FINDING THE BIG BANG

Cosmology, the study of the universe as a whole, has become a precise physical science, the foundation of which is our understanding of the cosmic microwave background radiation (CMBR) left from the big bang. The story of the discovery and exploration of the CMBR in the 1960s is recalled for the first time in this collection of 44 essays by eminent scientists who pioneered the work.

Two introductory chapters put the essays in context, explaining the general ideas behind the expanding universe and fossil remnants from the early stages of the expanding universe. The last chapter describes how the confusion of ideas and measurements in the 1960s grew into the present tight network of tests that demonstrate the accuracy of the big bang theory.

This book is valuable to anyone interested in how science is done, and what it has taught us about the large-scale nature of the physical universe.

P. James E. Peebles is Albert Einstein Professor of Science Emeritus in the Department of Physics at Princeton University, New Jersey.

Lyman A. Page, Jr. is Henry DeWolf Smyth Professor of Physics in the Department of Physics at Princeton University, New Jersey.

R. Bruce Partridge is Marshall Professor of Natural Sciences at Haverford College, Pennsylvania.

FINDING THE BIG BANG

P. JAMES E. PEEBLES,
Princeton University, New Jersey

LYMAN A. PAGE JR.
Princeton University, New Jersey

and

R. BRUCE PARTRIDGE
Haverford College, Pennsylvania

 CAMBRIDGE
UNIVERSITY PRESS

CAMBRIDGE UNIVERSITY PRESS
Cambridge, New York, Melbourne, Madrid, Cape Town, Singapore, São Paulo, Delhi

Cambridge University Press
The Edinburgh Building, Cambridge CB2 8RU, UK

Published in the United States of America by Cambridge University Press, New York

www.cambridge.org
Information on this title: www.cambridge.org/9780521519823

First published 2009

Printed in the United Kingdom at the University Press, Cambridge

A catalog record for this publication is available from the British Library

Library of Congress Cataloging in Publication data
Peebles, P. J. E. (Phillip James Edwin)
Finding the big bang / P. James E. Peebles,
Lyman A. Page, Jr., and R. Bruce Partridge.
p.   cm.
Includes bibliographical references and index.
1. Cosmic background radiation.   2. Big band theory.
3. Cosmology.   I. Page, Lyman A.   II. Partridge, R. B.   III. Title.
QB991.C64P44   2009
523.1–dc22
2008049448

ISBN 978-0-521-51982-3 hardback

To the memory of Dave Wilkinson
for his leadership in measuring
the fossil radiation

Contents

Preface

This is the story of a major advance in science, the discovery of fossil radiation left from the early stages of expansion of the universe – the big bang. Colleagues in informal conversations now only vaguely recalled led us to realize that this story is particularly worth examining because it happened in what was then a small line of research, and one that still is relatively simple compared to many other branches of physical science. That makes it well suited for an examination of how science actually is done, warts and all, in all the details – usually too numerous to mention – recalled by many of the people who did the work.

All the main steps in this story – the prediction, detection, identification, and exploration of the properties of the fossil radiation from the big bang – have been presented in histories of science. But these histories do not have the space (or the aim) to give an impression of what it was like to live through those times. We sense a similar feeling of incompleteness in many histories of science written by physicists, as well as by professional historians and sociologists. And there is a well-established remedy: assemble recollections from those who were involved in the work. An example in the broader field of cosmology – the study of the large-scale structure of the universe – is the collection of interviews in *Origins: the Lives and Worlds of Modern Cosmologists* (Lightman and Brawer 1990). We follow that path, but in more detail in a more limited line of research.

Early studies of the fossil radiation involved a relatively small number of people in what has proved to be a considerable advance in establishing the physical nature of the universe. This means we could aim for complete coverage of recollections from everyone involved in the early work who is still with us. We did not reach completeness: we suppose it is inevitable that a few colleagues would have reasons not to want to take part. We are fortunate, however, that almost everyone we could contact was willing to contribute recollections. All are well along in life now, but they have not slowed down; all had to break away from other commitments to complete their assignments. We are deeply indebted to the contributors for taking the time and trouble to make this collection possible, and for their patience in enduring the lengthy assembly of the book.

We are grateful to participants also for help in weeding out flaws in the introductory chapters, the collection of essays, the concluding chapter and Appendix which both treat what has grown out of the early work, and the Glossary that is meant to guide the reader through the story. We have also benefited from advice from those who started working in this subject more recently and have taken part in its growth into the present large and active science we outline in the concluding chapter. Their stories are important, but to keep the numbers manageable in the style of this book we had to impose a limit to recollections from people who were involved in this subject before 1970. That is when activity started gathering strength for the next leaps of technology and theory in increasingly large research groups.

Rashid Sunyaev was an invaluable guide to contacting contributors in Russia. We are grateful for help in the discussion in Chapter 3 of early measurements of the microwave radiation background from Eiichiro Komatsu and Tsuneaki Daishido, who led us to Haruo Tanaka's recollections of his work in Japan, from James Lequeux, who recalls early work in France, Virginia Trimble, who gives a picture of Gamow's thinking, and Jasper Wall, who led us to Covington's work in Canada. We have descriptions of the origins of the critical radiation energy spectrum measurements from Mark Halpern, Michael Hauser, and Ed Wishnow, and of the development of ideas on the distortion of the radiation spectrum from Ray Weymann. Ed Cheng helped us trace the origins of the WMAP satellite mission. We thank Steve Boughn, Josh Gundersen, Shaul Hanany, Gary Hinshaw, Norm Jarosik, Al Kogut, Paul Richards, John Ruhl, Suzanne Staggs, and Juan Uson for their help in entering and correcting the tabulation of experiments in Table A.3 in the Appendix, though of course all remaining errors are of our doing. We are grateful to Neta Bahcall, Joanna Dunkley, Brian Gerke, Toby Marriage, Jerry Ostriker, Will Percival, Bharat Ratra, David Spergel, Paul Steinhardt, and Ned Wright for help and advice on the cosmological tests; Michael Gordin for his instructions on similar collections of personal histories in other fields of science and on the lessons to be drawn from them; Mike Lemonick for help with his interview of David Wilkinson and his guidance to the art of communicating science; and Tatiana Medvedeva and Marina Anderson for their translations. Ned Conklin, Michael Fall, Masataka Fukugita, Martin Harwit, Michael Hauser, Malcolm Longair, Alison Peebles,

Bharat Ratra, and John Shakeshaft were particularly helpful guides to the presentation of the science and history of this subject, and to a substantial reduction of the error rate. They certainly do not share the blame for our remaining flaws of commission and omission.

Some steps toward the organization of this project ought to be recorded. Bernie Burke, Lyman Page, Jim Peebles, Alison Peebles, Tony Tyson, Dave Wilkinson, Eunice Wilkinson, and Bob Wilson met in Princeton on 9 February 2001, for an informal discussion over dinner of the story of the detection and identification of the fossil radiation. Wilson's written notes agree with Peebles' undocumented recollection of the general consensus that the story is worth telling. But we all returned to other interests. In a second attempt to get the project started, George Field, Jim Peebles, Pat Thaddeus, and Bob Wilson met at Harvard on 8 August 2003. This led to a proposal that was circulated to some 12 proposed contributors. (The number is uncertain because we did not keep records.) It yielded three essays – they are in this collection – but attention again drifted to other things. The third attempt commenced with a discussion between Bruce Partridge and Jim Peebles in September 2005 at the Princeton Institute for Advanced Study. That discussion led to a blunt actuarial assessment: if the story were to be told in a close to complete way it would have to be done before too many more years had passed. That generated the momentum that led to completion of the project.

We sent a proposed outline of the book with an invitation to contribute to 28 people on 7 December 2005. As one might expect, the outline for the book continued to change after that as we better understood what we were attempting to do. A more unsettling change is that although we had given the list of contributors careful thought, we continued to identify people who ought to contribute: we have in this book some dozen additions to the December 2005 list. A simple extrapolation suggests we have forgotten still others: we likely have not been as complete as we ought to have been. We hope those we inadvertently did not include will accept our regrets for our inefficiency. We hope all who did contribute to this book, in many ways, are aware of our gratitude.

Many of the figures were made for this book, whereas some were made by the contributors many years ago. Where we have reason to think a figure was published elsewhere and the rightsholder is not the contributor we have obtained permission to reproduce. We apologize in advance for any omissions in this procedure.

List of contributors

J. Richard Bond
Canadian Institute for
Theoretical Astrophysics
University of Toronto
Ontario, Canada

Stephen Boughn
Department of Astronomy
Haverford College
Haverford, PA, USA

Paul Boynton
Department of Physics
University of Washington
Seattle, WA, USA

Ronald N. Bracewell
STAR Lab, Stanford University
Stanford, CA, USA

Geoffrey R. Burbidge
Department of Physics
University of California
San Diego, CA, USA

Bernard F. Burke
MIT Kavli Institute for
Astrophysics and Space Research
Cambridge, MA, USA

Edward K. Conklin
Honolulu, HI, USA

Karl C. Davis
Richland, WA, USA

Andrei Georgievich Doroshkevich
Astro Space Center
Moscow, Russia

George F. R. Ellis
Mathematics Department
University of Cape Town
Cape Town, South Africa

John Faulkner
Astronomy and Astrophysics
Department
University of California
Santa Cruz, CA, USA

George B. Field
Harvard-Smithsonian Center for
Astrophysics, Harvard University
Cambridge, MA, USA

Martin Harwit
Cornell University
Washington, DC, USA

Paul S. Henry
AT&T Laboratories
Middletown, NJ, USA

David C. Hogg
Boulder, CO, USA

Michele Kaufman
Department of Physics, Ohio State
University, Columbus, OH, USA

David Layzer
Belmont, MA, USA

Malcolm S. Longair
Cavendish Laboratory, University
of Cambridge, Cambridge, UK

Jayant V. Narlikar
IUCAA, Pune, India

Igor Dmitriyevich Novikov
Astro Space Center, P.N. Lebedev
Physics Institute, Moscow, Russia

Donald E. Osterbrock
Lick Observatory
University of California
Santa Cruz, CA, USA

R. Bruce Partridge
Department of Astronomy
Haverford College
Haverford, PA, USA

P. James E. Peebles
Department of Physics
Princeton University
Princeton, NJ, USA

Arno Penzias
New Enterprise Associates
Menlo Park, CA, USA

Judith L. Pipher
Department of Physics and
Astronomy, University of
Rochester, Rochester, NY, USA

Martin Rees
Institute of Astronomy, Cambridge
University, Cambridge, UK

Peter G. Roll
Georgetown, TX, USA

Rainer K. Sachs
Department of Mathematics
University of California
Berkeley, CA, USA

John R. Shakeshaft
St. Catharine's College, University
of Cambridge, Cambridge, UK

Kandiah Shivanandan
Bethesda, MA, USA

Joe Silk
Department of Physics
University of Oxford
Oxford, UK

Yuri Nikolaevich Smirnov
Russian Research Center
"Kurchatov Institute"
Moscow, Russia

Kazimir S. Stankevich
Radiophysical Research Institute
Nizhny Novgorod, Russia

Robert A. Stokes
Versa Power Systems, Inc.
Littleton, Colorado, USA

Rashid Sunyaev
Max-Planck-Institut für
Astrophysik, Garching
Germany, and
Space Research Institute
Moscow, Russia

Patrick Thaddeus
Harvard-Smithsonian Center for
Astrophysics, Harvard University
Cambridge, MA, USA

Kenneth C. Turner
Carrollton, GA, USA

Robert V. Wagoner
Department of Physics, Stanford
University, Stanford, CA, USA

Jasper V. Wall
Department of Physics and
Astronomy, University of British
Columbia, Vancouver, Canada

Rainer Weiss
LIGO Group, MIT Kavli
Institute for Astrophysics and
Space Research
Cambridge, MA, USA

William "Jack" Welch
Department of Astronomy
University of California
Berkeley, CA, USA

David T. Wilkinson
Department of Physics
Princeton University
Princeton, NJ, USA

Robert W. Wilson
Harvard-Smithsonion Center for
Astrophysics, Harvard University
Cambridge, MA, USA

Arthur M. Wolfe
Department of Physics, University
of California, San Diego, CA, USA

Neville J. Woolf
Steward Observatory, University of
Arizona, Tucson, AZ, USA

Jer-tsang Yu
Office of the CIO, City University
of Hong Kong, Hong Kong SAR
China

1

Introduction

This is an account of the discovery and exploration of a sea of thermal radiation that smoothly fills space. The properties of this radiation (which we describe beginning on page 16) show that it is a fossil, a remnant from a time when our universe was denser and hotter and vastly simpler, a very nearly uniform sea of matter and radiation. The discovery of the radiation left from this early time is memorable because, as is often true of fossils, measurements of its properties give insights into the past. The study of this fossil radiation has proved to be exceedingly informative for cosmology, the study of how our universe expanded, cooled, and evolved to its present complicated condition.

The discovery of the fossil radiation grew out of a mix of lines of evidence that were sometimes misinterpreted or overlooked, and of ideas that were in some cases perceptive but ignored and in other cases misleading but entrenched. In the 1960s, it was at last generally recognized that the pieces might fit together and teach us something about the large-scale nature of the universe. We introduce the accounts of how this happened by explaining the lines of research that led up to the situation then. The story of what happened when the pieces were put together in the 1960s is told through the recollections of the people in the best position to know – those involved in the research. We have essays by most who took part in the recognition that this fossil exists, its properties may be measured, and what is measured may inform us about the nature of the physical universe. This did not happen all at once; nor was it done by a single person; nor was it always done knowingly. The collection of essays tell what happened in all the richness and complexity we suppose is typical of any activity that people take seriously.

The last part of this book describes how the developments in the 1960s led to the search and discovery of methods of accurate measurement of the properties of the fossil radiation and of methods of interpreting what is measured. This part of the story is told in a more orderly way – it is concerned with research directed to the solution of relatively well-posed problems – but it is no less rich. It shows how advances in technology and in the strategies of its application can dramatically increase our understanding of the world around us.

Look into the details of any other significant development in science and you are likely to find a story as rich and complicated as the discovery and exploration of the fossil radiation. Thus we offer this example of a particular advance of science as a lesson on the nature of the scientific enterprise. We can tell the story of the fossil radiation in finer detail than is usually done because this is a small slice of science, much of which played out not that long ago, with a relatively small number of actors. And because cosmology still is a relatively new science, it has not yet become exceedingly technical: we can explain the developments in words accessible to a nonspecialist who is willing to read carefully.[1] We believe this account is an instructive example for anyone who takes an interest in the nature of science and how it has led to our present understanding of the physical world.

The stories of search and discovery that scientists usually tell each other in books and scientific journals are much more schematic than what is presented here. Scientists as well as historians and sociologists complain about the distortions and simplifications that slight the wrong paths taken and understate the painstaking learning curves that experimentalists, observers, and theorists follow as they sometimes find better paths. But "tidied up" stories do serve a purpose in helping us keep track of the central ideas as well as reminding us that our subject does have a history. As a practical matter this is about the best scientists generally can do. Those who know what actually happened seldom are willing to take the time from research to tell it in detail; even if they did the rest of us would have little time to spare to read about it; and when we did we would find it difficult to pick out the threads that led to advances rather than dead ends. But it is important to have some examples that take the opposite tack: explore what happened in detail. This is our purpose in describing the discovery and exploration of the properties of the fossil radiation left from what we will term the "hot big bang."

The contributors to our set of recollections of what happened when the clues to the fossil radiation were put together in the 1960s have had a broad

---

[1] There are equations, for the pleasure of those who like them, but the equations that appear in the main text are not needed to understand the situation: the accompanying words are meant to convey the sense of the ideas. The more specialized mathematics and comments in footnotes and the Glossary are intended for specialists.

variety of careers. Some continued in this line of work after 1970, but many
have gone on to other things. Some were led to work on cosmology, the
study of the large-scale nature of the physical universe, by the elegance of
the issues: does the world as we know it last forever, or if not does it end
in fire or ice? Others were reluctant to get involved because the data one
could bring to bear on such questions were so exceedingly limited. Some were
drawn to cosmology by the challenge of making a particular measurement
or calculation. Others became involved by accident, not realizing that their
work would become important to the study of the expanding universe. We
have descriptions of what it was like to be a student then, or to be further
along into a career in science, along with accounts of how the contact with
this subject shaped careers and lives.

Our set of recollections cannot be complete because some of the actors are
no longer with us. That includes Yakov Zel'dovich, who led a research group
in the USSR that came close to the discovery of the radiation and, after its
discovery, contributed much to the exploration of its significance. We have
also lost Francesco Melchiorri, a pioneer in the use of bolometers to measure the radiation. In the USA losses include George Gamow, Ralph Alpher,
and Robert Herman. Their pioneering work in the 1940s and 1950s on the
thermal properties of the early universe is central to the history related in
Chapter 3. On the experimental side losses include Robert Dicke, Allan Blair,
and David Wilkinson. Bob Dicke suggested that Wilkinson and Peter Roll
search for this fossil radiation, using technology he had invented two decades
earlier. Al Blair with colleagues at the Los Alamos National Scientific Laboratory was one of the pioneers in the measurement of the fossil radiation
above the atmosphere. Dave Wilkinson, his colleagues and students, and
in turn their students, have played a leading part in the measurements of
the properties of the radiation, from the time of its discovery and continuing through to the two spectacularly successful satellite missions, Cosmic
Background Explorer (COBE) and WMAP, which have given us precision
measures that imply demanding constraints on the large-scale nature of the
universe. In England we have lost the pioneers of the steady state cosmology, Fred Hoyle, Hermann Bondi, and Thomas Gold, and a close associate,
Dennis Sciama. In the late 1960s Sciama became persuaded by the evidence
for a hot big bang, while Hoyle continued to lead the spirited exploration of
alternatives to the relativistic big bang cosmology. We do have recollections
by close associates; they are a valuable part of the story.

We are saddened by the loss of two contributors to the collection of essays.
Don Osterbrock, at the University of California in Santa Cruz, was among
the first to recognize evidence that most of the helium in stars is a fossil from

the early universe. This helium is closely related to the fossil radiation, but the observational indications are very different. Our explanation of his thinking commences on page 59; his recollections start on page 86. Ron Bracewell at Stanford University took an early lead in the development of the strategy for the measurements of the small departures from an exactly smooth sea of radiation. These measurements have proved to be exceedingly useful guides to how the concentrations of matter in galaxies and clusters of galaxies grew, in the process disturbing the radiation. His recollections begin on page 385. The technique he and his student Ned Conklin pioneered reappears in later generations of experiments. That is illustrated in Figure 5.6 on page 429. The recollections by our colleagues Don Osterbrock and Ron Bracewell, along with the other contributors to this volume, will edify generations to come.

Our guidance to contributors in the first round of invitations is summarized in the statement that

We invite your account of personal experiences. What did you know then about cosmology and what did you think of it as a branch of physical science? What issues of research or lines of thought led you by plan or serendipity to be involved with the idea of a primeval fireball (as it was then called)? What were your reactions to the discovery of the radiation, and what effect did the discovery have on your research?

We have made no attempt at documentation in these recollections, which we suspect would have been sparse compared to the density and complexity of the set of essays. We might have done better by going into the field to add interviews to the essays, and maybe even digging through notes and letters, though none of that is a practical plan for us. Lightman and Brawer (1990), in *Origins: the Lives and Worlds of Modern Cosmologists*, interviewed several of the people who contributed to these essays, and their questions are similar to ours, though not confined to as narrow a range of time and topic. They had the advantage of being able to ask a series of questions. But one may respond differently in an interview than to an invitation to write an essay, and we think we see the difference in the comparisons of what people who appear here and in *Origins* have to say. An analog of the follow-up question in an interview is the sharing of recollections of dates and events by some of our contributors. Apart from gentle hints, and a few corrections of well-documented points, we have not contributed to this interaction, or otherwise attempted to enhance the content or coherence of the essays.

The essays are informed by a considerable variety of philosophies of the theory and practice of science. To this must be added the variety of what the contributors happened to be doing in the 1960s, what they later considered worth recording in this volume, and what they happen to remember or are able to recover from fragmentary records. But in our opinion these recollections are the best feasible basis for an understanding of what actually happened and why. In science one seeks significant patterns in complex situations. We hope the reader will enjoy the opportunity of applying this tradition to the set of essays.

The research in the 1960s on fossils from the big bang grew out of what had happened earlier. In Chapter 3 we trace the histories of ideas and methods of measurement from early developments in the 1940s up to the general recognition in the 1960s that one may put these ideas and methods together. Our account of the science before 1960 is selective: we pay particular attention to those developments in cosmology that have proved to be relevant to the interpretation of a fossil from the early hot stages of expansion of the universe, the sea of radiation, along with a related fossil, the lightest of the chemical elements. This chapter concludes with a broader assessment of the state of the theory and practice of cosmology in the early 1960s: the observations and ideas that were more widely discussed and those that might have merited closer attention.

Our account of events leading to the situation in the 1960s is presented in the standard style for scientists that we mentioned earlier: we almost exclusively report what appears in the published scientific literature of the time (with a few exceptions that we hope are clearly apparent), and we present the development of our subject as a generally linear and orderly advance of knowledge. That is not the whole story by any means: we have omitted wrong steps that no longer seem relevant and all the other rough places that the essays are meant to illustrate. But, as we have remarked, this linear presentation is a well-tested and efficient way to present the main elements of the science. And because cosmology up to the 1960s was a small science, and only a small portion of that was concerned with fossils from the early universe, we have the space to explore the more interesting of the steps we now see were in wrong directions. This is important: mistakes are an inevitable part of advances in the enterprise of science.

There was an interplay of theory and practice in the science of cosmology leading up to the 1960s, including the first steps to the modern theory taken in the 1920s. But the scant observational basis allowed considerable and perhaps even unhealthy room for speculation undisciplined by observation. Even in the 1960s it was not at all unreasonable to doubt the progress toward checking ideas by piecing together an empirically based theory of the physical universe from our limited view in space and time. An example is in the foreword to the book *General Relativity and Cosmology* by Robertson and Noonan (1968). In the foreword the physicist W. A. Fowler wrote "Within its limitations special relativity is faultless. Whether this be true of general relativity remains to be seen. Cosmology is mostly a dream of zealots who would oversimplify at the expense of deep understanding. Much remains to be done – experimentally, observationally and theoretically. *Relativity and Cosmology* – Robertson's legacy made manifest by Noonan – surveys the fruit of past endeavors and is an almanac for the harvests to come."

When Fowler wrote this sensible assessment of the hazards of the enterprise of cosmology in the 1960s he may have been aware of the detection of the sea of radiation we now know is a fossil. (The detection is noted in this book, on page 390, but there is no mention of its possible significance for cosmology.) But in the mid-1960s Fowler was skeptical of the proposal that the radiation is a fossil from the past rather than something produced by processes operating in the universe as it is now. He was right to be cautious, and he was right also to caution that the use of Einstein's general relativity theory to describe the large-scale nature of the universe is an enormous extrapolation from the tests of this theory. At the time, experimental tests of general relativity were not very demanding, even on the length scale of the Solar System. If the observational and experimental basis for cosmology were as schematic now as it was in the 1960s, the discovery of the sea of radiation still would be an interesting development, but perhaps much less important to science than it has proved to be. That is because the measured properties of this radiation are a considerable part of the suite of evidence that now tightly constrains ideas about the large-scale nature of the universe, including stringent tests of aspects of general relativity theory applied on the enormous scales of cosmology. Fowler gave an accurate prediction of the present situation: much has been done, and it has yielded a rich harvest.

The counterpoint to the confusion of research on the frontiers of science is the development of webs of evidence that can become so tightly and thoroughly crosschecked that we can be confident they are good approximations to aspects of objective physical reality.[2] Chapter 5 shows an example of how an interesting issue, here the interpretation of the sea of radiation, can drive the development of new methods of measurement that build on earlier

---

[2] It is worth pausing to consider what is meant by this sentence. Research in physical science has made enormous progress by operating under the assumption that there is an objective physical reality that operates by rules we can discover, in successively improved approximations. The great advances of science reinforce the assumption: this is not an issue scientists generally consider worth discussing. The reality defined this way does evolve, of course. In quantum physics an isolated system may be in a definite state that does not have a real and definite energy until isolation is broken and a measurement forces the system to a real energy level. Here the older notion of reality is abandoned; we have a better approximation. The cosmology we are discussing is a physical science that operates by the standard and established conventions, including the highly productive working assumption of an objective physical reality, whose definition may evolve as we learn what questions we should be asking.

experience and teach us new things about the world around us. As experimentalists learned how to overcome the many obstacles to the spectacular precision of later measurements of the fossil radiation, they in turn drove theorists along their own learning curves on how to characterize the universe the measurements were revealing. The theoretical side of cosmology is guided by ideas of elegance, as is true of all physical science. But our ideas of elegance are informed by what observations and experiments teach us, and the ideas in turn inspire new observations.

By the beginning of the 21st century, at the time of writing this book, the interplay of theory and practice had produced a cosmology that passes a demanding network of experimental and observational tests. It is not practical to tell how this happened in the detail we could devote to the developments in the 1960s: too many people were making key contributions to too many lines of evidence. In Chapter 5 we return to the less realistic but more efficient linear style of presentation of Chapter 3 in describing what has been learned from precision measurements of the energy distribution of the fossil radiation and of the nature of its spatial distribution. This is supplemented by a tabulation in the Appendix of the series of experiments by which people learned how to make the measurements that so usefully characterize the radiation. A full account of how cosmology grew into the well-established science of the early 21st century would require tracing developments of other lines of evidence, some of which predate the idea of a hot big bang. We offer only the very condensed summary of this other work in Section 5.4. The course we have chosen leaves room instead for a closer study of how the science of the microwave radiation was done.

We have tried to make this worked example of science accessible to interested nonspecialists. We begin in the next chapter with explanations of the basic concepts of the established cosmology: what is meant by an expanding universe and a hot big bang, what can be said about the contents of the universe, and how the contents affect the history of its expansion. As we have mentioned, there are equations, but the text is meant to convey the sense of the discussion. The Glossary gives definitions of the jargon that appears in the essays and, inevitably, in the introductory and concluding chapters. The Glossary also is meant to serve as a guide to the somewhat complicated relations among ideas and issues. We offer references to the scientific literature for those who want to get into the really technical details. The citations are by the names of the authors and the date of publication, and the references to the literature are listed in the bibliography at the end of the book. The page numbers at the end of each reference in the bibliography serve as a supplementary index.

A gentler but still authoritative introduction to cosmology is in Steven Weinberg's (1977) *The First Three Minutes*. Helge Kragh's (1996) *Cosmology and Controversy* is a broader survey of the rich history of research in cosmology, and it is based on a broader variety of sources. We think of Kragh's style as intermediate between our more narrowly focused presentations in Chapters 2 and 3 and the full-blown details and complex panorama of recollections in the essays in Chapter 4. The reader will find that the essays are not fully concordant with these other accounts, careful though they are, or even with each other. Human events are complicated, and we have not sought to enforce a single vision of this example of research. Experts may find much of the science familiar, but unless they have long memories they would be well advised to look over Chapter 3, because the situation in cosmology in the early 1960s was very different from what grew out of it.

2

A guide to modern cosmology

The universe is observed to be close to uniform – homogeneous and isotropic – in the large-scale average.[1] That means we see no preferred center and no edge to the distribution of matter and radiation, and what we see looks very much the same in any direction. Stars are concentrated in galaxies, such as our Milky Way. The galaxies are distributed in a clumpy fashion that approaches homogeneity in the average over scales larger than about 30 megaparsecs (30 Mpc, or about 100 million light years, or roughly 1 percent of the distance to the furthest observable galaxies).

Space between the stars and galaxies is filled with a sea of electromagnetic radiation with peak intensity at a few millimeters wavelength and with spectrum – the energy at each wavelength – characteristic of radiation that has relaxed to thermal equilibrium at a definite temperature, in this case $T = 2.725$ K. This thermal radiation is much more smoothly distributed than the stars, but its temperature does vary slightly across the sky.[2] (The temperature differs by a few parts in 100,000 at positions in the sky that are separated by a few degrees.) The evidence developed in this book is that the radiation is a fossil remnant from a time when our expanding universe was much denser and hotter, and that the slight temperature variations were caused by the gravitational pull on the radiation by the increasingly clumpy distribution of matter in galaxies and clusters of galaxies.

We offer in this chapter a guide to basic ideas behind the interpretation of the radiation. We begin by explaining the concept of a universe that

---

[1] This situation is termed the "cosmological principle." It is an assumption that Einstein (1917) introduced and is now observationally well supported.

[2] The distributions of mass and this thermal radiation are seen to be close to homogeneous by the special class of "comoving" observers who are at rest relative to the mean motion of the matter and radiation around them. An observer moving with respect to this frame sees gradients in the distributions of matter and radiation. This definition of a preferred motion is not a violation of relativity theory, which of course allows observation of relative motion, here relative to the comoving rest frame defined by the contents of the universe.

is homogeneous and expanding in a homogeneous and isotropic way. Section 2.2 describes the meaning of thermal radiation and its behavior in this expanding universe. In the concluding section we present a list of the main known forms of matter and radiation in the universe as it is now. This inventory figures in the analysis of the properties of fossil remnants from the early stages of expansion of the universe: the thermal radiation and isotopes of the light chemical elements. The origins of ideas about these fossils in the 1960s are described in Chapter 3 and in the essays in Chapter 4.

2.1 The expanding universe

The expansion of the universe means that the average distance between galaxies is increasing. Figure 2.1 shows an early use of a model that helps illustrate the situation. Imagine you live in only two spatial dimensions on the surface of a balloon. Do not ask what is inside or outside the surface – you are confined to your two-dimensional space on the rubber sheet of the balloon. In your two-dimensional space you see a uniform distribution of galaxies: there may be local clustering, as we observe in the real universe, but the mean number of galaxies per unit volume (which in this example is an area) is the same everywhere. As the balloon is blown up the galaxies move apart. Another caution is in order here: the galaxies themselves are not expanding. An observer at rest in any galaxy sees that the other galaxies are moving away, at the same rate in all directions, as if the observer were at the center of expansion of this model universe. But an observer in any

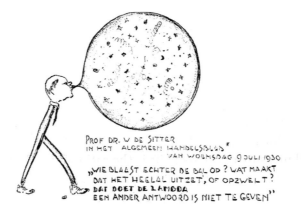

Fig. 2.1. A sketch of Willem de Sitter on the occasion of his explanation of the idea of an expanding universe in a Dutch newspaper in 1930. His body is sketched as the Greek symbol lambda, or λ, which represents Einstein's cosmological constant. As will be discussed, this constant was taken seriously then and came back into fashion.

other galaxy would see the same motion of general recession in all directions. The key point illustrated here is that this model universe is expanding but has no center of expansion: expansion is happening everywhere in the two-dimensional space. In the cosmology of our universe an observer in any galaxy in our three-dimensional space sees the same effect: the other galaxies are moving away.

A little thought about this expanding balloon model may convince you that an observer at rest in a galaxy sees that galaxies at greater distance r from the observer – measured along the balloon surface – are moving away at greater speed v. The recession velocity is proportional to the distance, following the linear relation

$$v = H_0 r. \tag{2.1}$$

The same argument, and this linear relation, applies to the expansion of the three-dimensional space of our universe.

Equation (2.1) is called Hubble's law, after Edwin Hubble (1929), who was the first to find reasonably convincing evidence of this relation. The multiplying factor, H_0, is called Hubble's constant.[3]

The speed v of recession of a galaxy is inferred from the Doppler effect. Motion of a source of light toward an observer squeezes wavelengths, shifting features in the spectrum of the source toward shorter – bluer – wavelengths, while motion away shifts the spectrum to the red, to longer wavelengths. The spectra of distant galaxies are observed to be shifted to the red, as if the light from the galaxies were Doppler shifted by the motion of the galaxies away from us. This is the cosmological redshift.

You will recall from the balloon model that in this expanding universe an observer in any galaxy would see the same pattern of redshifts, and hence also observe Hubble's relation $v = H_0 r$. It is of course a long step from the observation that the light from distant galaxies is shifted to the red to the demonstration that all observers in our universe actually see the same general expansion. But the proposition can be tested; that is one of our themes.

A numerical measure of the redshift is the ratio of the observed wavelength $\lambda_{\rm obs}$ of a spectral feature in the light from a galaxy to the wavelength $\lambda_{\rm em}$ of emission at the galaxy. In an expanding universe the ratio $\lambda_{\rm obs}/\lambda_{\rm em}$ of observed and emitted wavelengths is greater than one. Astronomers subtract unity from this ratio, defining the cosmological redshift z as

---

[3] In equation (2.1) H_0 often is called the "constant of proportionality." That can be confusing, because in the standard cosmology this factor of proportionality changes with time.

$$z = \frac{\lambda_{\text{obs}}}{\lambda_{\text{em}}} - 1. \tag{2.2}$$

Thus when the redshift vanishes, $z = 0$, the wavelength is unchanged.

The redshift z does not depend on the wavelength of the spectral feature used to measure it. That means we can define a single measure of the wavelength shift by the equation

$$1 + z = \frac{\lambda_{\text{obs}}}{\lambda_{\text{em}}} = \frac{a(t_{\text{obs}})}{a(t_{\text{em}})}. \tag{2.3}$$

The radiation was emitted from the galaxy at time t_{em} and received by the observer at the later time t_{obs}. The parameter $a(t)$ defined in this equation depends on time, but it does not depend on the wavelength, because we have observed that $\lambda_{\text{obs}}/\lambda_{\text{em}}$ does not depend on the wavelength. The parameter $a(t)$ serves as a measure of how the wavelength of radiation moving from one galaxy to another is changing now and has changed in the past.

Now let us consider how distances between galaxies change with time. As the universe expands the distance d between a well-separated pair of galaxies increases. Very conveniently, the theory says that the distance is stretched in the same way as the stretching of the wavelength of light moving from one galaxy to the other. That means the distance $d(t)$ between two galaxies – any pair of well-separated galaxies – is increasing as $d(t) \propto a(t)$. Thus we call $a(t)$ the expansion parameter.[4] When its value has doubled the mean distance between galaxies also has doubled. It follows that the mean number density of galaxies decreases as the universe expands, as

$$n(t) \propto a(t)^{-3}, \tag{2.4}$$

as long as galaxies are not created or destroyed.

In short, if we knew $a(t)$ we would have a measure of the history of the expansion of the universe. It is an interesting exercise for the student to calculate the rate of change of the distance $d(t)$ between a pair of galaxies in terms of $a(t)$; check that the result agrees with Hubble's law in equation (2.1); and find Hubble's constant H_0 in terms of the present values of $a(t)$ and its first time derivative. The rest of us may move on.

---

[4] To reduce confusion we urge the reader to bear in mind that our standard of length – be it a meter or a megaparsec – is fixed. Large-scale distances measured in terms of this standard are increasing. On the other hand, objects like ourselves or meter sticks are not expanding. A galaxy that is not accreting or losing matter is not expanding either. Its size is fixed by the gravity that is holding it together. The same is true of a gravitationally bound cluster of galaxies. The expansion parameter $a(t)$ describes the increasing distances between galaxies which are well-enough separated that we can ignore the local clumping of mass in galaxies and clusters of galaxies.

The standard cosmology of the early 21st century is based on Einstein's general relativity theory, the commonly accepted and successful theory of gravity. The use of this theory in the early days of cosmology was speculative, because there were few significant observational tests. But general relativity strongly influenced thinking, as follows.

In general relativity the rate of expansion of the universe changes as the universe expands. The gravitational attraction of the mass of the universe tends to slow its expansion. If the cosmological constant term mentioned in the caption in Figure 2.1 is present, and positive, then it tends to speed the expansion. The resulting acceleration – the second time derivative – of the expansion parameter $a(t)$ in equation (2.3) is represented by the equation

$$\frac{\mathrm{d}^2 a}{\mathrm{d}t^2} = -\frac{4}{3}\pi G\rho a + \frac{1}{3}\Lambda a. \tag{2.5}$$

Newton's constant of gravity is G and the mean mass density, averaged over local irregularities, is ρ. The minus sign in front of this mass density term signifies the gravitational effect of the mass: it tends to slow the rate of expansion of the universe. Einstein's cosmological constant appears in the last term. The style has changed here: people nowadays write it as an upper case Greek lambda, Λ, reserving the symbol λ for wavelength. (Note also that in Figure 2.1 the artist drew λ backward from the current convention, but in a style similar to Einstein's way of writing it.) If Λ is positive it opposes the effect of gravity. If Λ is positive and large enough it causes the rate of expansion to increase, or accelerate. The evidence reviewed in the last chapter of this book is that this is the situation in the universe now.

Einstein (1917) found that his original form of general relativity theory, without the Λ term, cannot apply to a universe that is homogeneous and, as he supposed, unchanging. You can see that from equation (2.5): if the universe were momentarily at rest then in the absence of the Λ term the attraction of gravity would cause the universe to start collapsing. That led Einstein to adjust the theory by adding the cosmological constant term, which he could choose so that the attraction of gravity and the effect of a positive Λ just balance: the right-hand side of equation (2.3) vanishes. That allows the static universe that made sense to him (since he was writing before Hubble's discovery). It takes nothing away from Einstein's genius to notice that he overlooked the instability of his model universe: a slight disturbance would reduce or increase the mass density ρ, and that would cause the universe to start expanding or contracting. (More generally, a local departure from exact homogeneity would grow and eventually make the universe much more clumpy than is observed.)

Aleksandr Friedmann (1922), in Russia, was the first to show that general relativity theory allows Einstein's homogeneous universe to expand or contract. He had the misfortune to do it a few years before there was a hint from astronomical observations that the universe is in fact expanding. Georges Lemaître (1927), in Belgium, rediscovered Friedmann's result and recognized that it meant Einstein's static universe is unstable. Lemaître also saw that the expansion of the universe might account for the astronomers' discovery that the spectra of galaxies are shifted toward the red, perhaps by the Doppler effect. Figure 2.1 shows de Sitter's explanation of Lemaître's idea. De Sitter is quoted as saying, "what causes the balloon to expand? That is done by the lambda. Another answer cannot be given." De Sitter is explaining Lemaître's idea that the universe was in Einstein's static condition, and that some disturbance had allowed the Λ term to push the universe into expansion.

Lemaître (1931) soon saw that the expansion could instead trace back to an exceedingly dense early state that he termed the "primeval atom." The evidence is that the universe did expand from a state that was dense, as Lemaître proposed, and hot. We will use the more familiar term for it, the hot big bang.[5]

It was soon recognized that the expansion of the universe does not require the cosmological constant, provided one is willing to live in a universe that expanded from a big bang. Einstein accordingly proposed that we do away with the Λ term. The physicist George Gamow (1970) quotes Einstein as saying that his introduction of Λ was his biggest "blunder." We might suppose Einstein meant that if he had stayed with his original theory, and kept to the idea that the universe is homogeneous, he could have predicted that the universe is evolving, either expanding or contracting. It is a curious historical development that Einstein's cosmological constant is back in style, for the reasons indicated in Chapter 5. The reasons are different from Einstein's original argument, but we imagine Einstein might not have been too disturbed by that. The cosmological constant was his invention, after all.

The names "primeval atom" and "big bang" are meant to indicate that, if the Λ term does not prevent it, general relativity theory predicts that there

---

[5] The evidence Mitton (2005) assembles is that Hoyle coined the term "big bang" in a lecture on BBC radio in March 1949. Mitton quotes Hoyle: "We come now to the question of applying the observational tests to earlier theories. These theories were based on the hypothesis that all matter in the universe was created in one big bang at a particular moment in the remote past." The connotation of a localized explosion is unfortunate – the theory deals with evolution of the near-uniform observable universe from a dense early state – but its usage is firmly established.

was a time in the past when the expansion parameter $a(t)$ in equation (2.3) vanished. The effect may be easier to see qualitatively by imagining the expansion of the universe running backward in time. The distances between galaxies are smaller in the past, and approach zero as $a(t)$ approaches zero going back in time. This means there was a time when the density of matter was arbitrarily large. If the effects of gravity and Λ are ignored then the recession speed v of a galaxy does not change, and in this case one sees from equation (2.1) (remembering that distance traveled is speed times time) that the distances between galaxies vanished at time H_0^{-1} in the past, or about 10 billion years ago.[6] This marks the moment of formally infinite density. It is conventional to speak of this moment as the beginning of the history of the universe as we know it, when $a = 0$. We include ourselves among the many who suspect that better physics to be discovered, perhaps within the concept of cosmological inflation, will remove this singularity, and teach us what happened "before the big bang," or "at the big bang," or whatever is the suitable term.

In the early 1960s another world view was under discussion. In the steady state cosmology proposed by Bondi and Gold (1948) and Hoyle (1948) matter is continually created – at a rate that would be unobservably small in the laboratory – and collects to form young galaxies which fill the spaces that are opening up as older galaxies move apart. The mean distance between galaxies – about 10 million light years (or about 3 Mpc) for relatively large ones such as the Milky Way – thus would stay constant. The universe on the whole would not be changing: there would be no singular start to the expansion and no end of the world as we know it. Einstein's (1917) original world model, taken literally, has no beginning or end of time either. But, if energy were conserved, all the stars would eventually exhaust their supplies of fuel and die, or if energy were not conserved and stars shone forever, space would become filled with starlight. The steady state cosmology offers an elegant solution: the expansion of the universe dilutes away the starlight and the dead stars, and continual creation supplies matter for unlimited generations of new stars. But this is not the way our universe operates. Part of the story of how that was established commences on page 51, where we consider the state of research in cosmology in the early 1960s. Chapters 4 and 5 describe what happened after that, and the role of the thermal radiation that fills space in teaching about the evolution of the universe. Let us consider now some properties of this radiation.

---

[6] For this reason H_0^{-1} is called the Hubble time, or the Hubble length measured in light travel time.

2.2 The thermal cosmic microwave background radiation

A warm body radiates; you can feel the thermal radiation from a hot fire. In a closed cavity with walls that are at a fixed temperature the radiation in the cavity relaxes to a spectrum – the intensity of the radiation at each wavelength – that is uniquely determined by the temperature of the walls. The time it takes for the radiation to relax to this thermal spectrum depends on how strongly the walls absorb and emit radiation. If the walls are perfectly absorbing – black – the relaxation time is comparable to the time taken by the radiation to cross the cavity. That suggested a commonly used name: blackbody radiation is radiation that has relaxed to thermal equilibrium at a definite temperature. The thin line in Figure 2.2 shows the spectrum of blackbody radiation at temperature

$$T_0 = 2.725 \, \text{K}, \tag{2.6}$$

above absolute zero. This is the thermal radiation – the cosmic microwave background radiation, or CMBR – that fills space.

Max Planck proposed the first successful theory for the spectrum of blackbody radiation in 1900; it was also the first step into the new field of quantum

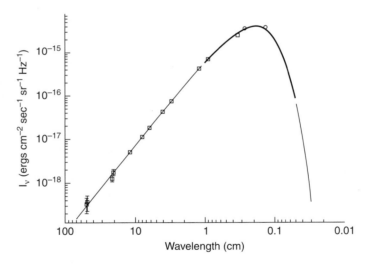

Fig. 2.2. The spectrum of radiation that uniformly fills space, with greatest intensity at millimeter – microwave – wavelengths. In this book this is termed the CMBR. The thin line in this figure is the theoretical Planck blackbody spectrum of radiation that has relaxed to thermal equilibrium at temperature $T_0 = 2.725$ K. The thick line running over the peak shows the measurements by the NASA COBE and UBC COBRA groups. They are not distinguishable in this figure. The symbols represent other measurements at more widely spaced wavelengths. This plot was made by David Wilkinson in 1992.

physics. Richard Tolman (1931) noticed that radiation in a homogeneous universe could relax to a thermal spectrum, if there were enough matter to absorb and reemit the radiation energy often enough to cause it to relax to equilibrium. In effect, the whole universe could be the blackbody "cavity." He also showed that the expansion of a homogeneous universe would cool the radiation. Most importantly, Tolman showed that once the radiation has relaxed to thermal equilibrium the expansion of the universe preserves the characteristic blackbody spectrum, with no further need for matter to promote or maintain thermal equilibrium. The expansion of the universe causes the temperature to decrease in inverse proportion to the expansion parameter in equation (2.3), that is,

$$T \propto a(t)^{-1}. \tag{2.7}$$

To summarize, blackbody radiation uniformly filling an expanding universe stays blackbody; only the temperature of the radiation changes as the universe expands. This is the essential signature. Since, as we now discuss, the spectrum of radiation filling our universe is close to thermal we have evidence that conditions were at one time right for relaxation to thermal equilibrium.

Figure 2.2 shows measurements of the intensity of the CMBR. It peaks at a microwave wavelength near 2 mm. The thick black line running over the peak shows measurements of the intensity at a densely sampled range of wavelengths. These measurements were made above the atmosphere, to avoid radiation from molecules in the air, independently from the National Aeronautics and Space Administration (NASA) COBE satellite (Mather *et al.* 1990) and from a UBC (University of British Columbia) rocket flight (Gush, Halpern and Wishnow 1990). The measurements are very close to – and not measurably different from – Planck's blackbody spectrum over a wide range of wavelengths.

The universe we see around us is close to transparent at wavelengths near the peak of this radiation. We know that because distant galaxies that are sources of radio radiation are observed at these wavelengths. This means that the universe as it is now cannot force radiation to relax to the distinctive thermal spectrum shown in Figure 2.2. And this means that the universe has to have evolved from a very different state, one that was hot and dense enough to have absorbed and reradiated the radiation, forcing it to relax to its blackbody spectrum. That is, contrary to the classical steady state cosmology, we have evidence that this cosmic microwave radiation is a fossil remnant from a time when our universe was very different.

One learns from fossils what the world used to be like. The fossil microwave background radiation is no exception: we have learned a lot from the close study of its properties. The evidence is that the thermal radiation played an important role in the history of the universe, including the thermonuclear reactions that produced light elements in the early stages of expansion and the dynamics of the growth of the mass clustering that we observe as galaxies and concentrations of galaxies. The study of both aspects, the radiation as a signature of what things were like and as a dynamical player in what happened, are recurring themes, in the recollections in Chapter 4 of research in the 1960s and in the subsequent developments described in Chapter 5 of the detailed measurements of the radiation and what the measurements have taught us. Our discussion of these themes begins with an inventory of other dynamical players: what does the universe contain in addition to the fossil thermal radiation?

2.3 What is the universe made of?

The world is full of many things, and we surely have discovered only a small fraction of them. But we do have credible evidence about what things are made of and about the amounts of the types of mass involved. Table 2.1 lists contributions to the total mass of the universe by some of the more interesting and important types of matter and radiation.[7] The numbers in the middle column, which usually are termed "density parameters," are fractions of the total. The last column lists the mean mass density in each component.[8]

People, planets, and stars are made of baryons – the neutrons and protons in atomic nuclei of the chemical elements – with enough electrons to keep the electric charge neutral. The mass in the inner parts of our Milky Way Galaxy is largely in baryons in stars. The same is true of the central parts of the other large galaxies. The outer regions of the galaxies contain baryons, mostly in the form of plasma, but there is more mass in dark matter, which is not baryonic. In the average over much larger scales the biggest contribution is shown as the entry for the first component in the table, dark energy. This is the new name for Einstein's cosmological constant, Λ.

The gravitational action of dark energy is illustrated in Figure 2.1. In general relativity theory the positive pressure of a fluid adds to the gravitational

---

[7] Fukugita and Peebles (2004) discuss the observational basis for these mass estimates and their uncertainties and also give estimates of the masses in a considerable variety of other components.

[8] The total mass density summed over all components is such that, in general relativity theory, space sections at constant world time are not curved. Spacetime is curved, but space sections at constant time have close to Euclidean geometry.

Table 2.1. *Cosmic mass inventory*

Category	Mass fraction[a]	Mass density, g cm^{-3}
The dark sector		
Dark energy	0.74	7.2×10^{-30}
Dark matter	0.21	2.0×10^{-30}
Thermal big bang remnants		
Electromagnetic radiation	0.00005	5×10^{-34}
Neutrinos	0.001	1×10^{-32}
Baryons[b]		
Diffuse plasma	0.042	4×10^{-31}
Stars	0.0022	2×10^{-32}
Stellar remnants	0.0005	4×10^{-33}
Atoms and molecules	0.0008	7×10^{-33}
Stellar radiation		
Electromagnetic	0.000002	2×10^{-35}
Neutrinos	0.000003	3×10^{-35}
Gravitational radiation from		
gravitational collapse	0.00000003	3×10^{-37}

[a]Energies have been converted to their equivalent masses.
[b]This includes enough electrons to make matter electrically neutral.
Source: Adapted from Fukugita and Peebles (2004).

attraction produced by the mass equivalent of its energy. Near the end of the life of a massive star the pressure grows large, and that contributes to the final violent relativistic collapse of its central parts to a black hole. But pressure can be negative: the tension in a stretched rubber band is in effect a negative pressure. This negative pressure slightly reduces the gravitational attraction produced by the mass associated with the energy of the rubber. Einstein's Λ acts like a fluid that has nearly constant energy density, and pressure that is negative. In this case the negative pressure is large enough in magnitude that its gravitational effect overwhelms the gravitational attraction of the energy (as opposed to the exceedingly small effect of the tension of a rubber band). The result is a contribution to the gravitational field that pushes matter apart.[9] The name "dark energy" comes from the intuition felt by many that Λ has something to do with an actual energy density, and that, like other forms of energy, Λ need not be exactly constant. But all we can say with confidence is that this term is needed to make sense of the evidence whose collection and analysis is the subject of Chapter 5.

---

[9] It is best left as an exercise for the student to see why this push has little or no effect on how the dark energy itself is distributed, and why the negative pressure allows the energy density in this component to remain nearly constant as the universe expands.

The second component in the table is dark matter. It acts like a gas of particles that move freely, apart from the effect of gravity. Fritz Zwicky (1933) seems to have been the first to notice the dark matter effect. As discussed in a little more detail in the next chapter (in footnote 13 on page 31), he found that the observed mass in stars in the Coma Cluster of galaxies (named for the constellation in which it appears in the sky) is much too small to gravitationally confine the motions of the galaxies deduced from the Doppler shifts of the galaxy spectra. It seemed unlikely that the cluster could be flying apart, because the distribution of galaxies near the center of the cluster is smooth and quite compact. But what might be holding the cluster together?

Zwicky's effect has since been found to apply to the other rich clusters: the cluster galaxies are moving too rapidly, and the plasma in the cluster is too hot, to be held by the gravity of the mass present in the galaxies. The same applies to the motions of stars and gas in the outer parts of individual galaxies outside clusters. The mass that is needed to hold clusters together, and to do the same for the outer parts of individual galaxies, used to be known as "missing mass." It is now termed "dark matter," but we still do not know what it is, apart from one clue. The evidence we will be describing is that the dark matter cannot be baryons, for that would contradict the successful theories for the origin of the light elements and of the properties of the CMBR. The evidence instead is that the dark matter is a gas of freely moving nonbaryonic particles. Discovering the nature of these mystery particles, and the nature of the dark energy – Einstein's Λ – is a wonderful opportunity for search and discovery by the generations after us.

The second category in the table is the thermal electromagnetic radiation and neutrinos left from the hot big bang. The radiation – the CMBR – has the spectrum shown in Figure 2.2. This radiation now contains about 400 thermal photons per cubic centimeter. The mass equivalent to the mean energy of one of these photons is so small that the radiation mass density adds only a trace to the total. But you will recall that the cosmological redshift (shown in equation 2.2) reduces the photon energy as the universe expands. In the early universe the thermal photons were energetic enough that their mass densities were the largest contribution to the total. (This is discussed in more detail in footnote 9 on page 29.)

The energetic photons in the early universe took part in the creation and annihilation of neutrinos by the reactions to be discussed in the next chapter. That would have produced a thermal sea of neutrinos. The number of neutrinos plus antineutrinos in each of the three families is now 3/11 times the number of thermal photons, or about 100 neutrinos per cubic centimeter at the present epoch. The present energy density is larger in these fossil neutrinos than in the radiation however, because the neutrinos have rest masses. (The experimental evidence that neutrinos have masses is clear, but the values of the masses are only loosely bounded. The number in the table for the present neutrino mass density is an order-of-magnitude estimate. But we can be sure there is not enough mass in the known families of neutrinos to serve as the dark matter: we need another kind of mystery particle.)

The third category is the baryons. The total mass density in this form is inferred from arguments that again are discussed through this book. The inference (but at the time of writing not a demonstration by detection) is that most of the baryons are in the form of diffuse plasma, the first entry in this category, because this amount of baryons in any other physically reasonable state would have been observed. There is a trace amount of this plasma in the disks of spiral galaxies such as the Milky Way. There is a larger amount in hotter plasma in clusters of galaxies, and a still larger amount in plasma gravitationally bound to the outer regions of individual galaxies. There also is a sea of diffuse plasma spread through the enormous spaces between the galaxies. The relative amount in the last two forms is not well established.

The second component in the baryon category in Table 2.1 is the mass in stars that are radiating energy by nuclear burning – the nuclear reactions that convert hydrogen to helium and heavier elements – in their central regions. The stars in the nearly spherical bulges of spiral galaxies such as the Milky Way formed when the universe was much less than half its present age. Most of the stars in elliptical galaxies, which have at most an inconspicuous disk, also are old. The stars in the disk of the Milky Way have a broader range of ages. Stars are still forming at substantial rates in the disks of spiral galaxies and in lower mass galaxies such as the Magellanic Clouds, largely out of the neutral atoms and molecules entered as the fourth component in this category. But the overall rate of star formation is markedly lower now than it was when the universe was half its present age. There is a large mass of baryons in diffuse plasma, but this plasma is cooling too slowly to supply baryons for ongoing star formation at the past high rate.

As the energy supply in a star is exhausted some baryonic matter is ejected in stellar winds and explosions and some is left in stellar remnants: white dwarfs, neutron stars, and black holes. The third component in the baryon category is an estimate of what has accumulated in these remnants. There are baryons in many other fascinating forms, including planets and people, but they are thought to amount to a very small fraction of the total, as indicated in the last entry.

The fourth category is the accumulated energy released by stars in electromagnetic radiation – starlight – and neutrinos. The larger amount of energy in neutrinos is a result of the copious emission accompanying the collapse of dying massive stars. These energy densities are averages over large scales. We receive more than average starlight (after correction for the Sun) because we are in a galaxy of stars, the Milky Way. This local energy density in starlight happens to be comparable to what is in the CMBR, but the two have little else to do with each other.

The fifth category is an estimate of the energy density in gravitational radiation produced during the formation of black holes by the gravitational collapse of mass concentrations or by the merging of black holes. Several of the contributors to Chapter 4 mention their interest in detecting this gravitational radiation, but that is another story.

As we have said, the tasks of discovering the physical natures of dark energy and dark matter are at the time of writing golden opportunities for research for future generations. One of our tasks in the rest of this book is to consider the lines of reasoning and observation that have led to the conclusion that we do have credible evidence that these dark components really exist. We begin in the next chapter with an account of the early development of ideas that led to the identification of two very helpful fossils from the early universe: the thermal CMBR and the isotopes of hydrogen and helium.

3

Origins of the cosmology of the 1960s

To understand the essays in the next chapter about what happened in the 1960s you have to appreciate the nature of research in cosmology then. To understand the nature of this research you have to consider its history. Figure 3.1 illustrates the major steps leading to one big advance in cosmology, the identification of the CMBR as a fossil remnant from the big bang. This figure was made by members of Princeton Gravity Research Group. David Wilkinson was its main author, he used it in lectures on cosmology starting in 1968, and it is a good illustration of his style. Another version was eventually published (in Wilkinson and Peebles 1983).

The figure maps relations among the topics we discuss in this chapter. The map is complicated because the story is complicated, but there are a few themes. We begin with the first of these, the development of the idea that the abundances of the stable isotopes of the lightest elements, hydrogen and helium, were determined by thermonuclear reactions in the early hot stages of expansion of the universe (with modest adjustments for what happened in stars much later). We consider next the line of thought that led Dicke to persuade Roll and Wilkinson to search for the CMBR. We then turn to the development of the means of detecting and measuring the properties of the radiation left from the hot big bang. We conclude this chapter with an assessment of what people were thinking and doing in cosmology in the early 1960s, at the start of the time surveyed by the essays in the next chapter.

3.1 Nucleosynthesis in a hot big bang

Hydrogen is the most abundant of the chemical elements (apart from places like Earth where the heavier elements have collected and condensed), helium amounts to about 25% by mass, and only about 2% of the baryon mass is in heavier elements. What produced this mix? In the 1930s people were

23

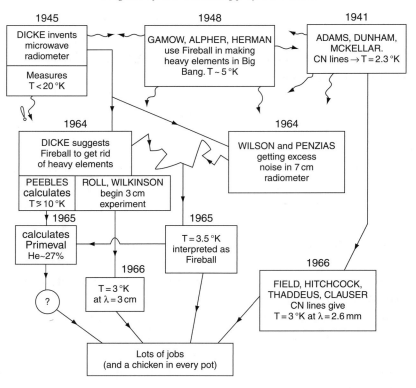

Fig. 3.1. This illustration of how the CMBR was and could have been identified was made in 1968 by David Wilkinson with other members of the Princeton Gravity Research Group.

exploring two main lines of thought, that the chemical elements might have formed in stars or in the early universe. The former was suggested by the growing evidence that the Sun and other stars radiate energy released by the fusion of atomic nuclei into heavier nuclei. One could imagine that the heavy elements produced in stars by this nuclear burning were ejected by stellar winds or explosions and that the debris formed new stars and planets. The other picture assumes that temperatures and densities in the early stages of expansion of the universe were large enough to have forced nuclear reactions among atomic nuclei that produced a mix of elements (that might have been adjusted by what happened later in stars). The later well-tested theory combines these ideas: the heavier elements originated in stars while most of the helium is a fossil remnant of the hot big bang, along with the thermal CMBR.

We review here the main steps in the development of the hot big bang part of this theory. Alpher and Herman (2001) describe the history and present recollections of the introduction of main features of the concepts by them and their colleagues in the 1940s and 1950s. Kragh (1996, Chapter 3: *Gamow's Big Bang*) presents helpful details about the events and the people involved in the research. The essays in Chapter 4 add to the story of what happened later, in the 1960s.

The earliest discussions of element formation in the big bang picture considered the idea that the relative abundances of the chemical elements and their isotopes might have been determined by relaxation to thermal equilibrium at some hot early stage of expansion of the universe. The concept can be compared to that of the blackbody radiation discussed in Chapter 2. We remarked that at thermal equilibrium the intensity of the radiation at each wavelength is determined by just one quantity, the temperature. At equilibrium the relative abundances of the elements and their isotopes would be fixed by two quantities, the temperature and the density of matter. The analysis by von Weizäcker (1938) showed that the situation has to be at least a little more complicated than that. He found that a rough fit to the observed pattern of abundances of the elements would follow if particle reactions generally ceased to be important – the pattern of element abundances were close to "frozen in" – when the expanding universe had cooled to a temperature of about $T \sim 2 \times 10^{11}$ K, but that residual reactions after that would have to have shaped the variation of the abundance from one atomic weight to the next in the middle part of the periodic table to that characteristic of a lower temperature, $T \sim 5 \times 10^9$ K. Chandrasekhar and Henrich (1942) repeated the analysis. Their more complete computations based on better data for the nuclear physics and element abundances indicated that the abundances of the heavier elements would had to have been frozen at about von Weizäcker's higher temperature, while the abundances of the lighter elements were determined by an approach to thermal equilibrium later, at about von Weizäcker's lower temperature. For our purpose the important thoughts are that the early universe might have been hot, expanded and cooled at a rate characteristic of thermonuclear reactions, and left an interesting variety of chemical elements.

The physicist George Gamow took the leading role in improving these thoughts. Gamow (1942) and Gamow and Fleming (1942) argued that the picture of near thermal equilibrium seems less plausible than a distinctly nonequilibrium process in an expanding universe. That might involve a "rapid breaking-up of the original superdense nuclear matter...Even in ordinary uranium-fission *a number of free neutrons are being emitted in each breaking-up process, and this number most probably increases in the case of the more violent fission of superheavy nuclei.* Neutrons produced this way will turn spontaneously into protons, and will contribute to a larger abundance of hydrogen" (Gamow 1942).

All these papers assumed we live in an expanding, evolving universe (in von Weizäcker's case one that is finite but large enough to include the most distant observed galaxies) but did not explicitly take account of the relativistic theory for the rate of expansion. Gamow (1946) took this important step. He remarked that in the early stages of an expanding universe the mass density would be large, and that would make the rate of expansion rapid:[1] "we see that *the conditions necessary for rapid nuclear reactions were existing only for a very short time*, so that it may be quite dangerous to speak about an equilibrium state" of the kind people had considered earlier. Gamow (1946) also noted that the positive electric charges of atomic nuclei tend to slow their fusion by pushing the nuclei apart, while the free neutrons he had mentioned earlier (Gamow 1942), which have no electric charge, react rapidly with protons and heavier atomic nuclei. That is wanted for a rapid build-up of the elements. One notices a roughly parallel development of ideas here and in the nuclear weapons program: both were thinking about neutron production and capture (a point Smirnov elaborates beginning on page 92).

Gamow's (1946) proposal was that the heavy elements were built by the "coagulation" of neutrons followed by nuclear beta decays that convert neutrons to protons (the decays being accompanied by the emission of electrons, or what is known as "beta" or "β" radiation). Ralph Alpher, who was Gamow's graduate student, made the coagulation idea more specific. In his doctoral dissertation (at The George Washington University, Alpher 1948a) the proposal is that the elements were built up by sequences of radiative captures of neutrons (that is, neutron capture accompanied by the emission of a photon, a quantum of electromagnetic radiation) and nuclear beta decays. In a preliminary report of this building-up idea, by Alpher, Bethe and Gamow (1948), Hans Bethe's name was added to produce an approximation to the first three letters of the Greek alphabet.

---

[1] The expansion rate has to be large to escape the strong gravitational attraction of the large mass density. One sees from equation (G.1) on page 518 that when the mass density is really large its value fixes the expansion rate, because the mass density term is by far the largest in the right-hand side of this equation. If the mass density is dominated by matter with relatively low pressure then when the density has dropped to the value ρ the model universe has been expanding for the time

$$t = 890\rho^{-1/2} \text{ s}, \tag{3.1}$$

where the value of ρ is measured in g cm^{-3}. If the mass density is dominated by radiation the time is three quarters of the value in this equation. The shorter time follows because the mass density in radiation falls more rapidly with the expansion of the universe than does the mass density in matter (for the reason in footnote 9 on page 29), and a larger earlier density requires a greater expansion rate to escape the stronger gravitational pull.

Alpher (1948a) and Alpher, Bethe and Gamow (1948) could cite a piece of evidence for their picture. They pointed out that in the building-up process the more readily an atomic nucleus of given mass and charge can absorb a neutron, to be promoted to a larger mass, the lower the expected cosmic abundance of that species of nuclei. And from measured nuclear reaction rates they concluded that if the elements were built up by exposure to a gas of neutrons that is hot – moving with velocities characteristic of a temperature of about 10^9 K – it would produce relative abundances of the elements that suggest a promising match to what is observed.[2] This encouraging result is still seen to apply in part of the periodic table, and the interpretation is the neutron capture building-up process, but transferred from the hot big bang to exploding stars, in what has become known as the "r-process." And basic parts of this building-up picture figure in the now well-tested theory for the origin of the lightest elements – the isotopes of hydrogen and helium – in the hot big bang. This was a memorable advance. But we must consider the introduction of several other important ideas.

Alpher (1948a,b) pointed out a problem with the theory. We remarked in footnote 1 that the relativistic big bang cosmology sets a relation between the mass density and the expansion rate. The condition that the rate of capture of neutrons produces a significant but not excessive amount of heavier elements sets another relation between the mass density and the expansion rate when the building-up reactions were at their peak. And there is a third condition, that the process must be completed in a few hundred seconds, before the neutrons have decayed.[3] Alpher showed that it is not possible to satisfy these three relations by the choice of two quantities, the characteristic matter density and characteristic expansion time when the building-up process occurred.[4]

---

[2] Alpher pointed out that if the velocities were much smaller than this then large rates of resonance capture by some nuclei would spoil the anticorrelation of element abundances and neutron absorption cross sections; if much larger the radiation would tend to break up the nuclei.

[3] In these exploratory discussions Gamow and Alpher generally left open the origin of the neutrons, and more broadly what was happening prior to the build-up of the elements. Alpher, Bethe and Gamow (1948) did note that the density of the universe might never have exceeded the density at element build-up, "which can possibly be understood if we use the new type of cosmological solution involving the angular momentum of the expanding universe (spinning universe)." On the origin of the neutrons, Gamow (1942) remarked that the fission of uranium produces free neutrons, and that the "rapid breaking up of the original superdense nuclear matter" might produce a large proportion of them. The simpler solution soon recognized is that if the universe had expanded from a very high temperature then heavy nuclei would have thermally evaporated and free neutrons would have been produced by the thermonuclear reactions to be discussed.

[4] In a little more detail the problem Alpher identified goes as follows. The condition that the build-up process produces an interesting but not excessive heavy element abundance is that the product $\sigma v n t$ is of order unity. Here σ is the radiative neutron capture cross section, v is the relative velocity of neutron and nucleus, n is the number density of nuclei, and t is

Alpher (1948a) also pointed to an effect that was missing in the calculation, which turned out to be the solution to his problem. The neutron gas was supposed to be hot (for the reason indicated in footnote 2), the hot matter ought to be accompanied by thermal blackbody radiation at the same temperature, and the mass density in this radiation would be much larger than the mass density in matter. But it was Gamow (1948a, with more detail in Gamow 1948b) who put these points together and solved the problem:[5] the large mass density in radiation speeds the expansion of the early universe so the build-up process can happen before the neutrons decay. This is the first analysis of the modern picture of the role of thermal radiation in element formation in the early universe.[6]

Gamow's argument begins with the thermal blackbody radiation present in a hot big bang. There would be a time, early enough in the expansion, when the temperature was high enough that the radiation would evaporate the atomic nuclei of any heavy elements, producing a gas of free protons and neutrons. As the universe expanded and cooled heavier elements could start to form in appreciable amounts, starting with captures of neutrons by protons to make deuterons (the nuclei of the stable heavy isotope of hydrogen). Each capture would be accompanied by the release of a photon (a quantum of electromagnetic radiation; at this energy usually written as γ) in the reaction[7]

$$n + p \leftrightarrow d + \gamma. \tag{3.2}$$

the expansion time, all evaluated during the build-up process. The measurements in nuclear reactors by Hughes (1946), as summarized by Alpher (1948a), indicate that the product σv is not very sensitive to v, and for the lightest nuclei amounts to $\sigma v \sim 10^{-19}\,\mathrm{cm^3\,s^{-1}}$. If the mass density is dominated by baryons then equation (3.1) gives $nt^2 \sim 10^{30}\,\mathrm{cm^{-3}\,s^2}$. It would follow from these two relations that build-up had to have occurred when the universe had been expanding for about $t \sim 10^3$ years. But that is absurdly large compared to the neutron lifetime, about 15 minutes.

[5] The Alpher, Bethe, and Gamow paper, which was submitted for publication on February 18, 1948, did not take note of the problem. They instead wrote that it "is necessary to assume" a much larger value of nt, that allowed the build-up process to happen at $t \sim 20\,\mathrm{s}$, well within the neutron lifetime. Alpher's thesis, which was accepted in April 1948, has a reasonable value of nt from the point of view of nuclear physics, and the consequent problem with the neutron lifetime. Gamow's paper, submitted on June 21, 1948, does not mention the problem but it presents the solution: take account of the mass density in radiation, which considerably increases the expansion rate. The published version of Alpher's dissertation, submitted July 2, 1948, states the problem but not Gamow's solution. It can take time to straighten out ideas.

[6] In the physical situation assumed in the calculations by von Weizäcker (1938) and Chandrasekhar and Henrich (1942) it is implicitly assumed that the baryons, being at near thermal equilibrium, are in a sea of blackbody radiation at the same temperature. We have found no one who took notice of the consequences of the presence of the mass density in this radiation prior to Gamow (1948a).

[7] Gamow's nonequilibrium calculation for this reaction assumes thermal equilibrium for the radiation but not for the relative abundances of the atomic nuclei. This is consistent: the interactions between radiation and electrons are fast enough to guarantee that the radiation remains very close to thermal, while the reaction in equation (3.2) is slow enough to break equilibrium.

The two-headed arrow means the reaction can go either way: a sufficiently energetic photon can break up a deuteron. Gamow noted that the critical temperature for the survival of deuterons, and hence the accumulation of appreciable numbers of them, is

$$T_{\text{crit}} \sim 10^9 \text{ K}. \tag{3.3}$$

At higher temperatures radiation breaks up deuterium as fast as it forms. When the temperature has fallen below T_{crit} the dissociation reaction going from right to left in equation (3.2) markedly slows because the cooler radiation does not have many photons energetic enough to break apart deuterons. This means deuterium starts to accumulate. As it does the deuterium can rapidly burn to helium by particle exchange reactions.[8]

It is essential for the consistency of this picture that at the critical temperature T_{crit} in equation (3.3) the total mass density is dominated by the energy of the thermal radiation that would accompany the hot plasma.[9] We know the radiation temperature, T_{crit}, when deuterons can start accumulating. The temperature tells us the mass density in radiation, which has to be very close to the total mass density. As we have noted, the mass density sets the expansion rate. The expansion time, shortened by the mass density in radiation, turns out to be comfortably less than the neutron lifetime, so neutrons could be available for the deuterium-producing reaction

---

[8] The most important reactions are

$$d + d \leftrightarrow {}^3\text{He} + n, \quad d + d \leftrightarrow t + p, \quad {}^3\text{He} + n \leftrightarrow t + p, \quad t + d \leftrightarrow {}^4\text{He} + n, \tag{3.4}$$

where tritium (t) is the unstable isotope of hydrogen that contains two neutrons. The reaction in equation (3.2) is slower, in conditions of interest here, because the electromagnetic interaction is weaker. That means the rate of equation (3.2) controls the rate of formation of helium by these reactions.

[9] At the present epoch the mass density in radiation is smaller than in matter, as is indicated in Table 2.1, but at the time of light element formation the mass in radiation was the largest component. This is because the energy of each photon, and its equivalent in mass, decreases as the universe expands, an effect of the cosmological redshift. The wavelength of a CMBR photon is increasing, as $\lambda \propto a(t)$, where $a(t)$ is the expansion parameter in equation (2.3) (and in equation 2.4 for the density and equation 2.7 for the temperature). Thus the photon energy is decreasing as $\epsilon = h\nu = hc/\lambda \propto a(t)^{-1}$, where c is the velocity of light and h is Planck's constant. The number densities of baryons and photons decrease as the volume of the universe increases, in proportion to $a(t)^{-3}$. Putting this together, we see that the mass densities in baryons and radiation vary as

$$\rho_m \propto a(t)^{-3}, \qquad \rho_r \propto T^4 \propto a(t)^{-4}, \tag{3.5}$$

as the universe expands. The middle part of the second expression is inserted as a reminder of a general relation: the energy density in thermal radiation at temperature T is $u = aT^4$, where a is Stefan's constant (not to be confused with the expansion parameter $a(t)$). When the temperature had fallen to T_{crit} (equation 3.3) the mass density would have been dominated by the radiation. This mass density sets the time elapsed, $t_{\text{crit}} \simeq 230$ s (from equation 3.1), from a really hot beginning to $T = T_{\text{crit}}$.

in equation (3.2). The matter density when the temperature has fallen to T_{crit} dictates how much deuterium could accumulate.

The density of neutrons and protons would have to have been large enough to allow production of an appreciable amount of deuterium, but not so much that it converts more hydrogen into heavier elements than is observed.[10] This consideration led Gamow to conclude that, if the neutron capture picture for element formation were right, then when the temperature in the early expanding universe had fallen to T_{crit} the number density of baryons – neutrons and protons – would have to have been $n_{\mathrm{crit}} \sim 10^{18} \, \mathrm{cm}^{-3}$.

Alpher and Herman (1948) took the next bold step: from the conditions required for element production in the early universe, predict the present temperature of the fossil radiation left from the hot early universe. When the temperature was T_{crit}, and elements heavier than hydrogen could start accumulating, Gamow had found an estimate of the baryon mass density n_{crit} that would allow production of a reasonable abundance of the heavier elements. The mass density and temperature drop as the universe expands, in the proportion $n \propto T^3$ (as one sees in equation 3.5). This means that when the temperature drops by a factor of 10 the mass density in matter drops by a factor of 1000. And one can similarly compute the temperature when n has dropped to its present value. Alpher and Herman found that, at present matter density,[11]

$$\rho_0 = 10^{-30} \, \mathrm{g \, cm}^{-3}, \tag{3.6}$$

the radiation temperature would be

$$T_0 \sim 5 \, \mathrm{K}. \tag{3.7}$$

In view of all the uncertainties this is strikingly close to what was measured many years later, $T_0 = 2.725 \, \mathrm{K}$.

One should not take the consistency of numerical values for the theory and measurement of T_0 too seriously, because there were problems with estimates of the mass density in equation (3.6). It was later learned that an error in the early estimates of the scale of distances to the galaxies introduced a

[10] As in the discussion in footnote 4, the Gamow condition is expressed as $\sigma_{\mathrm{crit}} n_{\mathrm{crit}} v_{\mathrm{crit}} t_{\mathrm{crit}} \sim 1$, where σ_{crit} is the radiative capture cross section for equation (3.2), v_{crit} is the relative neutron–proton velocity, and n_{crit} and t_{crit} are the baryon number density and the expansion time, all evaluated when the temperature is $T = T_{\mathrm{crit}}$.

[11] Alpher and Herman (1948) did not state this quantity. It is the mass density the group used in other papers, including Gamow (1946), Alpher (1948a,b), and Alpher and Herman (1949). Within rounding error it agrees with the indication in Alpher and Herman (1948) that the mass densities in matter and radiation are equal at $T = 600 \, \mathrm{K}$. There is the complication that their reported application of the Gamow condition at T_{crit} yields a matter density at T_{crit} that extrapolates to $T_0 = 20 \, \mathrm{K}$ at mass density $10^{-30} \, \mathrm{g \, cm}^{-3}$, not $T_0 = 5 \, \mathrm{K}$. Either there is an error in the paper or they adopted a present mass density well below the value they and Gamow generally used.

considerable overestimate of all the mass density measurements.[12] Also, it was known at the time that different methods yield quite different values for the mass density.[13] The point of lasting value is that in the hot big bang cosmology there is a relation between measurable conditions in the universe as it is now and conditions in the early universe when light elements could have been produced. The details of this relation have since been refined, as will be described, but this consideration by Alpher and Herman remains part of our standard cosmology.

There are several names to describe physical conditions in the early expanding universe, at the epoch of light element formation. In the published version of his thesis, Alpher (1948b) offered this opinion:

According to Webster's New International Dictionary, 2nd Ed., the word "ylem" is an obsolete noun meaning "The primordial substance from which the elements were formed." It seems highly desirable that a word of so appropriate a meaning be resurrected.

Alpher's ylem left us fossils, the light elements and the thermal cosmic background radiation (CMBR).

Gamow recognized that the thermal radiation in the ylem would be present after the early episode of element formation, and it would remain an important dynamical actor.[14] Alpher and Herman (1948) went further: they clearly stated that the thermal radiation would be present now. And

[12] Hubble's distance estimates were low by a factor of about 7.6. A mass density estimate from that time, corrected for the distance scale while leaving all other data unchanged, should be divided by the factor 7.6^2, which would divide the predicted present temperature of the CMBR by the factor $7.6^{2/3} \simeq 4$.

[13] The range of estimates of the mass density is an early indication of the dark matter problem. Hubble (1936) reported that the mean mass density is no less than about $\rho_{min} = 1 \times 10^{-30}\,\mathrm{g\,cm^{-3}}$ and may be as large as $\rho_{max} = 1 \times 10^{-28}\,\mathrm{g\,cm^{-3}}$. Gamow and colleagues used the lower value; Alpher (1948a) attributes it to Hubble (1936, 1937). It corresponds to the density parameter (defined in equation G.1) $\Omega_{min} = 0.002$. This number, which is independent of the distance scale, is comparable to the mass density in stars entered in Table 2.1. That makes sense, because Hubble's lower mass density used observations of the masses in the luminous parts of individual galaxies, which are dominated by the mass in stars and include most of the stars. Hubble's larger estimate, ρ_{max}, corresponds to density parameter $\Omega_{max} = 0.2$. It is comparable to the total mass density in matter entered in the table. This also makes sense, because Hubble based it on the mass per galaxy in clusters of galaxies, which he attributed to Smith (1936) (though Zwicky (1933) had made the point earlier). We know that clusters contain a close to fair sample of baryonic and dark matter, so the cluster mass per galaxy multiplied by the number density of galaxies gives a pretty good measure of the cosmic mean mass density. If Alpher and Herman had used ρ_{max}, it would have increased their estimate of T_0 by the factor $\sim 100^{1/3}$, which happens to about cancel the effect of the distance error. Let us notice, however, that the Gamow condition relates the CMBR temperature to the baryon mass density. The baryon mass is smaller than the total represented by Hubble's ρ_{max}, which is dominated by nonbaryonic dark matter, and larger than Hubble's ρ_{min}, which does not include the plasma in and around groups and clusters of galaxies.

[14] The paper Gamow (1948a) presents an estimate of the time – well after element formation – when the mass densities in matter and radiation were equal and, as Gamow recognized, the expanding universe became unstable to the gravitational growth of nonrelativistic concentrations of matter that eventually became galaxies.

in subsequent papers (Alpher and Herman 1949, 1950), they converted the present temperature to the present mass density in radiation, producing the first estimates of the third quantity in Table 2.1. Less easy to judge is whether they saw indications of the experimental methods to be described later in this chapter that might have been capable of detecting the radiation. Burke, on page 182, gives an assessment of the experimental situation.

The next important refinement in the theory is the process that fixes the relative number n/p of the neutrons and protons that enter the first step of element-building in equation (3.2). In the paper Gamow (1948a) the value of n/p is left open: Gamow was content to establish orders of magnitude. Hayashi (1950), on the other hand, recognized that in this hot big bang cosmology n/p may be computed from well-determined physics, as follows.

When the temperature in the early stages of expansion of the universe was above about 10^{10} K (and, by the argument in footnote 1 on page 26, the large mass density of the radiation caused the universe to have been expanding for about a second), the radiation was hot enough to produce a thermal sea of electrons and their antiparticles, positrons, and a sea of neutrinos and antineutrinos (ν and $\bar{\nu}$), mainly by the reactions

$$\gamma + \gamma \leftrightarrow e^+ + e^-, \qquad e^+ + e^- \leftrightarrow \nu + \bar{\nu}. \qquad (3.8)$$

These particles convert protons to neutrons and back again by the reactions

$$p + e^- \leftrightarrow n + \nu, \quad n + e^+ \leftrightarrow p + \bar{\nu}, \quad n \leftrightarrow p + e^- + \bar{\nu}. \qquad (3.9)$$

At temperatures above 10^{10} K the reactions drive the ratio n/p of numbers of neutrons and protons to its thermal equilibrium value,

$$\frac{n}{p} = e^{-Q/kT}, \qquad (3.10)$$

at temperature T.[15] Here $Q = (m_n - m_p)c^2$, where $m_n - m_p$ is the difference of mass of a neutron and of a proton, and k is the Boltzmann constant.

Hayashi found that as the universe expanded and cooled below 10^{10} K the reactions in equation (3.9) slowed to the point that the value of n/p froze,

---

[15] To be more accurate, we should note that the equilibrium value of n/p also depends on the lepton number, which is the sum of the numbers of e^- and ν particles minus the sum of the numbers of e^+ and $\bar{\nu}$. The reactions in equations (3.8) and (3.9) do not change the lepton number: its value had to have been set by initial conditions very early in the expansion of the universe. Equation (3.10) assumes the absolute value of the lepton number density is small compared to the number density of CMBR photons. A positive and large lepton number suppresses n/p, and a strongly negative lepton number increases n/p. This point figures in the cold big bang model we discuss beginning on page 35. The present observational constraints are consistent with the small lepton number assumed in equation (3.10). To be even more accurate we should take notice of the three families of neutrinos, but that does not figure in the history in this chapter.

and then n/p more slowly decreased as neutrons freely decayed to protons (by the last reaction in equation 3.9 going to the right). By the time the temperature had dropped to T_{crit} the ratio of neutrons to protons (n/p) would have fallen to $\sim$0.2. In the standard cosmology most of the neutrons present at this time combined with protons to form deuterons, and most of the deuterium burned to the heavy isotope of helium, ^{4}He, with a trace amount of the lighter isotope ^{3}He.

The paper Alpher, Follin and Herman (1953) presents a detailed application of Hayashi's idea. Their analysis of how the ratio n/p varies as the universe expands and cools is essentially the modern computation. Enrico Fermi and Anthony Turkevich (in work that is not published but is reported in Gamow (1949), ter Haar (1950), and in more detail in Alpher and Herman 1950, 1953) worked out the chains of particle exchange reactions that burn deuterium along with neutrons and protons to helium and trace amounts of heavier elements. These analyses essentially completed the formulation of all the pieces of what was much later established as the standard model for the origin of most of the isotopes of hydrogen and helium, along with the CMBR.

We can reconsider now the question in footnote 3 on page 27: what was the nature of the universe before the build-up of the light elements? The theory just described assumes the expansion traces back to temperatures above 10^{10} K, when the distributions of the radiation and the baryons are supposed to have been very close to spatially uniform – homogeneous even on small scales. The baryon density at that epoch is chosen to fit the observed light abundances. That is one way to determine the value at the present epoch listed in Table 2.1. (Another way is examined in Chapter 5.) The baryons have to have been created, but that is assumed to have happened still earlier, at much higher temperatures than we are considering. General relativity theory gives the rate of expansion of the universe. The early expansion is rapid, but at temperature 10^{10} K the exchanges of energy among particles and radiation are even faster. This means conditions then would have forced relaxation to thermal equilibrium, including the thermal ratio of neutrons to protons. Thus for the purpose of the theory of light element formation that commences at $T \sim 10^9$ K we need only these assumptions: we need not enquire about conditions at still earlier times. The question is fascinating, of course, and there are ideas: a favorite is the inflation picture (described in Guth 1997 and outlined on page 520). But for the story of the CMBR we need not consider the weight of evidence of whether inflation is a useful approximation to what actually happened in the exceedingly early universe.

We are very interested in the development of the empirical checks of these ideas on how the CMBR got its thermal spectrum (shown on page 16), and how the light elements formed. In the 1960s it was not at all obvious whether we would be able to find convincing tests, or, if we did, whether these ideas would pass the tests. People felt free – perhaps even compelled – to cast about for other ideas that might be philosophically attractive and perhaps better approximate reality. The debate over these alternatives is an important part of the story of how we arrived at the standard cosmology. We turn now to some of the ideas.

3.2 Nucleosynthesis in alternative cosmologies

The evidence developing in the 1950s was that the heavier elements were produced in stars. If so, might the stars also produce light elements? If that were so, helium production in a hot big bang could be a problem: it might produce too much helium. But that was easy to fix: adjust the prediction by adjusting the assumptions in the big bang model, or go to an alternative cosmology, the steady state picture for example. We review here some of the alternatives people were considering. The point to notice is that in 1960 the relativistic hot big bang model for the universe was not the obviously best possibility: there were other ideas that were arguably as elegant. We needed observations to show the way through the thickets of elegance. Our purpose in this book is to trace the development of a large part of the evidence.

Let us consider first what came of the proposal that the heavy elements were formed in the big bang along with the light elements. A problem with this idea is that there is no stable atomic nucleus with mass 5 (that is, a total of five neutrons plus protons). That means the abundant isotope of helium, with mass 4, cannot capture a neutron and then another one and subsequently decay to an isotope of lithium by the emission of an electron. This strongly suppresses the build-up of elements heavier than helium during the rapid expansion of the early universe. Alpher (1948b) remarks on the problem, and a like situation at mass 8, in the published version of his doctoral dissertation. The analysis mentioned above by Fermi and Turkevich failed to find a nuclear reaction that might carry significant nuclear burning in the early universe past the mass-5 gap. But Gamow (1949) noted a possible way out that is worth considering even though it proved to be wrong: false steps can be edifying.

Gamow's idea was that if the mass density in baryons when the temperature of the universe was T_{crit} were much larger than previously considered, and n/p were smaller (as, it was later realized, follows from Hayashi's 1950 analysis of equation 3.9), then after all the neutrons had combined with protons to form heavier elements a substantial fraction of the baryons would be left as protons, the nuclei of hydrogen atoms. That agrees with what is observed: hydrogen is the most abundant element. The larger matter density in the early universe would cause faster nuclear burning of deuterium, and perhaps that could push nuclear burning past mass 5 to make the heavier elements. We see another consequence that attracted no attention then: when the matter density had dropped to the present value the radiation temperature would have been much lower than in the hot big bang Gamow had introduced earlier.

Hayashi and Nishida (1956) presented an analysis of this idea. They considered the possibility that the baryon number density at temperature $T = 10^{10}$ K is at least a hundred million times what is assumed in Gamow (1948a) and Alpher and Herman (1948). That lowers the present temperature of the CMBR by a large factor, which Hayashi and Nishida would not have counted as a problem because the CMBR was not known. They took account of the helium-burning reactions

$$^4\text{He} +{}^4\text{He} \leftrightarrow {}^8\text{Be} + \gamma, \quad {}^8\text{Be} +{}^4\text{He} \leftrightarrow {}^{12}\text{C} + \gamma, \quad {}^{12}\text{C} +{}^4\text{He} \leftrightarrow {}^{16}\text{O} + \gamma, \quad (3.11)$$

which by then were known to be important in the evolution of stars after all the hydrogen in the central regions had burned to helium. In this "cool" big bang model universe, Hayashi and Nishida found significant production of carbon and oxygen. The deuterium abundance coming out of this model is much too small, according to what is now known, and the helium abundance is too large, though not by a large factor.[16]

This cool big bang universe produces helium in an amount that might not have seemed unreasonable at the time. It also produces a not insignificant amount of heavy elements. Layzer and Hively (1973) pointed out that the heavy elements produced in such a cool big bang might form dust grains that were able to absorb and reradiate starlight effectively enough to have produced the thermal CMBR spectrum out of starlight. Here is an example of an idea that is interesting but was not pursued, and as it happened later proved to be not viable. The light element abundances are wrong, and the picture cannot account for the relation between the large-scale distributions of matter and the CMBR that is discussed in Chapter 5.

Zel'dovich (1962, 1963a,b, 1965) proposed lowering the temperature all the way, to a cold big bang in which element production is left entirely to the

---

[16] That is because almost all the neutrons that survive to the time when the temperature has fallen to T_{crit} are burned to helium, and the value of n/p when deuterium starts accumulating is not very sensitive to the density of matter.

stars. He was led to this picture by his impression that the helium abundance in some stars is quite small.[17] His evaluation of the Gamow condition (in footnote 10 on page 30) led him to conclude that a hot big bang cosmology could only account for the low helium abundance if the present-day universe were hot, $T_0 \sim 30\,\mathrm{K}$ (at his estimate of the present baryon mass density, $10^{-29}\,\mathrm{g\,cm^{-3}}$; Zel'dovich 1963a), well above what was later observed. That is because a higher temperature today, at a given present mass density, implies a lower matter density at T_{crit}, which means fewer nuclear reactions that produce less helium, as Zel'dovich thought was required.[18] He argued that this high present temperature is unlikely because it would imply strong scattering of the thermal photons by fast-moving electrons, unacceptably limiting lifetimes of cosmic ray electrons.

Later developments on this issue are very relevant to our story. In a note added in proof in a paper published later that year Zel'dovich (1963b) stated that the latest data "indicate the temperature of intergalactic thermal radiation is below $1° - 0.5°\,\mathrm{K}$," which would add to his arguments for a cold big bang. Zel'dovich (1965) later mentioned the likely source of these data, Ohm (1961). The origin of Ohm's landmark paper, which actually could be read to suggest the presence of a sea of microwave radiation, is outlined in Section 3.5 beginning on page 44; Hogg describes the situation in more detail in Chapter 4 (beginning on page 70). Novikov (p. 99) explains why he and Doroshkevich, who were members of Zel'dovich's research group, were particularly interested in Ohm's paper. Novikov and Smirnov describe Zel'dovich's reaction to news of the identification of the CMBR. His reaction is illustrated also in Zel'dovich's letter to Dicke quoted on page 196. But this happened later: in 1963 Zel'dovich saw a good case for a cold big bang. It is worthwhile considering how he found what seemed to be a viable theory.

In Zel'dovich's cold model the very early universe contained equal number densities of protons, electrons, and neutrinos, all very nearly uniformly distributed, and cold, meaning the particle energies are as low as possible. This means one has to consider the effect of the exclusion principle that

[17] Zel'dovich (1963a) mentions evidence of stars with helium abundance Y as low as 2.5% by mass. He adds the careful statement (in the English translation) "We cannot make any estimate of the reliability of these results." But the paper proceeds on the assumption that Y is not more than about 0.1. Osterbrock (p. 86) describes the evidence known then that Y is larger than that.

[18] Smirnov's account of Zel'dovich's suggestion that he reanalyze element production in a hot big bang, and perhaps increase the challenge for a hot case, commences on page 94. The results, in Smirnov (1964), showed that the small primeval helium abundance he thought he should be aiming for could be accommodated in the hot big bang picture by lowering the matter density at a given radiation temperature, as Zel'dovich had proposed, but that would imply an unacceptably large abundance of deuterium. This is because at the lower densities Smirnov considered neutrons and protons combine to form deuterium, but the burning of deuterium to helium is incomplete.

limits the allowed number densities of electrons and neutrinos at a given energy. The high density of electrons in this cold early universe would cause the electrons to occupy all their available states up to a large energy. The energetic electrons would normally force themselves onto protons to make neutrons, by the first reaction going to the right in equation (3.9). But that is not allowed here because it would require the production of neutrinos, and in this picture all the neutrino states with the energy allowed by the reaction are already taken.[19] In this universe star formation would commence with nearly pure hydrogen. This is yet another interesting universe that proves not to be the one we live in.

Hoyle and Tayler (1964) knew that the helium abundance is large, and greater than seemed reasonable for production in stars. They too reconsidered the hot big bang model, but they also pointed to another possibility. In the steady state cosmological model the universe always has been as it is now: there would be no fossil helium. Hoyle and Tayler suggested that the helium could have been produced in the "little bangs" of very massive exploding stars. The evolution of temperature and density within a very hot exploding star is similar to the evolution in an expanding universe, so element formation is similar too. Worth noting here is that the energy released by the conversion of hydrogen to the observed amount of helium would produce radiation energy density comparable to what is in the CMBR.[20] Here is an elegant unified theory of the origins of helium and the CMBR, but it is yet another universe that we know is not ours. Like cool and cold big bangs, it cannot account for the measured properties of the thermal CMBR discussed in Chapter 5.

Still another alternative, which could eliminate fossil helium while leaving us with the fossil CMBR thermal radiation, was the idea that the laws of physics might change as the universe expands. An example of particular interest then (and now) is that gravity, which is weak now, might have been stronger in the past, making the early universe expand too rapidly to allow

---

[19] Another way to put this is that Zel'dovich assumed the lepton number mentioned in footnote 15 on page 32 is positive and large enough to force the equilibrium ratio of neutrons to protons at high density and low temperature to a value close to zero. Zel'dovich's idea of adjusting the cosmic lepton number can be extended to a hot big bang model; it changes the relation between the helium abundance coming out of the big bang and the CMBR temperature. The evidence now is that the lepton number is negligibly small (Steigman 2007).

[20] Suppose, for example, that 25% of baryon matter density $\rho = 10^{-29}\,\mathrm{g\,cm^{-3}}$, a value often discussed then, were burned from hydrogen to helium, with the conversion about 0.005 times the mass in helium to radiation (depending on what fraction of the released nuclear binding energy is lost to neutrinos, redshift, and maybe remnant black holes). Using the relation between temperature and blackbody radiation energy density in footnote 9 on page 29, we see that this energy is equivalent to radiation temperature $T = 6\,\mathrm{K}$. This line of thought is discussed further on page 58 and in the contributions by Faulkner (beginning on page 251) and Burbidge and Narlikar (beginning on page 267).

time for any appreciable build-up of the elements. This idea was inspired by the following consideration.

A measure of the relative strengths of the gravitational and electromagnetic interactions is the ratio of the gravitational and electric forces of attraction between an isolated electron and proton:

$$\frac{f_{\text{grav}}}{f_{\text{el}}} = \frac{Gm_em_p}{e^2} \sim 10^{-40}. \qquad (3.12)$$

The charges of a proton and electron are $+$e and $-$e, their masses are m_{p} and m_{e}, and G is Newton's gravitational constant. Since both forces vary in the same way with the separation of the particles, this ratio does not depend on the separation. Its small value – gravity is a very weak force compared to electricity – led Dirac (1938) to ask whether the strength of the gravitational interaction might be decreasing: maybe gravity is exceedingly weak now because the universe is very old. Alpher (1948a) mentioned the idea, and a consequence: if gravity were stronger in the past then the rate of expansion of the early universe would be larger than is predicted by general relativity theory (because stronger gravity then required a larger rate of expansion to escape the gravitational pull). That would affect the computations of element formation. Alpher quoted Teller's (1948) argument against the idea: if gravity were significantly stronger when Earth was young then the Sun would have been significantly hotter, making early life on Earth impossible. But beginning in the 1950s Pascual Jordan and Robert Dicke reconsidered Teller's argument and concluded that the observations might instead suggest that the strength of gravity is evolving. The important consequence of this line of thought for the purpose of our story is its effect on Dicke's thinking about the early universe.

Dicke felt that Dirac's proposal is an appealing illustration of another idea, which he, Dennis Sciama, and others termed Mach's principle.[21] Following earlier discussions, Ernst Mach (1883) had asked what determines the motion of a body that is moving freely and without rotation. It seemed unlikely to Mach that this free or inertial motion is an intrinsic world feature; he supposed rather that inertial motion is determined by motion relative to all the rest of the matter in the universe.[22] It seemed likely to Dicke that if

---

[21] Sciama (1959) and Dicke (1964) review their thoughts on what Mach's principle might mean for cosmology. These ideas still attract attention, but have not been fixed within a definite theory. The term Mach's principle accordingly means different things to different authors.

[22] Mach's arguments played an earlier role in the development of cosmology. Einstein considered them to be one of the guides to his general relativity theory: matter is the source term in the field equation that determines the geometry of spacetime, roughly what Mach and others had in mind. But the theory allows a universe in which there is an island of matter in a spacetime that is arbitrarily close to flat at arbitrarily great distance from the matter. A particle could escape

inertial motion were determined by what all the rest of the matter around us is doing then the same may be true of other aspects of physics, including gravity. Perhaps the thinning of the mass distribution around us as the universe expands causes the strength of gravity to decrease.

Jordan's thoughts about Dirac's proposal led him to the idea of an adjustment of general relativity to a scalar-tensor gravity theory in which the number in equation (3.12) decreases as the universe expands. The first version of this new theory is in Jordan (1952). Jordan (1962) summarizes ideas about possible observational consequences, largely on Earth's evolution. Dicke's reading of Mach's principle led to the exploration of the scalar-tensor theory in Brans and Dicke (1961). Dicke was taken with the idea that, in this theory, gravity in the very early universe could have been so strong that the universe was expanding so rapidly[23] that there was no production of elements heavier than hydrogen. That led to the comment in the letter from Dicke to Sciama quoted on page 199: at the time Dicke thought there is a good case for a hot big bang that left the fossil CMBR but no helium before nuclear burning in stars (Dicke 1968).

The Jordan–Brans–Dicke theory is another example of an idea that fascinates but fails, at least in its original intended application: we have tight experimental limits on any possible variation of the strength of the gravitational interaction or on many other conceivable departures from general relativity theory. Interesting ideas tend to be durable, however. This theory, and the idea that numbers such as the one in equation (3.12) may vary with time, continues to figure in debates about the physics of the very early universe.

In the 1950s and earlier it was logical to consider yet another departure from what had become conventional ideas: perhaps our universe of galaxies is not close to homogeneous. Perhaps the observed tendency of matter to

this island of matter, move arbitrarily far away, and yet retain its usual inertial properties. If the particle were large enough to house an observer with a gyroscope, the observer could determine whether the particle is spinning by referring to the motion of the gyroscope. But spinning relative to what? Einstein (1922a,b) noted that this situation is possible within general relativity theory, but he argued that if the universe were constructed this way "then Mach was wholly wrong in his thought that inertia, as well as gravitation, depends upon a kind of mutual action between bodies" (Einstein 1922a, p. 109). The problem is avoided if, as Einstein (1917) had proposed, matter uniformly fills space – apart from local irregularities. This picture of a homogeneous universe came to be known as the "cosmological principle." There are isolated island universes of matter, the galaxies. But the cosmological principle has proved to be a good approximation to the observed large-scale mass distribution. We do not know whether Einstein arrived at the right picture for the large-scale structure of the universe for the right reason.

[23] If this seems counterintuitive consider, as we have remarked earlier, that stronger gravity would more rapidly slow the rate of expansion, so the expansion rate would have had to have been larger to allow the universe to reach its present state. A more formal argument is in the first part of equation (G.1) increasing G when the mass density is large would make the expansion rate $\dot{a}/a$ larger than in standard physics.

be concentrated in galaxies, which are in turn found in groups and clusters of galaxies, extends in a hierarchy of clusters within clusters to the largest observable scales. Charlier's (1922) map of the distribution of the galaxies shows that this clustering hierarchy picture is a better fit to what was then known than Einstein's homogeneous universe. An alternative, in von Weizäcker's (1938) discussion of how the elements may have formed, is that the universe of galaxies is bounded and expanding into empty asymptotically flat spacetime. Observers tended to like these pictures because the galaxies are distributed in a decidedly clumpy way. Thus a report by Oort (1958) on the observational situation commences with the sentence, "One of the most striking aspects of the universe is its inhomogeneity." We remarked in footnote 22 that Einstein disliked the idea, but other theorists found it attractive: Charlier (1922, 1925) and Klein (1958) presented well-reasoned arguments in favor of large-scale departures from a homogeneous mass distribution. These arguments are worth reading, but they are not much heard now because the other side won by the weight of the evidence.

Oort (1958) remarked on one of the pieces of evidence: the counts of progressively fainter galaxies increase about as expected in a homogeneous universe. (The relation is shown in equation 3.18 on page 55). By the early 1960s the distributions of radio sources and the X-ray background radiation were observed to be close to isotropic across the sky. Radio waves and X-rays seem to propagate through intergalactic space without significant scattering. That means they could only be seen to be isotropic if we were in a special place, close to the center of the expanding cloud, which seems unreasonable, or else if the universe were close to homogeneous.[24] The network of evidence discussed in Chapter 5 shows that on the scale of the Hubble length mass density fluctuations amount only to a few parts in one hundred thousand. Einstein's picture of a reasonable universe, one that is close to homogeneous, was right.

These are examples of how elegant ideas may lead us astray or to aspects of reality. Let us consider next an arguably questionable idea that led to a decidedly interesting part of reality.

3.3 Thermal radiation from a bouncing universe

A big step toward sorting out all these ideas was the discovery of a fossil: the sea of microwave radiation that smoothly fills space. The chain of ideas and events that brought this radiation to the attention of the community includes the thought that our expanding universe might have bounced from

[24] The argument is given in more detail in Peebles 1971, p. 40.

a previous collapse. As we will discuss, the bounce might be expected to have filled space with thermal radiation.

The notion of a bouncing or oscillating universe certainly was not ignored. Lemaître (1933) had expressed the feeling that, from a purely aesthetic point of view, a universe that successively expands and contracts to exceedingly small size has "un charme poétique incontestable et faisaient penser au phénix de la légende." De Sitter (1933) was not so positive: "Personally I have, like Eddington, a strong dislike to a periodic universe, but that is a purely personal idiosyncrasy..." But he noted that a collapsing universe is unstable against the growth of departures from homogeneity, meaning different regions arrive at high density at different times. In a patch that does not become too dense most stars may avoid collisions; they may instead pass each other and move apart to join the new general expansion. Alpher, Bethe and Gamow (1948) suggested consideration of a bounce resulting from the net angular momentum of the universe. Wheeler (1958) put it that the bounce in an oscillating universe might be compared to "a glove which is turning itself inside out one finger at a time." Hoyle and Narlikar (1966) considered another variant: perhaps the universe is in a steady state overall, but "pockets of creation" set a part of the universe into a local oscillation. But the important notion for our purpose is Tolman's (1934) remark that a bounce could produce entropy, largely in the form of a sea of thermal radiation. Weinberg (1962) found a related result: neutrino emission and absorption in an oscillating universe would drive the distributions of low energy – and massless – neutrinos and antineutrinos to the form characteristic of thermal equilibrium. And at roughly the same time Robert Dicke (in an unpublished discussion that is described more completely in the essays in Chapter 4) made Tolman's picture of the production of thermal electromagnetic radiation during a bounce more tangible, as follows.

Dicke noted that the nuclear burning of four protons – the nuclei of hydrogen atoms – to form the nucleus of one helium atom in a star releases enough energy to produce roughly a million starlight photons. The burning of helium to heavier elements produces still more starlight photons. These starlight photons are shifted toward the red as the universe expands. If the expansion eventually stopped and the universe collapsed back to high density then during the collapse the starlight photons would be shifted toward the blue, to greater energy. If the blueshift were large enough then just a few blueshifted starlight photons would have enough energy to break apart each heavy atom, reducing it to protons. These protons would serve as fuel for nuclear burning in new generations of stars in the next cycle of expansion and collapse. The rest of the starlight photons would be thermalized, that is, turned into what we observe as the CMBR.[25] A few hundred bounces could make the observed energy density in thermal radiation out of starlight in a universe like ours, if the bounces conserved the numbers of baryons and photons.

Dicke had hopefully put aside, as a possibly minor nuisance, the developing evidence that general relativity is an incomplete theory of spacetime going forward in time to relativistic collapse to a black hole, and maybe incomplete also going backward in time to the big bang. Ellis (p. 379) recalls the gathering storms of the relativistic singularity theorems. The problem is still with us: general relativity cannot give a complete description of the arbitrarily remote past of our universe. But the general idea of a bouncing or quasiperiodic universe continued to attract interest (Steinhardt and Turok 2007). And it proved to be interesting enough in the 1960s that Dicke was able to persuade two members of his Gravity Research Group, Peter Roll and David Wilkinson, to build an instrument capable of detecting a sea of thermal microwave radiation.

News of the Roll–Wilkinson experiment reached Arno Penzias and Robert Wilson at the Bell Telephone Laboratories in Holmdel, near Princeton University. Hogg (p. 70) and Penzias and Wilson (pp. 144–176) recall the communications experiments that led to the detection of more microwave radiation than could be accounted for from known sources in and around their instruments. The essays in Chapter 4 recall how the news came to the attention of astronomers who saw that the radiation could account for the curious behavior of cyanogen molecules in the gas between the stars. The astronomers' puzzle and its resolution is our next topic.

3.4 Interstellar molecules and the sea of microwave radiation

We come now to methods of detecting the sea of microwave radiation, the CMBR, and we begin with interstellar molecules that serve as "thermometers." This provided a measure of the temperature of the radiation some two dozen years before its presence was recognized.

---

[25] In the language of thermal physics, Tolman (1931) had shown that the homogeneous and isotropic expansion of a universe filled with free thermal radiation is a reversible process: it conserves entropy. Tolman remarked that a bounce might be violent enough to be irreversible, producing entropy. Dicke gave an explicit example. Tolman's result follows in a free gas of particles with energy proportional to a power of momentum, as in photons or nonrelativistic particles, though the cooling rates as the universe expands are different. A rapid transition from a relativistic to nonrelativistic gas is irreversible. Yakubov (1964) computed the resulting entropy production in Zel'dovich's cold big bang model.

The function of a species of interstellar molecule as a thermometer follows from some results from quantum physics. The energy of an isolated object such as an atom or molecule has discrete – quantized – allowed values: it has a ground level with energy E_0, a first excited level with energy E_1, a second level at E_2, and so on. The energy levels of an object as large as a person would be fantastically closely spaced if we were able to truly isolate someone, but it can't be done: we interact – exchange energy – too strongly with our environment. The effect of the quantization of energy is clear and distinct on the much smaller scale of atoms and molecules, however.

In a dilute gas of molecules bathed in blackbody radiation at temperature T, absorption and emission of the thermal radiation causes the ratio of numbers of molecules in the first excited energy level and the ground level to relax to the value given by the equation

$$\frac{n_1}{n_0} = e^{-(E_1 - E_0)/kT}. \tag{3.13}$$

This has the same form as equation (3.10) for the thermal equilibrium ratio of numbers of neutrons to protons in the early universe, but here applied at much lower energies and temperatures and much later in the history of the expanding universe. Because the energy levels might be labeled by the spin angular momentum quantum number, this expression is said to give the spin temperature corresponding to a measured ratio n_1/n_0.[26]

The ratio n_1/n_0 for a species of molecules in interstellar space can be measured by comparing the strength of absorption of light from a background star by the molecules in the two energy levels. Starlight photons may be absorbed by a molecule in its ground level, with energy E_0, leaving the molecule in some highly excited level, with energy E_*. The photon has to supply the energy difference, $E_* - E_0$. From Planck's condition $E = h\nu$ we see that this absorption produces an absorption line at frequency $\nu_a = (E_* - E_0)/h$ in the spectrum of light from the star. A starlight photon with the lower frequency $\nu_b = (E_* - E_1)/h$ can be absorbed by a molecule in the first excited level, E_1, which again leaves the molecule at energy E_*. This produces a second absorption line, at frequency ν_b. The ratio of the amount of absorption at the two frequencies is a measure of the value of n_1/n_0. Since the energy difference $E_1 - E_0$ is known, equation (3.13) gives us a temperature. Thus, we have a thermometer.

There is the problem that the spin temperature measured by the ratio n_1/n_0 is determined not only by the effective temperature of electromagnetic

---

[26] It is conventional to use this spin temperature as a measure of the ratio n_1/n_0 even when the ratio is determined by energy exchanges that are not at all close to thermal equilibrium.

radiation bathing the molecules, but also by the interstellar particles that are colliding with the molecules we are studying and knocking them from one energy level to another. In effect the molecules are coupled to two heat reservoirs, radiation and interstellar particles, at different temperatures. The molecule cyanogen (CN, a carbon atom bound to a nitrogen atom) in interstellar space has two useful properties. First, it recovers quickly from collisions with particles. That means interstellar particles have relatively little effect on n_1/n_0: this thermometer is more sensitive to the temperature of the radiation than to the temperature of the interstellar matter. Second, the CN energy levels are well spaced for the measurement of radiation temperatures near that of the CMBR. The energy difference $E_1 - E_0$ for CN corresponds to the microwave wavelength 2.6 mm, which you can see is close to the peak of the spectrum in Figure 2.2. The spin temperature of interstellar CN thus provides a very convenient thermometer for the CMBR.

McKellar (1941) used equation (3.13) to translate observations of absorption of starlight by interstellar CN molecules in the two lowest levels to the spin temperature:

$$T \simeq 2.3 \text{ K}. \tag{3.14}$$

With hindsight, the inference we would draw is that interstellar CN molecules are bathed in radiation at about this temperature. But that was not suggested by McKellar or by Adams (1941), who made the measurements. Herzberg (1950) comments that this temperature "has of course only a very restricted meaning." The restriction he had in mind likely is that, as we have said, the excited levels of CN might be populated by particle collisions rather than radiation.

Astronomers are accustomed to dealing with complex situations that require them to remember and evaluate the possible significance of many curious things. In the early 1960s some knew that the observed excitation of CN by interstellar particles would require a curiously large collision rate. And after the proposed identification of the CMBR astronomers were quick to remember McKellar's spin temperature and recognize its possible relation to the hot big bang cosmology. How that happened is one of the threads running through the essays.

3.5 Direct detection of the microwave radiation

A direct detection of the CMBR uses a receiver that operates at some range of frequencies and an antenna that defines a beam, that is, the range of

Fig. 3.2. A Dicke microwave radiometer. From the left E. Beringer, R. Kyhl, A. Vance, and R. Dicke. Dicke is holding in front of the horn antenna a "shaggy dog," a good approximation to a source of blackbody radiation at room temperature. (From *Five Years at the Radiation Laboratory*, MIT, 1946.)

directions in the sky from which the radiation is received. The first two[27] direct detections, in experiments at the Bell Telephone Laboratories and then at Princeton University, had a feature in common: both used antennas shaped like a horn or funnel. A waveguide at the small end of the horn leads to a detector. The early example in Figure 3.2 shows the horn in the Dicke radiometer described by Dicke *et al.* (1946). A horn antenna can suppress radiation coming from directions well away from the main beam. That is important because the ground is a strong source of microwave radiation, and the system must be well shielded from this unwanted ground noise.[28] Hogg describes two horns used in the Bell communications experiments that detected the CMBR; these horns appear in Figures 4.1 and 4.2. Figures 4.22 and 4.23 shows the Princeton horn. The sizes are very different, but that is not important because the received energy flux from an isotropic sea of radiation does not depend on the horn size.[29]

[27] Other histories of this subject mention earlier experiments that may have detected the CMBR. We note on page 63 that we have not been able to substantiate any.

[28] Since the properties of horn antennas are important for the study of the CMBR we add here a few details. An example of the suppression of unwanted radiation incident from directions well away from the main beam of the horn is shown in Figure 4.13. The horns in Figure 3.2 and in the subsequent detections of the CMBR by the Bell Laboratories and Princeton groups have rectangular collecting areas. A reflector at the end of the Bell Labs horn brings radiation to a horizontal waveguide from a beam that may be swung across the sky. In communications applications using this Bel Labs design the waveguide is vertical and the beam horizontal. Burke (p. 179) compares this design to a scoop. Modern horns for measurements of the CMBR tend to have circular cross sections, as in a funnel.

[29] The solid angle of the antenna beam – the angular area in the sky from which radiation is directed to the detector – is inversely proportional to the collecting area of the antenna. The product of the solid angle and the collecting area determines the rate of collection of radiation energy from a source that is broader than the solid angle. This product is independent of

Unwanted radiation – noise – also originates in the receiver. The Bell Laboratories experiments used low-noise solid-state maser amplifiers developed there for the purpose of communication. The Princeton experiment used a detector with much larger noise, and they used a technique Robert Dicke pioneered to deal with it: rapidly switch the receiver between the antenna and a reference "load" that produces thermal radiation at a known temperature. The difference of the detector response subtracts the radiation originating in the detector and amplifier. The time average beats down the fluctuations in the difference. The Dicke radiometer thus yields a measurement of the difference $T_s - T_l$ between the wanted sky temperature T_s and the known load temperature T_l.

The low noise of the Bell Laboratories' receivers allowed the engineers to make useful estimates of the noise originating in the maser amplifier and add it to estimates of the noise from all other known sources of radiation: the atmosphere, the ground, the antenna, and the waveguide leading from the antenna to the amplifier. That sum could be compared to what was detected. A persistent discrepancy between what was detected and expected led Penzias and Wilson to take the final step: use a cold reference load to check the amount of radiation originating in the system (that adds to what is incident on the antenna). The results forced them to conclude that there is a source of radiation outside the system that they could not identify. Hogg, Penzias, and Wilson recall these events in the next chapter.

As this was happening Roll and Wilkinson, just 30 miles away in Princeton, were building an instrument to search for a possible sea of microwave radiation. They recall in the next chapter that Dicke had proposed this project to test the idea of a hot big bang; we reviewed his thoughts in Section 3.3. Roll and Wilkinson also used a cold load. The difference was that their instrument rapidly switched between sky and cold load, so as to compensate for the drifting level of the relatively large noise originating in the detector. Penzias and Wilson needed only to switch occasionally because their system noise was much lower.

Dicke had invented the instrument Roll and Wilkinson were building, a Dicke radiometer, as part of war research at the Radiation Laboratory at the

the horn size. The small horn in Figure 3.2 has poor angular resolution: Figure 3.3 shows the broad response of this radiometer to the relatively narrow warm objects. A larger collecting area gives better angular resolution, that is, sensitivity to compact sources. Figure 4.9 shows a dish reflector antenna, where incident radio or microwave radiation is directed by one or more reflecting surfaces – as in a dish – to a much smaller feed horn antenna leading to the detector. The size of a primary reflector can be made larger than the size of a horn, improving angular resolution. But that sometimes comes at the expense of poorer rejection of radiation incident from directions well away from the source, including ground noise, which can be a serious problem for observations of the CMBR.

Massachusetts Institute of Technology. An early application was the measurement of the emission – and hence absorption – of microwave radiation by the atmosphere, which at the time limited the push to develop radar at shorter wavelengths for better resolution.[30]

Figure 3.3 is Dicke's illustration[31] of the sensitivity of his radiometer. In this example, the reference load was at room temperature, and the switching was done by a wheel that swung the load into and out of the waveguide connecting the detector and the antenna. Later measurements used electronic switching and loads with temperatures that are colder and more closely matched to the CMBR temperature. A Dicke radiometer "sees" thermal microwave radiation wherever the horn is pointed, whether at the ground, or people, or the atmosphere. The strip-chart recording in Figure 3.3 shows the variation in the response when the antenna was pointed at the sky and at chimneys that were in use and so slightly warmer than their surroundings.

The top line in Figure 3.3, measured with the antenna scanning at an angle of 75° from the zenith, indicates a more or less uniform temperature of about 125 K. Variations in the temperature from one part of the sky to the other are small, less than about 10 K. This means the instrument used at MIT was capable of detecting temperatures as small as 10 K. Note also that the temperature measured well away from hot chimneys increases as the angle from the zenith is increased from 75° to 90°. Some of the increase in detected temperature at the larger zenith angle is the result of larger microwave emission by the Earth's atmosphere. This effect is the basis for a measurement of the radiation emitted by the atmosphere, as follows.

When the instrument is aimed closer to the horizon, it looks through a longer path through the atmosphere. The longer path length means the atmosphere produces more radiation along the line of sight.[32] By measuring how the temperature of the received radiation varies with the angular distance from the zenith, or with distance through the atmosphere, one

---

[30] Since microwave emission from the atmosphere is another important part of the story it is worth remarking here that if material at a nonzero temperature absorbs radiation then it also emits radiation. It is the balance of absorption and emission that produces blackbody radiation. Atmospheric absorption of radiation means that ground-based measurements of the microwave background have to deal with radiation produced by the atmosphere. Wilson, on page 163, and Wilkinson, on page 203, emphasize another point to bear in mind. The standard technique used by radio astronomers to measure the radiation received from an object outside the atmosphere is to compare the energy flux received when the detector beam is on the source to the flux received when the beam is directed to a point in the sky slightly off the source. The subtraction eliminates a good deal of the noise from the atmosphere as well as from the ground and detector. But this technique does not work for observations of the temperature of CMBR because it is uniformly distributed across the sky.

[31] The data were taken by Dicke in the summer of 1945. A redrawn version of this figure is in Lawson and Uhlenbeck (1950).

[32] In the approximation of the atmosphere as a plane-parallel slab of emitting material the detected atmospheric emission varies with the secant of the zenith angle.

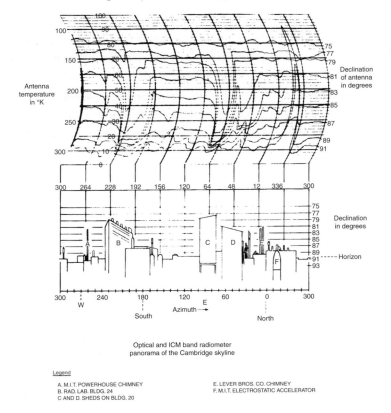

Optical and ICM band radiometer
panorama of the Cambridge skyline

Legend

A. M.I.T. POWERHOUSE CHIMNEY   E. LEVER BROS. CO. CHIMNEY
B. RAD. LAB. BLDG. 24   F. M.I.T. ELECTROSTATIC ACCELERATOR
C AND D. SHEDS ON BLDG. 20

Fig. 3.3. Illustration of the ability of a Dicke radiometer to detect thermal radiation. Below is a sketch of the skyline of Cambridge in 1945, seen from the roof of
Temporary Building 20 on the MIT campus. Above is a strip-chart recording of the
response of the radiometer to objects at different temperatures. Warmer objects
are recorded as deflections down on the chart, to higher temperature. The metallic
dome at F is cool because it reflects the beam from the cooler sky.

can extrapolate to what would be detected in the limit where the distance
through the atmosphere vanishes – where there is no atmospheric emission. This would be the observed temperature of space beyond the Earth's
atmosphere.[33]

Dicke *et al.* (1946) used this "tipping experiment" method to establish
that "there is very little (<20°K) radiation from cosmic matter" at the

[33] We should add that some of the variation in measured temperature shown in Figure 3.3 results
from what radio astronomers term "side-lobe pickup." An antenna does not produce a sharply
defined beam on the sky. Rather, there are subsidiary diffraction maxima, known as "side
lobes," that allow radiation to leak into the receiver from substantial angles away from the
direction of the main beam. As the zenith angle increases, these side lobes pick up more and
more radiation from ground that is at a temperature of about 300 K. Side-lobe pickup bedeviled
early attempts to detect the small variations in the CMBR temperature across the sky produced
by the lumpy mass distribution.

microwave wavelength (near 1 cm) they measured. Ironically, this paper appears in the same volume of the journal *Physical Review* as Gamow's 1946 letter on element formation in the early stages of expansion of a big bang cosmology. Gamow was not yet discussing a hot big bang, however – the first publication on that subject was in 1948 – so it is not surprising that neither paper made reference to the possible significance of the other. The bound Dicke and colleagues placed on how hot space might be is well above what would be expected in the cool big bang situation Hayashi and Nishida (1956) later analyzed. It is not far from the situation Gamow (1948a) proposed and Alpher and Herman (1948) calculated, however. And it is not far from what was later found to be temperature of the fossil radiation from the big bang, $T_0 = 2.725$ K. But the connection between what Dicke's radiometer can measure and what might be expected from a hot big bang was not noticed for another two decades.

In the course of research on other subjects in the 1950s and early 1960s measurements equivalent to the Dicke *et al.* (1946) tipping experiment were repeated, and the CMBR eventually detected and recognized. Detection happened first as a byproduct of research at Bell Telephone Laboratories on the development of low-noise maser amplifiers for communication systems (De Grasse *et al.* 1959; Ohm 1961; Jakes 1963).

Figure 3.4 shows a particularly detailed tipping measurement from this communications program. These data are from Project Echo, which demonstrated communication by microwave signals sent from the ground and reflected back to the ground by a satellite (a large balloon with a conducting surface). The paper on this measurement (Ohm 1961) presents the following numbers. When the Echo receiver was pointed to the zenith it detected microwave radiation equivalent to blackbody radiation at temperature $T_{\text{system}} = 22.2 \pm 2.2$ K at 2390 MHz (12.6-cm wavelength). When the instrument was tipped away from the zenith the system temperature increased because it was looking through more atmosphere. From that variation Ohm could estimate that the atmosphere contributed the equivalent of $T_{\text{atm}} = 2.3 \pm 0.2$ K in the direction of the zenith. This plus estimates of the radiation originating in the instrument and that entering the horn antenna from the ground was, in this experiment, estimated to total $T_{\text{local}} = 18.9 \pm 3.0$ K. The difference,

$$T_{\text{excess}} = T_{\text{system}} - T_{\text{local}} = 3.3 \pm 3.7 \text{ K}, \qquad (3.15)$$

is a measure of what might be entering the atmosphere from cosmic sources. This is a considerable improvement over the Dicke *et al.* (1946) measurement, $T_{\text{excess}} < 20$ K. The measured value in equation (3.15) is consistent

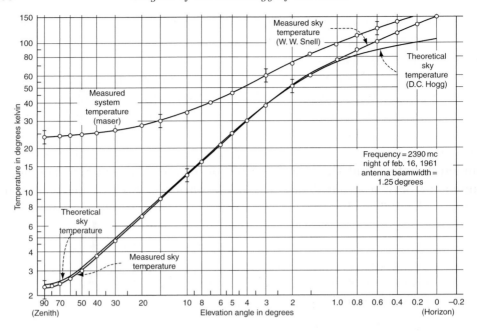

Fig. 3.4. A tipping measurement (Ohm 1961). The top curve shows the sum of the microwave radiation flux from the detector, ground, atmosphere, and whatever comes in from above the atmosphere. The lower curve is the result of subtracting estimates of what came from the instrument and ground. Reprinted with permission of Lucent Technologies/Bell Labs.

with the Alpher and Herman (1948) estimate of the Gamow condition (in equation 3.7), within the uncertainties. It is also close to the CMBR temperature (equation 2.6). But it is also formally consistent with zero. Hogg (p. 72) describes the situation in more detail.

The measurement was repeated in the Telstar Project that demonstrated transmission of a television signal from the ground to a satellite that reradiated the signal back down to the ground. Jakes (1963) reported that (at 7.2-cm wavelength) "The over-all system noise temperature was measured to be somewhat less than 17°K pointing at the zenith, which included about 4.5°K for waveguide losses, 2.5°K sky noise, 2.5°K for antenna side lobes and heat losses and 5°K for the maser." The sum and difference – which Jakes does not state – amounts to

$$T_{\text{excess}} = T_{\text{system}} - T_{\text{local}} = 2.5\,\text{K}. \qquad (3.16)$$

This is close to the central value of the range of estimates of T_{excess} from Project Echo, and again close to the CMBR temperature.

The consistency of central values of T_{excess} from the Echo and Telstar systems did not force attention to the idea that there might be a detectable sea of extraterrestrial microwave radiation: a reader of these papers could imagine that local sources of radiation had been slightly underestimated. That might be what Ohm (1961) had in mind in writing that "the '+' temperature possibilities of Table II" listing local noise contributions "must predominate." We know in hindsight that there was no need to assume the system temperatures in the Echo and Telstar systems were systematically underestimated, and that the contribution of radiation entering the horn antenna from the ground likely is an overestimate. All this became clear later in the 1960s when Penzias and Wilson added a low-temperature calibrator to the Telstar system, for the purpose we mentioned on page 46 and they explain in Chapter 4. That made the difference T_{excess} between what was detected and what was expected from the instrument, ground and atmosphere a clear and pressing issue for them. It then became a pressing issue for the cosmology community.

The next chapter presents recollections of Dicke's reaction to this issue that are consistent with the idea that when he asked Roll and Wilkinson to look for the CMBR he did not know the Bell experiments suggested excess noise. We have no evidence whether he was even aware of these communications experiments. But memories are complicated: Dicke's younger colleagues recall having to remind him that in 1946 he had published a measurement that placed a limit of 20 K on the CMBR temperature.

The broader reaction in the science community to this issue of excess noise was conditioned by the state of research in cosmology in the early 1960s. We consider this next.

3.6 Cosmology in the early 1960s

The book *Cosmology* by Hermann Bondi (in two editions, Bondi 1952 and 1960a) gives a good picture of research in this subject at the time they were written. Bondi reported the vigorous debate on the relative merits of the steady state and relativistic big bang cosmologies, and he assessed the state of observational tests of these and the other ideas then under discussion about the large-scale nature of the universe. He also painted a vivid picture of the role of the philosophies that explicitly or implicitly inform our approaches to theory and observation.

Bondi surveyed a broad range of fundamental issues about the basis for physical cosmology. Is the universe really close to homogeneous in the large-scale average? Though astronomers were not talking much about it, they knew that the nearby galaxies are distributed in a decidedly clumpy fashion. Are the redshifts of the galaxies really due to the expansion of the universe, as opposed to a "tired light" effect? Perhaps, as Zwicky (1929) had remarked, light tends to shift toward longer wavelength as it moves across the immense distances between the galaxies. If the redshift is in fact an effect of expansion, how do we know the universe is evolving, as opposed to the idea that continual creation of matter is keeping it in a steady state? If the universe is evolving were the laws of microscopic physics really the same now and in the remote past, when the universe is supposed to have been so very different? In particular, is gravity now and in the past well described by general relativity theory? If the relativistic cosmological model were a good approximation, what would be reasonable values for its parameters? Does it make sense to allow a possible role for Einstein's cosmological constant, Λ (with the property illustrated in Figure 2.1)? Einstein and de Sitter (1932) noted that a realistic model for a homogeneous universe requires the mass density term in equation (G.1), but the observations then did not require nonzero values for either of the other two unknowns, the space curvature and cosmological constant terms. The Einstein–de Sitter model makes the simplifying assumption that we keep just the one term we know is required. Einstein's feeling about the Λ term appears in the second edition of *The Meaning of Relativity* (Einstein 1945, p. 111), where he added the comment that the cosmological constant is "a complication of the theory, which seriously reduces its logical simplicity." In the chapter added to the second edition of his book Bondi (1960a) took note of the "outstanding simplicity" of the Einstein–de Sitter case. But in this case, and other high mass density solutions, the universe expanded from a state of enormously large density, as in Lemaître's primeval atom. What would the universe have been doing before that? Might we suppose the present expansion followed a bounce that terminated an earlier collapsing state? And if our universe is evolving how might it end, in a big freeze, as in the Einstein–de Sitter case, or a big crunch, as in higher density cases?[34]

This is a sobering list of issues,[35] but it certainly does not mean that the cosmology Bondi described was an empty science. People were assembling observational evidence, in part out of simple curiosity, in part driven by the goal of testing theoretical ideas, and the observations were in turn driving theoretical developments. We consider first lines of research that were largely

---

[34] Bondi (1952, 1960a) does not use the terms "big freeze" and "big crunch," and avoids also "primeval atom" and "big bang." In Gamow (1952) the phrase for the latter is the "big squeeze."

[35] A review of the current and accepted answers to most of these questions might commence with the discussion in Chapter 5 of this book.

inspired by the steady state concept, and then the state of ideas about fossils from a big bang.

3.6.1 The steady state cosmology and the cosmological tests

The steady state cosmology had a positive effect on research in cosmology in the 1950s and early 1960s by stimulating work on observational tests (many aimed at disproving the theory). It may have had the negative effect of distracting attention from other issues; one sees in the next chapter that the significance of measurements of the microwave background radiation was first recognized in Zel'dovich's group in the Soviet Union, where the steady state picture received little attention. But the picture was heavily influential elsewhere for reasons that merit our attention.

We noted the main ideas of the steady state cosmology on page 15: the universe is in a steady state of expansion, and continual creation of matter provides the material for the formation of young galaxies that fill the spaces between the older ones as they move apart, keeping the mean number density of galaxies constant.[36] This model was particularly well suited to the state of research in cosmology in the 1950s because it makes definite predictions that one might design observational programs to test. An example is the comparison of appearances of nearby galaxies with those observed at great distance. Distant galaxies are observed as they were in the past, because of the light travel time. In the big bang model distant galaxies, being younger, may be expected to look different from their nearby counterparts. In the steady state model distant and nearby galaxies are the same mix of young and old. That led Bondi (1960b) to state that if distant galaxies were observed to be systematically different from those observed nearby then "the steady-state theory is stone dead."

Here was an interesting opportunity: compare the appearances of nearby and distant galaxies, and perhaps find a critical test of ideas about the nature

---

[36] The model assumes homogeneity, isotropy, and a metric theory of spacetime. This means spacetime can be represented by the Robertson–Walker line element in equation (G.4) on page 526. Since the Hubble parameter $\dot{a}/a = H$ has to be constant in a universe in a steady state the expansion parameter has to scale as $a \propto e^{Ht}$. The physical curvature of a space section at fixed world time is $(aR)^{-2}$, and to make that constant we require $R^{-2} = 0$. The line element thus is fixed up to one locally measurable constant, H, to

$$ds^2 = dt^2 - e^{2Ht}(dx^2 + dy^2 + dz^2). \tag{3.17}$$

This happens to be one of the solutions de Sitter (1917) found (in a different coordinate labeling) for a universe that is empty except for Einstein's cosmological constant Λ, though there is no Λ in the steady state model. Equation (3.17) also is close to the situation in the present-day universe because the mass density is low and space curvature is small. In the limit of negligibly small mass density (where Λ in equation 2.5 is dominant) $H^2 = \Lambda/3$. Equation (3.17) also applies to the inflation scenario for the very early universe, but with a much larger value of H.

of our universe. There are apparently young galaxies nearby. Hoyle and Narlikar (1962) took that as an argument for the steady state picture, with ongoing creation of galaxies. Gamow (1954), on the other hand, made the now generally accepted point that the colors of most nearby galaxies are much the same, consistent with a close to uniform age of most of the stars in most of the present-day galaxies, and inconsistent with the broad mix of ages of galaxies in the steady state picture.

Another aspect of the age issue is that in a big bang cosmology the universe has expanded from densities and temperatures so large that stars could not have existed. That means the oldest stars have to be younger than the time taken for the universe to expand from high density to its present state. We noted on page 15 that this expansion time might naturally be expected to be about equal to the Hubble time, H_0^{-1}, where H_0 is Hubble's constant (defined in equation 2.1).[37] In the 1930s errors in estimates of distances to galaxies led to an underestimate of H_0^{-1} by a factor of about 8. That made it awkward to reconcile a big bang age of the universe with the radioactive decay ages of mineral deposits on Earth and in meteorites.[38] In the steady state cosmology there are galaxies of all ages, but that does not help much because the mean age of a galaxy is just one-third of H_0^{-1}. (It is an interesting exercise to show this, following the discussion in footnote 36.) As Gamow remarked, it is awkward to argue that the Milky Way is older than H_0^{-1}, as would be required to reconcile radioactive decay ages of minerals with the short Hubble time, because this galaxy looks like other nearby spirals, not much older. By 1960 the major errors in the galaxy distance scale had been identified, and Sandage (1958) had arrived at a larger value for H_0^{-1} (close to what was later established by the methods in Chapter 5 and the Appendix). In the second edition of *Cosmology*, Bondi (1960a) greeted this with the comment "it is not easy to appreciate now the extent to which for more than fifteen years all work in cosmology was affected and indeed oppressed by the short value" of the Hubble time H_0^{-1}.

Sandage (1961) concluded that with the new value of H_0^{-1} the big bang cosmology could be older than the oldest stars, but that the fit is tight and might require the postulate of a positive cosmological constant Λ (as

---

[37] In an expanding model universe with $\Lambda = 0$ and negligibly small mass density the time t_0 since the big bang is equal to the Hubble time H_0^{-1}. If $\Lambda = 0$ a significant mass density slows the expansion, making t_0 less than H_0^{-1}. In the Einstein–de Sitter model $t_0 = 2H_0^{-1}/3$. A positive Λ acts in the opposite way, increasing t_0. In the cosmology established by the beginning of the 21st century the effects of mass density and Λ about cancel, making $t_0 \simeq H_0^{-1}$.

[38] Hubble's (1936) distance estimates indicated $H_0^{-1} = 1.8$ billion years. Patterson's (1955) measurements of the decay of uranium to lead isotopes showed that Earth and the asteroids are about 4.5 billion years old.

discussed in footnote 37). Sandage anticipated what happened: the evidence collected in Chapter 5 convincingly shows the effect of a positive Λ. But the central point in the 1960s was that observations of distant galaxies and the measurements of isotope abundances accumulated by radioactive decay in minerals in the Solar System yield similar ages. The coincidence from such very different observations encouraged at least some to think that there might be something to this expanding universe concept. A set of coincidences of this sort from a considerable variety of measurements is presented in the summary of cosmological tests in Section 5.4.

Another influential – and controversial – opportunity to challenge the steady state cosmology was based on the counts of galaxies detected by radio telescopes. Some galaxies are very strong sources of radio radiation, so they can be seen at great distances, where the properties of spacetime can affect what is observed.

Consider first a simple case: suppose the universe is not expanding, has the flat geometry of Euclid, and contains a uniform spatial distribution of galaxies that are not evolving. Then the count, $N(>S)$, of galaxies that appear brighter than S (that is, S is the rate of arrival of radiation energy from the source per unit collecting area of the telescope) varies with S as[39]

$$N(> S) \propto S^{-3/2}. \tag{3.18}$$

This was a familiar and important relation. Hubble (1936) had shown that counts of galaxies as a function of their optical brightness S fairly closely follow this relation. Since the relation assumes a uniform distribution Hubble's counts encouraged the assumption that the distribution of galaxies is close to homogeneous.

By the mid-1950s it was becoming clear that some galaxies are sources of radio radiation strong enough to be detectable by radio telescopes. Ryle (1955) presented the early estimate in Figure 3.5 of how the counts of these objects vary with their radio brightness. The data in this early study indicated that the counts increase with decreasing brightness S more rapidly than in equation (3.18). In a big bang cosmology this need not be a problem. Since distant sources are seen as they were in the past, because of the light travel time, one may account for the large number of faint sources by

---

[39] In static flat space the energy flux density from a galaxy with luminosity L at distance r is $S = L/(4\pi r^2)$. The volume within this distance r is $V = 4\pi r^3/3$. If all galaxies had the same luminosity, and their number density were n, the number of galaxies brighter than S would be $N = nV = 4\pi n r^3/3 \propto r^3 \propto S^{-3/2}$, which is equation (3.18). Different galaxies have different luminosities. To take that into account separate the galaxies into luminosity classes. The law $N \propto S^{-3/2}$ applies to each class, so it applies to the sum over all galaxies. The expansion of the universe, spacetime curvature, and galaxy evolution all change this relation when S is small enough (r is large enough).

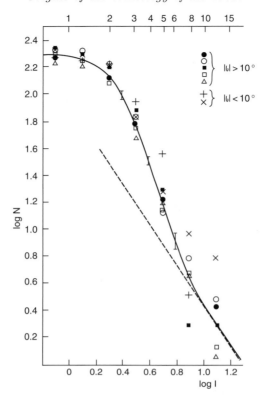

Fig. 3.5. The count N of radio sources brighter than S (marked I in this figure) in 1955. The dashed line is the relation in equation (3.18) (Ryle 1955). The excess counts between $\log I = 0.2$ and 0.8 were the subject of considerable debate.

supposing that in the past many more of the younger galaxies were strong sources of radio radiation. In contrast, in the steady state cosmology the number density of radio sources has always been the same. Here the departure from the assumption of a static Euclidean universe causes the counts $N(>S)$ to increase with decreasing S less rapidly than in equation (3.18). That is opposite to what was observed.

Ryle's (1955) paper was greeted with great interest and controversy: do radio source counts seriously challenge the steady state cosmology? Bondi's (1960a) comment in his second edition of *Cosmology* is limited to the statement that "Further work on this promising field may be decisive." For a succinct and authoritative historical account of this debate, see Longair (2006, p. 326). The influence of the issue on community thinking about cosmology is seen in the next chapter in comments by Burke (p. 178), Longair (p. 239), Rees (p. 263), Burbidge (p. 269), Narlikar (p. 272), and Shakeshaft (p. 290).

An important change in the situation in the mid-1960s was the discovery of quasars, some of which are detected as radio sources (hence an early name, quasi-stellar radio sources). It was soon seen that quasars are considerably more abundant at redshifts $z > 0.5$ than at low redshifts. The interpretation proposed then, and later well checked, is that quasars were more abundant in the past (Longair 1966; Sciama and Rees 1966). The use of Hubble's law (equation 2.1) to convert quasar redshifts to distances certainly could be questioned (Terrell 1964; Hoyle 1965; Hoyle and Burbidge 1966), but that has since been checked in a fairly direct way. When a quasar appears in the sky close to a galaxy at lower redshift it is observed that the gas in the galaxy produces absorption lines in the quasar spectrum. The quasar clearly is behind the galaxy, not in front of it. And that agrees with the conventional idea that larger redshift means larger distance. Cosmic evolution was a controversial issue in 1960. Now the measured low abundance of present-day quasars is a clear example of the effect.

A line of research in the early 1960s that foreshadowed another important advance was stimulated by Sandage's (1961) survey of how the large collecting area of the 200-inch Hale telescope might be used to test cosmological models. Sandage considered counts of galaxies as a function of brightness in visible light, an optical analog of Figure 3.5. But he concluded that with the technology and astronomy at hand the best way to distinguish between the steady state model and the family of big bang models was to measure the relation between brightness and redshift of the most luminous galaxies, all of which have close to the same intrinsic luminosity. This has come to be termed the "redshift–magnitude" relation (after the astronomers' measure of how bright an object appears in the sky, the apparent magnitude). Equation (2.1) determines the redshift–magnitude relation at low redshifts. Sandage was interested in what happens when the redshift is large, meaning the apparent recession velocity is comparable to the velocity of light.

The steady state cosmology predicts a specific redshift–magnitude relation. As we have remarked, that had the virtue of giving the observers something definite to aim for. The big bang model is much less predictive. It has free parameters, such as space curvature and Einstein's Λ, and it predicts that the galaxies observed at great distances are younger and so maybe systematically different from nearby ones. Sandage limited his proposed goal to the use of the redshift–magnitude relation to distinguish between big bang models and the specific steady state prediction. He cautioned that even this limited test will be "difficult and perhaps marginal." But it is worth noting that observations then reached to a galaxy at redshift $z = 0.46$, meaning the universe has expanded by the factor 1.46 since the light we receive left the galaxy (as expressed in equation 2.3). This is an impressively deep probe out in space and back in time.[40]

3.6.2 Light elements from the big bang

We have already discussed aspects of yet another topic of research in the early 1960s, the ongoing debate on the origin of the chemical elements. In the steady state cosmology matter is continually created. But Bondi opined that creation of matter with the observed relative abundances of the chemical elements and their isotopes would be bizarre. This led him to conclude that a demonstration that the elements could not be produced in the universe as it is now – presumably in stars – would be strong evidence against the steady state cosmology, while a demonstration that the elements did come from stars would reduce the list of arguments for a big bang. Bondi (1952), in his first edition, gave references to the work by Gamow, Alpher, and Herman on the theory of element formation in a hot big bang, but the discussion of this idea and of the idea that elements were formed in stars was brief. By the time of the second edition, Bondi (1960a) could report significant advances in the latter (Burbidge, *et al.* 1957; Cameron 1957). This important development was encouraging for the steady state philosophy. As Gamow (1956) noted, however, it was not necessarily a challenge to the big bang picture: one could image that nuclear reactions in stars only altered the abundances that came out of the big bang. But the abundance of helium offered a critical test.

An important, though at the time not widely appreciated, step in this direction was Burbidge's (1958) recognition that the helium abundance in the Milky Way is larger than might be expected from the rate of production of helium in known types of stars in the numbers indicated by the stellar luminosity of this galaxy. He noted that some galaxies emit large amounts of energy at radio wavelengths (making them detectable by radio telescopes, as we have discussed), and he asked whether the source of this energy might be the copious conversion of hydrogen to helium. He did not mention the possibility of helium production in a big bang. Gamow (1956) did: "the calculations in that direction, carried out by the present writer," (Gamow 1948b) "and later in some more detail by Fermi and Turkevich, ... lead to

---

[40] Applications of this redshift–magnitude test at the beginning of the 21st century use the light from supernovae rather than galaxies and reach expansion factors $1 + z$ close to 3. The apparently modest but deeply important increase in the distances the observations reach indicates that the redshift–magnitude relation is close to the steady state prediction discussed in footnote 36. The task of explaining why this is taken as evidence for the effect of the cosmological constant Λ illustrated in Figure 2.1, rather than as evidence for the steady state cosmology, is left to Section 5.4.

a value of the H/He ratio which is in good agreement with observational data." That result has lasted.

In the second edition of *Cosmology*, Bondi (1960a, p. 58) added this assessment of the issue of the origin of helium and the heavier elements:

Since it has also been shown that any hot dense early state of the universe could not have left us any nuclei heavier than helium, the origin of such nuclei is no longer a question of cosmology.

It might however be said that the abundance of helium may conceivably be greater than would be accounted for by ordinary stellar transmutation and so might have to be explained on a cosmological basis, but the evidence as yet is far too slight to merit serious consideration now.

North (1965), in *The Measure of the Universe: A History of Modern Cosmology*, presents a brief description of the hot big bang picture for element formation and concludes, with Bondi, that "The actual abundance of helium is still uncertain, however, and it may eventually be necessary to invoke some such explanation as Gamow's." Gamow, Bondi, and North did not document the measurements of the helium abundance. They are summarized in Osterbrock and Rogerson (1961).

Osterbrock and Rogerson reviewed measurements of the abundance of helium in the plasma around and between the stars based on observations of recombination line strengths. These measurements are considered unambiguous (at the accuracy needed for this purpose). They developed an estimate of the helium abundance in the Sun from models of its structure and estimates of its heavy element content (which determines opacity within the Sun, and hence the weight of helium needed to account for the observed solar luminosity). They concluded that the mass fraction, Y, in helium is considerably larger than the mass fraction in heavier elements, and that Y is not much different in the Sun (which we noticed on page 54 is 4.5×10^9 years old) from what is observed in the interstellar plasma. Osterbrock and Rogerson concluded:

It is of course quite conceivable that the helium abundance of interstellar matter has not changed appreciably in the past 5×10^9 years, if the stars in which helium was produced did not return much of it to space, and if the original helium abundance was high. The helium abundance $Y = 0.32$ existing since such an early epoch could be at least in part the original abundance of helium from the time the universe formed, for the build-up of elements to helium can be understood without difficulty on the explosive formation picture.

Their reference for the "explosive formation picture" is to Gamow (1949).

To our knowledge this is the first well-documented proposal for a relation between the theory and the observational evidence of a fossil from the early universe. It appeared in *Publications of the Astronomical Society of the Pacific*, a journal that was (and is) quite familiar to astronomers and even some physicists. But we have found no evidence that anyone took note of the significance of this paper for cosmology before the mid-1960s, after recognition of evidence for the detection of a related fossil, the CMBR.

It is worth mentioning some papers of the early 1960s that referred to Osterbrock and Rogerson (1961), and some that did not. Peebles (1964) used the Osterbrock and Rogerson estimate of helium abundance in a study of the structure of the planet Jupiter, but did not notice the big bang connection. This is recalled on page 190. O'Dell, Peimbert and Kinman (1964) added to the evidence for a large helium abundance in old stars, and took note of Burbidge's (1958) point that the production of this amount of helium in stars would require that galaxies were considerably more luminous in the past. This is acceptable in a big bang model, of course, but not in the steady state cosmology. O'Dell *et al.* referred to Osterbrock and Rogerson's paper, but did not mention the possibility of helium production in a big bang. In the previous section we noted the reanalyses of the theory of light element production in a hot big bang by Smirnov (1964) and by Hoyle and Tayler (1964). Smirnov did not know about the Osterbrock and Rogerson paper: he thought the abundance of helium in the oldest stars is relatively small, $Y < 0.1$ (p. 36). Hoyle and Tayler knew and documented the evidence that the helium abundance is larger than that, and that the big bang model could account for it, but their references to the literature do not include Osterbrock and Rogerson (1961).

3.6.3 Radiation from the big bang

The idea of another fossil, a remnant thermal sea of radiation that might be expected to accompany the production of helium in a hot big bang, was clearly expressed in the literature, including Alpher and Herman (1948, 1950).[41]

Osterbrock recalls (p. 88) hearing Gamow lecture at the University of Michigan in the summer of 1953. In the published version of these lectures, Gamow (1953a) described his ideas about element formation in a hot big bang. He presented his 1948 argument (based on the condition in footnote 10 on page 30) that, for production of a reasonable amount of deuterium that would mostly burn to helium, "about the right density" of baryons when the

---

[41] Gamow (1948a,b) expressed the same idea, but less directly. He pointed out that the radiation would be present after element formation, and that it could be an important dynamical actor in structure formation. This is discussed in Section 3.6.4. Gamow continued to emphasize the potential importance of the radiation, but his arguments were unfortunately confused by the error noted in the next footnote.

temperature was 10^9 K is 2×10^{18} baryons cm^{-3} (calculated from the numbers he gives). Under these conditions the temperature would have dropped to $T_0 = 7$ K at Gamow's estimate of the present baryon density (equation 3.6). This is close to what Alpher and Herman had obtained earlier from Gamow's condition (equation 3.7). Gamow did not write down this last step, however. Earlier in the lectures he discussed the idea of a sea of thermal radiation that cools as the universe expands, and he considered the effect of the mass equivalent of the radiation on the rate of expansion. He argued from a consideration of expansion time that when the mean mass density had dropped to 10^{-25} g cm^{-3} the radiation temperature would have dropped to 320 K. That extrapolates to present value $T_0 = 7$ K at his estimate of the present density. This is close to the Alpher and Herman value, and to what was later measured, but the calculation is unfortunately wrong.[42] The CMBR idea nevertheless is in these lectures, though we have not encountered anyone who noticed until much later.

An illustration of the low visibility of the CMBR idea is the absence of any mention of the idea of radiation from the early universe in Bondi (1952, 1960a). Another useful indication of what people were thinking comes from the proceedings, or published records, of international conferences. Most exchanges of ideas at these events tend to be in informal discussions, and the formal lectures are not always close to what appears in the proceedings, but the published versions do show what people considered worth recording. The proceedings of the Solvay conference, *La Structure et l'Evolution de l'Univers* (the eleventh in a distinguished series of meetings on major advances in physics), in Brussels in 1958; the 9th International Astronomical Union Symposium, *Paris Symposium on Radio Astronomy* (Bracewell 1959), in Paris in 1958; and the 15th International Astronomical Union Symposium, *Problems of Extra-Galactic Research* (McVittie 1962), in Santa Barbara, California in 1961, all include papers on issues in cosmology as well as on advances in the astronomy of radio sources, stars, and galaxies. In these volumes we find no mention of the idea that space might be filled with a sea of microwave radiation, perhaps one left from the early universe.

There are accounts of informal discussions of the idea. Alpher and Herman (2001) recall asking radio astronomers about the possibility of detecting the radiation. Tayler (1990), in his recollections of the work with Hoyle in 1964 on helium production, mentions their thoughts – which did not enter

---

[42] The calculation assumed space curvature in the expansion equation (G.1) becomes the dominant term just when the mass densities in matter and radiation are equal. That need not be so. Gamow (1953b, 1956) repeated this calculation and derived from it the present temperature, $T_0 \simeq 7$ K. But, as opposed to his now well-tested argument for light element formation, this calculation is not of lasting interest.

their paper – about remnant radiation from a big bang. Hoyle (1981) recalls discussions of the possible temperature of a sea of microwave radiation in conversations with Gamow and Dicke. In a very readable popular book, *The Creation of the Universe*, Gamow (1952) describes the cooling of a sea of thermal radiation in an expanding universe. He notes, as an example, that a universe with mass dominated by radiation cools to a temperature of $50\,\mathrm{K}$ at three thousand million years after the big bang, a common estimate then for the age of the universe. (The method of calculation is indicated in footnote 1 on page 26.) The calculation is not directly relevant, because the mass of our universe is not now dominated by radiation, but the pointer to a sea of thermal radiation is quite direct.

On the issue of detectability we have an account by Virginia Trimble of Gamow's encounter with Joe Weber, who had expertise in microwave technology.

Joe Weber was an amateur radio operator in his early teens and, at the time of the Sicilian invasion, was the skipper of one of the first submarine chasers to have a 6-cm radar (SC 690). As the war wound down, the Navy moved him to a desk job in Washington in electronic countermeasures, largely to descope the effort, but also to hand out some grants. When he decided to resign his commission (as lieutenant commander), several grantee organizations offered him jobs, but he accepted instead a full professorship of electrical engineering at the University of Maryland. The fall 1948 appointment was contingent on his obtaining a PhD in something quite soon, since his highest degree was a 1940 BS from the US Naval Academy.

Thus summer 1949 found Weber visiting Washington-area universities in search of a PhD project and advisor. One of the first places he visited was George Washington University, and one of the people he talked with there was George Gamow. "Do you have any interesting thesis problems?" Weber enquired. "What can you do, young man?" responded GG. "I'm a microwave spectroscopist," said JW. "No, I don't think of any interesting problems" concluded Gamow. So Weber went on to Catholic University, where he completed a 1950 PhD dissertation (Weber 1951) with Keith Laidler on the inversion spectra of normal and deuterated ammonia. Since Weber at the time knew about the technology for detecting faint radio signals, whether the story is funny depends on whether you think Gamow should have had radiation from the early universe in mind in 1948. It is, of course, a second-hand story, but I was married to Joe from 1972 until his death in the year 2000, and men, as you probably know, like to tell war stories. There is also a good one about the inhabitants of Tonga Tabu, following the sinking of the Lexington in the battle of the Coral Sea in May 1942.

For other aspects of this issue see Trimble (2006) and Burke's account on page 181.

Our conclusion is that the idea that space might be filled with thermal radiation left from the early stages of expansion of the universe – what we now term the CMBR – was "in the air" in the early 1960s. But it was less visible than other issues in cosmology, particularly the debates on the relative merits of the big bang and steady state scenarios.

The first ground-based measurements that, with the full benefit of hindsight, might be said to have offered a suggestive indication of detection of this microwave radiation were the Bell Laboratories Echo and Telstar communications experiments we discussed in Section 3.5. They are described in more detail in the next chapter.

There were earlier measurements that placed what were later recognized to be interesting upper bounds on the CMBR temperature. We have mentioned the Dicke *et al.* (1946) limit, $T_0 < 20$ K at wavelengths of 1–1.5 cm. They referred to "radiation from cosmic matter," not a cosmological model. Dicke (1946a) also used his radiometer to measure the radiation from the Moon and the Sun at 1.25 cm.

Covington (1950), in Canada, was studying bursts of radio radiation from the Sun at longer wavelength and the correlation of these bursts with radiation produced by disturbances of Earth's upper atmosphere (the ionosphere). That required an estimate of radiation from Earth's undisturbed atmosphere and beyond. Covington reported that at 10.7-cm wavelength this radiation is small, "not more than about 50° K." Within the experimental error, ± 25 K, this is not significantly different from zero, or from the temperature, 50 K, in the example cosmological model in Gamow's (1952) popular book. We have seen no notice taken of the latter coincidence.

Haruo Tanaka *et al.* (1951), at Nagoya University in Japan, using a square horn antenna at 8-cm wavelength, obtained a better constraint, that the incident radiation at zenith (including what is produced by the atmosphere) is no more than about 5 K. Tanaka (1979) offers this comment on their measurement.[43]

14 years before [Penzias and Wilson's discovery], we measured the temperature of sky at the wavelength of 8 cm, and estimated it to be between 0 and 5 K. ... The measurement [of the sky temperature at zenith] was made for an absolute calibration of the intensity of solar radio waves. ... Except for a parabolic reflector requiring accurate shaping, our instruments were all handmade: we obtained the necessary parts from the disposal goods of the army. ... Since we could not calculate a gain of the parabolic antenna, we built a pyramidal horn antenna whose length was 2 m. ... At that time A. E. Covington in Ottawa, Canada, had been observing the solar radio waves at 10.7 cm since 1947. He calibrated the solar flux using the sky temperature of 50 K. However it seemed too high for us, and we decided to measure the sky temperature by ourselves. ... I understand that 0–5 K and 3.5 ± 1 K are

---

[43] We are grateful to Eiichiro Komatsu and Tsuneaki Daishido for identifying the references to Tanaka's work and selecting and translating these excerpts from Tanaka (1979).

completely different values and meanings. However, had someone like Gamow or Dicke notified us of the significance of our measurements, it would not have taken us 14 years [to detect the CMBR]. This is a bit of regret.

Medd and Covington (1958) reported discussions with Tanaka, and an improved measurement, 5.5 K, with a probable error of about 6 K, at 10.7-cm wavelength. This is close, but it is also consistent with no background microwave radiation.

In the Soviet Union, Shmaonov (1957) reported a study of emission properties of a dish reflector antenna and receiver at 3.2-cm wavelength. Rapid switching to a reference load gave the instrument excellent stability. Our impression is that he could have detected the CMBR if he had thought to do it, though that would have required closer attention to the suppression of radiation from the ground and accounting for radiation from the atmosphere.

In France, measurements that placed a bound on the background radiation temperature at 33-cm wavelength are recalled by one of the authors, James Lequeux, who writes

In the winter of 1954–55, we measured the paraboloid dish antenna pattern of a former German "Würzburg" radar equipped with a 33-cm receiver built by Le Roux, that we used for mapping the Galaxy. This involved measuring the signal received from a remote transmitter while pointing the antenna in various directions. Then we calculated the contribution of the ground and the atmosphere to the antenna temperature as a function of the direction pointed by the antenna, and compared to observation (far from the galactic plane, of course). The observed antenna temperature was calibrated with blackbodies. Then we concluded that any contribution from the sky would be less than 3 K, and would be rather uniform. This is what is published in the *Comptes Rendus*; and signed by Le Roux.

Given our equipment, and in spite of careful measurements, it would have been foolish to claim a positive detection. Our remote antenna lobes were considerably stronger than those of the horn used by Penzias and Wilson. Thus in the *Comptes Rendus* paper we only claim an upper limit for the CMBR, admittedly close to the actual value, but only an upper limit.

In contemporary reports of these measurements, Denisse, Lequeux and Le Roux (1957) estimate that T_0 is less than about 3 K, while Delannoy *et al.* (1957) conclude (in a translation and commentary kindly provided by Lequeux) "Delannoy *et al.* write on page 236 of their paper, 'We may only conclude that the temperature of the sky at 900 MHz is certainly not larger than about 20 degrees Kelvin.' And the footnote says 'A stricter upper limit that was proposed [in Le Roux's thesis, 1956] underestimated the errors on the measurements of the antenna beam.'" We emphasize these comments because Le Floch and Bretenaker (1991) have suggested that this experiment may have yielded a detection rather than a limit. The evidence we have seen argues against it.

Jasper Wall (p. 280) describes an experiment in the 1960s that may have been capable of detecting the CMBR, if they had thought to do it. At the relatively long wavelengths of their observations, the radiation from our Galaxy is large and would have complicated the interpretation. But at the time of writing the art of measuring the CMBR energy spectrum at long wavelengths is still under development.

To summarize, our reading of the evidence is that the Bell Laboratories measurements (De Grasse *et al.* 1959; Ohm 1961) we discussed starting on page 49 were the first likely – though at the time not recognized – direct detection of the CMBR to be presented in the literature.

Doroshkevich and Novikov (1964), whose recollections begin on page 99, likely were the first to understand the importance of the early Bell Laboratories measurements for cosmology. This grew out of their study of the amount of electromagnetic radiation that would be expected to have accumulated from all known sources of radiation. They compared this to what was known about the brightness of the sky after elimination of radiation from local sources on Earth, in the Solar System, and in the Milky Way Galaxy.[44] That led them to make three important points. First, the cosmic radiation from known sources – starlight and radio-luminous galaxies – is minimum at wavelengths near 1 mm to 1 cm. Second, measurements near this minimum, at microwave wavelengths, "are extremely important for experimental checking of the Gamow theory," because the fossil radiation would peak up at these wavelengths.[45] Third, there already is a useful microwave measurement, from Ohm (1961).

[44] Others had been considering this. Shakeshaft (1954) had compared the measured mean brightness of the sky at radio wavelengths, $\lambda \sim 3$ m, to observed counts and radio luminosities of galaxies. He concluded that the mean radio sky brightness could be produced by the galaxies if, as was becoming clear, some are intense sources of radio radiation. Estimates of the mean energy density in intergalactic starlight, taking account of the shift to the infrared and the loss of energy by the cosmological redshift, were presented in increasing detail in Bondi (1952), McVittie and Wyatt (1959), and Sandage and Tammann (1964). Comments on the considerable challenge of measuring this cosmic infrared background radiation are, in increasing detail, in Baum (1956), Roach (1964), and Harwit (1964). Harwit's analysis of the problem of separating extragalactic starlight from the zodiacal light (sunlight scattered by, and absorbed and reemitted by, interplanetary dust) is recalled on page 329. Progress in the increasingly focused work of checking consistency of the measured sky brightness as a function of wavelength with what is known and conjectured about sources of radiation and their evolution is reviewed in Hauser and Dwek (2001).

[45] When the microwave background radiation was identified, at about one percent of the wavelength Shakeshaft was considering, the question he addressed naturally arose again: could this microwave radiation have come from sources in the universe as it is now? The coincidence Doroshkevich and Novikov noted, that the foreground radiation from known sources in galaxies is minimum at microwave wavelengths, aided recognition that the microwave background is

Doroshkevich and Novikov's reference for the "Gamow theory" is Gamow (1949). As it happens, that is the same paper Osterbrock and Rogerson (1961) cited in their comment about the "explosive formation picture" quoted on page 59. But Osterbrock and Rogerson discussed helium, while Doroshkevich and Novikov discussed radiation. Doroshkevich and Novikov were members of Zel'dovich's research group in Moscow. We noted (p. 35) that Zel'dovich (1963a) recognized the importance for cosmology both of helium and the microwave background radiation, but that his reading of the evidence at the time led him to conclude that the big bang likely was cold, not hot.

The recollections in the next chapter make it clear that some astronomers in the early 1960s remembered the evidence we have reviewed (p. 44) that the spin temperature of the interstellar molecule CN is surprisingly large, and that that suggested the presence of a microwave radiation background at a temperature of a few degrees above absolute zero. But the connection to the hot big bang picture seems to have been made only after the radiation had been recognized in direct detection.

3.6.4 Galaxy formation

There was one other – very indirect – hint to the fossil radiation, from large-scale structure: the concentrations of mass in galaxies and clusters of galaxies. In the printed version of the lectures by Gamow (1953a) that Osterbrock recalls hearing (pp. 60 and 88) Gamow proposed that

Only two observational quantities, the present density and age of the universe, together with the equations of general relativity and black-body radiation enable us to predict the formation of galaxies at a definite date in the past, and give a mass and radius of these galaxies which is at least comparable with observations.

Gamow based this statement on two arguments he had introduced earlier (Gamow 1948a,b) and explained in his 1953 lectures. First, the CMBR temperature and the present mass density set the matter temperature in the early universe (because at high redshift the matter was ionized and tightly thermally coupled to the radiation). The matter temperature sets the minimum size of a cloud of matter that gravity can hold together against the matter pressure that tends to drive the cloud apart. This minimum is termed the Jeans length and mass (after the mathematical astronomer James H. Jeans, who considered gravitational fragmentation of a gas cloud that might end up as a star cluster). Thus, it could set the scale of masses for galaxies.

a new phenomenon. And this minimum of the foreground radiation where the CMBR spectrum peaks up greatly aided the precision measurements discussed in Chapter 5.

Second, the gravitational growth of nonrelativistic mass concentrations such as galaxies in an expanding universe would have commenced when the mass density in the thermal radiation fell below the mass density in matter. Prior to that, the rapid expansion driven by the large mass density in radiation suppressed the gravitational growth of concentrations of matter (on scales less than the Hubble length). The mass density at the transition to matter-dominated expansion could have set the scale of mass density in galaxies.

Gamow's points were not much noticed. Bondi (1960a, p. 176) gave references to studies developing the idea that the present-day concentrations of mass grew by the attraction of gravity out of small departures from an exactly homogeneous mass distribution in the early universe, but he did not mention Gamow's two points. (The only reference we have found in the literature, apart from his immediate colleagues, is by ter Haar 1950, p. 129.) But Gamow's deeply fertile intuition again proved to be prophetic: his points, with modifications, are pieces of what later became the standard model for structure formation. His assessment quoted above may have seemed overly optimistic at the time, and we have no reason to think it could have inspired others to consider the idea of a sea of thermal radiation. But it is worth recording that these ideas were in the literature in 1953, though little discussed for another decade. The essays indicate how the role of the CMBR in structure formation came to the general attention of the community. The learning curve for how to measure the effects of structure formation and turn them into demanding tests of ideas about the large-scale nature of the universe is a subject for the next chapter.[46]

3.6.5 The situation in the early 1960s

Let us conclude with an overview of this wonderfully tangled situation. Some were aware that the abundance of helium is larger than seemed reasonable in the otherwise promising theory of element formation in stars, and some even recognized that a hot big bang could supply the helium. Others recognized that in a hot big bang universe space now could be filled with a thermal sea of microwave radiation at a temperature a few degrees above absolute zero. The spin temperature of interstellar CN molecules was known by some to be large compared to what might be expected from excitation by particle collisions,

---

[46] The related issues are the effect of the CMBR on the evolving distribution of the matter and the effect that has on the distribution of the CMBR. The basic principles were developed in the 1960s after the identification of the CMBR. We will discuss how gravity disturbs the radiation (Sachs and Wolfe 1967), the coupling of plasma and radiation (Peebles 1965), the dissipative decoupling (Silk 1967, 1968b), and the patterns that decoupling can leave in the distributions of matter and radiation (Peebles and Yu 1970; Sunyaev and Zel'dovich 1970c). Detection of these effects took some three more decades.

and some who knew that had considered the possibility that the molecules are excited by some source of microwave radiation. Others knew how to build a receiver capable of detecting microwave radiation at a temperature a few degrees above zero. And some of them had indications that communications receivers may have detected a near-isotropic component of radiation.

Why were all these pieces not put together until the mid-1960s? It was in part a matter of contingency: it can take time to notice relations among such a broad variety of considerations. But an important fact to bear in mind is that it was not obvious in the early 1960s that these are the relevant set of ideas. In a cold big bang universe, neutrino degeneracy could control helium production during the early stages of expansion. And what actually would happen during the early expansion of a hot big bang universe would depend on the laws of physics, which people suspected then, and still suspect, evolve with the expansion of the universe. The cosmological tests summarized in Section 5.4 show that this evolution could not have significantly affected the early production of helium, but that was not at all obvious in the early 1960s. The steady state cosmology also commanded attention. It was clear enough then that in a steady state universe one could postulate that helium is continually created along with the continual creation of neutrons, protons, and electrons. Perhaps better, helium could be produced in a hypothetical – but physically possible – class of massive exploding stars, and the energy released from the burning of hydrogen to the observed amount of helium could supply a significant sea of radiation.

The essays recall how the community found its way through this thicket of clues and conflicting ideas to the ones that led the way forward, in an iterative consultation of theory and practice. The process tends to be haphazard, and exciting, and on occasion exceedingly awarding. It has yielded deep advances in understanding of the world around us. The essays in the next chapter reveal the beginnings of such an advance, and Chapter 5 describes what grew out of it.

4

Recollections of the 1960s

Our plan of ordering the essays is to group them by topic, with chronological order within groups. The grouping is by the focus of the research, as indicated by the section headers. Since this focus tends to evolve with time, the result is that these recollections of what happened are presented in a roughly chronological order. For example, in the second half of the 1960s a first order of business on the experimental side was the test of how the energy of the CMBR varies with wavelength, and on the theoretical side it was the exploration of ideas about what a significant departure from a thermal spectrum might mean. These continued to be pressing issues at the end of the 1960s, but there was increasing interest in the experimental search for departures from an exactly isotropic distribution of the radiation, and in the development of the theory of the departures from isotropy that might be expected to accompany the known departures from an exactly homogeneous distribution of the matter. Thus we present the essays whose main focus is the spectrum before those largely concerned with the anisotropy of the CMBR.

Since many of the essays do not fit the headers our plan required arbitrary and debatable decisions on ordering. This is a realistic illustration of what was happening in the 1960s, of course. The confusion extends to the recollections: the stories are not complete and they are not always even consistent with each other. The reader, therefore, must be prepared for a distinct change of style from the linear – but we hope efficient – history of ideas in the previous chapter to the chaos of the real world of science.

4.1 Precursor evidence from communications experiments

4.1.1 David C. Hogg: Early low-noise and related studies at Bell Laboratories, Holmdel, NJ

The US National Academy of Engineering cites Hogg's election to the Academy for his "contributions to the understanding of electromagnetic propagation at microwave frequencies through the atmosphere." A native of Saskatchewan, Hogg's current interest is the composition of music.

A giant in radio science, Harald T. Friis[1] was head of the Bell Radio Research Laboratory in Holmdel. Having pioneered work on the superheterodyne receiver in the late 1920s, he played a key role in Karl Jansky's initial experiments and the beginning of radio astronomy in the early 1930s. His interests then turned to shorter wavelengths which eventually led to the construction of a nationwide microwave radio-relay system employing "horn-reflector" antennas that he patented with Al Beck. This antenna design is highly efficient and was used in all of the low-noise microwave systems to be discussed here. These remarks are made to indicate that high-quality equipment, designed for very practical purposes, can be used as a tool for first-class science. It is to the credit of the United States that the AT&T Bell Laboratories existed, allowing such broadminded interactive research to be done.

In the 1950s, not long after John R. Pierce had traveled to the UK, including Oxford, he brought Rudi Kompfner to Bell Labs. They asked me to calculate the thermal noise from the Earth's atmosphere over the microwave band. This noise level was needed for their calculation of the feasibility of microwave communication by reflection from an orbiting balloon (Pierce and Kompfner 1959). It was fortunate that some time earlier, with Arthur B. Crawford, I had measured the millimeter-wave absorption by the oxygen and water vapor in the sea-level atmosphere (Crawford and Hogg 1956). Thus the broadening constants for computation of "sky noise" were determined and that calculation was completed (Hogg 1959).

However, no sky noise measurements were available to corroborate this theory. Nevertheless, again fortunately, at that time Derek Scovil and Bob De Grasse at the Bell Laboratory, Murray Hill, NJ, were well along in developing microwave traveling wave solid state masers (TWM), with noise temperatures on the order 10 K (De Grasse, Shulz-Du-Bois and Scovil 1959). Again encouraged by Kompfner and Pierce, we, therefore, combined this maser and antenna to produce a "low-noise" receiving system.

---

[1] This is a good occasion to remind the reader that the Glossary is meant to serve as a guide, in this case to Friis' interview of Burke (p. 177) and to Friis' influence on the development of the technology that detected the fossil microwave radiation (pp. 159 and 162).

Here we discuss and compare three systems at various microwave frequencies, with emphasis on the technology and low-noise results that pointed the way toward a determination of the microwave cosmic background noise. In all three cases, the equipment was designed and built to demonstrate the feasibility of satellite communications, and the cooperation of NASA and the Bell Laboratories System Department were important factors.

The first low-noise microwave system (De Grasse *et al.* 1959) operated at 5.65 GHz. The antenna mount, constructed of wood, on the lip of Crawford Hill, allowed manual beam pointing in elevation only; a photograph is shown in Figure 4.1. The rectangular waveguide input to the TWM was fed via a rotating joint in the circular waveguide from the antenna. The output from the TWM preamplifier was then fed to a conventional superheterodyne. This combination resulted in a (zenith) system noise temperature of 18.5 K.

At that time I was invited by John Shakeshaft to Cambridge, England, and gave these results in Maxwell's lecture room at the Old Cavendish. A seminar also was given at the old McDonald physics building at McGill University, Montreal, Canada, where Ernest Rutherford did research on the alpha particle and helium.

Although this system was unsophisticated, it did serve as a prototype for the following two systems that were used for actual communications via satellites: Echo and Telstar.

The second low-noise microwave system (Ohm 1961) was designed and built specifically for the Echo satellite project at a frequency of 2.39 GHz, for receiving signals reflected from an orbiting balloon. The receiver design was engineered by Ed A. Ohm and the project engineer was W.C. (Bill) Jakes Jr. The antenna (Crawford, Hogg and Hunt 1961), a 20-ft aperture horn-reflector, is shown in Figure 4.2; design and construction was managed

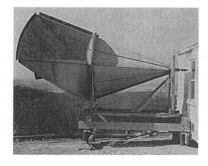

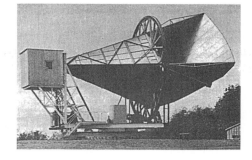

Fig. 4.1. The antenna in the first low-noise microwave system.

Fig. 4.2. Horn-reflector antenna used in the Project Echo experiment.

Table 4.1. *Summary of satellite communication systems*

Experimental communication system	1	2	3
Frequency	5.65	2.39	4.17
Estimated antenna back lobe and resistive noise	3.5	2.3	2.5
Zenith atmospheric noise			
Measured	2.5	2.3	2.5
Estimated	2.75	2.4	2.5
Zenith system noise			
Measured	18.5	21	17
Estimated	18.5	19	14.5

Note: Noise temperatures are in kelvin and frequencies in GHz.

primarily by Arthur B. Crawford and Henry W. Anderson. The 2.39-GHz measured radiation patterns and gains were found to agree well with theory (Crawford, Hogg and Hunt 1961); both azimuth and elevation pointing were available. The dual-channel TWM, provided by the Derek Scovil group at the Murray Hill Laboratory, received waves of both clockwise and counter-clockwise circular polarization, with a noise temperature of 7 K. Ohm carried out an exhaustive study of the uncertainties in the measured and esti-mated noise contributions. The overall system noise temperature (at zenith) was 21 K.

The third low-noise system operated at 4.17 GHz as a sensitive receiver for the Crawford Hill station of the Telstar Project (Jakes 1963). The antenna was the same as for the 2.39-GHz Echo experiment (Figure 4.2), but gain, beamwidths, and pointing characteristics were checked at the new shorter wavelength and found satisfactory when compared with theory. The TWM, with a 5-K noise temperature, was provided by the Scovil group to amplify both senses of circular polarization.

The "first-ever" live TV from Europe was obtained via Telstar with this receiver. The overall zenith system noise temperature was just less than 17 K. W. C. Jakes Jr. was project engineer. The main US ground station was at Andover, Maine.

Table 4.1 shows that the zenith atmospheric sky noise, measured by tip-ping the antenna beam in elevation, is within 0.25 K of the theoretical computation in the microwave band; the theoretical values are computed for average summer conditions of temperature and humidity (USA).

The table also shows that in no case does the estimated system tem-perature exceed the measured system temperature. Of course, none of the estimated system temperatures contain any contribution from the CMBR.

However, it is amusing that, in the case of system 1, the estimated is the same as measured; this indicates sizable uncertainties, probably in both quantities.

Many of the uncertainties in the noise contributions from the microwave circuitry can be avoided by switching between the antenna and a precisely calibrated cold load located near the antenna input *per se* as shown by Penzias and Wilson (1965a), at 4.08 GHz. With this improved measuring system they were able to deduce 3.5 K excess to the expected antenna temperature; this excess is interpreted as the CMBR.

The very low-loss switch used in these measurements is a treasure in microwave radiometry. It is made of a gently squeezed section of waveguide of circular cross section. Penzias and Wilson quote only 0.027 dB of loss for their switch; to my knowledge, it was first mentioned by George C. Southworth in his book on microwave technology and was first used in radiometry by Douglas H. Ring at K-band.

However, measuring the contribution of the lower hemisphere (back lobes) to the antenna temperature is quite another matter (for some estimated values, see Table 4.1). Ideally, one would measure the antenna radiation pattern over the lower hemisphere, measure the ground etc. radiation over that same hemisphere, and integrate the product of those two over the hemisphere. This radiation is comprised of both emission from the ground surface and reflection of sky temperature by that surface (Hogg 1968). Penzias and Wilson calculated a net contribution of just less than 1 K for the antenna *per se*. Apparently further research on antenna design, measurements and siting is called for.

As implied in the historical introduction, some equipment, designed and built for practical (economic) application, can impact scientific studies, provided the quality is good. An example of this is the fruitful use of electromagnetic and electronic equipments, designed for microwave satellite communications development, in pursuing the microwave cosmic background radiation.

Recently, the importance of science and engineering innovation to the USA has been emphasized in a proclamation by the President of the US, backed by a report issued by a panel supported by the National Academies, and chaired by Norman Augustine (2007). That there is fruitful feedback between the two is well exemplified by the exercise we have just discussed. The cosmologists and astronomers who carry on such research and innovation are to be commended.

4.2 Precursor evidence from interstellar molecules

4.2.1 Neville J. Woolf: Conversations with Dicke

Nick Woolf is Professor of Astronomy at the University of Arizona. He was a postdoc in the Princeton University Department of Astrophysical Sciences from 1962 to 1965. His current research interest is astrobiology.

I have these memories that tell me that I cost Bob Dicke the Nobel Prize.

One evening in the attic of Palmer, I think in early 1964, Bob turned to me and asked me whether there was any way to know the amount of the background radiation. He had already turned Roll and Wilkinson onto the topic, but I believe they had only just started.

I said "Well, there were your own measurements in 1946." He grunted. And I said, "and then there are the interstellar molecules."

He didn't say a word. "Oh," I thought, "I must have said something stupid" and I shut up. If I had said more, about the searches for excitation in iron and CN and the other stuff, I am sure he would have picked up on it and he would have been ahead of Penzias and Wilson – but that is the world of Might Have Been.

I also mentioned the molecules to George Field during this time, or slightly earlier, and George said something about that he thought they were excited by collisions. Later I asked him, and he said that he had tried a calculation around that time, but later realized that it had been wrong.

Finally, when I was at the Institute for Space Studies in 1965, Bob Dicke wanted Bill Hoffman and me to fly a balloon to detect the background radiation. Well, I knew that we were far from that level of precision, though in a couple of years later Bill did detect the 100-μm radiation from the galactic center. So I hurriedly diverted Bob to the molecules. And in the hurry of the moment I left him with a reference to McKellar's paper before he had measured the excited state. So Bob got Pat Thaddeus into the picture, and Thaddeus tracked down the literature – but this was all after Penzias and Wilson (1965a) had observed the background.

Anyway, once Bob knew of the excitation he visited me at the institute, and asked who was working on CN at that time. "Guido Munch" I said. "Call him, and ask if there is anything new," said Bob, so I picked up the phone and called Guido. I asked about the cyanogen, and Guido said "Are you working with George Field?" "No, why?" "Well George called me yesterday about this."

Later I found that at almost the same time Shklovsky gave a colloquium in Moscow on the same topic.

So that is the story of how one postdoc's hesitation lost Bob the Nobel Prize. And I believe it would be worth telling the tale, so that some other young person next time is not as hesitant as I was.

And of course, there it is in Herzberg's book about the temperature being 3 K, "but this number has no physical significance whatever" ... I quote from memory.

And like Gamow I have now moved into astrobiology.

4.2.2 George B. Field: Cyanogen and the CMBR

George Field is Senior Research Fellow at the Smithsonian Astrophysical Observatory, and was Director of the Observatory from 1973 to 1982. His current research interest is turbulence in astronomical settings.

My encounter with the microwave background began in 1955. I had come to Harvard as a postdoc, intending to search for intergalactic hydrogen by looking for 21-cm absorption in the spectrum of the radio source Cygnus A. While I was making the observations at the Harvard 28-footer, I studied the problem of the excitation of the upper level of the line, as that is crucial in calculating the absorption coefficient. I realized that it can be excited by fluorescence Lyman-α photons, by collisions with atoms or free electrons, or by absorption of 21-cm photons from whatever source. Ed Purcell and I calculated the collision cross section, I estimated the effect of Lyman-α radiation, and I proposed that we measure the continuum near 21 cm to get the radiation field that would excite the line. When I asked Doc Ewen how to measure the continuum, he said it could not be done at that time because it would require an absolute measurement whose zero point was known. So I extrapolated continuum maps at 21 cm to the coldest point and estimated 1 K. Clearly that was a lower limit, because without a zero point, there was no way to know what the coldest point represented. I published the result in 1959. Of course we now know that the zero point is 3 K.

In 1957, I joined the faculty at Princeton, where I had taken my PhD in astronomy in 1955. I knew about interstellar molecules from Lyman Spitzer, who was studying optical interstellar lines at Mt. Wilson as part of his research at Princeton on the interstellar medium. In fact, in my first published paper, in 1955, Lyman and I mentioned an unidentified line that appeared in our tracing that later was identified as interstellar CH^+. But I was particularly intrigued by a reference in Herzberg's (1950) book on diatomic molecules which stated that one of the lines of interstellar CN arose from a rotationally excited state ($J = 1$) in the ground electronic and vibrational state. The excitation temperature was estimated to be 2.3 K. This was unique in interstellar studies, and so with my experience with atomic hydrogen, I calculated the excitation to be expected from collisions and fluorescence radiation transitions. They failed by a large factor to account for the excitation. To calculate the effect of radiation at 2.6-mm wavelength, which might excite the molecule from the $J = 0$ to 1 levels, I needed two things: the permanent dipole moment of CN, in order to calculate the Einstein B coefficient, and the mean intensity of 2.6-mm radiation at the positions of the interstellar molecules.

The dipole moment had never been measured, so I estimated it from CO, for which it is 0.1 Debye, to be 0.05 Debye, enough to couple the excitation to the radiation field at 2.6 mm. Just as in the case of the 21-cm line, the mean radiation intensity, expressed as a radiation temperature, had not been measured either, but I convinced myself that from the CN observations themselves, 2.3 K was a good estimate. I wrote all this up, and concluded that there must be previously unrecognized source of radiation at 2.6 mm. I gave the paper to Lyman Spitzer to read. He thought it was too speculative to submit for publication, probably because the dipole moment, which determines the coupling to the radiation field, was only an estimate. All this took place before 1960. I recall that because I was then at the old Observatory on Prospect Street in Princeton, whereas we moved to a new building at that time.

One event that took place in the new building was a visit from Arno Penzias. I recall standing in the door of his office discussing his plans to observe 21-cm radiation emitted by atoms in intergalactic space. If the hydrogen is excited solely by the background radiation, no emission will be detected, as it is exactly cancelled by the absorption of the background. Thus we were led to think about the temperature of the background radiation. As I recall, Arno was not optimistic about the absolute measurements required.

The Dicke group was working on the Brans–Dicke (1961) theory of gravitation at Palmer Lab. I knew Dicke and Peebles, and recall attending a seminar there by Jim Peebles explaining his work on helium production in the big bang, of course in Brans-Dicke cosmology. I went up afterward and told Jim that colleagues of George Gamow, including Alpher and Herman, had done similar work. I think I knew at that time of the prediction of 5 K for the background radiation by Alpher and Herman, but I don't recall mentioning it. Moreover, it did not occur to me to mention my work on CN either, because I had not made the connection with the big bang.

I also recall that while teaching a course in Palmer I noticed a microwave horn out of the window, pointing to the vertical. It must have been Roll and Wilkinson's experiment, but again I did not make the connection.

Fast forward to 1965, when the discovery was published in *The New York Times* (Sullivan 1965). I missed it, perhaps because I was packing to move to Berkeley that summer. However, I soon learned about it from a call from Bernie Burke that I got in my Berkeley office. When he said "3 K" I at once realized that it could be the source of radiation that I had predicted in my work on CN before 1960. Unfortunately, my manuscript on the subject at that time is either lost or in cold storage.

Nevertheless, I thought maybe the CN data would be useful. I knew that to make the case I needed to find a value for the CN dipole moment. By a strange coincidence, it was hiding in my wastebasket. At the time, I was writing an article for the *Annual Reviews of Astronomy and Astrophysics*, and the editor had sent me proofs from another article as a guide to marking my own proofs. The article was on *The Spectra of Comets*, by Claude Arpigny, in which he discusses how to predict the emission spectra of molecules – including CN – using rate equations for level populations. One of the parameters is the dipole moment of the ground electronic state, which he had adjusted to fit the data. His number, 1.2 Debye, was not far from a more recently measured laboratory value, 1.4 Debye. Arpigny's dipole moment enabled me to calculate the coupling of the $J = 1$ rotational level of CN to the radiation field at 2.6 mm. I found that the coupling to radiation is stronger than to collisions or fluorescence by a large factor. Much stronger, even, than I had concluded before 1960, by the square of the dipole moment, a factor of 200. I knew then that we had a radiation thermometer at 2.6 mm.

Another coincidence occurred the same day. When John Hitchcock, a graduate student working in the next office, heard what I was doing, he came in and told me that at that moment he was working on observations of the rotational excitation of interstellar CN. He was reducing data that he had taken from six plates that George Herbig had taken of the spectrum of the star Zeta Ophiuchi at the wavelength of the interstellar CN line. Suddenly we had new data to which to apply the theory of excitation. Together with George Herbig we wrote an abstract of a paper for the 120th meeting of the American Astronomical Society, which was meeting in Berkeley (another coincidence). At that meeting, held December 28–30, 1965, we presented evidence that the background radiation follows a blackbody spectrum over the 28-fold wavelength interval from 7.4 cm to 2.6 mm. Our value of the temperature was given as 2.7–3.4 K (Field, Herbig and Hitchcock 1966).

John and I published two more papers on the subject (Field and Hitchcock 1966). One in *Physical Review Letters* gave a result of 2.7–3.6 K for Zeta

Persei, a star on the other side of the sky from Zeta Oph, and 300 pc distant from it. Thus the hypothesis that the radiation is universal passed the test. In a later paper in *The Astrophysical Journal* we considered the possibility that the spectrum of radiation is not blackbody after all, but as suggested to us by Nick Woolf, dilute blackbody at a higher temperature. We were able to rule out this hypothesis with reasonable certainty. It is interesting that the peak of the blackbody curve in frequency units is 1.7 mm. With our measurements at 2.6 mm, we were climbing the peak.

4.2.3 Patrick Thaddeus: Measuring the cosmic microwave background with interstellar molecules

Pat Thaddeus is the Robert Wheeler Willson Professor of Applied Astronomy, and Professor of Applied Physics, Harvard University, and Senior Space Scientist, Smithsonian Astrophysical Observatory.

Following Penzias and Wilson's great discovery, it was clear that measuring the microwave background near its peak intensity at a wavelength of 2 mm was the crucial observational test to demonstrate the blackbody spectrum of the radiation and its cosmological origin. Most of the energy of a 3-K blackbody lies in the vicinity of 2 mm, and because the energy in the CMBR is enormous on a cosmic scale – 100 times that of starlight when averaged over the great voids between galaxies – that is the observation which constituted the dagger at the throat of the steady state universe. Because the opacity of the terrestrial atmosphere increases greatly at short wavelengths, it seemed likely that a spacecraft observation was required, with all the expense, difficulty, and delay which that was likely to entail.

I had recently designed a small radio telescope for the Nimbus satellite to study the thermal emission of the Earth at a wavelength of 2 cm (chosen because that is a band where the contrast between water and land is large, and icebergs are readily distinguished against the surrounding ocean). I was therefore well aware of the technical and political problems – and the frustrations – which a spacecraft measurement imposed. I discussed these from time to time with my colleague William Hoffman at the Institute for Space Studies in New York City, who was working in collaboration with Neville Woolf on a small balloon-borne cryogenic telescope to conduct the first survey of the Milky Way in the far infrared. That instrument, which went on to discover how remarkably rich the Galaxy is in the far infrared, was the forerunner of Infrared Astronomical Satellite (IRAS) and the other far-infrared telescopes which have had such a large impact on space astronomy.

Woolf had been in New York working with Hoffman but had recently gone to the University of Texas, where he remained a member of our informal discussion group, and he was aware of my interest in the microwave background. From there in September 1965 he wrote me a letter which set me back, because it suggested that I might be wasting my time. "On a theoretical approach," he wrote, "I wonder what can be made from the case of the missing CN lines. Lines arising from a level only $3\,\mathrm{cm}^{-1}$ above the ground state are absent." This came as a severe shock and a disappointment, because it implied that there might be no background radiation at all in the millimeter-wave band, and that the Penzias and Wilson radiation – whatever it was – was not the faint blackbody remnant of the primeval fireball predicted by Gamow. An attempt to observe directly the short wavelength background could be a significant waste of time and money.

So I decided to look into the question of interstellar CN and the mechanisms which excite it in the interstellar gas. It didn't take long to discover that this widely studied radical, readily observed on the Sun, in comets, and in laboratory discharges and flames, was not stationary in the interstellar gas at all, but was rotating instead by just the amount expected from excitation by 3-K blackbody radiation. Observed as an extremely faint satellite, the R(1) line, to the R(0) line from the rotational ground state, this seemingly insignificant fact made almost no impression on astronomers when it was first observed by McKellar in 1940 as a barely perceptible absorption line on a high-resolution Mt. Wilson spectrum of the second-magnitude star Zeta Ophiuchi. But Gerhard Herzberg, the eminent molecular spectroscopist and a sharp-eyed observer who missed little, was aware of it – to the point of citing it on the penultimate page of his well-known monograph *Spectra of Diatomic Molecules*, whose first edition appeared in May 1950. There Herzberg quotes a rotational temperature of $2.3\,\mathrm{K}$, but proceeds to say that this excitation has "... only a very restricted meaning."

This conclusion was far from unreasonable at the time – it is probably what most astronomers and spectroscopists would have said had they taken the trouble to consider the matter at all. In 1940 only three molecules were known in the interstellar gas: CH, CH^+, and CN. As the heaviest of the three, and the one with the lowest frequency rotational transitions, CN was the most susceptible to excitation by purely local processes such as collisions with H atoms or resonant scattering (fluorescence) of starlight. Herzberg therefore reasoned that CN was the first interstellar molecule which one might expect to find rotationally excited in space, and its excitation was hardly surprising and had no general significance. It presumably varied from interstellar cloud to cloud, in response to the local density, intensity of starlight, etc.

But with the scant interstellar CN data which existed at that time and the poor signal-to-noise which characterized all of it, variation from cloud to cloud was very difficult to demonstrate. Astronomers would have been puzzled to discover that CN everywhere was excited by the same amount: just under 3 K. Was Herzberg right? was the crucial question.

About this time I took on for his doctoral research a versatile and energetic graduate student in the Columbia Physics Department, John F. Clauser, and together we undertook to do two things to see if CN really served as a good thermometer for the CMBR. The first was theoretical: to see if local processes were fast enough to explain the CN excitation. The second was observational: to obtain better CN data to better measure the CN excitation and the CMBR temperature, and to see whether the excitation was constant from cloud to cloud, as predicted by a universal mechanism.

Our theoretical calculations were hampered at the outset by the fact that the CN electric dipole moment – the parameter that determines the rate of rotational excitation by background photons and how tightly CN is coupled to the CMBR – had not been measured. So it had to be obtained indirectly. In our first paper, Clauser and I argued that the small value for the dipole moment of the first excited electronic state of CN, which had been obtained from pressure-broadening in a flame, implied that the dipole moment of the ground state was probably substantial, at least 1 Debye (in the standard unit of molecular dipole moments, the Debye, 10^{-18} esu), making CN a sensitive thermometer for the CMBR. That assumption turned out to be correct. Thanks to a communication from George Field, who we discovered was also pondering the question of the excitation of interstellar CN, we then learned that the Belgian astronomer Claude Arpigny had recently deduced the dipole moment of CN to be 1.1 D from cometary spectra, and that is what we adopted for our calculations. A few years later Thompson and Dalby observed the optical Stark effect of CN in the laboratory – a difficult experiment – and with it measured directly the CN dipole, and found it to be even larger: 1.45 D. So our indirect estimate turned out to be conservative: CN was even more tightly coupled to the CMBR than we had assumed, and was an even better radiative thermometer than we had supposed.

Clauser and I soon calculated that if the CN was located in a normal HI region, as generally assumed, the excitation by local processes, in particular collisions with hydrogen atoms and fluorescent excitation by background starlight, was quite slow compared to that from the CMBR, and it was therefore the CMBR which largely determined the excitation. We found that in an

HII region the situation was less clear-cut: collisions with charged particles, slow protons in particular, could be a significant source of CN excitation; but there was then – as today – little evidence that interstellar CN is found in HII regions, so that possibility could be discounted. Our conclusion was therefore, contrary to what Herzberg had assumed, that the excitation of CN was the work of the CMBR, and the rotational temperature of CN was a direct measure of the temperature of the background at the wavelength of the 1-0 rotational transition, 2.64 mm – quite close to the peak of the 3-K blackbody curve, which is the reason why our measurement was important. It implied that the enormous amount of radiant energy locked up in the CMBR which had escaped detection by the low-frequency measurements of Penzias and Wilson and other radio astronomers was really there, and could not be swept under the rug as the proponents of the steady state theory might have liked.

Our attempt to show that the CN excitation was constant from star to star was somewhat disappointing, because in the 1960s the number of bright stars with interstellar CN absorption was very limited, consisting of only two really good examples: ζ Oph, the second-magnitude star studied by McKellar, and ζ Per, a similar star in the opposite part of the Milky Way. We were nonetheless able to conclude from a total of ten stars with some evidence of interstellar CN that all the available data were consistent with invariant CN excitation, and the existence of background radiation at a temperature of 3 K or somewhat less. Today with modern echelle spectrographs and CCD detectors, one could demonstrate the constancy of CN excitation much better than in the 1960s, but with the remarkably accurate direct measurement of the CMBR spectrum by COBE, the question now is largely of historical interest.

Over the next few years we made an effort to obtain a really good measurement of the CMBR with CN and the other molecules then known to exist in the interstellar gas. This was well before the flood of interstellar molecule discoveries, which began with the discovery of ammonia and water by Townes and Welch and coworkers at Berkeley in 1968–1969. Radio lines of OH had been identified in 1963, but OH optical lines analogous to those of CN do not exist, and this widely distributed molecule unfortunately does not serve as a useful radiative thermometer for the background. Of the more than 130 molecules now known to exist in the interstellar gas or circumstellar shells, it is remarkable that none surpasses CN as a thermometer for the CMBR – or even comes close.

For a radio astronomer, obtaining the necessary time in the 1960s on a large optical telescope with a fast high-resolution spectrograph was far from easy. To get started, Clauser and I succeeded in begging time on the large McMath solar telescope at Kitt Peak, which had a very high-resolution spectrograph and had been used for a small amount of interstellar work, and with it we succeeded in obtaining a few rather ragged spectra of ζ Oph showing interstellar CN. These yielded a rotational temperature of about 3 K, but our data really represented little or no improvement over McKellar's pre-war observations at Mt. Wilson. It became clear after discussions with optical observers who worked at high spectral resolution that the instrument to use for a significant improvement was the Lick 120-inch telescope, which thanks to George Herbig had a very high-resolution Coudé spectrograph that was fast and efficient.

To obtain a substantial block of time on this telescope as a visiting faculty member of the University of California, I moved with my wife and two children to Berkeley for the spring quarter of 1968, and was granted then and over the summer a substantial block of time on the 120-inch telescope for the CN project by Albert Whitford, the Lick Director. With another New York student, Victor Bortolot, I ultimately obtained over 30 very high-quality Coudé spectra on IIao emulsions, carefully baked to enhance sensitivity, at the highest available resolution of the spectrograph, 1.2 A/mm; most were 5 mm wide to store a large amount of information on each exposure, which typically required several hours. Back in New York we digitized these spectra and added them together numerically to obtain the interstellar spectrum of ζ Oph with a signal-to-noise not previously achieved. I was fortunate to be tutored in the art of high-resolution Coudé spectroscopy by two masters, George Herbig, then on the faculty at Santa Cruz, and Gene Harlan, a fastidious and exacting member of the Lick technical staff.

During one of our Lick runs we were paid an impromptu visit by George Mueller, director of the Apollo Program, and Wernher von Braun, director of the Marshall Space Flight Center and the chief architect of the Saturn V launch vehicle developed for the Moon landing. They were contemplating a telescope of the Lick class in orbit and wanted to see what one actually looked like. As a prototype for a space telescope, the Lick 120-inch with its long focal ratio was a bizarre choice, and one is thankful that this visit did not sour NASA on a big telescope in space for good. Mueller soon left, but von Braun, who knew something about optics and had an affection for telescopes, decided to stay the night. I put him to work in the bowels of the Lick Coudé guiding the telescope, keeping the light from the star streaming down the entrance aperture of the spectrograph. He was an interesting companion for a long observing night, and a good talker. It was not long before this – about 1960 – that von Braun published his memoir *I Aim at the Stars*, which a well-known comedian (Mort Sahl) said might have been better called *I Aim at the Stars, and Sometimes Hit London*, and it was about the time Norman Mailer in his book *Of a Fire on the Moon* described von Braun as looking like "the head waiter in the largest hofbrau house in Heaven." It was an unusual evening at the Coudé focus of the 120-inch.

Our synthesized spectrum represented well over 100 h of 120-inch observation. With it we determined the rotational temperature of CN toward ζ Oph to be 2.99 ± 0.06 K, which with small corrections for optical depth and collisional excitation yielded 2.78 ± 0.10 K for the temperature of the CMBR at $\lambda = 2.64$ mm. It is gratifying that this result obtained by photographic spectrophotometry in 1972 is within 0.6σ of the COBE temperature published 18 years later.

Our synthesized final spectrum covered not only the violet band of CN, but also, as Figure 4.3 shows, the stronger interstellar bands of CH and CH^+, and it yielded also a marginal detection of $^{13}CH^+$, which furnished the first observational evidence for carbon-13 in the interstellar gas. The $^{13}C/^{12}C$ ratio, which we obtained, was consistent with the terrestrial value, $1/89$, and with the many measurements later made of this important isotopic ratio.

Several years later I took on as a postdoc in radio astronomy John Mather, a brilliant student of Paul Richards. For a research topic John took up the recently discovered SiO masers, but, true to his training in Richards' laboratory, his real interest was in the microwave background and the challenge of its short wavelength spectrum. The first proposal for a Cosmological Background Radiation Satellite was written in my office in New York. Already present at our first meeting to draft a proposal to NASA for a spacecraft experiment were some of the key COBE players, including Weiss, Wilkinson, Hauser, and Silverberg. Mather was so good at both the political and the technical requirements of this enterprise that when he later took a job at the Goddard Space Flight Center, the project followed him with my support. The discovery of polyatomic molecules in space was underway, and that promised the kind of science which was closest to my heart.

My excursion into optical astronomy was short, but it left an indelible impression. After a many-hour exposure, the thrill of holding up a developed photographic plate in the darkroom to the light and seeing the faint, barely perceptible absorption line of excited CN, knowing that it was a fingerprint of the universal radiation filling all space, once as brilliant as the surface of the Sun, was an aesthetic and intellectual pleasure which I have

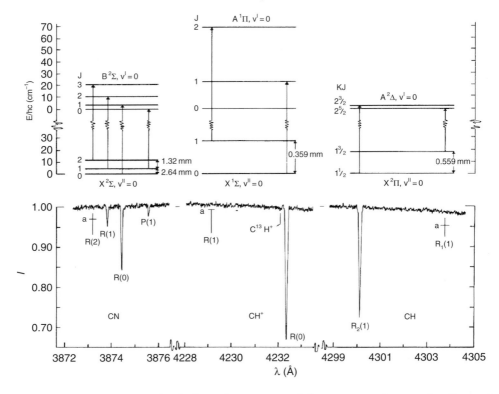

Fig. 4.3. The spectrum of ζ Oph in the vicinity of the strongest interstellar lines of CN, CH, and CH$^+$, from our first 25 Lick spectra, ca. 1968 (Bortolot, Clauser and Thaddeus 1969). The R(1) and P(1) lines of CN are a direct result of the CMBR, and provide a measure of its intensity at $\lambda = 2.6$ mm near the peak of the 3-K Planck curve. Directly above each line the transition is shown on a diagram of the relevant energy levels. "a" indicates the strength of the missing lines if the intensity of radiation in interstellar space at $\lambda = 1.32$, 0.359, and 0.059 mm were that obtained by Shivanandan, Houck and Harwit (1968) from a rocket flight which suggested a large amount of short wavelength background radiation. ©1969 American Physical Society.

never again experienced in research – a shock of recognition perhaps comparable to that felt by Rutherford when he saw the back-scattered alphas and realized that our entire picture of the structure of matter was wrong. The now largely obsolete techniques and paraphernalia of optical photographic spectroscopy – the baking and cutting of the big glass plates to fit the curved plate holders, the development of the plates in total darkness in the darkroom with its characteristic pungent smell – provided pleasures which I am afraid few astronomers today will enjoy.[2]

---

[2] My early work on measuring the CMBR with interstellar molecules is summarized in Thaddeus and Clauser (1966), Bortolot, Clauser and Thaddeus (1969) and Thaddeus (1972). Two other

Although the excitation of interstellar CN is remarkably simple, determined almost entirely by the photons of the microwave background, the general question of molecular excitation in the interstellar gas is still very much with us, with fascinating ramifications. In general, a molecule in space is subject to simultaneous excitation by three thermal or quasi-thermal reservoirs: the microwave background at 2.725 K, collisions with the ambient gas at a kinetic temperature typically between 20 K and 200 K, and background starlight, highly dilute radiation at 5000–10,000 K. There are astronomical regimes where each of these reservoirs predominates, and others where the competition between them can produce striking departures from thermal equilibrium: population inversion and maser amplification in OH and H_2O, and refrigeration below 3 K in formaldehyde, so that absorption against the CMBR is observed in the absence of any localized background source, an observation which would have been quite unintelligible before the discovery of the CMBR. A questionable but highly interesting example may be provided by the interstellar diffuse bands, the several hundred unidentified interstellar features which date from as long ago as the 1930s. It is widely thought that the width of these bands is the result of rapid radiationless transitions, but other interpretations are possible. The bands seem to be formed preferentially in low-density regions of the interstellar gas, and they are probably produced by strong electronic transitions with large f-values. Under these conditions, rotational excitation by starlight is a possible cause of the width of the diffuse bands, a mechanism which we considered early on to explain the excitation of interstellar CN, but rejected because of the small f-value of the CN optical transitions.

No one in the 1960s realized how profoundly the discovery of the microwave background would enhance our knowledge of the universe – realized that we were on the eve of great events which would alter cosmology beyond recognition in little more than a generation. (Possibly Dicke, our most far-seeing teacher, sensed this.) Comparable in its impact to Watson and Crick's model of the structure of DNA, the discovery of the background started a revolution, which has yet to run its course. To me as a practitioner of small science, it is a pleasure to realize that much of this revolution (obviously not all) has been done on the cheap, by small teams, in the classical way in which experimental science has been pursued since Jansky, Rutherford, Faraday, and before.

papers based on our Lick high-resolution optical spectra are *Probable Detection of Interstellar* $^{13}CH^+$ (Bortolot and Thaddeus 1969) and *Weak Interstellar Lines in the Visible Spectrum of* ζ *Ophiuchi* (Shulman, Bortolot and Thaddeus 1974).

4.3 Precursor evidence from element abundances

4.3.1 Donald E. Osterbrock: The helium content of the universe

Donald Osterbrock played a leading role in the study of AGNs. He is author of the influential book, Astrophysics of Gaseous Nebulae and Active Galactic Nuclei *(1989), and coauthor, with Gary J. Ferland, of the greatly expanded second edition (Osterbrock 1989; Osterbrock and Ferland 2006). At the time of his death in January 2007 he was Professor Emeritus of Astronomy and Astrophysics at the University of California, Santa Cruz.*

I have never done any research in cosmology, but as an onlooker I have been interested in it for many, many years. I was inclined toward science from boyhood, partly no doubt because of my father's background as an engineering professor, and my mother's as a chemistry assistant in an industrial laboratory in Cincinnati, where both of them, my brother, and I were all born and grew up. My high school had an excellent library, and in it, and also in books from our local public library, I read a lot about astronomy. I had a small amateur-made reflecting telescope with an alt-azimuth pipe mounting, and could look at the poor images that it produced of the Moon, planets, and bright nebulae like the Ring and Orion. My father took me to occasional meetings of the local amateur astronomical society when a famous professional came to town, and I remember especially Harlow Shapley and Otto Struve.

I graduated from high school six months after Pearl Harbor, and in another seven months I was in the Air Force, training to be a weather observer. On a troopship from Honolulu to Okinawa, we proceeded by way of Eniwetok, Guam, and Saipan, and on the way I first saw the star Fomalhaut and the Southern Cross. After the war ended, I was able to enter the University of Chicago under the so-called GI Bill of Rights, and in three years completed a bachelor degree in physics, and a master in astronomy and astrophysics. Chicago had the best faculty in physics and astronomy in the country at that time, in my opinion, and I was especially inspired by courses in quantum mechanics, taught by Gregor Wentzel, and nuclear physics, by Enrico Fermi. There were no active cosmologists there, but I attended colloquia by George Gamow, on what we call the "big bang" today, but he called the "ylem-theory" then, and by Maria Goeppert-Mayer on her new interpretation of the so-called "magic-number" nuclei in terms of nuclear shell structure with strong spin–orbit and spin–spin coupling. These two colloquia seemed quite reasonable to me and, I noticed, to nearly all the professors who were there too.

Then, for three years at Yerkes Observatory I again had excellent teachers, especially Struve, S. (Chandra) Chandrasekhar, W. W. Morgan, Bengt Strömgren and Gerard P. Kuiper. All of them taught us about stars, nebulae, and galaxies, even Kuiper, although he also lectured on the Solar System, on which he had begun working during and just after the war. I did my thesis with Chandra, on the gravitational interaction between stars and cloudy interstellar matter, which we would call "giant molecular clouds" today. For two or three weeks in the summer of 1951, I went to a "summer school," organized by Leo Goldberg at the University of Michigan. I wanted to hear the lectures of George C. McVittie, on hydrodynamics of interstellar clouds (though Chandra advised me not to go – he said he had already taught me more than McVittie knew on the subject!). The "school" was held in the old Detroit Observatory building at the UM campus, where all the professors and grad students had offices. I believe I was the only student from outside UM who attended, and so I shared an office with McVittie and with David Layzer, who had just joined the faculty there that summer, with his fresh Harvard PhD degree.

Once a day we all got together in the main room of the observatory, to have coffee and talk about astronomy. In those conversations McVittie and two of the older professors, Dean McLaughlin and Freeman Miller, were scathing in their remarks on Fred Hoyle's steady state theory of cosmology, involving continuous creation of matter. McVittie was a classical mathematical cosmologist, and I had soon seen from his lectures that Chandra had been right. He had little if any physical insight, and his criticisms of Hoyle's ideas were ridiculous, I thought. Basically, he said continuous creation just couldn't happen, and McLaughlin and Miller chimed in as his conservative claque.

After I completed my PhD at Yerkes in 1952, I was fortunate to be appointed a postdoc at Princeton for a year. There I worked out the internal structure of red-dwarf stars, which turned out to have deep outer convective zones, but radiative centers with the main energy production by the proton–proton reaction. I had learned of the problem in Strömgren's stellar-interiors course at Yerkes, and he encouraged me to follow it up at Princeton. Martin Schwarzschild and his students were working on red-giant stars, and he helped me tremendously in my work. Lyman Spitzer, the head of the astronomy department, asked me to teach the stellar atmospheres graduate course the second semester I was at Princeton, so he could spend full time on his research on deriving energy for peaceful uses from controlled nuclear reactions, called Project Matterhorn at that time. I was glad to teach the course; there were only four grad students in it: Andy Skumanich, Jack Rogerson,

George Field and Leonard Searle. As they all had long and successful careers as research astrophysicists, I can't help thinking that at least I didn't hinder them in this first course I ever taught.

Hoyle came to Princeton that year as a visitor, working with Schwarzschild on the structure and evolution of red-giant stars for two or three months. Fred's office was next to mine, in the quiet rear of the old observatory building, and we often discussed his research and mine. He was extremely hard working, brilliant, and knew a lot of astrophysics. I was impressed by Hoyle, and although he was not doing cosmology there at that time, I still had an open mind on it. We never discussed cosmology, so far as I can remember. Hoyle was all business on red giants there, as I was on red dwarfs, and those were the two subjects we talked about.

After one year at Princeton, I was appointed to the faculty of Caltech's then very new astronomy and astrophysics department, headed by Jesse Greenstein. My wife and I drove west in the summer of 1953, stopping for a month at Ann Arbor for a second astrophysics summer school, again organized by Goldberg. This one was much more successful than the earlier one, with Walter Baade and Gamow the two main lecturers, backed up by Ed Salpeter and Kuiper for shorter series of talks. About 30 grad students, postdocs, and young faculty members were there. I was most interested in learning from Baade, but Gamow's lectures, mostly on his cosmology, were quite good. He was always humorous, but with plenty of good ideas. By that time in his life he was a fairly heavy drinker, but it never seemed to mar his thoughts nor his lectures.

Baade was a fantastically inspiring lecturer, and I was glad indeed to have him and Rudolph Minkowski as my chief mentors in Pasadena. At that time the Caltech and Mt. Wilson (now Carnegie) astronomers shared the 200-in and 100-in telescopes, and I worked largely on nebular spectroscopy, with some forays into emission-line galaxies, but never into cosmology. There were too many interesting things for me to do with objects in our own and nearby galaxies. Hoyle came to Caltech two or three times while I was there, mostly to work with Willy Fowler and Geoff and Margaret Burbidge, who came there on visits, on nucleosynthesis in stars. Fred was a visiting professor for one quarter, lecturing on the same subject, and I sat in on most of his lectures. But I never discussed cosmology with him then, nor heard him discuss it with others around the astrophysics lunch table in the faculty club, except to utter an occasional disparaging remark about the "big bang."

From Caltech I went to the University of Wisconsin in Madison with Art Code, to help him build up a full-size graduate astronomy and astrophysics department there. Again, I continued largely observational research there with our smaller telescope, using its excellent photoelectric scanner which made it highly effective for nebular problems.

Then in 1960–1961 I had a Guggenheim Fellowship to go back to Princeton on leave, this time as a visiting fellow at the Princeton Institute for Advanced Study, where Bengt Strömgren had recently become the professor of astrophysics, "the man who got Einstein's office." Among the other visiting fellows then were Anne Underhill, who had worked with Bengt at Yerkes and in Copenhagen, Su-shu Huang, another Yerkes PhD, and Hong-Yee Chiu. We had weekly astronomy lunches with Spitzer and Schwarzschild, and Field and Rogerson, who had come back as assistant professor and research associate, respectively, and others. These were held in a faculty cafeteria upstairs in Firestone Library, not as spacious or well appointed as the IAS dining room that was built later, but still quite a step up from the aluminum-sided diner on Nassau Street where we had gone in 1952–1953.

I think Martin suggested to Rogerson and me that we review the status of the helium abundance in the objects we knew best: the Sun, on which Jack had done a lot of research while a Carnegie postdoctoral fellow at Mt. Wilson, and gaseous nebulae, with which I was familiar. I had seen Rogerson often in his two years in Pasadena, and we were good friends.

The helium abundances in nebulae were simple; we used the measurements of the Orion HII region and several planetary nebulae, made by my first PhD thesis student at Caltech, John Mathis, who had also calculated the relations between line-strength and abundances of helium and hydrogen. These were supplemented by somewhat later theoretical calculations by Mike Seaton. Our results were that the helium to hydrogen ratio was very nearly the same for planetary nebulae (mean value $N(He)/N(H) = 0.16$), and for the Orion nebulae ($N(He)/N(H) = 0.15$). They contradicted the idea that the helium content in our Galaxy might have increased with time, from when the stars had formed that were at present in the planetary nebula stage (then estimated as 5×10^9 years ago) to today.

For the Sun we used absorption-line strengths Rogerson had measured for weak [OI] lines in the Solar spectrum to determine the relative abundance of oxygen as a representative of the heavy elements (usually called "metals," an especially poor term for all the elements heavier than helium, in my opinion!) to hydrogen. Then from the relative abundances to oxygen of all those heavy elements, often described in earlier years as the "Russell mixture," but using more recent compilations, we derived the abundance ratio by mass, $Z/X = 6.4 \times 10^{-2}$. In this notation X, Y, and Z represent the fractional abundance, by mass, of hydrogen, helium, and heavy elements (where the helium fraction is $Y = 1 - X - Z$).

The other relation we used for the Sun was derived from a series of Solar interiors models that Ray Weymann had recently calculated at Princeton under the guidance of Schwarzschild. These new models were then current state-of-the-art, taking into account a shallow outer convection zone, an intermediate, unevolved radiative zone, and a large inner radiative but hydrogen-burning region, in which the results of nuclear processes over 5×10^9 years had affected the variation of hydrogen and helium content with distance from the center. Energy production was mostly but not entirely by the proton–proton reaction, and there was no central convective core. These were the best models then available, but in addition I liked them personally because Ray had been the brightest and best undergraduate student I had taught at Caltech, and also because his models took into account revisions and extensions of my early research on red dwarfs by Nelson Limber, my close friend from Yerkes days. Nelson had also gone on to Princeton as a postdoc after me.

The well-observed Solar radius, luminosity and mass gave $X = 0.67$, $Y = 0.29$, $Z = 0.04$ for the original abundances in the Sun, at its formation 4.5×10^9 years ago. This set of abundances is not quite the same as we had derived for the planetaries and the Orion nebula had given, but well within the estimated error, we believed. In the end the best overall fit we adopted was $X = 0.64$, $Y = 0.32$, $Z = 0.04$, essentially unchanged for the past 5×10^9 years. Our evidence was that that the helium abundance in the Sun is essentially the same as the results mentioned above for planetary nebulae and the Orion nebula based on the very straightforward recombination-line theory for H^+ and He^+.

Although many of the numerical values have been revised slightly on the basis of better measurements and improved theoretical interpretations of nebular and Solar spectra, our conclusion has remained unchanged. The abundance of helium in our Galaxy, and presumably in other galaxies as well, had changed little from their earliest days. Most of the helium must have been formed in the big bang. Personally, I could have accepted the idea that both helium and hydrogen had been created together in a steady state universe, but evidently Hoyle, Hermann Bondi, and Tommy Gold could not, nor could other later theoretical cosmologists.

Rogerson and I had done our paper because Schwarzschild suggested it at the time. I don't remember why he thought it was important, but I don't think it was for cosmology. Certainly I did not have that idea in my mind back then. I was interested in it chiefly because Martin seemed to me so uncertain about what the helium abundance was in stars near the Sun. He had used various abundances for it in his early stellar interiors and evolution papers with students, postdocs, and visitors at Princeton as collaborators. Looking back now (I didn't realize this at the time), he had even used $Y = 0$ (no helium at all)! This was heresy to me, as all grad students at Yerkes were indoctrinated from early on with the interpretation of the spectral sequence as basically a temperature sequence in stellar atmospheres, all with the same abundances in them, with luminosity as a secondary criterion, but only a very few minor abundance variations which Morgan, Keenan and Kellman (1943) had noted in bright stars, and Nancy Grace Roman (1950) had found more in somewhat fainter ones in her postdoctoral research. It was evident that helium was much more abundant than anything else except hydrogen from the great strengths of its lines in hot stars, though we didn't know just how abundant it might be. All the astronomers I talked with in 1953–1958 at Mt. Wilson and Palomar Observatories had the same general idea, I believe.

Only Martin did not have it in 1952–1953, and he didn't seem to in 1960–1961, although maybe he was just pretending, to convince Jack and me to prove it. I now realize that Schwarzschild had calculated those models with $Y = 0$ to compare with earlier calculations by Hoyle and Lyttleton (1942). The assumption $Y = 0$ agreed with Hoyle's interpretation of the steady state theory. As I mentioned above, I could have accepted continuous creation of both hydrogen and helium if that fitted observational data. Perhaps by that time, 1961, Hoyle was already semiconvinced that continuous creation was dead because he knew from his contacts with American observational astronomers that Y does *not* equal zero anywhere in our Galaxy. But I may be wrong, and I do not want to put words into his mouth or in Martin's either!

In addition to Burbidge *et al.* (1957), three early theoretical papers that I know of had treated the expected helium abundance in our Galaxy as a result of nuclear reactions in stars. Burbidge (1958) estimated its increase with time from the approximately known luminosity of the whole Galaxy, Maarten Schmidt (1959) formulated and calculated an early "closed-box" model, and Mathis (1959) carried out a somewhat less exhaustive one. All three assumed that the initial helium content was zero, and built up gradually with time, as a result of nuclear processing in stars and return of matter to interstellar space from evolved stars, but all three found, in one way or another, that this hypothesis would not work, although they did not put it that directly. None of these authors considered how the heavy-element content might have increased; that was still an unknown process.

When the CMBR was discovered in the 1960s, I readily accepted it as a confirmation of the big bang picture. I believed, and still believe, in following the observational evidence, as long as it was based on sound theoretical interpretations. However, I think it is a great mistake to trust any detailed numerical values, derived from observational measurements, too far. The theory is always too simple to match reality "exactly." For instance, I have heard lectures, and seen cosmological papers, in which values of X and Y derived from nebular spectrophotometry are quoted and used to three significant figures. Observers are often overly optimistic in stating their probable errors, and theorists who use them can be even more so. But in addition all the available calculations of the HI and HeI emission-line intensities that I know are based on simplified model nebulae, either with one "mean" temperature and one "mean" electron density, or on models in which local means, varying only with distance from the photoionizing star or stars, are used. Yet direct images of nebulae show that down to the finest resolution we have been able to achieve to date, even at excellent seeing-sites on high desert mountains or from space with the Hubble Space Telescope, fine structure, "filaments," and "clumps" are present in nebulae. No doubt these contain a range of densities, temperatures, and excitation conditions down to very small scales. The "mean" values may not represent these conditions to high accuracy, as many current papers are showing. As our understanding of the effects of fine structure, and also perhaps of hydromagnetic heating of nebular gas, improves, the precision of the derived relative abundance will also increase.

4.4 The path to the hot big bang in the Soviet Union

4.4.1 *Yuri Nikolaevich Smirnov: Unforgettable Yakov Zel'dovich*

Yuri Smirnov is a Leading Research Scientist at the Russian Research Center "Kurchatov Institute," Moscow. He is coauthor (in collaboration with Victor Adamskii, Yuri Babaev, Andrei Sakharov, and Yuri Trutnev) of the 100-megaton thermonuclear bomb; he took part in the test of its half-power version on October 30, 1961. He was one of the initiators and participants of the program for deep seismic sounding of Earth's crust with the help of underground nuclear explosions for the accelerated revealing of the prospective regions containing oil, gas, and other minerals. He participated in the preparation of 14 "peaceful" explosions; in 11 of them (for deep seismic sounding) he was State Commission Vice-Chairman. His research interests include atomic energy and the history of the Soviet atomic project.

There is a deep analogy between physical processes which take place inside nuclear bombs (especially thermonuclear bombs) and stars. It was soon realized by the designers of nuclear weapons. That is why the newcomers who just came to Sarov, the Russian nuclear center, were sometimes told, "we research astrophysics here." There were good reasons for that.

For instance, the thermonuclear reaction inside a ball-shaped hydrogen medium can be characterized by the mass and radius. Variation of these parameters over a wide range of values changes the process out of recognition: from an explosion in the case of a bomb to stationary burning in the case of a star.

At the same time, the comparison of a "bomb" and a "star" is relative. Thus, for example, the principal reaction inside stars is $p+p$ (the interaction of protons) whereas in a thermonuclear bomb reactions with deuterium and tritium, $d + d$ and $d + t$, dominate. Or again, in the case of nuclear burning in a star the retentive force is gravity, while in a bomb it is the compression produced by explosives or radiation.

This is why theorists who participated in the Soviet atomic project commonly became experts in astrophysics or cosmology. Thus it was natural that David Frank-Kamenetskii (1959), who left Sarov as long ago as 1956, published his well-known monograph *Physical Processes in Stars* and a series of papers on the origin of elements in the universe as early as 1959. He was one of the most brilliant theorists to collaborate with Yakov Zel'dovich.

Nevertheless it was Ya. Zel'dovich, who, as a chairman of the physics seminars, performed a quick turn from elementary particle physics to relativity theory and cosmology in 1961–1962. Andrei Sakharov, Nikolay Dmitriev, Andrei Doroshkevich, Michail Podurets, Sergey Kholin, Valery Yakubov, and others were influenced by nobody but him in their studies of astrophysics and cosmology in Sarov. I didn't escape the common lot as well.

For me, a 24-year-old colleague of Sakharov, who had just returned from the test of a 50-megaton superbomb on October 30, 1961, it was extremely interesting to join that renewed seminar. That was especially because Zel'dovich equalized starting positions of all participants and definitely carried young people by the choice of subjects. Along with the chief we synchronously began to study chapters on general relativity in the book *The Theory of Field*, by Landau and Lifshitz (1960), and, following Zel'dovich's proposal, reported on them in the seminar.

Discussions gave rise to questions which turned into tasks. Soon publications of the participants of the seminar began to appear. And as the chairman accelerated the pace he shortly became an authoritative leader in this branch of physics that was new for him.

Professional discussions with Yakov Zel'dovich were remarkable for their dynamism and expressiveness. Having familiarized myself with his paper *The Initial Stages of the Evolution of the Universe* (Zel'dovich 1963a), which was just published in the journal *Atomic Energy*, I discovered some inaccuracies and dropped in at his office. Right off the bat we found ourselves in front of the blackboard, and I barely followed the dance of a piece of chalk in the hand of the master. Remark ... calculation ... another remark ... rejoinder ... another one. And suddenly he said with satisfied smile: "You beat the academician!" Then he walked up to the desk, opened a notebook, and asked me to read a manuscript of his latest work to check whether I was interested in it. I was interested.

When he was leaving for Moscow for the traditional two weeks in February 1963, we failed to meet, and he left the note shown in Figure 4.4 with an assignment for me. He watched over me patiently and with curiosity until the task was completed.

When the work was done, I proposed to publish it as joint authors. Yakov Zel'dovich strongly objected: "No! This is your paper!" He knew the manuscript and he criticized it (in black and white!) and gave several suggestions how to improve the text. It was an objective, impressive, and unforgettable lesson for me.

The paper I published (Smirnov 1964) differs from others, since the calculations were performed in the framework of the Gamow–Alpher–Herman–Hayashi theory. The nuclear reactions in question were as follows:

$$
\begin{aligned}
& n + p \leftrightarrow d + \gamma, && d + t \rightarrow He^4 + n, \\
& d + d \rightarrow He^3 + n, && He^3 + n \rightarrow t + p, \\
& d + d \rightarrow t + p, && He^3 + d \rightarrow He^4 + p.
\end{aligned}
\tag{4.1}
$$

The seventh reaction took into account the decay of a neutron into a proton, electron, and antineutrino.

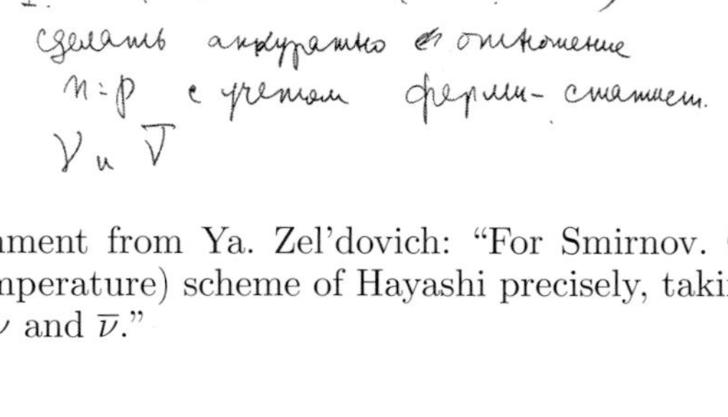

Fig. 4.4. An assignment from Ya. Zel'dovich: "For Smirnov. Calculate the ratio n:p in the (high-temperature) scheme of Hayashi precisely, taking into account the Fermi statistics of ν and $\bar{\nu}$."

Table 4.2. Prestellar abundances

ρ_1	H	He^4	d	t	He^3
10^{-8}	92	5	3	0.01	0.03
10^{-6}	66.6	33.3	0.1	$\sim 2 \times 10^{-5}$	$\sim 2 \times 10^{-6}$
10^{-4}	63.5	36.5	0.3×10^{-3}	$\sim 10^{-7}$	$\sim 2 \times 10^{-7}$

In fact, these calculations were the first to be published on the early evolution of the universe that took account of the tritium channel, and thus they attracted attention. They resulted in the composition of the prestellar matter of the universe in Table 4.2.

The abundances are expressed as a percentage of the mass. Following Zel'dovich, I took the total mass density in all types of relativistic particles to be four times the mass density in electromagnetic radiation. In my computations the mass density in nucleons is much smaller. In the first column of the table, ρ_1 is the mass density in nucleons at a conveniently chosen temperature for the calculation, $T_1 = 4.3 \times 10^8$ K. If $\rho_1 = 10^{-6}$ g cm^{-3}, the case in the middle row in the table, and the present baryon mass density is 10^{-30} g cm^{-3}, the present temperature in this model is about 4 K.

Zel'dovich's assignment was completed. According to his idea, my paper should "work" in favor of the "cold" model published by him in *Atomic Energy*, and should argue against Gamow's "hot" model. Thus my paper concludes

The theory of a "hot" state for prestellar matter fails, then, to yield a correct composition for the medium from which first-generation stars formed: for $\rho_1 \leq 10^{-6}$ g/cm^3, several percent of deuterium is obtained, in conflict with observation [9], while for $\rho_1 \geq 10^{-6}$ g/cm^3, too high a He^4 content is found.

The reference is to Zel'dovich (1963b).

After the discovery of the CMBR (at temperature close to what would be implied by the middle row of the table) this conclusion lost its significance and became the only drawback of the paper. This is why, even after 1965, Zel'dovich (1966), Sakharov, Peebles (1971), Sciama (1971, p. 169), and others mentioned and quoted my paper as an additional illustration of the efficiency of Gamow's "Primeval Fireball." Thus in his paper *Symmetry of the Universe* Sakharov (1968, p. 85) wrote

One of the typical conclusions of the Primeval Fireball is the formation of a sufficient amount of helium as a result of thermonuclear reactions (up to 33% of mass, according to the calculations by Fermi and Turkevich USA) in the early prestellar stages of the expansion of the universe. It is interesting that those results were not published, since they involved the data on thermonuclear reactions with tritium, which were considered confidential at that time (the fifties). In succeeding years those calculations were reproduced by Smirnov (USSR) and Peebles (USA).

To be just, let me mention that during the work on the paper I realized the significance of the value of the He^4:p mass fraction as a parameter that might be compared with the observations, and thus could influence the choice of the model of the universe. After the discovery of the CMBR in 1965 the situation was enriched: there appeared to be a chance to compare the value of the temperature of the expanding universe, calculated in the framework of Gamow's model, with its observable magnitude. But it was not that simple, since even six years later Peebles (1971, p. 127) wrote

The close agreement of the Gamow–Alpher–Herman temperature of the residual radiation from the "Ylem" with the effective temperature $T_0 = 2.7\,\mathrm{K}$ of the modern candidate for the "Primeval Fireball" is an impressive result, but unfortunately it is somewhat delicate for use as an argument in support of the Primeval Fireball hypothesis.

Yakov Zel'dovich strongly believed in his hypothesis. Naturally his paper in *Atomic Energy* appeared in the special issue in commemoration of Igor Kurchatov. Zel'dovich was an inexhaustible source of enthralling tasks in fresh arrangement.

Sakharov, who published at least two papers in the framework of Zel'dovich's model, was interested in the idea of the "cold" model as well. He emphasized in his *Memoirs* (Sakharov 1990, p. 246) that this work provided "a kind of psychological 'warmup' that made possible my subsequent papers of the 1960s."

When the CMBR was discovered, the "cold" model in its original form collapsed. However, it is a fact that Zel'dovich devoted himself not only to the development of the "cold" model, but also to the general physics of the relativistic big bang cosmology, and thus was right on the crest of the thrilling fundamental research and one step away from the triumph. At the time he was out of favor with fortune, and decisive progress fell to others' share.

According to Lev Altshuler,[3] Yakov Zel'dovich felt this keenly: "One day he stopped by our place. He was entirely dispirited. He said that from the point of view of general physical considerations he had every reason to adduce his hypothesis of the 'cold' Universe. But indeed, he said, 'if the Universe were hot there should be relict cosmic microwave background radiation!

---

[3] Lev Vladimirovich Altshuler (1913–2003) was a veteran of the Soviet atomic project.

And I should have explained to experimentalists what they should observe and how'."

On March 3, 1984, Zel'dovich wrote in his autobiographical afterword (Zel'dovich 1985, p. 445; English translation in Sunyaev 2004, p. 352)

At the beginning of my astrophysical activity, I was bothered by habits acquired in the course of my applied activities. An astrophysicist should pose the questions: how is nature constructed? What observations provide the possibility of elucidating this? However, I formulated the problem more like this: how would it be best to construct the universe, or a pulsar, in order to satisfy given technical conditions – forgive me, I meant to say direct observations? This is how the idea of a cold universe arose, and my idea of a pulsar as a white dwarf in a state of strong radial oscillations. As justification for these ideas, I can only say that I was never stubborn about my errors. Apparently, on the whole, my activity – scientific and propagandistic – has been useful.

Serving to truths of science, he has acknowledged defeat with dignity. And having acknowledged it he advocated with all his passion the epoch-making significance of the discovery of the CMBR for the concept of the evolution of the universe.

I remember him immediately doing brilliant reports on this advance in overcrowded auditoriums of the Institute for Physical Problems, on the International Conference on blast waves in plasma, which took place in Novosibirsk Akademgorodok in 1967, speaking to mathematicians and attracting their attention to cosmological problems. He soon published a review, *The "Hot" Model of the Universe* in the journal *Uspekhi Fizicheskikh Nauk* (Zel'dovich 1966). There he referred to the discovery of the CMBR as a result of paramount importance for astronomy.

And again there was an increasing stream of publications!

Yakov Zel'dovich was one of the creators of the first Soviet atomic bomb. The successful test of it on August 29, 1949, liquidated the atomic monopoly of the USA. Zel'dovich contributed a lot to atomic defense subjects for the Soviet Union. He originated the physical basis of the internal ballistics of solid fuel missiles. He passed down more than a vast scientific heritage: he gave rise to schools of thought in chemical physics, hydrodynamics, combustion theory, nuclear physics, elementary particle physics and astrophysics.

Tsukerman and Azarkh (1994, p. 144) emphasized that

What always struck one about Zel'dovich was his indefatigable scientific energy, his lively interest in everything new, his extraordinary versatility, and his intuition ... Zel'dovich's influence on his students and colleagues was striking. Thanks to him they often discovered within themselves capacities for creative work which might otherwise have gone partially, or even totally, unrealized ... Zel'dovich was ... a man of absolute honesty, self-critical, willing to recognize when he was wrong and others were right ... When he succeeded in doing something substantial or overcoming a methodological difficulty through some elegant manoeuvre, his joy was child-like.

When the CMBR was discovered it opened new directions for research in cosmology. It was through Zel'dovich's leadership that physicists in the Soviet Union were ready to do research in these new directions in much the same way as their colleagues in the USA. I have described the many aspects of his leadership elsewhere (commencing on page 103 in Sunyaev 2004); here I mention a brief summary.

No doubt a central example is his organization of the Sarov "youth" seminar dedicated to astrophysical and cosmological problems. Having left Sarov for Moscow he immediately assembled a new collective of young devotees under the same banner. Nowadays Zel'dovich's alumni and progeny are well-known experts in their branches of science and work not only in Russia but also in other countries. In his seminars, and in all his activities, he was an enemy of scientific inertia: he purveyed "hot news" from the most broad variety of branches of physics and from the physical societies of the country. According to the apt statement of Victor Adamskii, Yakov Zel'dovich never "held his knowledge and ideas inside: he splashed them out soon." As Zel'dovich said of himself, in science he was attracted not so much to the cascade of discoveries already made, but rather to the evident, wide incompleteness of a theory. During contacts with him there always appeared a pleasant sensation of a touch with the front line of research, at the very edge of it, where not so much solution as an avalanche of questions dominates. He never parted with his small slide rule, which he wielded masterly, and with a thick writing-pad where he performed his calculations. He never lost the chance to discuss scientific problems he was interested in with a visitor. It was thanks to him that his collaborators and colleagues experienced the delight, were within arm's reach from the wonder of discovery, of the CMBR; it happened in our sight. As every generously gifted person, he was simple, "regular," and convenient, but not always and not for everybody. While recalling this exceptional person one realizes that time will swallow up the details and particular streaks and will keep his true and majestic scale.

In conclusion let me offer the following thought.

Penzias and Wilson (1965a) rehabilitated the "Primeval Fireball" hypothesis of Gamow, when they discovered isotropic noise with a temperature of about 3 K. By that time P. J. E. Peebles together with R. H. Dicke, P. G. Roll and D. T. Wilkinson, who already were purposefully looking for the CMBR, shortly realized and explained what their luckier colleagues had observed.

So, the unexpected "noise" in a supersensitive horn antenna resulted in an outstanding discovery, which was awarded with the Nobel Prize. It was basically akin to the discovery of spontaneous fission of uranium by Georgiy Flerov and Konstantin Petrzhak, who also fought against the "background" in their supersensitive ionization chamber in the late 1930s.

Those common things for astrophysicists and nuclear physicists and the correspondingly interdisciplinary similarity were neither unique nor accidental. They are much more meaningful. If one recalls the 1930s, one can see how generously and unexpectedly these two areas supplied each other.

Lev Landau published three papers: *On the Theory of Stars* (Landau 1932), *Internal Temperature of Stars* (Gamow and Landau 1933), and *On the Origin of Stellar Energy* (Landau 1937).

Hans Bethe (1939) discovered (independently of other authors) the proton–proton and carbon–nitrogen cycles of thermonuclear reactions which are the sources of energy for stars on the main sequence. He won a Nobel Prize for that work 30 years later. It was the first Nobel Prize in astrophysics.

Oppenheimer and Volkoff (1939) performed the first computation of the structure of a neutron star, and Oppenheimer and Snyder (1939) predicted the existence of black holes.

In 1939 to 1940 Yakov Zel'dovich and Yuliy Khariton published three papers, which became the basis for the modern physics of nuclear reactors and energetics.

And what were the results?

Oppenheimer became the head of the research laboratory in Los Alamos, where the American atomic bomb was built, and Bethe became the head theorist of the group.

The picture was similar in the Soviet Union. Yu. Khariton was appointed supervisor for the Soviet atomic bomb project in Sarov, the Russian nuclear center, and Ya. Zel'dovich was the head theorist. G. Flerov and L. Landau were direct members of the atomic project.

It is not surprising, if one recalls the beginning of my story.

4.4.2 Igor Dmitriyevich Novikov: Cosmology in the Soviet Union in the 1960s

Igor Novikov served as head of the Department of Relativistic Astrophysics at the Space Research Institute and head of the Department of Theoretical Astrophysics at the Lebedev Physical Institute, both in Moscow, and then as director of the Theoretical Astrophysics Center of the University of Copenhagen. He is now at the Niels Bohr Institute.

The beginning of my scientific career in the middle of the 1960s coincides with the events that caused the astrophysical community to become aware of the real existence of the CMBR.

I outline here the development of the situation at this period in the Soviet Union. It so happened that I played a part at this stage of the story. This description relies on my recollections, published and unpublished material of my colleagues, and special recent discussions with participants of the events. I have used also some material from the books Zel'dovich and Novikov (1983) and Novikov (1990), and from the paper Novikov (2001).

I would like to remind you that at the beginning of the 1950s the theory of an expanding, indeed an evolving, universe with the beginning of time at some finite period ago was practically forbidden in the USSR. There was a postulate that only an eternal universe without directed evolution as a whole is compatible with the materialistic attitude.

Only at the beginning of the 1960s did the first serious discussions and publications on the physics of the expanding universe became possible. At that period some important works on the structure of the cosmological singularity and gravitational instability of the expanding universe were published by E. Lifshitz, I. Khalatnikov, and I. Novikov. But these works did not discuss the physical conditions in the early universe. In the early 1960s Yakov Zel'dovich began turning his attention to cosmology. He very quickly became one of the greatest cosmologists of the last century.

In 1962 Zel'dovich published a paper (Zel'dovich 1962) in which he modernized the cold universe scenario. According to this scenario at the initial stage of the evolution of the universe the matter consists of a mixture of protons, electrons, and neutrinos in equal amounts and the entropy is low. Then at high density (on the order of nuclear density) and at zero temperature neutrinos and electrons form a degenerate relativistic Fermi gas. The process of interaction of protons with electrons with the formation of neutrons and neutrinos is forbidden since the neutrino states that are energetically obtainable in this process are occupied. Upon expansion such a substance remains pure cold hydrogen. It was assumed that all other elements were generated much later, in stars. According to this model the CMBR radiation should not exist in our epoch.

Zel'dovich's hypothesis was widely discussed in the USSR. One of the main reasons for the hypothesis was some indication in the literature at that time that the helium mass fraction in the oldest stars is much less than 20%.

Probably this means that the primeval helium mass fraction is essentially smaller than predicted in the theory of the hot universe, $\approx 25\%$. Zel'dovich believed also that according to the hot universe theory the matter density of the CMBR should be of the order of the modern average nucleon density. In this paper his conclusion was: "These deductions are incompatible with the observations."

Curiously, Ya. Zel'dovich at that period, as well as the originators of the hot model, was mainly interested in the integral properties of the relic radiation (CMBR) – its density, pressure, and temperature – but not its spectrum.

Here my story begins. At that time I had just completed the postgraduate course at Moscow University; my science adviser was Professor Zel'manov. My adviser was mostly interested in the mechanics of motion of masses in cosmological models when no simplifying assumptions are made about their uniform distribution. He was less interested in specific physical processes in the expanding universe. At that time, I knew almost nothing about the hot universe model.

Not long before the end of my postgraduate term, I was attracted to the following problem. We know how different types of galaxies produce electromagnetic radiation in different ranges of wavelength. With certain assumptions about the evolution of galaxies in the past, and having taken into account the reddening of light from remote galaxies owing to the expansion of the universe, one can calculate the present distribution of the integrated galactic emission as a function of wavelength. In this calculation, one has to remember that stars are not the only sources of radiation, and that many galaxies are extremely powerful sources of radio waves in the meter and decimeter wavelength ranges.

I began the necessary calculations. Having completed the postgraduate term, I joined the group of Professor Ya. B. Zel'dovich; our interests focused mostly on the physics of processes in the universe.

All calculations were carried out jointly with A. Doroshkevich, who I met when I joined Zel'dovich's group. We obtained the calculated spectrum of galactic radiation, that is, of the radiation that must fill today's universe if one takes into account only the radiation produced since galaxies were born and stars began to shine. This spectrum, shown in Figure 4.5, predicted a high radiation intensity in the meter wavelength range (such wavelengths are strongly emitted by radio galaxies) and in visible light (stars are powerful emitters in the visible range), while the intensity in the centimeter, millimeter and some still shorter wavelength ranges of electromagnetic radiation must be considerably lower.

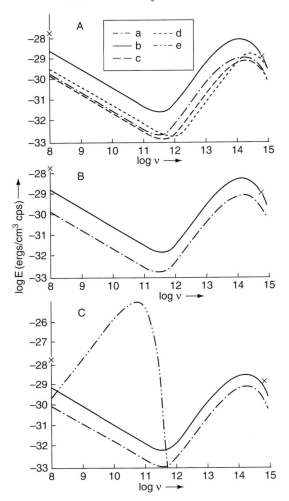

Fig. 4.5. From Doroshkevich and Novikov (1964). Spectrum of the metagalaxy. Curves (a)–(d): the integrated radiation from galaxies under several assumptions about the cosmology and the evolution of the galaxies. Curve (e): equilibrium Planck radiation with $T = 1\,\mathrm{K}$. Crosses denote experimental points. ©1964 American Institute of Physics.

Since the hot and cold universe scenarios were eagerly discussed in our group (consisting of Zel'dovich, Doroshkevich, and myself), the paper that Doroshkevich and I prepared for publication added to the total the putative radiation surviving from the early universe if it indeed had been hot. This hot universe radiation was expected to lie in the centimeter and millimeter ranges and thus fell into the very interval of wavelengths in which the radiation from galaxies is weak! Hence, the relic radiation (provided the early universe had been hot!) was predicted to be more intense, by a factor of many thousands or even millions, than the radiation of known sources in the universe in this range of wavelengths.

This background could, therefore, be observed! Even though the total amount of energy in the microwave background is comparable with the visible light energy emitted by galaxies, the relic radiation would be in a very different range of wavelengths and thus could be observed. Here is what Penzias (1979a) said about our work with Doroshkevich (Doroshkevich and Novikov 1964) in his Nobel lecture:

The first published recognition of the relic radiation as a detectable microwave phenomenon appeared in a brief paper entitled "Mean Density of Radiation in the Metagalaxy and Certain Problems in Relativistic Cosmology" by A. G. Doroshkevich and I. D. Novikov in the spring of 1964. Although the English translation appeared later the same year in the widely circulated "Soviet Physics - Doklady", it appears to have escaped the notice of other workers in this field. This remarkable paper not only points out the spectrum of the relic radiation as a blackbody microwave phenomenon, but also explicitly focuses upon the Bell Laboratories 20-ft horn-reflector at Crawford Hill as the best available instrument for its detection!

Our paper was not noticed by observers. Neither Penzias and Wilson, nor Dicke and his coworkers, were aware of it before their papers were published in 1965; Penzias told me several times that this was very unfortunate.

I want to mention a strange mistake related with an interpretation of one of the conclusions of the Doroshkevich and Novikov (1964) paper. Penzias (1979a) wrote: "Having found the appropriate reference (Ohm 1961), they [Doroshkevich and Novikov] misread its result and concluded that the radiation predicted by the 'Gamow theory' was contradicted by the reported measurements." Also in the paper Thaddeus (1972) one can read: "They [Doroshkevich and Novikov] mistakenly concluded that studies of atmospheric radiation with this telescope (Ohm 1961) already ruled out isotropic background radiation of much more than 0.1 K."

Actually in our paper there is not any conclusion that the observational data exclude the CMBR with the temperature predicted by the hot universe. We wrote in our paper: "Measurements reported in [14] [Ohm 1961] at a frequency $\nu = 2.4 \cdot 10^9$ cps give a temperature 2.3 ± 0.2 K, which coincides with theoretically computed atmospheric noise (2.4 K). Additional measurements in this region (preferably on an artificial Earth satellite) will assist in final solution of the problem of the correctness of the Gamow theory."

Thus we encouraged observers to perform the corresponding measurements! We did not discuss in our paper the interpretation of the value 2.4 K obtained by Ohm with help of a technology developed specially for measuring the atmospheric temperature (see discussion in Penzias 1979a).

Below I will tell more about our discussions with some radio astronomers in the USSR.

At that time the possibilities for communications with our foreign colleagues were very restricted. I learned about the discovery of the CMBR radiation at a conference in London in the summer of 1965. When I was back in Moscow I informed Professor Ya. Zel'dovich.

At the first moment when I told Ya. Zel'dovich about the discovery he obviously did not remember the details of my paper with Doroshkevich and started to scold us that we had not included in our paper the figure with the predicted spectrum of the CMBR. When I immediately showed him the corresponding figure in the reprint of our paper he started to scold us for the absence of the effective propaganda of our paper.

It was clear that this discovery means the strict proof of the hot universe. This discovery was widely discussed among Soviet physicists and astronomers. Zel'dovich abandoned his hypothesis of the cold universe and became an ardent proponent of the theory of the hot universe. In his letter to Professor Dicke he wrote on September 15, 1965 (unpublished and kindly provided to me by J. Peebles): "I am not more so cock-sure in my cold universe hypothesis: It was based on the assumption that the initial helium content is much smaller than 35% by weight. Now I understand better the difficulty of helium determination."

Zel'dovich began active work on the hot model even before the discovery of the CMBR. V. M. Yakubov, a collaborator of Ya. Zel'dovich, repeated the earlier calculations of Hayashi (1950) of the process of the nucleosynthesis in the hot model. These calculations were much simpler and more transparent and based on new knowledge of the weak processes. These calculations were set forth in the paper Zel'dovich (1965). Thus the process of the big bang nucleosynthesis (BBNS) became known and understandable for us.

When I started to work on these notes (autumn, 2003) I asked Professor V. Slysh (Lebedev Physics Institute, Moscow) for his recollections of the events of that period in the group of Professor I. Shklovsky at the P. K. Shternberg State Astronomical Institute in Moscow. He told me the following. In 1965 just after learning about the discovery of the CMBR I. Shklovsky asked V. Slysh to find in the Institute Library papers, published around 1940, concerning the interstellar absorption lines in the spectrum of the light coming from the star ζ Ophiuchi; the absorption was caused by the molecule CN. I. Shklovsky himself was a specialist in the physics of the interstellar medium and remembered the papers. V. Slysh had found the papers by Andrew McKellar (1940, 1941). In these papers McKellar concluded that these lines (in the visible part of the spectrum) could have arisen only if the light was absorbed by rotationally excited CN molecules. The rotation must be excited by radiation at a temperature about 2–3 K. In the paper McKellar (1940) wrote: "Effective temperature of the interstellar space ...< or = 2.7 K." In another paper McKellar (1941) wrote that the "Rotational temperature of interstellar space is about 2 K."

On the basis of this information I. Shklovsky wrote and published a paper Shklovsky (1966) where he declared that this temperature was the temperature of the whole universe rather than the temperature of only the interstellar medium as it had been declared by McKellar.

We are not yet through the chain of missed opportunities that plagued the discovery of the relic radiation.

Let us return to the question about the technical feasibility of detecting the cosmic microwave background radiation (CMBR). At what time did this become possible? Weinberg (1977) writes: "It is difficult to be precise about this, but my experimental colleagues tell me that the observation could have been made long before 1965, probably in mid-1950s and perhaps even in mid-1940s." Is this correct?

In the autumn of 1983, Dr. T. Shmaonov of the Institute of General Physics, Moscow, with whom I was not previously acquainted, telephoned me and said that he would like to talk to me about things relevant to the discovery of the CMBR. We met the same day and Shmaonov described how, in the middle of the 1950s, he had been doing postgraduate research in the group of the well-known Soviet radio astronomers S. Khaikin and N. Kaidanovsky: he was measuring radio waves coming from space at a wavelength of 3.2 cm. Measurements were done with a horn antenna similar to that used many years later by Penzias and Wilson. Shmaonov carefully studied possible sources of noise. Of course, his instrument could not have been as sensitive as those with which the American astronomers worked in the 1960s. Results obtained by Shmaonov were reported in 1957 in his PhD Thesis and published in a paper (Shmaonov 1957) in the Soviet journal *Pribory i Tekhnika Eksperimenta* (*Instruments and Experimental Methods*). The conclusion of the measurements was: "The absolute effective temperature of radiation background ... appears to be 4 ± 3 K." Shmaonov emphasized the independence of the intensity of radiation on direction and time. Errors in Shmaonov's measurements were high and his 4-K estimate was absolutely unreliable, but nevertheless we now realize that what he recorded was nothing other than the CMBR. Unfortunately, neither Shmaonov himself, nor his science advisors, nor other astronomers who saw the results of his measurements knew anything about the possibility of the existence of the relic radiation and so failed to pay the results the attention that they deserved. They were soon forgotten. When Doroshkevich and I, having completed our calculations, were calling in 1963 and 1964 on several Soviet radio astronomers with the question "Do you know any measurements of the cosmic background in the centimeter and shorter wavelength ranges?" not one of them remembered Shmaonov's work.

It is rather amusing that even the person who made these measurements failed to appreciate their significance, not only in the 1950s – this is easy to explain – but even after the discovery of the microwave background by Penzias and Wilson in 1965. True, at that time Shmaonov was working in very different field. His attention turned to his old results only in 1983, in response to semiaccidental remarks, and Shmaonov gave a talk on the subject at the Bureau of the Section of General Physics and Astronomy of the USSR Academy of Sciences. This event took place 27 years after the measurements and 18 years after the publication of the results of Penzias and Wilson (1965a).

Shmaonov's observations were reanalyzed by N. Kaidanovsky and Yu. Pariĭskĭ (1987). Their conclusions were: "Thus, one can conclude that the contribution of the relic radiation [CMBR] was $T_{rel} = 2 \pm 1\,K$. Of course the result is rough but unambiguous." In autumn 2003 I discussed the situation with both N. Kaidanovsky and Yu. Pariĭskĭ. They confirmed this conclusion. Kaidanovsky emphasized that at the time of the measurements (1955–1956) the reality of Shmaonov's results was out of any doubt, but unfortunately there was not any theorist who could tell them the possible interpretation.

Fate takes unexpected and tortuous turns. Nevertheless, the entire story is very instructive. To hit upon a phenomenon is not yet equivalent to discovering it. One has to realize the significance of the find and give the correct explanation. A combination of circumstances and sheer luck does play a role here – no doubt about it. Nevertheless, success does not come by accident. Success requires lots and lots of work, vast knowledge, and persistence in the work itself and in bringing the results to the attention and recognition of others.

In conclusion I want to say that just after the discovery of the CMBR we in the group of Professor Zel'dovich started to work on the theory of the origin of galaxies and large-scale structure of the universe (see for example Doroshkevich, Zel'dovich and Novikov 1967) and on other physical processes in the hot model. Analogous works started in the groups of Professor V. Ginzburg, Professor I. Shklovsky, and other groups in the USSR.

4.4.3 Andrei Georgievich Doroshkevich: Cosmology in the 1960s

Andrei Doroshkevich is a Leading Research Scientist at the Astro Space Center, the P. N. Lebedev Physical Institute of the Russian Academy of Sciences, Moscow. His research interests include cosmology, galaxy formation, and the relic radiation background.

For cosmology the 1960s were a very active period. Over these years the rapid development of observational techniques offered many discoveries, including the relic radiation (the CMBR), X-ray sources, and quasars. At the same time significant progress had been achieved in the understanding and description of the nature of black holes. These discoveries have demonstrated the complex nature of the universe composed of a set of objects with a wide variety of properties and evolutionary histories.

This progress immediately rearranged cosmology and transformed it to the physical theory focused on the interpretation of new phenomena and the construction of corresponding physical models. These models have opened up new fields of application for classical and quantum physics in exotic situations such as the early stages of the evolution of the universe and objects with superstrong pressure, magnetic and gravitational fields. Theoretical analysis of such problems had begun already in earlier publications of Landau, Oppenheimer, Volkoff, Snyder, Tolman, Gamow, Alpher, Herman, and some others. However, only at this period did the application of abstract theoretical constructions to observed objects become possible.

For cosmology as a whole the discovery of the CMBR was most important. In this context the key role must go to G. Gamow, who formulated the hypothesis of the hot universe using very limited observational estimates of the chemical composition of old stars. Unfortunately his expectations of the direct observations of the CMBR were not so promising and they were not realized over an extended period after his predictions.

The development of Soviet cosmology at this period is well reviewed here by I. Novikov. I can only repeat that in 1963–1964 Novikov and I began to work in Zel'dovich's group and were interested in manifestations of the properties of galaxies. Thus, we calculated the background radiation produced by galaxies and found that it is quite small in the millimeter wavelength range. (It is interesting that this inference remains valid up to now.) For comparison we plotted in the same figure the Planck spectrum of the relic radiation for the temperature $T = 1\,\mathrm{K}$. From this figure it was obvious that the CMBR can be successfully observed by high precision measurements at suitable wavelengths. The main problems were linked with the technical limitations and reduction of atmospheric noise. Unfortunately, owing to very limited contacts between Soviet and Western astronomers, our publication (Doroshkevich and Novikov 1964) remained unknown for many years.

In the 1960s the very limited available information about the discovered phenomena stimulated construction of a wide set of cosmological models, the majority of which are forgotten now. Nonetheless, some of such models can be restored in the light of the new observational evidence in favor of weak large-scale anisotropy of the universe. Of course, such models must be close in main features to the now accepted ΛCDM cosmology which quite well describes the main observations. Beyond that point, according to presently discussed simple models of cosmological inflation, the primordial anisotropy, global rotation and magnetic field catastrophically decayed. This means that there are problems that call for further observational and theoretical investigations.

However, the list of the parameters used for the analysis of the intensity and polarization of the CMBR can be extended and, for example, include the possible contribution of small-scale initial entropy perturbations and possible complex composition of the dark matter component. Such analysis strongly restrict amplitudes of these "hidden parameters" but cannot reject them. At the same time, it can noticeably change the derived parameters of the ΛCDM cosmological model.

These comments are called on to demonstrate a succession of many basic ideas in cosmology which have been discussed in the sixties and remain relevant up to now.

4.4.4 Rashid Sunyaev: When we were young ...

Rashid Sunyaev is Director of the Max-Planck Institute for Astrophysics, Garching, Germany, and Chief Scientist, Space Research Institute, Russian Academy of Sciences, Moscow, Russia.

Meeting and working with Zel'dovich

I became Yakov Zel'dovich's student in March 1965, half a year short of the news of the discovery of the CMBR by Penzias and Wilson (1965a) reached us in Moscow. At that time Zel'dovich (Figure 4.6) was putting together his group at the Institute of Applied Mathematics that he had joined a year before. There were already several people in the group, including Igor Novikov and Andrei Doroshkevich.

For over a year I was the junior member of the team. Before my first meeting with Zel'dovich I had never heard about him. I was proud that in 1963,

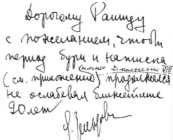

Fig. 4.6. Yakov Zel'dovich on the left, Aleksandr Solomonovich Kompaneets on the right. The inscription was written by Zel'dovich on a copy of the book *Relativistic Astrophysics* (Zel'dovich and Novikov 1967). The inscription reads: "To dear Rashid, with wishes that the period of Sturm and Drang (see Addition VIII) continues without weakening for the nearest 90 years. Ya. Zel'dovich."

despite intense competition, I, a student of the fourth year, was admitted to the Elementary Particle Physics Chair of my university – PhysTech (the Moscow Institute of Physics and Technology). This Chair was based at the Institute for Theoretical and Experimental Physics – at that time a research center with a proton accelerator, nuclear reactor, more than two thousand employees, and an excellent theoretical department.

The majority of students in our group were looking forward to getting involved with elementary particle theory. We took numerous courses and passed a lot of exams, but instead of the welcome moment of being admitted to the ranks of theoreticians we were offered more and more courses and exams. It was only later that I found out that the Theory Department did not have vacancies for PhD students. We were considered a reserve of knowledgeable physicists for the Institute's experimental groups that had no shortage of PhD student positions.

I remember my first conversation with Zel'dovich. He told me that he was involved in astrophysics, and that would be the topic of my dissertation. I responded that I had over a year left to finish my diploma, and only then could I take a PhD course. Unexpectedly to me, Zel'dovich said that it had been many years since he stopped advising graduate students, and that he was just selecting the best university graduates to work for him. I did not expect such a reaction, but for whatever reason, Zel'dovich, having a change of heart, all of a sudden winked with his right eye and said: "Let us try, but don't get offended. If you fail, then you will need to finish your diploma somewhere else." After hearing this I threw off reserve and repeated what my department chair had said, that astrophysics was a useless science, and I said that I would prefer to work on elementary particle theory. Zel'dovich smiled and promised me that after I helped him solve one or two of the problems in astrophysics that he was most excited about we would both go back to studying elementary particles. I would like to note right away that after I had published, together with Yakov Borisovich (YaB), the very first two articles in the *Astronomical Circular* (a timely publication of Sternberg Astronomical Institute in the Russian language only) I never went back to my desire to study field theory. Thanks to Zel'dovich! He never reminded me of those words said during our first conversation. At the conclusion of the meeting, he told me that my excellent grades and oral answers to his questions were not sufficient and he would ask me to solve a couple of simple problems and present solutions in a few days. After I had passed this test as well, he said that he would try to transfer me to his group.

For some reason Zel'dovich liked the fact that I had learned a lot about elementary processes not only in the interaction of relativistic particles, but also in the radiation processes in a rarefied plasma and collision of electrons with atoms and ions, recombination, level excitation, and ionization. He also liked that on PhysTech written exams in mathematics I usually scored almost a hundred percent and that I had an offer to transfer to mathematics.

The majority of the people he had hired by that time for his group at the Institute of Applied Mathematics were interested mostly in general relativity and hydrodynamics. He put me on physical cosmology, suggesting that I focus on methods of helium detection in intergalactic gas. This task was very interesting and I soon realized the importance of the two peaks in the temperature-dependence of the emissivity of a hydrogen–helium plasma. Unfortunately these first results were initially published only in the *Astronomical Circular*. Nowadays hardly anybody knows the publication existed. Later Andrei Doroshkevich joined me in this research. Together, we published a preprint version of our paper and, almost two years later, our paper (Doroshkevich and Sunyaev 1969) was published in *Soviet Astronomy Journal*. We showed a two-peak curve, at sixteen thousand and a hundred thousand degrees, and we discussed the role of Compton cooling of intergalactic gas. By that time Ray Weymann's article on the same topic was well known in the West. I remember Zel'dovich telling me, after an issue of *The Astrophysical Journal* with Weymann's article started circulating in our department, that we had to insert a reference to it into our article's proof sheets.

I write about these first works, virtually unknown to anybody, only because they later led me to understand the role of Comptonization in the formation of distortions in the relict radiation (CMBR) spectrum and the effect of Lyman-α resonance line photons, which happened to be so important in the problem of cosmological hydrogen recombination. It was at that time (back in 1965) when Zel'dovich advised me to read an article written by Kompaneets in 1956.

Yu. N. Smirnov describes in his essay in this book how YaB taught himself and his theoretical group in the Sarov (Arzamas-16) nuclear center, how they became proficient in general relativity, and learned astrophysics and cosmology.

Zel'dovich liked to write reviews. This was his way to learn. It permitted him to add his own opinion when describing results of other scientists, to describe his own (often significantly simpler) way to derive important results and to add original details. In 1965 I read practically all the reviews in cosmology and astrophysics he had written at the beginning of 1962 and published in the journals *Soviet Atomic Energy*, *Soviet Physics Uspekhy*, *Advances in Astronomy and Astrophysics*, and in the collected articles and reviews, *The Issues of Cosmogony* (which was not translated into English). Outstanding reviews (Zel'dovich 1965) written before the CMBR discovery, and immediately after (Zel'dovich 1966), became the basis of cosmology chapters of the excellent books published together with I. D. Novikov. The first, *Relativistic Astrophysics*, appeared in Russian in 1967 and became the manual for many young (and not so young) scientists. Its 5500 copies sold out immediately. It is a pity that the English translation of these books was greatly delayed despite efforts by Kip Thorne, David Arnett, and Gary Steigman. One reason was the rapid development of astrophysics and cosmology: the authors were too busy adding descriptions of fresh results to the book.

The review of the hot model of the universe (Zel'dovich 1966) is written in beautiful Russian. It is really a pity that the translation into English was of low quality. Now it is possible to read a significantly improved English version of this review in the book of selected works (Zel'dovich 1993). It is very impressive to see how much YaB knew about the hot model at the moment CMBR was discovered. It is a great surprise to see how different in subject and style are the three reviews (Zel'dovich 1966 on the hot model, Zel'dovich 1965 on global properties and instabilities in the universe, and Zel'dovich 1968 on the Λ term) and how much was known at that time. However, our knowledge now is much deeper and more detailed, and practically all theoretical models and assumptions are confirmed or checked by the observations and detailed simulations that later became possible.

Meeting Yakov Borisovich completely changed my life. From the time I met him up to 1970 my life was full of festivity. Nearly every day he would call me in the morning (or I would call him, when I spent the night in

Dolgoprudnyi, at PhysTech) and ask what was new with me, tell me his own news, and make an appointment to meet. After his death in December 1987 it became emotionally difficult for me to come, at the invitation of his daughters, to the building on the former Vorobiev Road, to the room where he worked and to which he invited everybody – where I was at least two thousand times during two decades of working with him.

I arrived in Zel'dovich's group at a time that was happy for him, and indeed for all astronomy. In Russia many remember the All Moscow Astrophysics Seminars at the Shternberg Astronomical Institute in 1965 to 1968, when they were supervised jointly by Zel'dovich, Vitaliy Ginzburg, and Iosif Shklovsky, exceptional people in their behavior, scientific level, activity, and scope of interests; with Solomon Pikelner prepared at any minute to step up and explain the incomprehensible to physicists; with the young L. M. Ozernoi, A. G. Doroshkevich, I. D. Novikov, N. S. Kardashev, V. S. Imschennik, V. G. Kurt, Y. N. Efremov, V. F. Shvartsman, L. P. Grishchuk, G. S. Bisnovatyi-Kogan, and A. M. Cherepashchuk actively participating; with Evgeny Lifshitz, Isaak Khalatnikov, or Andrei Sakharov showing up occasionally. I remember the conference hall at the Shternberg Institute, usually packed to the limit, with doors open on both sides, behind which clustered the latecomers for whom there weren't enough seats. Many of the young people there later became the cream of Soviet astrophysics.

Before and after the CMBR discovery

When *The Astrophysical Journal* issue with the article on the discovery of the CMBR by Penzias and Wilson (1965a), and a Princeton Group article on the importance of this discovery, reached Moscow, and was received, as far as I remember, by Shklovsky, it naturally caused a sensation and a heated discussion at a seminar in the Shternberg Institute. Iosif Shklovsky immediately termed it "relict radiation." This term has become firmly rooted in Russian science and popular science literature.

I also remember that YaB expressed regret at the seminar that he had held to the cold model of the universe until the discovery of the relict radiation. It was striking how he immediately acknowledged the correctness of Gamow's point of view and delightedly praised it, which was not so simple at that time (as Gamow had fled the USSR in the 1930s, which was unforgivable). In general, YaB has done a lot to ensure that the names of Friedmann and Gamow and their great contributions to cosmology are appropriately reflected in the USSR scientific literature.

He loudly bemoaned his own mistake, when he made an attempt to explain (including in a published paper) the newly discovered phenomenon of radio pulsars as the effect of processes on white dwarfs. At that time, he told me that his mistake was the result of his many years of work in weapons research: it was sometimes necessary during the course of a single night to make a decision about what to do the next day. There had to be only one final option, and it had to be realistic, simple, reliable, and the most economical. Mistakes were not permitted. Nature, he convinced himself time and time again, could allow itself things that seemed at that time absolutely unexpected or improbable, although, from the point of view of physics, those also were entirely natural solutions. In his later years, he, who was absolutely certain that the neutrino has zero mass, radically changed his point of view and used to repeat: everything is allowed that is not forbidden. Not everybody can survive such a radical change in perceptions and start working based on a new philosophy.

YaB was ready to acknowledge the hot model. He was used to accepting the results of straightforward and beautiful experiments and, long before the discovery of Penzias and Wilson, while realizing the importance of Gamow's predictions[4] on the existence of the 5- to 6-K relict radiation, supported and recommended for publication in *Soviet Physics Doklady* the work of Doroshkevich and Novikov (1964) on the compilation of radio observations of background radiation.

To him, the appearance of a new point in the background radiation spectrum was a long-awaited, however unfavorable, experimental fact. He could not but acknowledge this experimental fact and therefore had to agree right away that our universe was hot. During the previous four years, when getting ready to be transferred from Arzamas-16 to Moscow, and during the first two years in Moscow, Zel'dovich, intensely and with great interest, worked on the theory of physical processes in the expanding universe within the frameworks of both cold and hot models. The latter was important in order to find predictions, which were supposed to contradict current and future experimental data and observations.

I remember the first improvised seminars after the discovery of the relict radiation, when YaB talked, as if it were obvious, about the dipole anisotropy as a means of measuring Earth's velocity, and about the presence of a unique reference system that could be defined by this radiation.

---

[4] Zel'dovich knew about every paper by Gamov and his colleagues. At his request librarians at Arzamas-16 had searched everywhere for all of Gamow's old papers. Gamow also knew about Zel'dovich's papers and was following them. I remember how proud I was when a thick package of reprints of his hot model papers reached me. This was in 1968, and this was his reaction to my first papers and preprints I had mailed him. Not too many great scientists knew my name at that time. Every reprint had a stamp with an image of "Gamov's dacha" and several handwritten words.

I remember how I first heard from him at these seminars about the quadrupole anisotropy component in an anisotropic universe, and the unavoidable existence of angular fluctuations of the radiation. Only a year later, reading fresh issues of *Nature*, *Physical Review Letters*, and *Astrophysical Journal Letters*, did I understand that there is a level of knowledge starting from which such things became evident, and cosmologists of this level in Cambridge, Princeton, and Moscow could seize this simultaneously. And then I realized for the first time that to be alongside YaB meant, as a minimum, to be at a world-class level.

As a person who had been denied travel abroad, YaB was long certain that if he were ever allowed to go to some major conference in the West something important would come of it. For almost 20 years, he put it that he had "only sub-orbital velocity:" he was allowed to travel only to Eastern Europe.

In 1967, he was able to get me – his postgraduate student – approved to join a group of young scientists going to the General Assembly of the International Astronomical Union (IAU) in Prague. That was my first and next-to-last trip abroad for more than 12 years of being banned to travel out of the country. Naturally, I remember this meeting well. The mood of the people, carried in the air, was unusual to us, and clearly heralded the imminent arrival of the "Prague Spring." But I also remember YaB, fully busy, escorted by his students and fellows to meet various celebrities (and these meetings were scheduled one after the other). At them I first saw Ed Salpeter, Margaret and Geoffrey Burbidge, Herbert Friedman, the young and active Riccardo Giacconi, George Field, and Dennis Sciama. Professor Sciama was memorable also for being accompanied by a student who later became well known – Martin Rees. I also met Joe Silk for the first time. During these short meetings, YaB managed to say something about his latest work, ask several specific questions, primarily on experimental data, and also let his young students talk, though some spoke English with difficulty. He led us to a swimming pool, where he, as always, swam pleasurably and talked about a movie he had seen the day before. At intervals, he could be seen sitting on a sofa and rapidly writing several lines in his large, expressive handwriting, usually forwarding his notes to I. D. Novikov to be included in the next book.

In Moscow Zel'dovich behaved essentially in the same way: he was usually not late for appointments, of which he had many each day, he regularly went skiing or walking, and he was writing something in thin school notebooks nearly every free minute. Everything he wrote he tried to give away immediately for inclusion in a book or a paper, or simply to be used as food for thought. Sometimes this was a prepared text, sometimes only equations or estimates with questions between them. I have kept many such sheets. However, he quickly understood that I was "not a writer" and very disorganized. As a result, I was left primarily with questions and sometimes equations.

It was striking how he could instantly concentrate, settle himself comfortably, fall silent, and rapidly write. Then, completely unexpectedly, he would come out of this state and begin a conversation on a new topic. If the papers were not given to someone on the spot, they were intended for somebody else, most likely for books. He also valued experimental data, and he was pleased with his pupils' works that predicted straightforward experimental tests.

YaB carried a colossal burden. While we were working in close collaboration with him during 1966 to 1973, he at the same time wrote reviews, worked on a book on low-energy nuclear physics, reissued *Higher Mathematics for Beginners*, *Elements of Applied Mathematics* (jointly with Myshkis), very popular books then, and, together with Yury Raizer, worked on a wonderful, serious book, on the physics of shock waves and high temperature hydrodynamic phenomena (Zel'dovich and Raizer 1966), all the while continuing research on the theory of combustion. Two or three times a week he would come to the Institute of Applied Mathematics for several hours, quickly report something on a blackboard, or set up a small improvised seminar, and then have a short conversation with some of his young colleagues.

If initially (in 1965 to 1966) I failed to show him something new, then for the next few days he practically ignored me until I could tell him that I had original and, from my point of view, interesting news. Quite often on the next day he would call me or ask a secretary to find me in a dormitory and invite me to come to his home or to meet with him at one of the institutes where he had other business to attend. He himself worked almost without breaks, and he wanted at least the same from me. At the same time he knew how to, and liked to, rest: he liked to drive a car, travel all over the Soviet Union, and did not forget every year to spend a month on the Black Sea in his beloved Crimea. But even there he worked a lot and sent me letters or sometimes summoned me to Crimea to talk and to find out what I had come up with.

We worked at a mathematical institute where such well-known mathematicians as Mstislav Keldysh, who soon became President of the Academy of Sciences of the Soviet Union, Israel Gelfand, and many others worked at that time. There I began working with outstanding people – Timur Eneev, Ilya Sobol, Vladimir Gol'din, Nikolay Kozlov, Boris Chetvertushkin, and

Lev Pozdnyakov, who remained my friends for life and taught me a lot. YaB approved of this collaboration. Thanks to his recommendations I got acquainted also with the leading specialists in atomic processes – Leonid Vainshtein and Israel Beigman of the Lebedev Institute of Physics, with whom I continue to discuss problems of common interest.

Malcolm Longair in Moscow

In 1968 Malcolm Longair, a young Scottish (as he made clear to us right away) scientist from Cambridge came to Moscow for an internship in an exchange between the Academy of Science of the Soviet Union and the Royal Society. Prior to that, he had graduated from a Russian language course in England. For us it was a huge event – Malcolm happened to be a kindly and a pleasant person to socialize with. He was a good conversationalist and managed to have a great relationship with all the scientists he happened to meet at the Lebedev Institute of Physics and the Shternberg Institute. He kept company with such giants as Vitaly Ginsburg, who had invited him to the USSR, Yakov Zel'dovich, and even Victor Ambarzumian, who invited him to visit his Byurakan Observatory in Armenia. It was interesting for all of us to observe how he behaved and thought, to listen to him talking about experimental results of Martin Ryle's group and Tony Hewish's radio telescope that detected pulsars. It was most interesting when he talked about the cosmological evolution of radio sources.

I was banned from travel abroad; however, somehow Zel'dovich managed to arrange for me to be present during his meetings with foreign scientists visiting Moscow and to participate in international conferences in Soviet territory. (I need to note that I never worked on defense issues and never had access to any special secrets.) Malcolm and I used to meet and talk after seminars at the Shternberg Institute, then I dared to walk him to the Lebedev Institute, and by the end of his stay we would often go for walks in Moscow. I even used to invite him to my tiny room in a shared small Krushchev-time apartment that Zel'dovich "pushed through" for me. I was very proud of this room of my own, and Malcolm never allowed himself to explain that this pride was groundless.

In those days young scientists of my rank "were not recommended" to see foreigners without supervision of somebody who was more "experienced and tested." I knew that people responsible for security at the Institute of Applied Mathematics and the Academy of Sciences were perfectly aware of my meetings with Malcolm and his regular phone calls to my home (he never called at work). But nobody demanded that I immediately stop these meet-ings. Sometimes they jokingly asked why we were walking around as a couple, but they treated with respect the articles written jointly with Malcolm when I would bring them for finalizing and sending out to "GLAVLIT" (Main Administration for Literary and Publishing Affairs). Apparently there was some sort of an arrangement behind my back, or maybe it was done to demonstrate to the Royal Society that exchange scientists visiting the Academy of Science had a real opportunity for collaboration.

In a year, Malcolm and I coauthored several articles. We thought about CMBR angular fluctuations due to the presence of the foreground of extragalactic radio sources (Longair and Sunyaev 1969). I learned a lot from him. That was not only in acquiring skills to write articles in a completely different style than was customary in Soviet journals, but also in presenting papers and formulating my ideas with precision. I will not even mention how he helped me improve my quite mediocre (wretched, to be honest) English. It was much later that I learned that Malcolm did a huge favor for us. After coming back to Cambridge, he visited several leading universities around the world and at seminars there reported what was going on in Moscow in the field of cosmology along with explanations of the major results obtained by Zel'dovich and his group. I am grateful to Malcolm for suggesting to students, immediately upon his return to Cambridge, that they should seek to detect a decrease in the brightness of the relict radiation in the direction of clusters of galaxies, what was later called the "SZ effect". He heard with interest and told others about our work with YaB on angular fluctuations of the relict radiation and the baryon oscillations and on cosmological hydrogen recombination. It was a learning experience for me to work with him on a review of cosmic background radiation (CBR), from radio to γ-rays, and an attempt, never finished, to write a two-volume book on physical cosmology and extragalactic astronomy under the comprehensive name *The Matter and Radiation of the Universe*. I have to admit that I am to be blamed for this failure. However, I shall always remember how Malcolm piously followed the well-known motto of the Soviet postrevolutionary poets – "not a day without a line." Unfortunately I was not capable of such deeds. I would rather stay up at night fiddling with my equations, thinking over better solutions and clearer graphs. If not for insistence, and, sometimes, strictness, on the part of Zel'dovich, many of the results of these nightly vigils would have never been published.

Why did we publish in these particular journals?

In those far-away years it was not possible to publish a paper, even in a Soviet journal, before it was reviewed by a special expert commission at the Institute. Naturally, this practice caused delays in publication. Later this would become an issue between the journals themselves and GLAVLIT – a special agency responsible for ensuring that no state secrets and no suspicious information, from the government point of view, got into the open press. At the beginning of the 1990s a lot was written about GLAVLIT, so my words are targeted only at Western readers who did not follow the internal Soviet Union problems disclosed during the years of Glasnost and Boris Yeltsyn's rule.

It took even more time to obtain a permit to publish a preprint or send an article to a foreign journal. An article that had passed an expert commission at the Institute had to be turned over to GLAVLIT and then sit and wait for permission for the article to be sent abroad. Normally that took from one to three months. Translation into English was a big problem for us. Very often, our translations done with help of professionals received negative feedback due to unreadable English. It was next to impossible to publish an article in a journal of such a level as *The Astrophysical Journal*. People who were responsible for mailing articles abroad (and they were the ones who handled all our foreign correspondence) knew that one needs to pay to be published in such journals, and just didn't let these envelopes go through.

That is why Zel'dovich advised me to send a series of articles on the relict radiation to the journal *Astrophysics and Space Science*, as Zdenek Kopal, a publisher from Manchester, offered not only to publish our articles in two languages – English and Russian – which was politically correct, but also took it upon himself to translate them. Similarly, Martin Gordon's visit to Moscow resulted in the publication in *Comments on Astrophysics and Space Physics* of a number of articles, which, as we both thought, were quite important. In 1969, I mentioned our problems with publishing our results in a conversation with Stuart Pottash, who at that time was one of the editors of *Astronomy and Astrophysics*. Sometime later, Stuart sent letters on official letterheads to YaB and myself with an offer to send him articles that he was ready to publish for free, if the referee reports were satisfactory. We then started publishing our work in this journal as well.

It is natural that the bulk of our articles were printed in Soviet scientific publications. Unfortunately, for the majority of the Soviet journals, it took over a year or a year and a half for an article to be published. Sometimes, Solomon Pikelner of *Soviet Astronomy* managed to publish our group's original articles significantly faster, but sometimes even he failed to do it. The best (but not all) Soviet journals were translated into English. However, translation caused another delay of at least eight to nine months, and the translations were available only at the libraries of the big science centers.

That is why Zel'dovich told me a few times that I had to identify a problem and solve it at least a year prior to my Western colleagues, as they were writing in their mother tongue and they didn't have our problems with translation, slow mail, and GLAVLIT.

He considered preprints as one of the possible solutions to this problem and he went to a lot of effort to get them published regularly at the Institute of Applied Mathematics. While time still was wasted on procedures at the Institute, as well as on waiting for GLAVLIT's approval and mailing, at least there was no need to wait for a year while an article was put on hold by a journal's editorial staff. Also, it was possible to work on translation while procedures at the Institute and GLAVLIT were taking place. Most of the results that I mention below were first published as preprints and sent out to the leading world groups and libraries long before they were published in the journals.

Zel'dovich was very well aware that in his country in the field of cosmology he was equal to none, and that only in the West could the beauty of his theoretical work be appreciated. He used to tell me: "You compete with the whole world, as you practically don't know personally anybody you compete with, and you don't know their intellectual potential and abilities." For a person of his class it was obvious that in the Soviet Union the sciences of mathematics and theoretical physics were outstanding. At the same time, the gap between the United States and the Soviet Union in the area of experimental astrophysics was rapidly increasing despite the country's huge investments in a 6-m optical telescope, the radio telescope Ratan-600, and space research. Not a single great astrophysical discovery of the 1960s – the relict radiation, quasars, radio pulsars, compact X-ray sources, the cosmological evolution of radio sources and quasars – was done in the USSR. And this increased YaB's desire to tell the best observers in the world, as early as possible, about the new ideas generated by his group.

Cosmological recombination

In the spring of 1966 I got interested in CBR in different spectral bands. It was at this time that I became friends with Dima (Vladimir) Kurt, who, with the interplanetary probe "Venera," was observing the solar Lyman-α line scattered by interplanetary hydrogen. Dima used very simple detectors with filters, which made it possible to measure the very intense Lyman-α line brightness using a broad filter (1050–1340 Å) simultaneously with a broadband of wavelengths longer than Lyman-α (in the range 1225–1340 Å). We hoped to detect this line produced by intergalactic plasma and displaced by redshift (Kurt and Sunyaev 1970). It was this that later led me to the idea of external zones of ionized hydrogen in galaxies. It is natural that hydrogen in these zones could have been ionized only if there were a cosmic ionizing radiation background. And, the other way round, the lack of a cutoff in the distribution of neutral hydrogen and the emission measure in these zones allowed an estimate of the extragalactic ionizing radiation background (Sunyaev 1969), something impossible to measure in any other way. It was very important to me that at that time Ed Salpeter valued this work.

In September 1966 I gave a talk at the All Moscow Astrophysics Seminar in the Sternberg Institute where I mentioned hydrogen recombination in the universe, and I noted that, according to the Saha equation, it occurred at redshift $z \simeq 1300$. Dima Kurt came up to me after the seminar and asked: "And where are the Lyman-α line photons emitted by the recombining atoms and displaced by the redshift?" I immediately explained that because of the high-specific entropy of the universe there would be a relatively small amount of these photons, and that they come to us in a range of wavelengths that was observationally inaccessible at that time. But Dima's question sunk into my mind. The optical depth in the Lyman-α line is huge. The cosmological redshift caused these photons to very slowly escape through the long wavelength wing of the line. With Dmitry Varshalovich of the Ioffe Institute in Leningrad we solved an integral radiative transfer equation for the spectrum of Ly-α radiation from intergalactic gas (Varshalovich and Sunyaev 1968). At that time I was interested in the spectrum of this line, although Kurt's detectors were unable to resolve it. I told YaB about our solution and about Kurt's question. YaB advised me to try to compute the line profile by the method of characteristics, and also to look up an article by Kipper (1950) on the rate of two-photon decay of the 2s level of hydrogen.

To my surprise, the very slow two-photon decay rate $(8.1\,\mathrm{s}^{-1})$ was (with the cosmological parameters accepted at that time) more likely to happen than escape of Lyman-α photons through the wing of the Lyman-α line. Knowing the lifetime of the 2s level it was not hard to calculate the strongly delayed hydrogen recombination process, and to find the effect of emission by the two-photon decay on the relict radiation spectrum in the Wien region. Another implication was that the residual ionization could not be lower than 10^{-4}. A result that astonished me was that Compton heating kept the electron temperature close to the CMBR temperature, and before redshift $z \simeq 150$ electron collisions transferred enough heat to the hydrogen to keep it at the radiation temperature. Only at lower redshifts did the hydrogen cool more rapidly, in accordance with the difference in adiabatic indexes of matter and radiation. These results were obtained fairly fast, but I, as always, was slow in perfecting the article. I was naive to suppose that it would occur to nobody else to check how recombination differed from the classical Saha thermal equilibrium solution.

When all results of the Zel'dovich, Kurt and Sunyaev (1968) paper had been obtained, Zel'dovich told me that Iosif Shklovsky from the Shternberg Institute and Igor Novikov from our group were going to the United States to attend the Texas Symposium in 1967 and that I had to immediately formulate all our recombination research results concisely and clearly on a couple of pages and submit them to Shklovsky and Novikov. The world had to know about our results! Almost two years after I learned about the results of this action, when I saw Jim Peebles' (1968) article in *The Astrophysical Journal*, in which he, referring to Shklovsky's remarks at the Texas Symposium, obtained the same results that we were so proud of.

Energy release and distortions of the CMBR spectrum

At that time publications were appearing saying that thermalization of starlight by intergalactic dust might imitate the observed spectrum and high isotropy of the relict radiation. (The accuracy of the spectrum measurements was not very high, and many astronomers didn't doubt that there could be large deviations from a blackbody spectrum.) It was clear that it was necessary to study in detail the physical processes that could affect the CMBR spectrum: is it at all possible to create the blackbody radiation during the evolution of the universe?

I was very surprised, when making the simplest estimates, how incredibly small the Rosseland optical depth of the universe was due to bremsstrahlung. It became large only at $z > 10^8$, due to the appearance of electron–positron pairs.

I looked for traces of bremsstrahlung emission of intergalactic gas. In 1967 I managed to demonstrate that in the history of the universe a period of neutral hydrogen was unavoidable, judging by the absence of measurable low-frequency bremsstrahlung in the decimeter range as well as by the upper limit of possible nuclear energy release due to all the existing baryons. The mathematical task was reduced to finding the extremum for a simple functional. YaB liked the result although he thought the answer to be evident. He recommended this article for publication in the *Reports of the Academy of Sciences of the USSR* (Sunyaev 1968). This journal was considered very prestigious, as only Academy members could publish there or recommend articles for publication. Unfortunately it is little known in astronomical circles, even the part of it translated as *Soviet Physics Doklady*.

The outstanding paper by Gunn and Peterson (1965) made it clear that intergalactic hydrogen must be highly ionized at $z \leq 2$. With Gennady Sholomitsky we thought about the role of Thomson (that is, nonrelativistic) scattering of the CMBR by free electrons and the resulting polarization. At large z (before recombination) the optical depth for this process was very high, but it does not change the spectrum. But after listening to lectures at ITEPh I remembered that Compton scattering by near-relativistic electrons changes the photon frequency and creates photons (double Compton). I had already in my hands an article on the subject by Aleksandr Kompaneets (1956), and a book, *Quantum Electrodynamics* by Akhiezer and Berestetsky, still one of my favorites, on the bookshelf in my dormitory room.

The Kompaneets equation

I learned a lot from Kompaneets' article. (It made even more interesting my occasional meetings with Professor Kompaneets.) At different times YaB would tell me by bits and pieces how they managed to guess at this equation that describes the evolution of the spectrum of radiation exposed to the frequency shifts of Compton scattering by electrons in a hot plasma.

Only the exact form of a diffusion operator in a standard Fokker–Planck equation was calculated. The terms responsible for the recoil effect and induced Compton scattering were found using an elegant trick, based on the condition that the equation predict no distortion of a blackbody spectrum at full thermodynamic equilibrium. Now I know that this is a fairly standard method.

It was amazing to know that even Lev Landau and the famous mathematician Israel Gelfand were engaged in the work of solving the Kompaneets equation at the request of Zel'dovich's group in Arzamas-16. Several times Zel'dovich told me, with sadness, about the role of Sergey Dyakov in the work on the equation. He drowned in the Moscow river in a suburb of Moscow. (Kompaneets referred to Sergey Dyakov and to all the scientists mentioned above in the last lines of his article. It is written there that the "problem was formulated by Zel'dovich.")

Zel'dovich told me that attention was drawn to this beautiful equation during the analysis of the highest plasma temperature that could be reached in a hydrogen bomb explosion. In the early 1950s, it was found that radiative and other mechanisms did not allow the temperature to significantly exceed a million degrees, so the Compton cooling could be neglected. It became clear that the Kompaneets equation was useless for weapons research, and in 1955 these results were declassified.

I wanted to take a look at the original description of the derivation of the equation, obtained at the end of the 1940s. In his article, Kompaneets refers to a report of the Institute of Chemical Physics of the Academy of Sciences published in 1950, which had been declassified for a long time and was located in the Library of the Russian Academy of Sciences. Unfortunately, the report, which I was shown, didn't have a single word on the derivation of the equation and contained only an attempt to find energy losses of plasma due to Comptonization in a cylindrical geometry.

It is natural that the giants who worked on the Kompaneets equation solutions in those far away years were not interested in a time-dependent problem, which is critical for the uniform expanding ionized gas of the early universe. Reading Kompaneets' paper again one sees that the author and his colleagues were not interested in the effect of Comptonization on the radiation spectrum. Their main goal was to find the cooling rate of hot plasma due to this effect, and to bremsstrahlung that provides low-frequency photons for Comptonization. In astrophysics, the complete solution of the Kompaneets equation has not been found, but we managed to find beautiful and simple solutions in particular cases, when it was only a diffusion operator related to Doppler broadening that matters, or only recoil or induced scattering effects dominated (see a review by Pozdnyakov, Sobol and Sunyaev 1983).

Physics of Fluids published an article by Ray Weymann (1965) with an equation identical to the Kompaneets equation, referencing a Livermore Lab report. In 1966 Weymann published numerical calculations for the distortion of the relict radiation spectrum when extremely large energy is released in the plasma. The journal with this article reached us in the spring of 1967, but by that time we, together with YaB, were fully engaged in the work on analytical solutions describing spectrum distortions in useful limiting cases. I was really lucky that nobody before was interested in analytic solutions of the type we found.

Two major types of distortion of the CMBR spectrum

The energy exchange between electrons and radiation due to Comptonization (that is, redistribution of the photon energies in the process of multiple Compton scatterings on hot electrons) is characterized by the parameter

$$y = \int_{t_{\min}}^{t_{\max}} \frac{k(T_e - T_r)}{m_e c^2} \sigma_T N_e(z) \, c \, \mathrm{d}t. \tag{4.2}$$

In our article (Zel'dovich and Sunyaev 1969) we found analytic solutions describing spectrum distortions both at small y values and at any value of

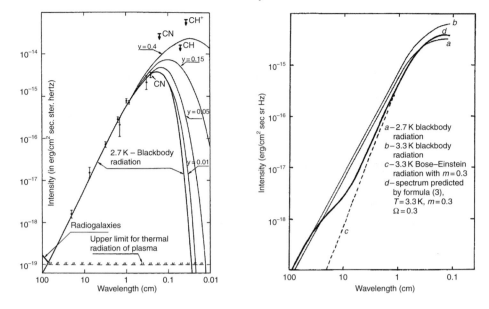

Fig. 4.7. The y-parameter distortion of the CMBR spectrum is illustrated on the left and the μ-distortion on the right.

this parameter when the recoil effect may be neglected. These y-distortions allowed us to relate any energy release in the universe at $z < 5 \times 10^4$ to the deviation of the CMBR spectrum from blackbody. The left-hand graph in Figure 4.7 shows that the deviation is especially strong in the shorter wavelength Wien part of the spectrum, where, at that time, observations showed a significant excess of the background spectrum over a blackbody radiation curve.

Zel'dovich realized the importance and the beauty of these solutions, but, as I have noted above, it took a long time to be published in Western journals. He decided to send the first edition of the article to the *Soviet Physics Uspekhy* review journal. There is a reference to our never-published paper in this journal in a wonderful article by Zel'dovich and Shakura (1969) on spherically symmetric accretion onto a neutron star, where our solution is used to obtain the radiation spectrum of a shock wave in the vicinity of a neutron star surface. After Zdenek Kopal's visit to Moscow, Zel'dovich changed his mind and recommended that I send the article to *Astrophysics and Space Science* for publication. The publication of the article was significantly delayed, to March 1969.

In a subsequent article (Sunyaev and Zel'dovich 1970a) we demonstrated that earlier energy release (at $z > 10^5$) led to another effect: Bose–Einstein distortion or μ-distortion of the relict radiation spectrum. This is shown in

Figure 4.7 (from Sunyaev and Zel'dovich 1970b). In scattering by free electrons that preserves the number of photons the radiation spectrum relaxes to the form characteristic of a boson gas with nonzero chemical potential μ.

But bremsstrahlung (and double-Compton emission) can create photons at low frequencies, which Comptonization can raise to higher energies, lowering μ, and even relaxing back to a true blackbody spectrum. We demonstrated that any energy deposited at redshift $z > 10^7$ would leave no trace in the observed relict radiation spectrum. Later ($10^5 < z < 10^7$) energy release would lead to a Bose–Einstein spectrum distortion.

At the end of the 1960s, Hannes Alfvén visited Moscow (I was present during his conversations with Zel'dovich) and argued for the possibility of annihilating matter and antimatter, which would inject energy that could affect the CMBR spectrum. Leonid Ozernoi and Artur Chernin (1968) were actively developing a turbulent model of the universe, which could also lead to significant energy release. Nowadays decay and annihilation of different elementary particles is widely discussed, including those which possibly are connected with dark matter.

Observations conducted at the end of the 1960s allowed, and sometimes indicated, large spectrum distortions. However, even then unexpected experimental facts emerged. I remember a seminar by I. S. Shklovsky and V. I. Slysh on McKellar, who, back in the 1940s, evaluated the excitation temperature of a CN molecule in the intergalactic medium by the observation of absorption in the ultraviolet doublet line of this molecule in the spectra of bright stars. This measurement of the relict radiation temperature at wavelength 2.54 mm, and similar data on CN^+ and CH molecules obtained by Thaddeus, Field, and others were for quite a while the most important constraints on the CMBR spectrum in the Wien region. Only a really great instrument, COBE/FIRAS (let's also mention a Canadian rocket experiment, Gush, Halpern and Wishnow 1990) much later provided extremely rigid upper limits for both y and μ (Mather *et al.* 1990) and, consequently, energy release in the early universe, thus considerably limiting the freedom of action for theoreticians.

The CMBR and clusters of galaxies

The history of the "SZ" effect in directions toward clusters of galaxies goes back to the discussions of the problem of the search for "missing" matter in clusters of galaxies. I do not remember now who told me first about the existence of a paradox of "missing" mass in clusters formulated by Fritz Zwicky in the 1930s. Most likely it was Samuil Kaplan, a professor from Gorkii University.

I began to read papers about clusters of galaxies and very soon strongly believed that there is a lot of hot intergalactic gas inside clusters because the limits to the amount of neutral hydrogen there were very strict. It was simple to show that Thomson scattering of isotropic CMBR photons on electrons in this gas leaves no trace in the angular distribution of CMBR. When I told this to Zel'dovich, he began to laugh and said only "Tindal effect: smoke from the chimney in a fog."

A beautiful and unexpected result came when the change of the photon frequency due to Comptonization on the hot electrons was taken into account. The CMBR brightness in the directions toward clusters had to decrease at centimeter and millimeter wavelengths. The corresponding increase of the brightness in the submillimeter band did not excite anybody at the time. This behavior of the CMBR brightness toward clusters was a direct consequence of the solution (Zel'dovich and Sunyaev 1969) of the Kompaneets equation with small y.

Obviously I told Zel'dovich. However, he at that time was not interested in individual objects and their properties: the universe as a whole was his main interest. Therefore, I had no reaction from him during or after this conversation. Nevertheless, he was not against it when I included this topic in one of my talks in the Sternberg Institute seminars, or I spoke about this prediction with radio astronomers. My thoughts about the CMBR brightness change toward the directions of clusters of galaxies are included in the paper Sunyaev and Zel'dovich (1970c). The case of Coma cluster of galaxies was considered. This was at the end of the paper, which YaB usually had no time to read. I had little doubt that he would object to the comments on the effect of clusters of galaxies on the CMBR. I was sure everything was correct. It was known that differential observations to detect this effect are much easier than absolute measurements. I learned this from conversations with Aleksandr Salomonovich, Yury Pariĭskiĭ and Kazimir Stankevich. It allowed me to dream that the effect would be observed earlier than the global spectral distortions due to energy release. That has proved to be the case.

In 1971 the UHURU spacecraft confirmed that the nearest rich clusters of galaxies are very bright X-ray sources. Scientists participating in the discussion in the Sternberg Institute All Moscow Astrophysics Seminar had very different opinions about the origin of this X-ray emission. The majority of them liked the model of inverse Compton scattering of CMBR photons on ultrarelativistic electrons accelerated in the bright radio sources or coming from shock waves in numerous supernova remnants in the galaxies belonging to the cluster. YaB and I were happier with the thought that this might be the result of free-free emission of the hot intergalactic gas. Such a model was proposed by George Field. It was necessary to look for independent methods that could prove the existence of intergalactic gas in clusters. This time, already at the end of 1971, Zel'dovich began to request that I write a separate paper on the thermal effect on the CMBR toward clusters of galaxies. Then he accused me as usual that I am not "a writer" and finally dictated to me himself the first page of the paper that was submitted to *Comments on Astrophysics and Space Physics* (Sunyaev and Zel'dovich 1972). YaB was right as usual; it was important to write a clear separate paper.

The peculiar motion of a cluster was also able to cause a significant decrease (or increase) of the CMBR brightness just due to the Doppler effect influencing photons scattered by electrons moving together with the whole cluster relative to the unique frame in which the CMBR was isotropic. We both knew about the "kinetic effect" at that time and this is distinctly mentioned in Sunyaev and Zel'dovich (1972). However, our detailed paper on the "kinetic effect" (Sunyaev and Zel'dovich 1980) was published much later.

I remember well my first talks when I tried to describe the thermal effect to physicists. For me the most interesting fact was that this effect if observed would permit a demonstration that the CMBR originates at redshifts greater than the redshift of the cluster itself. Some physicists and astronomers at that time still had doubts about the cosmological origin of the CMBR. I was excited that the spectral dependence of the effect and its amplitude were independent of redshift. This was really unusual; it opened a way to observe very distant clusters. Nevertheless, the audience was usually much more interested in the possibility of proving the existence of hot intergalactic gas in clusters and disproving the hypothesis of a nonthermal origin of the X-ray emission due to inverse Compton scattering on relativistic electrons. It is rather difficult to say that the reaction to my first talks was enthusiastic. The information about the physics of the effect was perceived silently and with obvious distrust. However, there were exceptions. Yury Pariĭskiĭ began attempts to observe the effect in the direction of the Coma cluster of galaxies after several conversations with me in 1970 and 1971. A rumor of detection of the effect (Pariĭskiĭ 1972) was one of the reasons why Zel'dovich started to press me to publish the "separate paper" about the effect.

Gas in clusters has a small Thomson optical depth. The rare single scatterings of a small fraction of the photons crossing the cluster defined the change of the CMBR brightness. I was very curious and wanted to find the kernel of the integral equation which could lead to the Kompaneets equation after a Fokker–Planck type expansion. It turned out to be not a simple task. Practically ten years went by before the kernel describing the profile of the line after single scattering was ready for publication (Sunyaev 1980; see also Sazonov and Sunyaev 2000, where the kernel taking both Doppler and recoil effects into account was obtained). This kernel allowed one to find the spectral dependence of the effect in single-scattering approximation. Naturally this spectrum fully coincided with the spectrum obtained by solving Kompaneets equation.

Here I am glad to write words of gratitude to the radio astronomers who spent a significant part of their lives to detect thermal effect and to make it possible to use these measurements for the purposes of cosmology. I should mention John Carlstrom, Mark Birkinshaw, and Francesco Melchiorri first.

Small-scale fluctuations of the CMBR

I remember being very impressed by Joe Silk's articles (1968b,c) where he, for the first time, wrote about small-scale damping[5] of density perturbations due to radiative viscosity, and he indicated the simplest equation for adiabatic perturbations in the radiation temperature,

$$\frac{\delta\rho_m}{\rho_m} = 3\frac{\delta T}{T}. \tag{4.3}$$

Of course, everybody was familiar with this relation, but in Silk's article it was used to predict the level of expected angular fluctuations of the relict radiation. For me it was easy to agree that it had been exactly like this until the moment of recombination, when there would be mixing and blurring of radiation fluctuations on small scales. This simplest and beautiful equation was valid only under the approximation of instantaneous recombination.

Some time in the fall of 1968 I had come up with an approximate analytic solution describing hydrogen recombination. This allowed me find an analytic expression for what is now called the "visibility function" that describes the zone from which photons, freed by recombination, come to us without further scattering. Later this derivation was included in Sunyaev and Zel'dovich (1970c), which reached the journal in June 1969. According to the WMAP satellite, the redshift where the peak of the visibility function is located, and its effective width differ by just a few percent from the value obtained in those early years.

---

[5] I should mention that I earlier heard about the importance of viscous damping from Andrei Doroshkevich, but Silk's papers reached us before Andrei finished his work.

Our interest in the amplitude of primary CMBR fluctuations was increased by the statements by radio astronomers (first of all, by Yury Pariĭskiĭ) that the measured fluctuations were substantially weaker than what Silk had predicted. We needed a more realistic method of estimating the amplitude of the angular fluctuations. That was easy to do. Baryons together with electrons were moving due to the growth of density perturbations at the last stages of recombination. Scattering of photons on moving electrons was changing their frequencies (and temperature), due to the Doppler effect. The resulting change of the radiation temperature was defined by the simple equation:

$$\frac{\partial(\delta\rho/\rho)}{\partial t} = -\nabla \cdot u, \quad \frac{\delta T}{T} = \int_0^\infty \frac{u_1(z)}{c} e^{-\tau(z)} \frac{d\tau}{dz} dz, \qquad (4.4)$$

where u_1 is the projection of the velocity along the direction of the ray. Taking into account the analytic expression for the "visibility function," we obtained the expected value for the amplitude of angular fluctuations on the angular scale of tens of minutes of arc at the level of 2×10^{-5}. This is close to the value measured later by Boomerang, Maxima-2, WMAP, and many ground-based experiments.

Baryonic oscillations and acoustic peaks

I very much liked a "duck's beak" diagram Zel'dovich once drew on a blackboard in my presence. It described the evolution of density perturbations at different length scales; a version is shown in Figure 4.8. I had read in different articles and reviews that in the expanding universe (as in any object affected by Jeans' instability) density perturbations on scales smaller than a Jeans wavelength had to behave like acoustic waves. I remember a very interesting conversation on this topic with Lev Gurevich, almost blind at

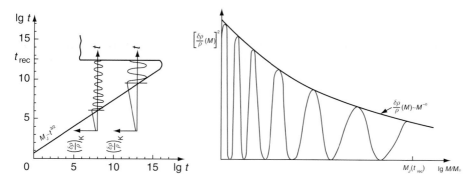

Fig. 4.8. Illustrations of the origin of the quasiperiodic oscillations of the baryon and radiation distributions (from Sunyaev and Zel'dovich 1970c).

that time, at his "dacha" near Leningrad in the summer of 1967. Zel'dovich just loved to explain the results obtained within the framework of the general theory of relativity in a customary Newton language. At the same time all of us knew that the behavior of acoustic waves during the radiation-dominated era of the universe was analyzed by Evgeny Lifshitz (1946) in his classic work. We knew that the universe was a unique object at the early stages of its expansion. The velocity of sound was close to the velocity of light, and therefore a Jeans wavelength was just a bit smaller than the horizon scale. In the very early universe observationally interesting wavelengths were larger than the Jeans wavelength. They became smaller than the Jeans length as the universe expanded. And at recombination the baryons decoupled from the radiation and the Jeans wavelength for the baryons sharply dropped.

In the very early universe only the growing mode of density perturbations survived, meaning all acoustic waves were launched with the same phase. The sharp decoupling cut off fluctuations of different wavelengths at different phases. This led to a wonderful prediction of the quasiperiodic dependence of the perturbation amplitude after recombination on the length scale. This effect now is called the "baryonic acoustic oscillations." (The presence of dark matter complicates this picture, but not the physical essence.) This quasiperiodic behavior in the distribution of baryons and electrons and their velocities, naturally, had to be reflected in the quasiperiodic behavior of the angular distribution of CMBR temperature. (See Figure 2 in Zel'dovich, Rakhmatulina and Sunyaev 1972, where an attempt was made to show how to use a spherical harmonic expansion for the demonstration of acoustic peaks. This effect was discovered by the Boomerang, Maxima-1, and WMAP experiments.) In Doroshkevich, Zel'dovich and Sunyaev (1978) the Silk damping was taken into account.

YaB liked to get cozy and vet article manuscripts. Unfortunately, my typewritten drafts were very long and he was never able to read the whole text. With "a touch of a master," he would go through the first several pages and briefly look through the figures. However, he used to edit abstracts of the articles many times. I remember how upset I was when, in the abstract of our article (Sunyaev and Zel'dovich 1970c), he crossed out my words on the importance of observing the quasiperiodic scale-dependence of the amplitude of the CMBR angular fluctuations. He wrote that the effect was very small and could hardly be observed. To calm me down he said that the physics of the phenomena described in the paper was beautiful, and the article needed to be published. Another time, after a long conversation with A. D. Sakharov, who used to visit Zel'dovich quite often at the Institute of

Applied Mathematics, Zel'dovich told me that A. D. Sakharov was nervous and changing every day his opinion on an important, however far from scientific, issue, and that we needed to call the quasiperiodic dependence in our article "Sakharov oscillations." That would be a great joke! I tried to protest that I had never heard anything similar to our results from Sakharov, and that we would have to have some reference. YaB responded: "Let's refer to Sakharov's (1965) article on a cold universe, where every second paragraph starts with the words 'As Zel'dovich told me.' "

Sergei Shandarin told me once that if Zel'dovich ever thought that these baryonic oscillations and acoustic peaks would be observed and become of importance then he would never have made such jokes.

The situation around the name of Andrei Sakharov in the second half of 1969 and early in the 1970s continued to deteriorate. It was obvious that any joke about him was inappropriate. Zel'dovich did everything possible to keep the name of Sakharov alive in Soviet astrophysical literature, mentioning in particular again and again "Sakharov oscillations" in his papers and books. This completely changed the initial meaning of "Sakharov oscillations" as Zel'dovich had introduced them. Now after practically 40 years it is possible to write how and why this term originated.

Terms are functions of time, but acoustic peaks and baryonic oscillations will be present on the sky for the next billions of years.

Epilogue

In the four decades since the events I have described cosmology has made unbelievable progress: much of what seemed impossible then has been studied in detail and measured with a high degree of accuracy. WMAP spacecraft (after Boomerang and Maxima-1 balloon flights) made a most significant contribution to the revolution in observational CMB cosmology. I remember my last conversation with David Wilkinson at the Kennedy Space Flight Center on Cape Canaveral after the WMAP launch. David recalled an old question of mine: why MAP was going to be placed in a low orbit, while for the Relikt-2 project Igor Strukov and Dmitry Skulachev chose the great opportunity of launching to the second Lagrangian libration point (Strukov and Skulachev 1991). This possibility was reviewed in detail at the Space Research Institute in Moscow by Pavel Eliasberg and Ravil Nazirov's group. At that time David asked me to tell Alan Bunner at NASA's Headquarters about this plan. It was amazing that David managed to find time to thank me and (indirectly Strukov and Skulachev) for this advice, when WMAP was already on the way to the Lagrange point.

I regret that limits of time and space do not allow me to say much about Relikt-1, the first spacecraft to scan the sky to look for CMBR angular fluctuations. Everything including the detectors was designed and produced in the USSR. The first map of the microwave sky brightness was obtained by this experiment. Emission from the plane of the Galaxy was mapped in detail. The amplitude and direction of the CMBR dipole component were measured, and strict limits placed on CMBR angular fluctuations, before the launch of COBE (Klypin *et al.* 1987; Strukov *et al.* 1987; Skulachev and Skulachev 1988; and Klypin, Strukov and Skulachev 1992).

The Strukov and Skulachev group was very close to the great result later obtained by COBE/DMR (Smoot *et al.* 1992). All that was missing was the presence of at least one additional frequency channel and a modest factor in detector sensitivity – or mission lifetime. But this is the point of view of a person who was not knowledgeable in the nuances of the experiments.

It would be a pleasure to write about Semen Gershtein and Zel'dovich's (1966) idea on the upper limit for the muonic neutrino mass based on cosmological data; about the upper limit on the amount of neutrino species, obtained by Viktoriy Shvartsman (1969); on Andrei Sakharov's seminal articles and ideas including his thoughts about the nature of baryon asymmetry in the universe (Sakharov 1967); on Georgiy Zatsepin Vadim Kuz'min (1966) with their prediction of the high energy cutoff in the ultrahigh cosmic ray spectrum due to the presence of CMBR (Greisen–Zatsepin–Kuz'min limit); and about an influential article by Vitaly Ginzburg and Leonid Ozernoi (1966), where they for the first time raised the issue of mechanisms of heating and ionization of the intergalactic gas. It would be possible to write about many other things, but this is enough. It suffices to say that I remember that as a happy time when everything turned out well and every day gave us something new and unexpected.

4.4.5 Malcolm S. Longair: Moscow 1968–1969

Malcolm Longair carried out his postgraduate studies at the Mullard Radio Astronomy Observatory of the Cavendish Laboratory, University of Cambridge, from 1963 to 1967. He spent the academic year 1968–1969 as Royal Society Exchange Fellow to Moscow. Subsequently, he was Astronomer Royal for Scotland and Director of the Royal Observatory Edinburgh from 1980 to 1991. He returned to Cambridge in 1991 as Jacksonian Professor of Natural Philosophy and from 1997 to 2005 was Head of the Cavendish Laboratory.

I was fortunate enough to be in Moscow during the period September 1968 to July 1969 under the Royal Society – USSR Academy of Science Exchange Fellowship scheme when the Moscow Group led by Zel'dovich made many of their fundamental contributions to cosmology. For me, this was an incredible period of study when I came into daily contact with physicists and astronomers who had been brought up in a rather different scientific and political tradition from that of Western scientists. I have already given an account of how I came to spend this period as a postdoc working with Zel'dovich and his colleagues in my essay *Encounters with Zel'dovich* in the volume celebrating his life and work, *Zel'dovich: Reminiscences* (Sunyaev 2004). This essay concerned my interactions with him and his colleagues from 1968 onward.

In this present essay, my aims are quite different. The intention is to give some impression of the circumstances under which Zel'dovich and his colleagues carried out their ground-breaking research in astrophysics and cosmology. All my friends and colleagues in the USSR carried out their research in a rather constrained social, political, and scientific environment. The fact that they were so successful can be attributed to the leadership and dynamism of an outstanding group of senior scientists.

Astronomy and cosmology up to the death of Stalin

The history of the late 1960s cannot be appreciated without some understanding of the historical background to astronomical research in the early years of the Soviet Union and the post-War period up to the death of Stalin. Before the October Revolution of 1917, Russian astronomy had enjoyed a considerable international reputation. Founded in 1839, the Pulkovo Observatory at St. Petersburg grew into a world-class organization under the successive directorships of F. G. V. Struve and his son Otto Struve. The Moscow University Observatory, later to become the Shternberg Astronomical Institute, was founded in 1830 and was primarily intended for teaching. Distinguished work on variable stars was carried out by its director Vitold K. Tseraskii and his wife Lidiya P. Tseraskaya in the latter years of the 19th century. At the outbreak of World War I, the Russian empire, which included observatories at Kazan, Tashkent, and Kharkov, was undoubtedly a major force in world astronomy.

Matters deteriorated rapidly after the outbreak of World War I. Following the October Revolution of 1917 and the devastating Civil War which followed, living conditions became very harsh and contacts between the Observatories and with the West were broken off for several years. It was only in the early 1920s that communications were restored with the international community. Some astronomers were actively involved in the Civil War. For example, Pavel K. Shternberg, Director of the Moscow University Observatory, had been a long-standing member of the Bolshevik party and was involved in the artillery barrage which supported the Red Guard troops who stormed the Kremlin in 1917.

An example of the impact of this scientific isolation of the new formed Soviet State was that it was not until the early 1920s that Aleksandr A. Friedman had access to Einstein's papers of 1915 on general relativity. He and his colleagues then began an in-depth study of the implications of this work. This led to Friedman's (1922, 1924) famous papers which established what are now known as the standard Friedman world models of modern cosmology. Friedman was employed by the Main Geophysical Observatory at Leningrad where his principal responsibilities and researches were in meteorology and the atmospheric sciences. His death from typhoid in 1925 was a tragic loss.

Initially the astronomers were optimistic that the future of astronomy would be placed on a firmer funding basis as part of a planned socialist economy, rather than depending upon the generosity of private benefactors such as Lick and Carnegie, as was the case in the USA. Stalin's victory over his rivals in the 1930s resulted in the drive to replace bourgeois pre-revolutionary specialists by unqualified political appointments, and astronomy suffered badly. This was at least partly ideologically driven as part of the philosophy of dialectic materialism, which was enshrined as official Soviet policy for all aspects of Soviet life, including the sciences. As expressed by Bronshten and McCutcheon,[6]

Scientists were now expected to denounce all theoretical trends in modern science that were incompatible with Soviet ideology. Not to do so made one vulnerable to accusations of idealism and anti-Soviet sympathies.

Those who were perceived to continue carrying out research according to the traditions of the pre-revolutionary era were in considerable danger. Dissension broke out at the Pulkovo Observatory and this contributed to the purge of Leningrad astronomers during the period of Stalin's Great Purges of the late 1930s. Many astronomers were arrested. Boris P. Gerasimovich, Director of the Pulkovo Observatory, was executed in 1937 and Boris V. Numerov, the director of the Leningrad Astronomical Institute, had the same fate in 1941. Similar devastation occurred at the Tashkent Observatory. At the Pulkovo and Shternberg Observatories, A. D. Drozd and

---

[6] I obtained much valuable information about this period from the October 1995 edition of *Journal for the History of Astronomy*, Volume 26, which was devoted to Astronomy in the Soviet Union, eds. Doel and McCutcheon (1995).

A. A. Kancheev were appointed Directors of the respective Observatories, neither of them having the stature or qualifications to run these observatories, but they were Communist Party members who were fully committed to the Soviet "cultural revolution."

Matters did not improve markedly immediately after World War II. Just as the Zhdanov Committee had condemned Shostakovich for "formalism" in his musical compositions, and genetics ceased to exist as a discipline following the Lysenko affair, other sciences including astronomy were subject to political intervention. In 1951, the Soviet authorities convened an all-Union Conference on Cosmogony with the specific objective of criticizing cosmological ideas which were not consistent with dialectic materialism. This conference was similar to earlier conferences on biology, chemistry, psychiatry, and neurology which undoubtedly held back progress in these disciplines. The 1951 conference was highly critical of many aspects of Western cosmology. For example, the singular beginning of the universe according to big bang cosmology was presented as religiously inspired. Some distinguished astronomers contributed to the attack on Western values. As stated by Doel and McCutcheon (1995),

Western astronomers, already aware of Lysenko's disruption of Soviet genetics, became anxious about the intellectual autonomy of Soviet astronomy, particularly as V.A. Ambartsumian, P.P. Parenago and B.A. Vorontsov-Velyaminov published polemical attacks criticising the capitalist ideology of Western astronomers and its injurious effects on theoretical astrophysics.

Although a similar conference was planned for physics, it escaped these political strictures for the very practical reason that Stalin needed to develop nuclear weapons to counter the threat of the West. The Manhattan project in the USA had its parallel in the USSR, where many of the very brightest physicists contributed to the efforts to develop the atomic and hydrogen bombs. Although the Soviet physicists had access to Western data thanks to the efforts of spies within the US–UK–Canadian collaboration, a huge amount of original science was carried out and the Soviet hydrogen bomb used a different principle from that developed in the USA.

The Soviet effort was led by Yulii B. Khariton, Igor V. Kurchatov, and Yakov B. Zel'dovich and they were supported by a galaxy of great Soviet physicists including Lev D. Landau, Andrei D. Sakharov and Vitali L. Ginzburg, and the great mathematician Israil M. Gelfand. These scientists worked under the greatest pressure to succeed in the development of nuclear weapons and they carried out physics of the very highest quality in pursuit of this goal. A beautiful example is the work on the derivation

of the Kompaneets equation which was completed as early as 1949 through
their joint efforts but which only appeared in the literature in 1956 under
the name of Aleksandr S. Kompaneets (1956) who was given permission to
publish this research. The contributions of Zel'dovich, Landau, Gelfand, and
Dyakov are acknowledged in the published paper. The physics which came
out of these studies was to have a profound impact upon Zel'dovich's future
studies in cosmology.

The Soviet nuclear program was a great success but it led directly to
the Cold War which was to dominate relations between the USSR and the
West for many years to come. The leaders of the program were awarded the
highest honors for this work. Zel'dovich was awarded the Order of Lenin,
named three times Hero of Socialist Labour and four times awarded the State
Stalin Prize. The enormous contributions by the theoretical and experimental physicists afforded them a rather special status within the Soviet system
and physics did not suffer political intervention to anything like the extent
of other disciplines.

Another problem concerned the persecution of particular communities
within Soviet society. Campaigns were waged against "rootless cosmopoli-
tans," meaning largely those of Jewish origin. This was a potential threat to
a number of prominent members of the physics and astronomy community.
The result of all these strictures was to isolate the Soviet astronomers from
the rest of the international community. The meeting of the IAU planned
to be held in Leningrad in 1951 was cancelled under pressure from Western
astronomers.

Topics such as special and general relativity and cosmology were therefore potentially dangerous areas for study. Yet, the flame was kept alive by
a few individuals. Among the most important contributions made through
this period was Evgeny M. Lifshitz's profound and influential paper (Lifshitz
1946) on the development of small perturbations in the expanding universe.
In so far as the authorities might have been able to understand this great
paper, which involved perturbations of the spacetime metric in general relativity, Lifshitz's conclusion that the large-scale structure of the universe
could not have come about through the growth of infinitesimal fluctuations
in an expanding universe, because they grow as a power law rather than
exponentially, must at least have seemed politically correct.

Another key figure was Abram L. Zelmanov, who worked on the mathematical theory of general relativity at the Shternberg Institute throughout
his career. He was a shy and physically weak man but he had exceptional
qualities as a knowledgable scientist and a good teacher. He was accused of
having a "lack of publications," of not being "scientifically active," "disjoint

from real life," and so on. He was dismissed from his position at the Institute but the accusations were milder than they might have been. He was loyal to the system and a member of the Communist Party and so was given back his post at the Shternberg Institute where he remained for the rest of his life. There, he provided a healthy scientific environment and made his expertise available to anyone interested in cosmology. Besides the distinction of his own work, he was the teacher of the next generation of physicists and mathematicians who would become specialists in general relativity. These were to include Igor D. Novikov and Leonid P. Grishchuk.

A third key figure was Vladimir A. Fok, who contributed to many different areas of physics, particularly to quantum mechanics and relativity. He was a remarkable survivor through the turbulent years of the 1930s to 1960s when he did his greatest work. From our present perspective, his most important contribution was his book *Theory of Space, Time and Gravitation* (Fok 1964), which was to begin the revival of research in these areas in the 1960s. Interestingly, the book contains not only a sound exposition of the special and general theories of relativity but also what Freeman Dyson describes as "... an eloquent and at times somewhat polemical plea for an unorthodox interpretation of Einstein's theory of gravitation."

Stalin died in 1953 and the subsequent Khrushchev and Brezhnev eras saw a relaxation of the constraints on astrophysical and cosmological research, but the restrictions on contacts with the West remained in place. Most important for our story is the fact that, although the relaxation came as a great relief to everyone, the political system bequeathed by Stalin and the means of enforcing the restrictions on Soviet life remained in place. In particular, the system set up by the Committee for State Security (KGB) during the Stalin era remained omnipresent in Soviet life and was governed by the same suspicion of individuals which had pervaded the Stalinist era. There is no question but that the USSR was a police-state in which the KGB had essentially unlimited power to make life difficult if one did not follow the written and unwritten rules.

It is also important that what I have described above was all relatively recent history when I arrived in Moscow in 1968. I knew essentially nothing of this history at the time. I had, however, been inspired by my meetings with Ginzburg in Cambridge and by the work in cosmology which I had heard described at international conferences. I was therefore very pleased that Ginzburg agreed to act as host when I obtained the Royal Society – USSR Academy of Sciences exchange fellowship to spend the academic year 1968 to 1969 in Moscow. As it turned out, I actually had much stronger scientific collaborations with Zel'dovich and his colleagues than with Ginzburg's colleagues. This proved to be an extraordinary learning and research experience. I also had, and continue to have, a deep love of Russian music and literature – I wanted to experience the culture at first hand. Fortunately, I had been very well briefed by the Royal Society about what the KGB might get up to in order to demonstrate the perfidy of the West and/or to obtain blackmail material to be used at some time in the future when one might be in a position to be seriously embarrassed by their revelations.

The Moscow Scene

The theme of this collection of essays is cosmology and the big bang and so I will concentrate upon the astrophysicists and cosmologists whose work was most influential in these areas. Let me emphasize the great privilege it was to be on friendly terms with many of those listed below, despite the fact that contact with a long-term visitor from the West carried the potential of making difficulties for them. I had nothing but goodwill and lasting friendships with many of them.

The principal players in the story and their institutes were as follows:[7]

- *Lebedev Physical Institute, Moscow*
  Vitali L. Ginzburg, Sergei I. Syrovatskii, Leonid M. Ozernoi, Gennadi V. Chibisov, Tigran A. Shmaonov, Andrei D. Linde, and David A. Kirzhnits.
- *Shternberg Astronomical Institute, Moscow University*
  Iosef S. Shklovsky[*], Nikolai S. Kardashev[*], Solomon B. Pikelner, Leonid P. Grishchuk, Vladimir G. Kurt[****], and Abram L. Zelmanov.
- *Institute of Applied Mathematics, Moscow*
  Yakov B. Zel'dovich, Rashid A. Sunyaev[**], Andrei G. Doroshkevich, Igor D. Novikov[**], and Alexei A. Starobinsky[***].
- *Kapitza Institute of Physical Problems, Moscow*
  Evgeny M. Lifshitz.
- *Landau Institute of Theoretical Physics, Chernogolovka*
  Isaak M. Khalatnikov and Vladimir A. Belinsky.
- *Ioffe Physical Institute, Leningrad*
  Lev E. Gurevich, Artur D. Chernin and Dmitri A. Varshalovich.

---

[7] * Moved to the Space Research Institute (IKI) in 1972. ** Moved to the Space Research Institute in 1974. *** Starobinsky was a student at Moscow University, supervised by Zel'dovich. **** Kurt joined IKI in 1967 but was housed in the Shternberg Institute until the new IKI building was completed in the early 1970s.

- *Pulkovo Branch of the Special Astronomical Observatory, Leningrad*
  Yuri N. Pariĭskiĭ.
- *Department of Applied Mathematics, Gorkii State University*
  Samuil A. Kaplan.

It is apparent that the cosmological scene was dominated by the institutes based in Moscow. There were regular visitors from Leningrad and other astronomical centers throughout the USSR, in particular from the Observatories in the Crimea, the Special Astrophysical Observatory which ran the 6-m optical telescope at Zelenchukskaya, from the RATAN-600 radio telescope, from the Tartu Astronomical Institute in Estonia, from the Observatories in Armenia and Georgia, and so on. Undoubtedly, however, the cosmological scene was dominated by Moscow and, in particular, by Zel'dovich and his colleagues.

There are a number of important differences in the way science was carried out in these institutes as compared with the UK or the USA. First of all, these were all state-funded and state-controlled organizations. Although some of the scientists lectured at Moscow Physical-Technical Institute, these were primarily research institutes. Thus, the environment was not that of a University Department and University scientists made little contribution to cosmological studies.

A consequence of the institutes being state-controlled was that there was necessarily a strong KGB presence in all of them. Cooperation with the KGB was necessary for everyone to a greater or lesser extent and this had to be regarded as just a part of normal life. Identity cards were needed to enter these institutes. There was, however, a very different degree of formality associated with these institutes. During my stay in 1968–1969, I was involved with the first three in the above list.

- The *Institute of Applied Mathematics* was a completely closed Institute, which was involved in aspects of the nuclear program and other high security disciplines. Zel'dovich had his base in this Institute. My colleagues indicated to me that I should not be seen in the vicinity of this Institute and to this day I do not know where it is – I did not even want to ask for obvious reasons.
- The *Lebedev Physical Institute* was the principal Physics Institute in Moscow and Vitali L. Ginzburg was the host for my visit of 1968–1969. The Lebedev Institute was a very powerful physics institute indeed and had a remarkably strong liberal culture. After his exile and return to Moscow, Andrei Sakharov remained a member of the Lebedev Institute,

which continued to offer him an intellectual base despite being on difficult terms with the authorities.

- The *Shternberg Astronomical Institute* was the most open of all the institutes and this was the place to meet colleagues, particularly at the seminars run in alternate weeks by Zel'dovich and Shklovsky.

The scientific atmosphere in all these institutes was at the very highest level, particularly in the area of theoretical physics and astrophysics. The seminars at the Shternberg Institute were always memorable since all Moscow astrophysicists would make a point of attending these each week and the discussions were very lively indeed. What impressed me most was the depth of physical understanding which all the participants displayed. It is difficult to imagine a more lively intellectual atmosphere. All the pioneering work of the Moscow astronomers was discussed and thrashed out at these seminars. Of course, there were tensions – relations between Ginzburg and Shklovsky were not good, as is apparent from Ginzburg's (2001) reminiscences.

There were, however, strong impediments to the normal practice of research. Besides the constraints already discussed and the memories of the difficult years of the recent past, one of the biggest problems was access to Western journals. The institutes subscribed to all the major journals such as *The Astrophysical Journal* and *Nature*, but the originals could not be put on the library bookshelves. All advertisements had to be cut out of the journals, as well as all articles in journals such as *Nature* which might contain items not in accord with the political views of the authorities. These bowdlerized versions of the journals were then reprinted and arrived at the libraries. This entailed typically a six-month delay or longer before the journals became available to the Soviet scientists. In fact, one or two early copies did arrive in Moscow by quite unofficial routes thanks to the generosity of certain Western astronomers and these were pored over voraciously by all those who could get access to them.

Just as difficult were the problems of publication in the Western literature. All papers which were ready for submission to the scientific journals had to go through a rather long approval process before they could be sent off. For papers to be published in the Soviet journals, such as *Astronomicheskikh Zhurnal*, *JETP*, or *Soviet Physics Doklady*, reasonably prompt publication could be assured, but the system made it essentially impossible for papers to be sent to *The Astrophysical Journal* or *Monthly Notices of the Royal Astronomical Society*. Zel'dovich and some of the most senior and distinguished astronomers were able to overcome these problems, but the vast majority of research was significantly delayed in reaching the attention of Western astronomers. Rashid Sunyaev and I were able to publish three joint papers in Western journals while I was in Moscow, but this was because I was in a special position under the Royal Society–USSR exchange agreement. Translations of the major Soviet journals began to appear in the West in the late 1950s and these were of considerable help in disseminating the Soviet research, but these always appeared at least six months after their publication in the Soviet Union.

The other means of disseminating Soviet research to Western scientists was through international conferences. Foreign travel was, however, very restricted and those who were allowed to travel to major meetings had to be accompanied by representatives of the KGB. Zel'dovich and Sunyaev were not allowed to leave the Soviet bloc until the 1980s. In Zel'dovich's case, this was undoubtedly due to his involvement in the nuclear weapons program. Sadly, it meant that he was not able to take up the Honorary Doctorate awarded by Cambridge University because of his inability to come to Cambridge in person. In Rashid's case, it was almost certainly largely due to the very productive collaborations we had during my various visits to Moscow. As a result, Zel'dovich and Sunyaev were never able to present their work in person, or benefit from face-to-face discussions with Western scientists working in the same field, unless they could meet in the Soviet bloc.

As a result of these travel problems, those who were permitted to travel abroad had to present the new results on behalf of those who couldn't. I remember being very impressed indeed by the researches which Igor Novikov presented in January 1967 at the New York "Texas" Symposium. He had to summarize a vast amount of innovative work of outstanding quality. This work had not yet become available in the scientific literature and there were concerns among my Moscow colleagues about the means of establishing priority for the origin of new ideas when Western scientists could publish the same material very much more rapidly in Western journals. Although I personally believe this was not really a problem, the perception that it might be was always there because of the severe restrictions on publication and travel.

The way round these problems was to organize meetings within the Soviet bloc. Interestingly, although the Soviet Union withdrew from essentially all International Scientific Unions after World War II, it remained a member of the IAU throughout the period of the Cold War. As a result, Soviet astronomers were able to attend officially approved meetings of the IAU. It was, however, only the privileged few of the best astronomers who were able to travel abroad. Therefore, the organization of IAU meetings within the USSR was one of the most important ways of developing communications with the West. I was able to help with two of these. Zel'dovich entrusted me with organizing the scientific program of the 1973 IAU Meeting in Krakow, Poland entitled *Confrontation between Cosmological Theories and Observational Data*. For the first time many of the very greatest minds from the USSR and the West were able to meet under what would be considered normal international conditions anywhere else. The same may be said of the 1977 meeting on *The Large Scale Structure of the Universe* which Jaan Einasto and I organized in Tallinn in Estonia.

Surveillance by the KGB was simply a part of life. I was certainly aware of being the subject of surveillance but it did not impact my science program. Later, when I went back to the Space Research Institute to work with Rashid in the mid-1970s, all our discussions were monitored and an "interpreter" sometimes was present in case we exchanged written notes. Socially, one had to be careful, but I was still able to go regularly to the Bolshoi Teatr, to the jazz sessions at the Cafe Pechora, and to the Teatr Taganka where some of the most liberal plays were performed. Indeed, on one occasion a KGB officer was helpful in enabling me to hear the Russian Orthodox Easter services.

Finally, one important aspect of the Soviet system was the way one progressed up through the scientific hierarchy. Although the prices of all basic commodities for living were kept artificially low, any luxury goods were very difficult to come by. Appointment as a corresponding or full member of the USSR Academy of Sciences not only brought prestige, but also a very significant increase in salary and, just as important, privileges to use special shops selling goods that were normally impossible to buy and improved accommodation. As a result, promotion within the system carried with it a very significant improvement in lifestyle. As can be imagined, rivalries for promotion were intense.

Zel'dovich and his colleagues

This is the background against which Zel'dovich and his colleagues carried out their ground-breaking research in astrophysics and cosmology. Zel'dovich left the nuclear program in 1963, but even before then he developed a strong interest in astrophysics and cosmology. Within the Institute of Applied Mathematics, he put together a very powerful team of some of the very best young theoretical physicists to tackle problems which were to lead to the revival of interest in general relativity and cosmology. Igor Novikov was among the first to join him and, as Igor has recalled in his reminiscences, his job was to provide the necessary expertise in general relativity which was not then a strong weapon in Zel'dovich's armory. He was joined by Andrei Doroshkevich and later by Rashid Sunyaev, who was Zel'dovich's research student.

The science they carried out is dealt with elsewhere in this book and is now a matter of record. Zel'dovich had the enormous advantage of understanding the physics of ultra-high temperature plasmas in great detail and was a coauthor with Yuri P. Raizer of the definitive monograph on the subject (Zel'dovich and Raizer 1968). Paralleling the experience in the USA, the nuclear program made possible much better estimates of the nuclear cross sections needed to predict the primordial abundances of the light elements and, although the details were not published until very much later, they made excellent estimates of the primordial abundances and how they depended on physical conditions in the early universe.

Zel'dovich was a hard taskmaster and would expect the highest performance from his younger colleagues. They had to be on call at essentially any time of the day or night to come to his apartment in Vorobeyevsky Prospekt to discuss astrophysics and cosmology. He worked very fast and kept a very strong interest in all the works of his colleagues. The breadth and depth of his researches were quite staggering. He began to publish papers in astrophysics and cosmology only from the age of 50 onward, quite different from the profile of all other great astrophysicists. But this was possible because of his very deep involvement and commitment to physics and research throughout an extraordinarily eventful lifetime. It is not an exaggeration to say that Zel'dovich was the driving force behind the enormous development of astrophysical cosmology in the Soviet Union. Furthermore, it had to be done essentially in isolation from the rest of the international physics and astrophysics community.

Concluding remarks

These are not easy topics to write about because my Moscow colleagues were all living and working under constraints quite unfamiliar to Western scientists. While the period under discussion is now becoming better understood, it is not obvious to me that we will ever really get to the bottom of many important issues. Symptomatic of this was the hope and expectation during the 1990s that, once the state archives had become more publicly available after the collapse of the Soviet Union, a clearer picture of what had actually happened would become available. As soon as the first archives were opened, however, it became obvious that they could cause embarrassment to many individuals and the archives have been closed again. It is not at all obvious that the real story will come out until after many of those involved have passed away.

This concern applies equally to trying to disentangle how Zel'dovich and his colleagues were able to overcome all the impediments of the system and produce science of the very highest quality. For me, all the astrophysicists on my list of key players are heroic figures in that they overcame the somewhat unnatural constraints on normal scientific discourse and carried out wonderful and original science. I owe them more than can be adequately expressed in words for the scientific enlightenment they so generously offered me and for their lasting friendships.

4.5 Detection at Bell Laboratories

4.5.1 Arno Penzias: Encountering cosmology

Arno Penzias shared the 1978 Nobel Prize in Physics with Robert W. Wilson for their discovery of the CMBR. He joined the staff of Bell Laboratories after graduate school and remained there until retiring as Vice President of Research and Chief Scientist. He then joined New Enterprise Associates, a Silicon Valley venture capital firm, where he advises emerging companies in the fields of information technology and alternative energy sources.

My first serious brush with cosmology came in 1958, when Charles Townes accepted me as one of the students in his radio astronomy group at Columbia University. My project was to be the first maser-based astronomical study of 21-cm line emission from neutral atomic hydrogen. At that time, the only known source of radio line radiation, neutral atomic hydrogen, had by then been studied by several groups of observers. Since my system would yield an order of magnitude improvement in sensitivity over the best systems available to other radio astronomers, it seemed to me that I could extend trails already blazed in interesting directions. All I had to do was pick the most interesting body of prior work.

The choice of this observing project stemmed from a review of the then current radio astronomy literature, most notably a special (January 1958) issue of the Proceedings of the Institute of Radio Engineers devoted entirely to radio astronomy. From the first, I was most taken by an article (Heeschen and Dieter 1958) that addressed an interesting puzzle: clusters of galaxies appeared to contain more mass (determined from dynamical studies) than could be accounted for by the sum of the masses of their constituent objects. According to the data reported in this article, that discrepancy could be accounted for by the large amounts of neutral atomic hydrogen observed within each of the clusters investigated by the authors. Having selected an H I survey of clusters of galaxies as my target, I proceeded to design the maser preamplifier and other components that I would need, to create a low-noise radiometer for my radio telescope – an 85-ft parabolic antenna owned by the US Naval Research Laboratory (NRL). (In practice, that meant installing my equipment on a mount that could only be reached by a 40-ft scaffold, and servicing it with cryogenic liquids. Small wonder then, that I later became so attracted to the cozier geometry of the Bell Labs' horn-reflector antennas.)

In order to stabilize my system against gain fluctuations, I employed a scheme in which the maser input was switched between the antenna's feed horn and an attenuator immersed in the same helium bath that cooled my maser preamplifier. When finally completed and installed on the radio telescope, the system performed perfectly, yielding scans across the sky with unprecedented sensitivity, limited only by the thermal noise expected from the system. To my dismay, however, my data showed little, if any, trace of the hydrogen the literature promised. I made further searches at longer integration times to improve sensitivity, but found nothing more than traces of continuum radiation from individual galaxies. By then, my time on the telescope had run out, leaving me with enough to qualify for my PhD, but far less data than I had hoped for.

With my degree, and hands-on knowledge of how best to apply cryogenic radiometry to microwave radio astronomy, I applied for a temporary job at Bell Radio Research Laboratory, an organization in which David C. Hogg was then a leading contributor, and began working there in the late spring of 1961. At that time, this group's satellite communications research infrastructure made it the best place to continue my project and bring it to a more satisfactory conclusion. In his essay, Dave Hogg describes that project in the broader context of the work which surrounded it, together with accounts of the first glimpses of the CMBR.

When I arrived at Bell Labs early in May of 1961, the 20-ft horn-reflector was still being used in the last stages of the Echo satellite project (Figure 4.9). In the interim, preparing for my planned project left me with time to complete the write-up of my thesis, and to initiate a search for line emission from interstellar OH radicals, using the same horn-reflector that Dave Hogg and his collaborators (De Grasse *et al.* 1959) had used in the pioneering 5-cm studies recounted in his section.

During this time, I also helped my engineering colleagues by applying radio astronomy techniques to solve a series of technical problems – starting with devising a way to calibrate the pointing accuracy of satellite receiving systems by tracking radio astronomy sources as they moved across the sky.

Fig. 4.9. The 20-ft "home-made" horn-reflector antenna in the foreground designed by A. B. Crawford (head of the Bell Labs department I joined in 1961) served as the Lab's receiving station for Echo, the world's first communications satellite project. Unlike this horn antenna, whose response pattern is tightly grouped about its main beam, the commercial 60-ft (diameter) parabolic antenna, shown in the background, picks up an appreciable amount of radiation from the ground, via spillover of the (downward-facing) feed horn located at its focus.

In the pointing project, I made use of the fact that Bell Labs experimental satellite receiving systems were designed to function as radiometers as well as receivers – so as to provide a convenient means of measuring each system's sensitivity (normally expressed in units of equivalent noise temperature), as well as a way of monitoring atmospheric attenuation. As a result of this work, most early commercial satellite receiving systems were also configured to operate in a radiometric mode. In that way, operators could use celestial radio sources as reference objects for antenna pointing as well as measuring overall sensitivity. This practical work allowed me to stay connected to the work going on around me, even though the majority of my time continued to be spent on radio astronomy.

In the meantime, the 13-cm Echo receiver was removed from the 20-ft antenna and replaced by a 7-cm receiver – the wavelength employed by Telstar, the follow-on satellite project to Echo – thereby delaying my access to that antenna until shortly after Telstar's successful launch in July of 1962. At that point, the Holmdel horn, and its new ultra low-noise 7-cm traveling wave maser, became available for radio astronomy – subject only to the concurrence of local management: Rudi Kompfner, the director of our Laboratory. All I had to do was give a seminar-like talk outlining the research topics that seemed most interesting. Reasons to use the 7-cm system before moving to 21 cm seemed almost self-evident. Two-wavelength measurements of astronomical objects (most notably our own Galaxy) with the same instrument would yield valuable spectral information. This stroke of good fortune came at just the right moment. A second radio astronomer, Robert Wilson, came from Caltech on a job interview and was hired. In addition to finishing our separate projects, we set to working together early in 1963.

At that time, Bob was also working with Dave Hogg, who had come up with a novel way of measuring the effective collecting area of the Andover antenna (AT&T's primary satellite ground station). The idea was to measure our 20-ft horn by means of a helicopter-borne source, use that calibration to measure the absolute flux of strong "radio stars," and then use the antenna temperature obtained with the Andover antenna with those sources, to determine the collecting area of that antenna (Hogg and Wilson 1965).

This addition to our program appears to have left an indelible mark on the folklore of cosmology. Once the relatively elaborate helicopter data had been collected, we were unable to modify the antenna in any way, until the related flux measurements of discrete radio astronomy sources (intended as intermediate flux standards) had been completed. As a result, while we were able to evict a pair of band-tailed pigeons from their preferred resting place in the throat of our antenna, removing all signs of their prior presence had to be deferred for several weeks after the start of our observations. Once we had measured the flux densities of Cassiopeia A (Cas A) and the other discrete radio sources whose absolute fluxes we wished to establish as calibration objects for future use, we cleaned the throat of bird droppings and found, as expected, no measurable increase in antenna efficiency, and only a minor diminution in antenna temperature.

In putting our radio astronomy receiving system together we were anxious to make sure that the quality of the components we added were worthy of the superb properties of the horn antenna and maser that we had been given. We began a series of radio astronomical observations, including the ones that I had proposed so as to make the best use of the careful calibration and extreme sensitivity of our system. Of these projects, the most technically challenging was a measurement of the radiation intensity from our Galaxy at high latitudes. In particular, we needed to resolve the uncertainty surrounding the seeming extraneous sources of system noise encountered by several of our Bell Labs colleagues, and described in Dave Hogg's section (beginning on page 70).

This multi-year endeavor, which resulted in our discovery of the CMBR, is described in detail in Bob Wilson's (1979) article on the subject. Briefly, we spent most of 1963 converting the horn to radio astronomy. A mechanically based coordinate converter which allowed us to move the antenna in right ascension and declination, the cold load, a carefully built switch, and back end electronics were the main items that we added.

Since we planned to depend on our "cold load" as a noise standard (Penzias 1965), I decided to first design the microwave device I wanted, and then worry about how I might cool it. Clearly, I would use an absorber immersed in liquid helium, and connected to its (room temperature) output flange by a waveguide. Instead of the plated stainless steel generally used in cryogenic microwave spectroscopy, I opted for a meter-long section made of the well-behaved high-copper brass alloy used in AT&T's microwave radio towers because of its low attenuation. In addition to thinning the walls of the waveguide by machining away material from its outside surfaces, to reduce its thermal conductivity, I added a series of gas baffles to allow evaporating liquid helium to cool the transmission line as this gas flowed upward toward the vacuum pump connector. Calibrated thermistor diodes, attached to each of the baffles as well as other key points along the waveguide, allowed us to monitor its temperature profile – thereby allowing us to calculate the noise temperature at its output flange to greater accuracy.

Owing to the large thermal mass and size of the Dewar flask which contained the cold load, each day's fill consumed the contents of a 25-l helium container. Since each such fill lasted through a full day and night of observation, we were almost always ready to quit working well before our helium ran out. Remarkably, our local carpenter shop – headed by Carl Clausen, a long-time employee who had built the antenna that Karl Jansky had used in the 1930s – managed to build the 20-ft horn for a mere $20,000. On our part, Bob and I almost certainly spent more than that amount on liquid helium during the years we used that antenna for our observations.

Those observations began in late May of 1964 – with us working to collect data, while also tracing possible sources of the excess antenna temperature which proved to be the CMBR. By then, it seemed unlikely that the excess temperature was due to measurement errors, since three independent measurements had yielded similar results. Was it then due to the receiver, the antenna, or something outside the maser systems themselves? Our first observation exonerated the receiver. Figure 4.10 contains readings from each of the cold load's 11 thermistors, together with a temperature reading from a thermometer attached to a variable attenuator which connected the cold load to one of the two input ports of our waveguide switch.

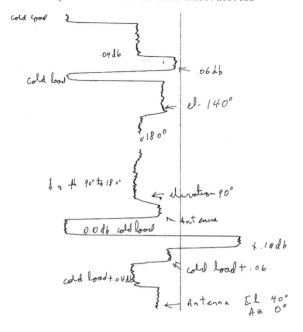

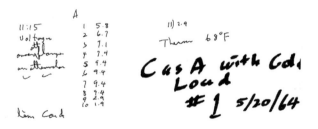

Fig. 4.10. Record of our first 7-cm observation. Data were recorded on rolls of paper by means of a Leeds and Northrop chart recorder, with the output voltage of a detector located at the output of our radiometer plotted as a function of time. Because of the inherent stability of our system, the phase-sensitive detection used in a Dicke radiometer was not used. The switch was turned manually between the antenna feed and the reference arm (denoted as "cold load," along with the setting on the variable attenuator described in the text).

The attenuator – a standard Western Electric component with its resistive absorber replaced by a much less lossy material – had a range of 0.12 dB, or about 10 K when used at room temperature.

As can be seen from the chart, the antenna temperature at 90° elevation was observed to be ∼3 K (∼0.04 dB) hotter than the noise temperature of our cold load. We knew from our prior calibration that our cold load had an output temperature of about 5 K with the attenuator set at zero – with its precise temperature determined by the physical temperature distribution of the connecting waveguide, and calculated from the thermistor readings. Since the atmosphere's contribution to the antenna temperature at zenith was about two degrees less than the physical temperature of the liquid helium bath in which the cold load's absorber was immersed – together with the fact that the antenna throat was expected to introduce roughly the same small amount of noise as that due to waveguide in the reference arm – we knew immediately that the excess noise temperature must be coming from outside our apparatus.

By the late fall of 1964, we had made all the absolute flux measurements we needed, and had exhausted an extensive list of possible terrestrial noise sources, as well as known astronomical sources. What to do? We wanted to publish our result, but were hesitant about writing a stand-alone paper. In those days, a considerable fraction of the radio astronomy literature was taken up with spurious results, and we didn't want to run the risk of having our first joint publication to be cited as totally wrong. We therefore decided to include our detection of excess temperature as a section in one of the other papers then in preparation – but fate intervened.

In December of 1964, Bernie Burke and I met at an American Astronomical Society meeting in Montreal, exchanging accounts of our work and promising to keep in touch. He called me a few weeks later (late February, as I remember it) to tell me of a talk he had heard about (from Ken Turner) – saying that there was "a guy from Princeton" with a theory predicting "ten degrees at X-band" (radio engineering jargon for the microwave band around 3-cm wavelength). Bernie's mimeographed copy of Jim Peebles' preprint arrived in my office a few days later. Sure enough, the abstract contained a prediction of 10-K radiation, confirming what Bernie had told me over the telephone. I was happy to find a theoretical explanation for our puzzling phenomenon, even though I wasn't sure that the general model described in the paper was necessarily the right one. I don't remember paying much attention to the details of the cosmological theory, other than that it mentioned a cyclical universe model, apparently proposed by Bob Dicke, who had organized an experimental search for this phenomenon at 3-cm wavelength.

I immediately picked up the phone, and was soon speaking with Bob Dicke – catching him in the middle of a meeting with Jim Peebles, Peter Roll, and Dave Wilkinson. Rather than saying that he would call me back after his meeting, as I thought he would, he and I began a conversation that lasted for some considerable time as I told him about our discovery, and the additional work that we had done in this connection in the months that had followed it. At the end of our conversation, I invited him to come and have a first-hand look at our apparatus and data, resulting in a visit by Dicke, accompanied by Peter Roll and Dave Wilkinson, to Crawford Hill shortly thereafter.

As soon as the group arrived, Bob Wilson and I brought them to the horn antenna where all five of us managed to squeeze into our control cab in order to give our visitors a first-hand look at our equipment. Bob Dicke looked over what we had, asked a few questions, nodded, and agreed that we had a real result. From there, we moved to a conference room in our main building, where I gave a presentation explaining the motivation behind this portion of our work in the context of our galactic continuum project. Apparently, I assumed more understanding of radio astronomy than the group possessed at that time, because Peter Roll remembers me talking about M31 (the nearby galaxy in Andromeda) and him thinking that we were interested in the sky background in order to aid in our measurements of that galaxy's emission. At that time, I understood that they knew more about their areas of expertise than we did, but it didn't occur to me that the inverse of my assumption (that they knew less about radio astronomy techniques than we did) could be true as well.

This latter situation became clearer when Bob and I paid a return visit to Peter and Dave's lab a short while later. Thanks to phase-sensitive detection, they had effectively eliminated the effects of random noise in their measurement. But they had done less well with systematic uncertainties – especially with their cryogenic noise standard. In particular, I remember the waveguide being covered with frost and condensed water where it emerged from a metal flange atop their liquid helium bath. To me, they seemed to be making many of the mistakes that I had made in my first encounter with such problems in my thesis experiment, and had solved in designing the "cold load" and related apparatus for our 7-cm system. On the other hand, they might just have underestimated the precision they would need, expecting a more intense level of background radiation than the level we had detected. Either way, I went through some of the ways in which they might improve their design details – an area that we hadn't touched upon during their visit to our facility.

Earlier on, toward the end of his visit to Crawford Hill, I remember discussing publication with Bob Dicke and suggesting a joint paper. For his part, Dicke refused immediately, leading me to then propose a pair of back-to-back papers in *The Astrophysical Journal* – the same place that our Cas A and galactic continuum papers were soon to go.

Our paper (Penzias and Wilson 1965a) consisted of a bare-bones account of our measurement – together with a list of the possible sources of interference we had eliminated – along the lines of what I would have included, had this result been a section of our paper on the 7-cm galactic continuum (Penzias and Wilson 1966). As a result, we submitted the write-up without a single mention of astronomy. We only added a sentence – stating this phenomenon could not be accounted for in terms of sources known to exist in the present universe – some days after we had sent the original version off to the journal. By the time our correction arrived however, the editor had already accepted the original version for publication. Not wishing to withdraw the paper, and replace it with a revised copy, we accepted the editor's offer of including that sentence as a "note in proof."

Notwithstanding the rapid acceptance of our paper, actual publication of the pair of CMBR papers was held up until July 1, with the issues themselves mailed out in the early fall. In the interim, another form of publication took over on May 21st, with a front-page *New York Times* article headlined: "Signals Imply a 'Big Bang' Universe." Walter Sullivan (1965), then the dean of American science writers, apparently had a "mole" in *The Astrophysical Journal* editorial office. At that time, Sullivan was hoping to get an early look at an expected submission by Allan Sandage, whom he thought was then about to report observations of particular cosmological significance.

The article reported our discovery and the prediction at Princeton, noting that: "It is clear that Dr. Dicke, and others would like to see an oscillating universe come out triumphant. The idea of a universe born 'from nothing' in a single explosion raises philosophical and well as scientific problems." At the time, however, the likelihood of resolving such cosmological issues seemed remote to me. My first reading of Jim Peebles' preprint had linked it to the cyclic model in my mind (and only later made the Gamow connection) even though Jim didn't have a strong connection to oscillation. Jim seemed to favor a cold early universe – more like the one I heard about from David Layzer later that same year. In those days, each of the principal cosmological theories seemed to be as much about personal preferences as it was about data, at least as far as those of us outside the field could tell. After all, it had taken until the mid-1950s for the Hubble age of the universe to catch up with the age of the oldest stars.

But then, the link between theory and data began to strengthen markedly once the *Times* article appeared. Most importantly, unexpected confirmation appeared from an unexpected direction, in the form of a trio of independent analyses by George Field and John Hitchcock (1966), Pat Thaddeus and Paul Clauser (1966), and Iosef Shklovsky (1966) – each inferring a 3-K temperature of the CMBR at millimeter wavelengths, and all making use of published optical spectra which indicated an otherwise puzzling excitation of interstellar radicals.

Ironically, George Field and I had discussed the optical CN data and its possible connection to radio astronomy, albeit in an entirely different context. In writing up my thesis, I had found myself faced with puzzling theoretical issues I couldn't figure out on my own, so I sought help from George, who was still at Princeton in those days. Some time later, I sought George's help again in connection with a search for line emission from interstellar OH radicals. In both cases, excitation of the emitting gas came up as an issue, and I recall discussing McKellar's CN observations with him, although our memories differ a bit. I recall George mentioning it during our OH discussion, while George remembers it taking place in connection with intergalactic hydrogen. Nonetheless, George made that connection for me with respect to my spectral studies, and later connected the CN excitation phenomenon to the CMBR. For my part, I didn't. While I adopted an estimate of 2 K as the lower limit of radiative excitation for OH radicals (Penzias 1964), I assumed that this "radiative excitation" was due to starlight, that is, confined to wavelength regions much shorter than the one associated with the 17-cm and 21-cm lines studied in my observations.

I realized my oversight a short while after *The New York Times* article appeared, when I visited Pat Thaddeus in his office. As Pat greeted me with "There's another way of measuring the . . . ," I glanced down and saw Herzberg's book on the table in front of him. The pieces of the puzzle were coming together faster than I could have imagined just a few weeks earlier.

As for another piece of the puzzle, the connection to the prior work done at Bell Labs, I remember being astonished to learn about Doroshkevich and Novikov's (1964) linking of Ohm's (1961) report of his 13-cm noise measurements to the CMBR implications stemming from the "Gamow Theory." At a time when being "plugged in" usually meant being on key colleagues' preprint lists, keeping up in astronomy generally depended on participating in the informal exchanges that marked life in academic departments – something that Bell Labs couldn't be expected to provide for its radio astronomers.

Fortunately, I soon found such a connection, when Lyman Spitzer invited me to give a colloquium sponsored by Princeton's astronomy department. From that time on, I became an increasingly active participant in the science and teaching of that department – a relationship that lasted well into the 1980s.

Other than the single pair of March 1965 visits already touched upon, Bob and I had little direct contact with the members of Dicke's group during the remainder of that year. In the meantime, Bob and I made the two additional 7-cm CMBR measurements described in his section, confirming our original result in both cases. By the time Peter Roll and Dave Wilkinson reported the results of the 3-cm measurements made with their reworked system, the following January (Roll and Wilkinson 1966), they had evidently solved the problems we had noticed in their earlier attempt, judging from the fact that their result produced "the right answer" – matching our 7-cm values, the earlier Bell Labs results at 5 cm and 11 cm, and the work done (on what was by then being called the "3-K radiation") at 2.3 mm from the CN results.

In the meantime, the connection with Gamow's earlier work, and the predictions that stemmed from it, gained increasing attention in the scientific community – in my case, via a personal letter from George Gamow himself. This letter (in Figure 4.11, misdated 1963, for some reason) begins by thanking me for sending him my "paper." Since this could only have referred to a

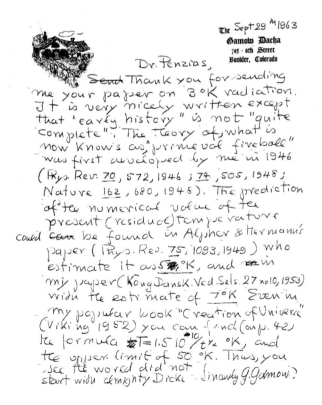

Fig. 4.11. Gamow's letter.

preprint of our 1965 article in *The Astrophysical Journal*, which included a sentence connecting our findings to the accompanying article by Dicke *et al.* (1965), I assumed that someone else had sent him a mimeographed copy. In those pre-computer days, a dedicated organization in Bell Labs distributed copies of papers submitted for outside publication by means of the same system used for internal technical memos, a kind of paper-based "Google" that employees could search by looking through an index based on authors and topics, and then "downloading" content for delivery via our company's internal mail service. In addition to the copies that Bob and I sent to colleagues, some of our colleagues in the Physical Sciences Division likely sent copies to some of their friends as well.

As I noted earlier, the connection to the predictions cited in Gamow's letter had been made by others even before the events recounted above. As time went on, and the agreement between theory and data grew stronger, many of us began to wonder why the measurements involved hadn't been attempted earlier. In my case (Penzias 1979a), I went so far as to attribute Ralph Alpher's apparent endorsement of Gamow's position (that bolometric measurements of the relict radiation would be confused with other sources of radiant energy, as outlined in a 1948 letter to Alpher and Robert Herman) as a demonstration of his having overlooked the possibility of microwave measurements. As I learned from Bernie Burke only recently, however, Alpher had indeed queried at least one radio astronomy group about the possibility of making microwave measurements like those recounted in the present volume, but was told that it couldn't be done (p. 182).

With the results of present-day CMBR measurements judged significant enough to be taught even in some high schools, it may be hard for contemporary readers to imagine a circa 1950 radio astronomer turning down such an "opportunity." Nonetheless, a more careful look at the state of radio astronomy in those days makes such a turn-down far more understandable. First of all, there were no idle radio astronomers. The first few radio observatories were just being set up, and almost anything they did would break new ground – at least as long as the rudimentary equipment they used worked well enough to produce useful data. Given such circumstances, together with the amount of effort a CMBR measurement would have required, it is not hard to imagine someone being likely to consider such an undertaking outside the realm of possibility.

Thanks to experience and improved techniques however, CMBR measurements began to look almost easy just a few years after the initial report of our discovery. In 1967, for example, Dave Wilkinson and his coworkers reported a trio of highly consistent CMBR measurements done at three different wavelength regions (Wilkinson 1967; Stokes, Partridge and Wilkinson 1967). Given the speed and precision of this work, it is understandably easy to overlook the fact that the same group's first 3-cm measurement took the better part of two years from start to submission for publication. Moreover, that 3-cm project had far better resources than any that would have been available 15 or so years earlier – along with the additional advantage of experience gained from familiarity with a successfully completed project similar to theirs.

Small wonder then, that a potential CMBR experimenter would have balked at anyone proposing such an undertaking back in the 1950s – especially with no more incentive than what was then a tenuous link between an unproven theory and hypothesized data. Under such circumstances, it is not hard to imagine a radio astronomer of that year saying "it can't be done," nor is it hard to imagine the subsequent frustration felt by George Gamow and his colleagues as the events of 1965 began to unfold.

By early 1966, Bob and I had completed our observations with the 7-cm system, and installed a newly-built 21-cm system in its place. Our CMBR measurements at this new wavelength went smoothly, and we were able to report that result later the same year (Penzias and Wilson 1967). Here, for the first time, we found "company" in the form of a similar measurement made by Howell and Shakeshaft (1966), allowing us to compare the results of two independent measurements at the same wavelength. Since the raw data (the sum of the CMBR and galactic radiation) in the two measurements differed by only 0.2 K, the combined result yielded an accurate determination of our Galaxy's spectral behavior – one of the items on my earlier research agenda. While we continued our 21-cm Hi studies for another year or so, our CMBR studies had come to an end. In its place, we began a long-term effort aimed at following the CMBR's companion thread in cosmology – the origin of the elements – by studying the chemical and isotopic composition of interstellar space.

In this endeavor, Bob and I once again moved to a new wavelength range – this one centered on the atmospheric window which stretches from 75–150 GHz (4–2 mm). In contrast to the small handful of hyperfine lines available to microwave radio astronomers, the then still-unexplored millimeter-wave portion of the astronomical spectrum encompasses a rich variety of molecular rotation lines. Fortunately, several of the key components required for such work had been developed for communications research purposes. With much help from Charles Burrus, one of our Bell Labs colleagues, Bob and I assembled a millimeter-wave receiver. Completed in the spring of 1968, I carried it to a precision radio telescope owned and operated by the National Radio Astronomy Observatory at Kitt Peak, Arizona, for preliminary continuum observations. Until we introduced our receiver, millimeter-wave observations with that telescope had been limited to bolometric measurements. Following the success of our continuum work, and the subsequent installation of an National Radio Astronomical Observatory (NRAO)-built spectrometer "back end," we – together with a number of collaborators from other institutions – discovered and studied a number of interstellar molecular species, thereby revealing the rich and varied chemistry which exists in interstellar space.

Since that time, millimeter-wave spectral studies have proven to be a particularly fruitful area for radio astronomy, and are the subject of active and growing interest, involving a large number of scientists around the world. The most personally satisfying portion of this work for me was using molecular spectra to explore the isotopic composition of interstellar atoms – thereby tracing the nuclear processes that produced them. Most notably, our discovery of the first deuterated molecular species found in interstellar space (Wilson *et al.* 1973) enabled me to trace the distribution of deuterium in the galaxy. This work (Penzias 1979b) provided the first direct evidence for the cosmological origin of this unique isotope, which by then had earned the nickname "Arno's white whale" among my observing colleagues. Of all the nuclear species found in nature, deuterium is the only one whose origin stems exclusively from the explosive origin of the universe. Because deuterium's cosmic abundance serves as the single most sensitive parameter in the prediction of CMBR, these measurements provided strong support for the "Big Bang" interpretation of our earlier discovery.

4.5.2 Robert W. Wilson: Two astronomical discoveries

Bob Wilson shared the 1978 Nobel Prize in Physics with Arno Penzias for their discovery of the CMBR. Wilson is a Senior Scientist at the Harvard–Smithsonian Center for Astrophysics and Technical and Computing Leader of the Sub-Millimeter Array Project.

As a child I acquired an interest in electronics from my father. I also learned from him that I could take apart almost anything around the house, probably fix it, and then reassemble it successfully. In my high school years I fixed radios and later television sets for spending money and built my own hi-fi set. Thus when I enrolled at Rice University, I declared a major in electrical engineering. During my freshman year I switched to physics after realizing that much of the EE course work would be in power engineering. Having read my father's copies of *Review of Scientific Instruments* I realized the physicists had the interesting instruments (good toys). At Rice and later at Caltech I had two formal courses in electronics for physicists. My earlier interest had prepared me to enjoy and absorb this material thoroughly. My senior thesis at Rice was centered on designing and building a current regulator for a high-field magnet in the low-temperature physics group. These early experiences, especially the trouble-shooting skills I learned in my high school days, have served me well while fixing many problems with radio telescopes.

I entered graduate school in the physics department at Caltech in 1957 after receiving my B.A. in physics at Rice earlier that year. I had no clear idea of what I wanted for a thesis topic. During my first year I became friendly with David Dewhirst who was visiting from Cambridge University and was using the original Palomar Sky Survey plates in the basement of the astronomy building for identifying 3C radio sources (the Third Cambridge Catalog, Edge *et al.* 1959). After David learned of my interest in instrumentation as well as physics, he suggested that I consider working with the new radio astronomy group which John Bolton had formed. There was the added enticement that they wanted to make maser amplifiers for the telescopes. The original Owens Valley Radio Observatory 90-ft antennas were nearing completion and it was an ideal time to join such a group.

My thesis was intended to be interferometric observations with these antennas at the 21-cm hydrogen line. I built the local oscillator and other parts of the receiver system for those observations. That project stretched out and my actual thesis was based on an intervening project John Bolton had started me on – making and interpreting a map of the plane of the Milky Way at 960 MHz (Wilson and Bolton 1960). We used one of the two Owens Valley 90-ft antennas before interferometric observations started. I used load switching against a liquid nitrogen-cooled load and scanning or drifting from the west to the east across the Milky Way. I covered up to about 20° either side of the plane of the Milky Way – enough that the radiation was falling off very slowly at the edges of my map. Having no better reference, I took the edges of my map to be zero. Since we are inside the Milky Way, it was clear to me that this technique only worked because the Milky Way is very thin compared to its diameter. I knew I did not have a true zero reference for my map. It is interesting in retrospect that I added 2.8 K to my observations to improve the comparison to a lower frequency survey in analyzing the radiation from the galactic plane into thermal and nonthermal components (Wilson 1963).

My only cosmology course at Caltech was taught by Fred Hoyle. While I had not had a course in general relativity, Hoyle's lectures did not require an understanding of the tensor math which general relativity is based on. Philosophically, I liked his steady state theory of the universe except for the fact that it relied on untestable new physics.

After a one-year postdoc at Caltech doing 21-cm line and polarization interferometry, I took a job at Bell Labs' Crawford Hill Lab. A major attraction there was the 20-ft horn-reflector antenna, and the promise that Arno Penzias and I could use it for radio astronomy. A second reason I was favorably inclined toward Bell Labs resulted from the help they had given the radio astronomy group. They had offered Caltech the opportunity to send someone to work in the group which had designed TWMs and make a pair for the observatory. TWMs were the lowest noise receivers at that time. I had hoped to be the person to go, but because I needed to finish my thesis, Venkataraman Radhakrishnan was chosen to go to Bell Labs. I worked closely with him to put the masers to use and developed a very positive opinion of the people and the working atmosphere at Bell Labs.

In the late 1950s, plans were made to start working on communication satellites at the Bell Labs' Crawford Hill site. John Pierce (1955) had had a long-time interest in communication satellites resulting from his science fiction writings. The first satellite tests were planned with NASA's Echo balloon. It was known that the return signal from Echo would be very weak because a sphere scatters the incoming radiation in all directions. While reading a paper by John Pierce describing the parameters required for a satellite system, Rudi Kompfner had the idea of using a traveling wave maser. Derek Scovil and his group at Murray Hill (De Grasse, Shulz-Du-Bois and Scovil 1959) had developed them for a high-sensitivity military radar. They worked at liquid helium temperatures and had a noise temperature of a few kelvin. Even after making a room-temperature connection to it, one could have a receiver with a noise temperature of 10 K or less.

It was natural to combine a TWM with a horn-reflector antenna. The horn-reflector was invented at Holmdel by Al Beck and Harald Friis for use in a microwave relay system. In addition to turning the corner between the waveguide going up a tower and the horizontal communication path, the horn-reflector has the distinct advantage that when two of them are put back-to-back on a tower and have a very weak signal coming in on one side, a strong regenerated signal can be transmitted from the other side without interference. Its front-to-back ratio is very high. The corollary of this is that a horn-reflector put on its back will not pick up much radiation from Earth and will be a very low-noise antenna. Therefore, Art Crawford

Fig. 4.12. The 20-ft horn-reflector with its parabolic reflector on the left and cab on the right. Since the cab does not tilt, almost any kind of receiver can be conveniently put at the focus of this antenna (apex of the horn). It is clear that the horn shields the receiver from the ground, especially when it is looking up.

built the large (20-ft aperture) horn-reflector pictured in Figure 4.12, to be used with a TWM to receive the weak signals from Echo (Crawford, Hogg and Hunt 1961).

Figure 4.13 shows a polar diagram of the gain of a smaller horn-reflector antenna compared with the gain of a theoretical isotropic (uniform response) antenna. If we put an isotropic antenna on a field with the 300-K ground down below and zero degree sky up above, we expect it to pick up 150 K; half of its response comes from the ground. The response of the horn-reflector is more than 35 dB (a factor of about 3000) less responsive to the ground than the isotropic antenna. So one would expect less than a tenth of a kelvin for the ground pickup from the horn-reflector.

In December of 1962 I went on a recruiting trip to Bell Labs. Of the groups I was interviewed by, I was most interested in the Radio Research Lab at Crawford Hill. I met Arno Penzias there and he showed me his OH experiment and the 20-ft horn-reflector. At that time, he had been there a year and a half. We had much more time to talk a week later at the winter American Astronomical Society meeting, where I gave a talk. He was clearly trying to get me to join him at Crawford Hill. Setting up and carrying out an observing program with the horn-reflector was certainly a job better done by two people than by one.

We were very different people and, as it turned out, had complementary skills. We made a good team for that job. Arno was as garrulous as I was

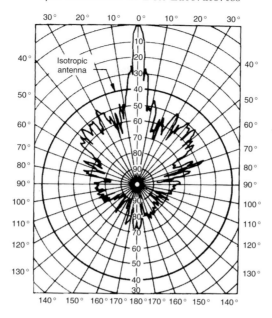

Fig. 4.13. Polar gain pattern of a horn-reflector microwave antenna. The radial units are dB and the gain is normalized to 0 dB at its peak.

reserved. He was interested in the big picture and tended to think of ways to most effectively use the resources at hand. I tended to be shy, persistent at getting all of the details correct, and liked to do things myself and with my own hands. As a graduate student Arno had built a maser amplifier and made observations with it. While I had had some experience with the maser from Bell Labs, I had worked much more on the back-end signal-processing electronics and antenna control at the Owens Valley Observatory. We were both intent on making accurate measurements.

Crawford Hill in 1963 offered a remarkable environment for us to work in. Several of the people there had been at the original Bell Labs building in Holmdel in the 1930s when Karl Jansky discovered extraterrestrial radio waves. They included our first department head, Arthur Crawford (unrelated to the family for which Crawford Hill was named). Crawford Hill, although part of the research arm of Bell Labs, was somewhat more mission-oriented than the rest of research. Long-haul communications was their primary focus. George Southworth had developed the waveguide at Holmdel in the 1930s. They had been a strong force in designing the first microwave relay system. Many of the Members of Technical Staff (MTS) had started there before the technology of microwaves was developed and had contributed to its development. There was a strong curiosity about new things and a feeling that new fields should be understood rather than exploited for the easy solutions which might be found. Bell Labs had "written the book" on many new fields and writing comprehensive books was still an ongoing endeavor. The Bell Labs merit review system rewarded good research and recognized the value of cooperative and interdisciplinary work, something which seemed to be missing at many universities. I could find experts on many subjects at Crawford Hill or other parts of Bell Labs who were happy to help.

The Crawford Hill building was built to house the original Holmdel group when the land they had occupied since the 1930s was taken over for the big lab at Holmdel. They moved in 1962. The front part of the building has a long hall with MTS offices on the front side and laboratories on the backside. Nearly every experimental MTS had a lab and often a technician to help him build things. The back part of the building had an extensive machine shop, a three-person carpenter shop, and a well-supplied stockroom. The machinists had a lot of experience building microwave components and were used to working from hand sketches rather than formal drawings. The head of the carpenter shop, Carl Clausen, had built Jansky's original antenna in the 1930s. When I arrived they were building a replica from the original drawings for the NRAO in their spare time. These resources were available to us with little evidence of limitations from accounting.

At that time, there was no computer at Crawford Hill. Mrs. Curtis Beatty, a mathematician who had come from Murray Hill, would either write and run Fortran programs for us or take care of the complexities of running our programs on the Holmdel or Murray Hill computers. She would often fix small errors by changing the assembly code to avoid the cost of running the Fortran compiler again.

It is reported that Karl Jansky, in common with many others of his era, had built a measuring set as his first job. There were "standard Holmdel measuring sets" in many of the labs whose design probably dated from the 1940s, but were logically derived from Friis' design which Jansky had used. They were very simple but effective and were capable of measuring with 0.01 dB accuracy over tens of decibels in the microwave bands which were used for communication. I was to use these extensively in building and measuring components for our receiver for the 20-ft horn-reflector.

One might ask why two young astronomers wanted to work with such a small antenna as the 20-ft horn-reflector with its collecting area of only $25\,\text{m}^2$. While other radio observatories all had much larger antennas, we knew it had very special properties. First, it is a small enough antenna that one could measure its gain very accurately. It was necessary to be only about a kilometer away to be in the far field for making an accurate gain measurement. And that, in fact, had already been started by David C. Hogg (Hogg and Wilson 1965).

The TWM amplifiers, which were available at several frequencies, would make this small antenna sensitive enough for work even with small diameter sources. For sources which were large enough to fill its beam, it would have been the most sensitive radio telescope in existence at the time. The other important thing is that we expected to be able to account for all of the sources of noise and make absolute brightness measurements. Radio astronomers don't often understand the background temperature when they do the usual on–off experiment (subtracting a measurement pointing away from the source from the measurement on the source), but the 20-ft horn-reflector offered the possibility of absolute temperature measurements. My interest in that possibility, of course, came directly from my thesis work at Caltech with John Bolton.

Soon after I went to Bell Laboratories, the 20-ft horn-reflector was released from the various satellite jobs it was doing. It had been designed for the Echo experiment which required operation at 13-cm wavelength, but it had later been used to receive a beacon from the Telstar(R) satellite. Thus when Arno and I inherited it, there was a 7.3-cm maser receiver on it (Tabor and Sibilia 1963). At that time it had a communications receiver with three low-noise amplifiers connected in series which a radio astronomer would find hard to believe. The maser was followed by a low-noise nitrogen-cooled parametric amplifier which was followed by a low-noise traveling wave tube amplifier. The gain stability was unbelievably bad. Our jobs were to turn all of this into a radio telescope by making a radiometer, finish up the gain measurement, and then proceed to do some astronomy projects.

We thought about what astronomy we ought to do and laid out a plan that would take a few years. The first project was an absolute flux measurement of Cas A, the brightest discrete source at that wavelength, as well as several other bright sources. We were planning our radiometer so that we could know its sensitivity to 1 or 2% accuracy based on physical temperatures we could measure.

Shortly after arriving I had joined Dave Hogg to make an accurate gain measurement of the 20-ft horn-reflector. Putting these together would let us measure the standard astronomical calibration sources more accurately than had been done before. This would be a service to both radio astronomers and the Bell System (and anyone else buying satellite Earth stations). The sensitivity of an Earth station could be accurately and easily checked by measuring its signal-to-noise on one of our calibrated radio sources.

I planned to follow up on my thesis by taking a few selected cuts across the Milky Way Galaxy and then confirm the spectrum of some of the sources that I had looked at. Next we wanted to check our ability to measure absolute temperatures so we could look for a spherical or halo component of the radiation from the Galaxy. Extrapolating from a lower frequency, we did not expect to see any galactic halo at 7-cm wavelength. We wanted to prove that when we did try to make such a measurement, we got a null result. After doing these projects, our plan was to build a 21-cm receiver similar to our 7-cm receiver. We already had the maser in hand. We would then make the halo measurement and do a number of 21-cm line projects including reworking Arno's thesis of looking for hydrogen in clusters of galaxies.

At one point during that time John Bolton came for a visit so we laid out this plan of attack and asked his opinion. He said that the most important thing to do in that list is the 21-cm background measurement. He thought that it was an unexplored area and something that we really ought to do.

By the time I joined Bell Laboratories, Arno had started making a liquid helium-cooled noise source (cold load) (Penzias 1965). Figure 4.14 is a drawing of it with an odd perspective. There is a piece of standard Bell System 90% copper 4-GHz waveguide, which runs from the room-temperature output flange down inside the 6-inch diameter Dewar to the absorber in liquid helium. About halfway down, the waveguide is thinned to reduce its heat conductivity, and finally there is a carefully designed absorber in the bottom. There is a sheet of Mylar in the angled flange near the bottom which keeps the liquid helium out of the upper part of the waveguide and makes a smooth transition from gas to liquid. Some holes in the bottom section allow the liquid helium to surround the absorber and there was no question of the physical temperature of the absorber itself. The heat flow down the waveguide which otherwise would have boiled the liquid helium rapidly has been

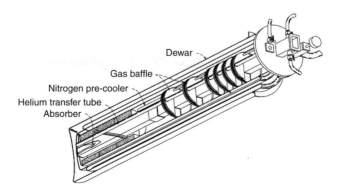

Fig. 4.14. The cold load for our radiometer.

taken care of by the baffles. They exchange heat between the cold helium gas leaving the Dewar and the waveguide. We realized that we had to know the radiation from the walls of the waveguide, so there is a series of diode thermometers on the waveguide for measuring its physical temperature distribution. We calculated the radiation of the walls using these temperatures and the measured loss in the waveguide.

When we first transferred the contents of a 25-l Dewar of liquid helium into the cold load, it would fill up to a high level. We calculated the radiation temperature at the top to be ∼5 K – just eight-tenths of a kelvin above the temperature of the liquid helium. After 15 hours or so (we usually ran down before the helium did), the liquid helium level would be down near the absorber and we would calculate the flange temperature to be about 6 K. Comparing it to the horn-reflector, the change agreed within something like a tenth of a degree over that period, so we felt we had a reasonably good calibration of what was going on in our cold load.

While Arno was doing that, I set up the radiometer shown in Figure 4.15 (Penzias and Wilson 1965b). As with most of our astronomical equipment

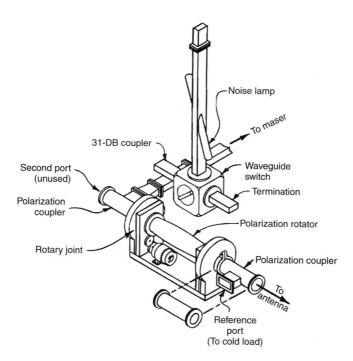

Fig. 4.15. The switch and secondary noise standard of the radiometer used for our measurements of the flux density from the radio source Cas A and the CMBR. The noise injected by the noise tube and its coupler was calibrated in three ways against thermal sources.

at Bell Laboratories, this is somewhat unusual. The 20-ft horn-reflector was fitted with an electroformed throat section which made a smooth transition from the square tapering horn to the circular waveguide which had been used in the Echo receiver. After a waveguide rotary joint, a second electroformed waveguide made the transition to circular 4-GHz waveguide. We decided to use this in a switching scheme which Doug Ring and others at Crawford Hill had used in the past. It takes advantage of the fact that two orthogonal polarizations will pass through circular waveguide. The polarization coupler near the antenna couples the signal from the reference noise source into the horizontal polarization mode traveling toward the maser and allows vertical polarization from the antenna to go straight through. The polarization rotator is the equivalent of a half-wave plate. It is a squeezed piece of waveguide with two rotary joints; another polarization coupler at the back picks one polarization off and sends it over to the maser.

By rotating the squeezed waveguide, we could switch between the reference noise and the antenna. An important aspect of this radiometer design is that except for the unused port, all ports of the waveguide were terminated at approximately the same low radiation temperature. Thus small reflections would not have a large effect. We adjusted all parts of the system to be well matched, however, and the unused port could be opened to room temperature with no effect on measurements. In addition, I added a motor to turn the squeezed waveguide to switch between the antenna and the reference noise source at 10 Hz. This, combined with a phase-sensitive detector I constructed, formed a "Dicke Switch" which was useful when measuring weak signals. After stabilizing the room temperature and all of the components of our system, the stability was so good that we usually just rotated the squeezed section by hand and recorded the receiver levels on a pen recorder.

Figure 4.16 shows a picture of the actual installation. The rotary joint that allowed the horn-reflector to turn while the receiver stayed stationary in the cab is at the right edge of the picture and the polarization rotator is on the left. An adjustable 0.11-dB attenuator seen at the bottom of the picture connects the cold load, which is below the picture, to the reference port of the switch. It could add well-calibrated additional increments of noise. The top of the maser is seen above the polarization coupler for the reference port and its massive magnet is hidden from view. The relatively large, strong cabin of the 20-ft horn-reflector, which does not tilt with respect to gravity, allowed us the freedom to build our receiver almost as though it were in a lab room and be with it during observations. The ease of working in the cabin undoubtedly contributed to our success. As graduate students

Fig. 4.16. Our radiometer installed in the cabin of the 20-ft horn-reflector.

Arno and I had both attached masers cooled with liquid helium to conventional antennas in which the focus tilted with elevation angle. We very much appreciated this arrangement.

Before we started making measurements with this system, there had been careful measurements of horn-reflectors with TWMs at Bell Laboratories. First, before going to the trouble of building a 20-ft horn-reflector, Dave Hogg had been asked to calculate the "sky noise" in the microwave band (Hogg 1959). To confirm his calculations the antenna and maser groups had put together a test system (De Grasse *et al.* 1959). They had a 6-GHz maser and a small horn-reflector antenna. They hooked the two up with a calibrating noise lamp and saw that indeed they got a system temperature of 18 K, which was very nice, but they had expected to do a little better. You see in Figure 4.17 that contrary to the expected value of less than 0.1 K for ground pickup from the antenna, they have assigned 2 K to it. They assigned 2.5 K for atmosphere and 10.5 K for the temperature of the maser. The makers of the maser were not very happy with that number. They thought they had made a better maser than that. However, within the accuracy of what the whole group knew about all the components, they solved the problem of making the noise from the components add up to the measured system temperature by assigning additional noise to the antenna and maser. Arno had used this horn-reflector for his OH project and was aware of the extra 2 K that had been assigned to it. One of the reasons that he built the cold load was to improve on their experiment.

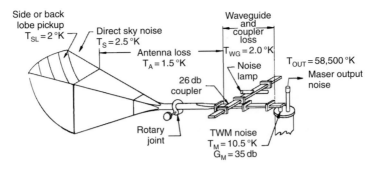

Fig. 4.17. The assignments of contributions to the system noise temperature in the De Grasse *et al.* (1959) radiometer.

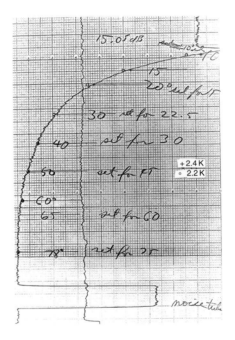

Fig. 4.18. A measurement of the radiation from the atmosphere by a tipping experiment with the 20-ft horn-reflector. The small circles and crosses are theoretical fits of the atmosphere to the measurements and show excellent agreement with the measurements.

This group had measured the atmospheric radiation (sky noise) by the same technique that Dicke had first reported on in 1946 (Dicke *et al.* 1946). Figure 4.18 shows a chart of such a measurement Arno and I made with the 20-ft horn-reflector. It shows the radiometer output as the antenna is scanned from the zenith (90° elevation angle) down to 10°. This is a chart with power increasing to the right, and shows what the power out of the receiver did. The circles correspond to the expected change if the zenith sky brightness is 2.2 K and the crosses to 2.4 K. You can see that the curve is a very good fit to the expected values down to at least 10° elevation. A well-shielded antenna makes an accurate measurement of the atmospheric radiation very easy.

After the 20-ft horn-reflector was built and was being used with the Echo satellite, Ed Ohm, who was a very careful experimenter, added up the noise contribution of all the components of the system and compared it to his measured total. In Figure 4.19 we see that from the sum of the components he predicted a total system temperature of 18.9 K, but he found that he consistently measured 22.2, or 3.3 K more than what he had expected (Ohm 1961). However, that was within the measurement errors of his summation, so he did not take it to be significant.

Our first observations with our new system were somewhat of a disappointment because we had naturally hoped that the discrepancies I have mentioned were just errors in the experiments. Figure 4.20 is the first measurement with our receiver. At the bottom and top, the receiver is switched to the antenna and in between to the cold load. The level from the antenna at 90° elevation matched that from the cold load with 0.04 dB of attenuation (about 7.5 K total radiation temperature). At the bottom I recorded measurements of the temperature-sensing diodes on the cold load.

TABLE II – SOURCES OF SYSTEM TEMPERATURE

Source	Temperature
Sky (at zenith)	$2.30 \pm 0.20°$K
Horn antenna	$2.00 \pm 1.00°$K
Waveguide (counter-clockwise channel)	$7.00 \pm 0.65°$K
Maser assembly	$7.00 \pm 1.00°$K
Converter	$0.60 \pm 0.15°$K
Predicted total system temperature	$18.90 \pm 3.00°$K

the temperature was found to vary a few degrees from day to day, but the lowest temperature was consistently $22.2 \pm 2.2°$K. By realistically assuming that all sources were then contributing their fair share (as is also tacitly assumed in Table II) it is possible to improve the over-all accuracy. The actual system temperature must be in the overlap region of the measured results and the total results of Table II, namely between 20 and 21.9°K. The most likely minimum system temperature was therefore

$$T_{system} = 21 \pm 1°\text{K.*}$$

The inference from this result is that the "+" temperature possibilities of Table II must predominate.

Fig. 4.19. Ed Ohm's (1961) tally of instrumental noise. Reprinted with permission of Lucent Technologies/Bell Labs.

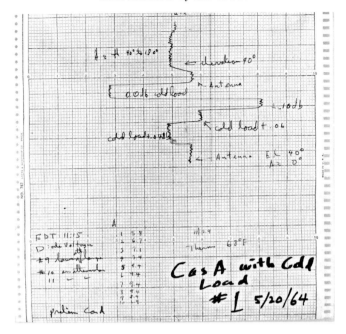

Fig. 4.20. Record of our first 7-cm observation.

That was a direct confrontation. We expected 2.3 K from the sky and 1 K from the absorption in the walls of the antenna, and we saw something that was obviously considerably more than that. It was really a qualitative difference rather than just quantitative because the antenna was hotter than the helium reference and it should have been colder. But we knew that the problem was either in the antenna or beyond. Arno's initial reaction was "Well, I made a pretty good cold load!" The most likely problem in such an experiment is that you do not understand all the sources of extra radiation in your reference noise source, but it is not possible to make it have a lower temperature than the liquid helium.

It initially looked like we could not do the galactic halo experiment, but at that time our measurements of the gain of the antenna had started (Hogg and Wilson 1965) and we wanted to go on with the absolute flux measurements before taking anything apart or trying to change anything. We ended up waiting for almost nine months before doing anything about our antenna temperature problem; however, we were thinking about it all that time.

We thought of several possible explanations of the excess antenna temperature. Many radio astronomers at the time thought the centimeter-wave atmospheric radiation was about twice what we were saying. That would have gone a long way toward explaining our problem. However, the curve for the zenith angle dependence in Figure 4.18 indicates that we were measuring the atmospheric absorption and emission correctly. It turned out later that the centimeter astronomers had applied refraction corrections to their measurements of radio sources in the wrong sense. John Shakeshaft finally straightened this out (Howell and Shakeshaft 1967a).

Since Crawford Hill overlooks New York City, perhaps man-made interference was causing trouble. Therefore, we turned our antenna down and scanned around the horizon. We found a little bit of superthermal radiation; but, given the horn-reflector's rejection of back radiation, nothing that would explain the sort of thing that we were seeing.

Could it be the Milky Way? Not according to extrapolations from low frequencies. The galactic poles should have a very small brightness at 7-cm and our actual measurements of the plane of the Milky Way did fit very well with the extrapolations.

Perhaps it was a large number of background discrete sources. The strongest discrete source we could see was Cas A and it had an antenna temperature of 7 K. Point sources extrapolate in frequency in about the same way as the radiation from the Galaxy, so they seemed a very unlikely explanation.

That left radiation from the walls of the antenna itself. We calculated nine-tenths of a kelvin for that. We took into account the actual construction of the transition between the tapering horn and the circular waveguide of the radiometer, which is the most important part. It was made of electroformed copper and we measured waveguides of the same material in the lab to determine the loss under real rather than theoretical conditions.

We had to wait some time to finish the Cas A flux measurement, but in the spring of 1965, almost a year later, we had completed it (Penzias and Wilson 1965b). The Earth had almost made a complete cycle around the Sun and nothing had changed in what we were measuring. We pointed to many different parts of the sky, and unless we had a known source or the plane of the Galaxy in our beam, we had never seen anything other than the usual antenna temperature. In 1962 there had been a high-altitude nuclear explosion over the Pacific which had greatly increased the amount of plasma in the van Allen belts around the Earth. We were initially worried that something strange was going on there, but after a year, the population of the van Allen belts had gone down considerably and we had not seen any change.

There was a pair of pigeons living in the antenna at the time, and they had deposited their white droppings in the part of the horn where they roosted. So we cleaned up the antenna, caught the pigeons in a havahart trap, and put some aluminum tape over the joints between the separate pieces of aluminum that made it up. All of this made only a minor improvement.

We were really scratching our heads about what to do until one day Arno happened to be talking to Bernie Burke about other matters. After they had finished talking about what Arno had phoned him for, Bernie asked Arno about our Galactic halo experiment. Arno told him about our dilemma of excess noise from the horn-reflector, that the Galactic halo experiment would not work, and that we could not understand what was going on. Bernie had heard from his friend Ken Turner about Jim Peebles' recent colloquium at Johns Hopkins where he described calculations of microwave radiation from a hot big bang. Bernie suggested that we get in touch with Dicke's group at Princeton. So of course Arno called Dicke. Dicke was thinking about oscillating big bangs which he concluded should be hot. After a discussion on the phone, they sent us a preprint and agreed to come for a visit. When they came and saw our equipment they agreed that what we had done was probably correct. Afterward our two groups wrote separate letters to *The Astrophysical Journal* (Dicke *et al.* 1965; Penzias and Wilson 1965a).

We made one last check before actually sending off our letter for publication. We took a signal generator, attached it to a small horn, and took it around the top of Crawford Hill to artificially increase the temperature of the ground and measure the back lobe level of the 20-ft horn; maybe there was something wrong with it. But the result was as low as we expected. So we sent the letter in!

Arno and I were very happy to have any sort of an answer to our dilemma. Any reasonable explanation would probably have made us happy. In fact, I do not think either of us took the cosmology very seriously at first. I had come from Caltech and had been there during many of Fred Hoyle's visits. Philosophically, I liked the steady state cosmology. So I thought that we should report our result as a simple measurement; after all the measurement might be correct even if the cosmology turned out not to be!

The submission date on our paper was May 13, 1965, and a few days later on May 21 my father was visiting us as part of a business trip. As was typical for him he woke up before I did and went for a walk. He came back with a copy of *The New York Times* which had a picture of the 20-ft horn-reflector and an article by Walter Sullivan entitled "Signals Imply a 'Big Bang' Universe" on the front page (Sullivan 1965). Besides being a very satisfying experience, this awakened me to the fact that the world was taking the cosmology seriously.

At the time of our paper, the spectrum of the CMBR was only determined by our measurement and an upper limit at 404 MHz which was dominated by galactic radiation. This was only enough to rule out ordinary radio sources. Soon after our result became known, George Field (Field, Herbig and Hitchcock 1966), Pat Thaddeus (Thaddeus and Clauser 1966) and Iosif Shklovsky (Shklovsky 1966) independently realized that the absorption of stellar optical radiation by CN in interstellar clouds, which had been known since 1940, could be used to measure the radiation temperature in those clouds. The measurements of those three groups indicated about 3 K for the radiation temperature at 2.6-mm wavelength. The Princeton group (Roll and Wilkinson 1966) completed their first measurement at 3.2 cm by the end of the year. Arno and I repeated our 7.35-cm measurement with a smaller horn-reflector with consistent results. We then installed a 21-cm receiver on the 20-ft horn-reflector and made a measurement (Penzias and Wilson 1967), which was consistent with Howell and Shakeshaft's 21-cm measurement (Howell and Shakeshaft 1966) made about the same time. In approximately a year there were seven measurements consistent with a 3-K CMBR, but it would be more than a decade before the spectrum was proven to be a blackbody spectrum rather than graybody, and thus definitively from the early universe. The details of these early measurements are covered by other essays in this chapter.

Looking back, it is a bit surprising how quickly our results were accepted among the astronomers I talked to. It probably helped that the steady state theory was failing to fit observations and Bell Labs had a reputation for doing good science. There were only a couple of occasions where I was challenged about the correctness of our measurements. More often, paradigm changes of that magnitude are resisted much more by established scientists.

It is interesting to compare the equipment Arno and I used to that which Roll and Wilkinson designed for the purpose of detecting the CMBR. Theirs had a large amount of symmetry between the path to the sky and that to the helium reference source, just the sort of thing a physicist would design. Ours required very careful measurement of the loss in the separate paths for making the comparison, but the high-sensitivity receiver and high-gain antenna had advantages in measuring the radiation from the Earth's atmosphere and in looking for and rejecting interference and foreground radio sources. We could make a measurement with a tenth of a kelvin accuracy in a second whereas they had to integrate a long time for that accuracy.

The ability to make meaningful tests in a short time can be invaluable when working with equipment which is not doing what you expect. In short, I think that our equipment inherited from other Bell Labs projects was ideal for finding something unexpected, but similar to what we were looking for, and theirs was more suited to a high-accuracy measurement. With hindsight, we should have explored the degree and larger scale isotropy of the CMBR more carefully before moving to 21 cm. Analyzing the records made for flux measurements of a number of sources on one day we were able to put a limit of 0.1 K on the large-scale anisotropy (Wilson and Penzias 1967). We could have made a measurement on a one degree to tens of degrees in angular scale, which would have been the most accurate for several years.

In 1966 Roy Tillotson, who had succeeded Rudi Kompfner as the director of our laboratory (an organizational unit of several departments at Bell Labs), told us two things which I still remember. First, he told us to identify and preserve our first record which showed the CMBR. Second, he reminded us that we had agreed (probably as a result of Congress having created COMSAT and taken the international satellite business away from AT&T) to each spend half time on radio astronomy and do things for the Bell System in the other half. Therefore, since we had been doing astronomy almost full time for several years, we should make good on the second part of the bargain. Over the next several years Arno and I continued to do 21-cm measurement with the 20-ft horn-reflector, but we were also involved in projects more directly targeted to communications.

For the first such project, Arno and I set up a propagation measurement at 10.6-μm wavelength between Crawford Hill and the Holmdel building a couple of miles away using one of Kumar Patel's first high-power CO_2 lasers. It was hoped that one could communicate over short distances in the far infrared much more readily than in the optical and near infrared. Dave Hogg had shown that those wavelengths were highly attenuated over the same path in foggy weather. Alas, 10 μm was much better, but not nearly good enough to be practical. It was, however, fun to convert the parts we got into a reliably operating laser.

I set up a small radio telescope to automatically track the Sun every day and measure its brightness as a way to explore the possibility of using bands at 1-cm and 2-cm wavelengths for domestic satellite communications. I showed that those bands were useful except during very heavy rains. I also found that if one were willing to have two Earth stations 5 or 10 miles apart one could work around the heaviest rain cells. I did this using fixed pointed radiometers which measured the radiation of Earth's atmosphere from which I calculated the attenuation. A somewhat longer wavelength band is currently used for direct broadcast satellite TV.

I was having considerable success and fun with the millimeter-wave propagation experiments and was drifting toward working more of the time on them, but we also continued our 21-cm work, especially with Pierre Encrenaz, a Princeton graduate student at that time.

Then in 1968 Arno suggested using a millimeter-wave receiver based on Schottky barrier diodes with NRAO's recently completed 36-ft antenna on Kitt Peak. Charlie Burrus, who was just down the hall from us, had developed the diodes and mixer assembly for a millimeter-wave (pre–optical fiber days) broadband communications system. This initial experiment demonstrated the feasibility of this effort, but produced little in the way of new science. We left that 90-GHz receiver for NRAO to use in developing the antenna. Two years later Sandy Weinreb of NRAO offered to provide a spectrometer and frequency control equipment for the 36 ft. We returned with a higher frequency Burrus receiver. Arno talked Keith Jefferts (a Bell Labs atomic physicist interested in millimeter-wave spectroscopy) and me into integrating the Bell Labs receiver into an NRAO receiver box that would fit at the focus of the 36-ft antenna. We would then go back to look for carbon monoxide in interstellar space. At one point in this process, Keith remarked that Arno had the two best technicians at Bell Labs wiring the receiver for him.

The payback came when Keith and I joined Sandy at Kitt Peak to get it all working. After several frustrating days, Sandy had to leave, but the next day we got it all tenuously working and put it on the antenna. I asked the telescope operator to point to the BN/KL region of the Orion Nebula where two nebulae which are bright, one in the near infrared and the other in the far infrared, would be in our beam. I was watching the rather crude output of the spectrometer when some of the center channels increased from their somewhat random previous outputs. The operator confirmed that we had just reached the source. I asked him to go off the source and the channels went back down. Thus in a few seconds, using a system which was hundreds of times less sensitive than the one on the 20-ft horn-reflector, we discovered carbon monoxide in an interstellar cloud. I had picked the BN/KL source because it was the source in our list of candidates which was overhead at the time, but it turned out that it is the strongest CO source in the sky. Arno arrived the next day to find that the key discovery had been made (Wilson, Jefferts and Penzias 1970).

The carbon monoxide and other simple molecules that we and others have found since can be thought of as stains which allow us to measure the structure and dynamics of the interstellar molecular clouds. The clouds are so cold that their main constituent, hydrogen, doesn't radiate. The radiation from simple molecules has shown that these dense molecular clouds exist, star formation is active in them and they are common in galaxies. Since that time, a large number of astronomers have worked on understanding the physical and chemical conditions in these clouds and the formation of stars within them. For several years after the discovery, Bell Labs gave Burrus diodes to other observatories and taught other groups how to make them.

This discovery changed the direction of my career. We spent five exhilarating years exploring interstellar clouds and discovering new molecules and their isotopic variants with our receivers and the 36-ft antenna at Kitt Peak.

I then became project director for the 7-m antenna. It was designed to do millimeter-wave astronomy when the weather was good and satellite propagation measurements at 1-cm and 2-cm wavelengths in weather bad enough to affect that band. We then had almost two decades of additional studies of molecular clouds and the cores around young stars which are embedded in them. The Crawford Hill astronomy group grew to include several additional people at its peak. Later the astronomy effort became less relevant to AT&T's need to prosper in the post-divestiture days and therefore declined. The Sub-Millimeter Array which I am working on now is an aperture synthesis array that spends most of its time observing radiation from the simple molecules and dust in these star-forming regions.

This work has taken me much closer to the origin of Earth and perhaps the organic molecules from which life originated, as opposed to the universe. On that larger scale, however, I have found the beautiful spectrum of the CMBR measured by COBE, and the evolving page full of accurate numbers derived from its fluctuations, immensely satisfying.

4.6 The Bell Laboratories–Princeton connection

4.6.1 Bernard F. Burke: Radio astronomy from first contacts to the CMBR

Bernie Burke is the William A. M. Burden Professor of Astrophysics, Emeritus, at the Massachusetts Institute of Technology.

Let me start out with some personal background. When I was a graduate student at MIT, 1950–1953, working in Woody Strandberg's microwave spectroscopy laboratory, I was exposed to radio astronomy through three routes. Woody had known Martin Ryle when he was posted to TRE Malvern (Telecommunications Research Establishment) during the war, as the Radlab representative. He worked with Martin on countermeasures – he said that the tension had been tremendous, and the radar people at TRE were "burnt out." He had heard about the use of Michelson interferometry by Martin, and of the Lloyd's mirror interferometer at CSIRO Radiophysics, and thought they were an excellent example of using cleverness instead of brute force to do radio astronomy. Woody had known Taffy Bowen; he also had known Hanbury Brown and Richard Twiss. The director of MIT's RLE, where I was working, invited Taffy to come to MIT to give three lectures on radio astronomy: one was about the Sun, one about the Moon, and the third about everything else. I was impressed, and was further impressed in the same year, 1951, when I heard Ed Purcell describe the discovery of the 21-cm hydrogen line at the joint Cambridge Monday-night physics colloquium. Through Woody's contacts, I also did some of my thesis work in Charlie Townes's lab at Columbia. It was a rival lab, but the atmosphere was wonderfully open; Charlie showed a new gadget that he was working on, called a "Maser," and explained how it worked.

As the end of my thesis work approached, I had to find a position, and I interviewed for a job with Harald Friis at Holmdel, the Bell Laboratories field station. He emphasized that they were in the telephone business. Radio astronomy was never mentioned. I met and got to know Bob Dicke, a good friend of Woody's, and knew about his K-band radiometer measurements on the roof of Building 20. The famous picture of Bob holding the "shaggy dog" in front of the radiometer horn was well known, and his derivation of the atmospheric K-band absorption that degraded K-band radar was also well-known. I probably knew about his upper limit to the cosmic background (Dicke *et al.* 1946), but its future connection to radio astronomy did not make much of an impression at the time.

I tried to obtain a Fulbright fellowship with Martin Ryle, but that did not work; I then found out that Merle Tuve, director of the Department of Terrestrial Magnetism (DTM) of the Carnegie Institution of Washington, was starting a radio astronomy effort. I had met Merle at an MIT summer study on undersea warfare (the Hartwell project) in 1950, so I contacted him at the DTM, received a postdoc offer, and joined the fledgling program in September 1953. Merle had imported Graham Smith from Ryle's group at the Cavendish, and my education in radio astronomy began. Our first big project was the 22-Hz Mills cross, and two years later Ken Franklin and I discovered Jupiter's radio bursts. This continued a long tradition of making a discovery in radio astronomy, but not the discovery that the radio telescope had been designed for. I got to know Grote Reber, a marvelous person to talk with, and gained an appreciation of his ability. Fred Haddock called him "not a scientist, but a scientific pioneer," which captures his maverick quality. Reber's (1958) personal account of his motivation and work is a masterly description of how pioneering science is done. There is a curious historical note that can be added. Edwin Hubble, the founder of modern observational cosmology, was taught in elementary school by Grote Reber's mother!

It should be remembered that the state of astronomy in the 1950s was quite different from today. There was an unresolved discrepancy between the Hubble age of the universe and the age of the Earth. There had been a few identifications of radio sources and the two brightest had just been identified: Cygnus A and Cas A, resolving the fierce controversy that had raged, led by Tommy Gold, who maintained that most were extragalactic, and opposed by Martin Ryle, who maintained that most were in our own Galaxy. This had been followed by the bitter controversy between Ryle, who maintained that the 2C source counts disproved the steady state universe, and Bondi, Gold, and Hoyle, who (quite correctly) maintained that the survey was so flawed that it did no such thing.

Now, back to the DTM. In the fall of 1953, Jesse Greenstein and Merle Tuve were at work, arranging a symposium to be held at the Carnegie headquarters in Washington, with the evident intent of instigating a resurgence of radio astronomy in the USA. Jesse was chairman of the National Science Foundation's (NSF) advisory committee on astronomy, the first such group that the NSF had convened, and he could be sure of close attention from that fledgling organization. Merle was well connected throughout the government, and the two, despite some philosophical differences, had considerable influence in official circles. They gathered together an outstanding group of participants, who assembled at the Carnegie headquarters on P Street under the aegis of Vannevar Bush in January 1954. The group included Lee DuBridge, president of Caltech, Leo Goldberg from the University of Michigan, Ed Purcell and Bart Bok from Harvard, Rudolph Minkowski and Walter Baade from Mt. Wilson, the optical astronomers who identified Cygnus A and Cas A, Bernard Mills from CSIRO and Graham Smith, both of whom had provided the accurate positions of radio sources that were needed for the identifications, John Hagen from the NRL, John Kraus from Ohio State, and Charlie Townes from Columbia, who was in the process of inventing the maser and the laser. There were the cosmologist Fred Hoyle, Bob Dicke and Lyman Spitzer from Princeton, Henk van de Hulst, Taffy Bowen, and many other prominent researchers. Lloyd Berkner, who would play a key role in establishing the NRAO, attended; he was president of AUI (Associated Universities Incorporated, a nonprofit corporation composed of representatives from nine northeastern research universities). John Firor and I, along with several of the young people from the Washington area, were also invited. For me, it was a grand introduction to the bright lights of physics and astronomy, and a broad-ranging tutorial in astronomy. The real purpose of the tutorials, however, was aimed at the NSF and Department of Defense officials who attended. Here was a new field of science, clearly related to various national interests, demanding attention from those who were funding science in the USA.

Although radio astronomy was being pursued at the NRL, Ohio State, Cornell, as well as at the DTM plus a fledgling group at Harvard, it was at the Washington Conference that American radio astronomy moved to join Britain and Australia as a major power in radio astronomy. Things moved fast. Caltech, Berkeley, Michigan, Illinois, and Stanford all began major projects. The NRAO was established, and while the 140-ft telescope project writhed in agony, the 300-ft telescope was started as a stopgap measure. The result was that less than two years later, observations began with the 300-ft transit telescope, which was built, as John Findlay put, for the price of sugar – 68 cents per pound. I was an early user, and made a map of the entire visible sky at 234 MHz, including an absolute brightness calibration. The cosmic component had no place in my thinking, for I was pursuing the question of the galactic radio halo, which was much more flattened than the Cavendish measurements implied. The results were published in the Carnegie Year Book, but the map itself was never published. Otherwise, the early 1960s were an eventful time. Otto Struve left the NRAO, an unfortunate case of capping an outstanding career with a conspicuous failure, and he died shortly afterward. Joe Pawsey, from CSIRO Radiophysics, agreed to take his place, but he was stricken by a fatal brain tumor and never took office. In this critical time for the NRAO, David Heeschen was named interim director; in fact, he had been the intellectual leader of the observatory from the beginning, even though he did not have the authority to influence major policy issues such as the finishing of the 140-ft telescope. Meanwhile, the search for a permanent director continued, unsuccessfully, and Heeschen was appointed director in 1962, making official what had been, in fact, the case since the NRAO was founded.

A new direction in radio astronomy was developing at the Bell Labs Holmdel station. The director was now Rudy Kompfner, a physicist with a broad range of vision. As in the case of Karl Jansky, the project started as a system to help plan for telecommunications. The Bell Labs knew that they had to look into satellite communications, and were performing scatter experiments on the Echo satellite, a simple aluminum-foil sphere. For sound engineering reasons, they wanted to develop the best possible low-noise antenna/receiver system, and calibrate it carefully. The frequency was 2390 MHz; the antenna was a shielded horn (it looked like a sugar-scoop), and the low-noise amplifier was a state-of-the art ruby maser. Their results were published by Ohm, the project leader, in the July 1961 *Bell System Technical Journal*, where they reported that separately the total noise of the system from all components was 18.90 ± 3.00 K (Ohm 1961). The total noise, measured on the sky, was 22.2 ± 2.2 K, and this meant that the microwave background of the sky was undetectable. It is said, however, that their initial measurements had smaller error bars, and an implied background temperature of 3.3 K was observed repeatedly, but the engineers talked each other into assigning larger error bars. Whatever the actual facts were, they missed the discovery, and their main fault, as engineers, was that there was an unknown source of noise that they did not pursue. My own contact with this work was almost nil – the case of the dog that did not bark in the night? I would say that I was aware of Ohm's work, but I had not seen the article in BSTJ, and I believe that the discrepancy went unnoticed by my colleagues.

For the young American radio astronomers of that time, it was a marvelous era. We all knew one another, and that included the graduate students. Arno Penzias was a student of Charlie Townes, who, to complete his PhD thesis, had taken his low-noise maser receiver to the NRL and installed it on their 50-ft dish. The entire Washington astronomy community was close-knit, helped by the quarterly community meetings that the Naval Observatory hosted. I knew the graduate students at Caltech, including Bob Wilson. I think that we took it as a good omen when the two decided to go to work at the Bell Labs field station at Holmdel.

Arno Penzias joined the Bell Labs in 1961, and was followed a year later by Bob Wilson. Here I have a personal story to tell. In late 1964 I shared an airplane journey with Arno (my recollection it was to Montreal), and I asked him what his plans were at Holmdel. He said that he was going to determine the absolute brightness of the sky at C-band. I said that he would certainly find it was so low that it was undetectable, based on the synchrotron spectrum, and he said yes, he knew that, but it had never been measured and the equipment at Holmdel was the best in the world for that purpose. Arno remembers that I mentioned the earlier upper limit set by Dicke at the Radlab. I don't remember that explicitly, but since I was aware of the measurement, it is entirely possible that I did.

Two years later, in 1965, a colleague at the DTM, Ken Turner (a PhD from Dicke's lab at Princeton), told me about an interesting colloquium that he had attended at the Johns Hopkins Applied Physics Lab. Jim Peebles, a theorist working in Dicke's group, said that there was good reason to suspect that if the "big bang" cosmology was correct, there should be a remnant microwave glow in the sky, the redshifted remnant of the time when the hot gas recombined at a redshift of about a thousand, and Dicke's group were in the process of measuring it. At the DTM, we had a lunch club, with the staff taking weekly turns as cook. It is my recollection that, on the very same day that Ken told me the news, the telephone rang during lunchtime (it may have been a day later, but the interval was very short). I was called to the telephone, and it was Arno, calling about some side issue, possibly about URSI (Union Radio-Scientifique Internationale) matters, and after we finished our business, I asked Arno "How is that crazy experiment of yours coming?" Arno replied "We have something we don't understand." I then said "You probably should call Bob Dicke at Princeton to discuss it." Arno called Princeton, talked to Bob while he was meeting with his group, and the rest is history. Penzias and Wilson received the Nobel Prize, quite deservedly, but it is a shame that it was not shared with Dicke. He shares the distinction of many friends who might have become Nobel laureates, were well deserving of the honor, but who were passed over.

Another footnote story can be told about how Bell Labs profited in a practical way from the CMBR discovery. Rudy Kompfner told me that the space relay system that Ohm's work had been designed for needed a reliable calibration system for the relay stations in the field. The engineers planned to launch a calibration satellite to do the job, when Arno and Bob pointed out that there were radio sources already in the sky that could calibrate the system, with no cost to Bell Labs! Shortly afterward, the engineers were contemplating an 8-mm telephone relay system, but again a calibrator was needed, especially to get a statistical record of atmospheric attenuation. Again, a calibration satellite was proposed, and again Arno and Bob pointed out that the Sun could serve as the calibrator, again at zero cost! The radio astronomy program saved Bell Labs several hundred million dollars in satellites that were not needed.

Many discoveries have precursors, and the discovery of the CMBR has some history of that sort. Joe Weber at the University of Maryland told me this example. He served in the USA Navy in the Pacific in World War II. After Joe resigned his commission in 1948, his expertise earned the offer of a professorship at the University of Maryland, provided he get a PhD. That led to an interview by George Gamow, who was in the midst of his calculations of a nuclear "big bang" at the beginning of the universe. Gamow asked him "Young man, what do you do?" Joe answered, "I'm a microwave physicist." Gamow replied "I'm sorry, but we don't have anything suitable for you at George Washington." It does not seem likely that Gamow had seriously considered how the relic radiation might be detected.

Gamow may have missed an opportunity, but his two younger associates, Ralph Alpher and Bob Herman, may not have. After Joe Weber told me his story about his interview with Gamow, he continued with a second story that illustrates how major discoveries have antecedents, might-have-beens, that for one reason or another did not happen. This was true for pulsars, and particularly for the Crab pulsar; both the radio and optical discoveries had failed precursors. Joe's story is that Alpher and Herman visited the NRL, almost certainly in 1948 or 1949, to see if detection of the microwave background was a possiblity. In radio astronomy, NRL was the only show in town at that time, and it is likely that they talked to the head of the radio astronomy group, John Hagen. He told them that the experiment was too hard, so they did not pursue the matter. I asked Ed McClain, one of their talented young engineers, if he had ever heard from Hagen about the experiment, and he said that he certainly had not. I remember Ed saying "That's odd, because John had a very good nose for new science." I doubt that Joe Weber's story is incorrect, because the Washington radio community was a close group, where everybody knew everybody, but it is, nevertheless, a second-hand story. Might the NRL group have succeeded? Hagen had a powerful team, including Ed McClain, Connie Mayer, Fred Haddock, and Russell Sloanaker, all of them talented microwave engineers and good physicists, familiar with the latest microwave technology. On the other hand, they had concentrated mostly on Solar radio astronomy, a strong source where sensitivity is less important, but receiver stability is vital. Nevertheless, they were clearly interested in fainter radio sources because, in 1950, they persuaded the Navy to fund a steerable 50-ft dish, which they placed on top of the central building at NRL (a terrible location). A simple calculation shows that they might have been able to do the experiment: their crystal mixers had a double-side-band noise temperature in the range 2000–3000 K. Typically, their IF bandwidth was 10 MHz, so a Dicke radiometer would have had an rms fluctuation of about 1 K for a one second integration. Connie Mayer, in particular, was meticulous in calibrating radiometers, and all of them knew about the hazards of atmospheric and stray radiation. They had access to liquid helium, so they could make cold loads. I conclude that they had a good chance of being successful if NRL gave them the resources. The conclusion, though, is clear: the experiment was not pursued, and it joins the long list of lost opportunities in science. Arno Penzias and Bob Wilson, on the other hand, were at the right place, at the right time, and their work is a model of how forefront science should be done.

A brief coda is in order. Kipling's ditty

As the dog returns to its vomit,
and the sow returns to its mire,
and the burnt fool's bandaged finger
goes wobbling back to the fire …

comes to mind, for after I returned to MIT in 1965, the lure of the CMBR pulled me in. I had a talented graduate student, Marty Ewing, and along with Dave Staelin we hatched a plan to measure the CMBR at a shorter wavelength. We chose 9 mm, because atmospheric transparency is good there, and we chose White Mountain, east of the Owens Valley, as the obvious site. Nello Pace of Berkeley had established a high-altitude physiology lab there, so there were electric power, living facilities, and road access. Common sense said that it had to be, at 12,400-ft altitude and east of the Sierras, an uncommonly good place to do the experiment. The Princeton group evidently thought so too, and so in the summer of 1967, side-by-side, we measured the CMBR. The leaders of the Princeton team were Dave Wilkinson and Bruce Partridge, and we became friendly competitors. They were carrying out their measurements at three wavelengths, a better experimental design, but at least we confirmed their results, using a different calibration technique. We used a Dicke radiometer, switching against a liquid helium cold load, and calibrated the overall system by using a helium-cooled "shaggy dog" (actually, a shaggy egg-crate) (Ewing, Burke and Staelin 1967). Dave Wilkinson's group calibrated by using the same reflector to look at the sky and to look down at an egg-crate in a bath of liquid helium (Stokes, Partridge and Wilkinson 1967).

There is a final twist to the story. In about 1970, there was a rocket experiment that tried to measure the CMBR temperature above the Planck maximum, and they found that the sky was much too hot. I doubted the result (which turned out to be caused, not by a hot universe but by hot rocket gases), and sought out Rainer Weiss, a colleague at MIT and a friend of many years. Rai is a great experimental physicist, at the time doing fancy things with lasers, and I think I communicated my enthusiasm. A balloon-borne radiometer was obviously the way to go. In addition to sending data by a radio link, Rai wanted an on-board recorder, and this I borrowed from my close colleague Al Barrett, who had been carrying out balloon radiometer observations of the Earth and its atmosphere for some time. Al was reluctant to lend his precious gear, and on the first flight the wrong squib was fired, and the experiment fell 100,000 ft to the Earth. Al's recorder was among the casualties. I told Al that Rai would buy him a better recorder, and the next balloon experiments worked: the CMBR still showed 3 K beyond the peak. Otherwise, I had little to do with the experiment (Lyman Page was the graduate student who helped Rai with the heavy lifting) but some years later, Rai paid me the ultimate compliment: "Bernie, you wrecked my lab."

A more extended history is given in my article, *Early Years of Radio Astronomy in the U.S.*, Burke (2005).

4.6.2 Kenneth C. Turner: Spreading the word – or how the news went from Princeton to Holmdel

Ken Turner has done research in radio astronomy at the Arecibo Observatory, Puerto Rico, and served as Program Officer for Extragalactic Astronomy and Cosmology at the USA National Science Foundation. His current interests include the study of psychology.

After finishing up my PhD at Princeton in 1962, I was awarded a Post Doctoral Fellowship at the Department of Terrestrial Magnetism of the Carnegie Institution of Washington. We were located in Northwest Washington, DC, and I was working with Bernard F. Burke learning radio astronomy, mostly related to the study of neutral hydrogen, and utilizing a 60-ft radio telescope at nearby Derwood, Maryland.

At Princeton I had been a member of the Dicke group investigating the experimental foundations of general relativity and any other cosmological or gravitational effect we could think up. Jim Peebles was also a member of the group and a good friend, so when I heard that he was going to give a talk at the Johns Hopkins Applied Physics Laboratory in Baltimore I made it a point to attend. Jim outlined the current activities of the group, which included an experiment to look for the red-shifted primordial radiation of the "big bang," which was expected to peak in the microwave region. Although this had been predicted by Gamow and Alpher some 20 or so years before, that prediction was "lost in the literature" and was unknown to the Dicke group at the time.

I was much taken by the idea of this experiment, and when I returned to the DTM I told Bernie Burke all about it. A short time afterward he was visiting Arno Penzias at Bell Laboratories, and, as I recall the story, Arno had told him that they were trying to make an absolute calibration of the big horn antenna there and were having trouble accounting for the last few degrees of noise temperature they had measured. At that point, Bernie told Arno and Bob Wilson, who was working with Arno on the experiment, about the background radiation that the Princeton group was tooling up to look for.

Arno and Bob immediately saw the implication of their "difficulty" and published their discovery of the radiation of the "primeval fireball," a phrase coined by John Wheeler to characterize the effect Peebles had predicted from his calculations of the conditions thought to prevail in the very early universe.

4.7 Developments at Princeton

4.7.1 P. James E. Peebles: How I learned physical cosmology

Jim Peebles has been at Princeton University since 1958 and is now Albert Einstein Professor of Science Emeritus.

I arrived in Princeton in 1958 from the University of Manitoba as a graduate student intending to study particle physics. At Princeton Bob Dicke somehow saw that I was much better suited to work on his new research interest, gravity physics.

Dicke had recently changed directions from research in quantum optics and precision measurements in atomic physics to the study of the physics of gravity. At the time we had an elegant theory, general relativity, but very limited tests. Dicke set out to improve the situation. By the time I arrived a considerable number of people were working with him, including undergraduate and graduate students, postdocs, and junior faculty. Bob Moore, who had been two years ahead of me at the University of Manitoba and was one of Dicke's graduate students, brought me to a meeting of Dicke's Gravity Group. I was fascinated by the variety of topics under discussion, and intimidated by how much everyone knew. Dicke, in particular, seemed to have a ready and well-informed assessment of every issue that arose. But he was drawing from a deeper well of understanding of the physics of the real world than anyone else I have encountered.

Bob Dicke encouraged me to join the group. I wrote a doctoral dissertation under his direction, on constraints on the time-variability of the strength of the electromagnetic interaction (Peebles 1962). Bob's motivation was his fascination with Mach's principle, which might be read to say that as the universe evolves so do the laws of physics. I was fascinated by all the evidence one could bring to bear, from the laboratory to geology and astronomy. My evident lack of interest in Mach didn't seem to bother Bob: I stayed on as his postdoc and evolved into a member of the faculty.

I learned about the general relativity theory solution for a homogeneous and isotropic expanding universe as part of preparation for the physics department graduate general examinations. I remember feeling a little surprised that people might consider this a serious model for the real world rather than one of the over-simplified problems you solve in exams, along with the acceleration of a frictionless elephant on an inclined plane. My textbooks on general relativity and cosmology, Landau and Lifshitz (1951) *Classical Theory of Fields* and Tolman (1934) *Relativity Thermodynamics and Cosmology*, present beautiful theoretical physics but little phenomenology. When we were graduate students Ken Turner introduced me to a book that has more phenomenology, Bondi's (1960a) *Cosmology*. I don't remember what I thought about this book at the time, apart from being shocked by the steady state cosmology: they just made this up. But I felt much the same about the relativistic big bang cosmology. Bob Dicke led me to see that cosmology then was a real physical science, with meaningful – if sparse – connections of theory to experimental physics and observational astronomy. By the end of the 1960s I had learned that there are many good things to say about the physical science of cosmology, including the steady state model, and I wrote a book about it, *Physical Cosmology* (Peebles 1971).

I don't remember much about the Gravity Group meeting at which Dicke explained why we might want to look for a sea of blackbody radiation that nearly uniformly fills space. But I think it was at this meeting that he gave an explanation that sticks in my mind for why the radiation would cool as the universe expands. He invited us to imagine placing a box with perfectly reflecting walls in the sea of radiation, with the same radiation temperature inside and out. The walls are expanding with the general expansion of the universe. They have no effect on the radiation (at wavelengths small compared to the box size) because for every photon that approaches the box from outside and is reflected there is on average an interior photon that bounces off the wall to replace it. I think I remember his concluding remark: we all know that radiation is cooled by the adiabatic expansion of the cavity. It was obvious to Bob that the spectrum remains thermal as the radiation cools. I don't remember whether he explained that. I convinced myself of it by a variant of the argument that is presented in the glossary under the CMBR energy spectrum.

Bob invited Peter Roll and David Wilkinson to build a Dicke radiometer to look for this radiation. His casual remark that I might look into the theoretical implications of the outcome of the experiment set the direction for my career. Great people can do things like that.

What was Bob thinking? I know he liked the idea of an expanding universe, and I remember his inviting us on more than one occasion to consider what the universe might have been doing before it was expanding. The answer he liked was that the universe was collapsing following a previous cycle of expansion. He instructed us on the production of entropy – largely in a sea of blackbody radiation – during the bounce, and on the role of the radiation in the thermal dissociation of the heavier elements produced in stars in the previous cycle. (The argument is reviewed in Section 3.3.) I believe his proposal to Roll and Wilkinson was meant to test this idea; I don't think the possibility of distinguishing between the big bang and the steady state cosmologies was a serious consideration. My scarcity of recollections of Bob's comments about the steady state philosophy, apart from his dislike of the passionate debates about it, leads me to suspect that this line of thought simply did not interest him. In those days the very limited fund of empirical evidence allowed us a lot more freedom in following our instincts in the search for clues to the nature of the universe.

I have some notes about what I was doing following Bob's invitation to think about the physical implications of the search for the fossil radiation. But I rarely put dates on the notes, so I can only say for sure that by the fall of 1964 I was making progress on two ideas. One was that thermonuclear reactions during the early rapid expansion of a hot universe, when the radiation temperature was $T \sim 10^9$ K, could produce appreciable amounts of helium and deuterium. The other was that when the temperature was greater than about 3000 K matter would have been thermally ionized and radiation drag on the plasma would have strongly affected the growth of the clustering of mass we observe now in galaxies and concentrations of galaxies. In 1965 I learned that much of the first idea had already been worked out. I think my first clue was Dicke's instruction to look up a paper by Hoyle and Tayler (1964). I don't know how Bob knew this paper. In that same year my second colloquium on what I was doing led to the connection between the Princeton search for a sea of microwave radiation and the problem of unexpected noise in a Bell Laboratories microwave radiometer.

I presented my first colloquium on this subject at Wesleyan University in Connecticut on December 2, 1964. Henry Hill, a former member of Dicke's Gravity Group, invited me. He wanted to explore the possibility of my moving to Wesleyan. I was impressed by the faculty, and particularly remember Thornton Page for his instructions about astronomy. But I don't remember any feedback about cosmology, and nothing came of the job idea.

In the colloquium I showed the two graphs in Figure 4.21. The curves in the panel on the left are examples of thermal spectra. The hotter one would have about the energy density of the Einstein–de Sitter cosmological model. (The mass in this model is such that the universe in effect is expanding just at escape velocity, with no cosmological constant.) The symbols show measurements or upper bounds on the cosmic radiation energy density across a broad range of wavelengths. It was known then that space is filled with a near uniform sea of X-ray to γ-ray radiation. The amount of energy in this form is much less than the equivalent of the observed mass in stars in galaxies. There was an upper bound on the cosmic mean energy density at optical wavelengths. Now we have a measurement of the accumulated amount

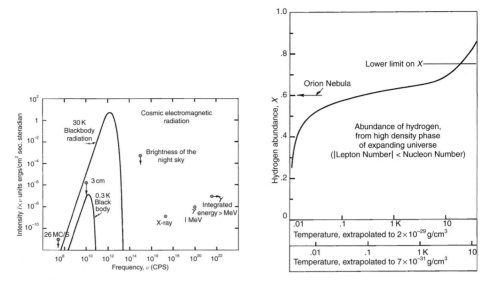

Fig. 4.21. Measures of the background radiation spectrum and calculations of the helium abundance in December 1964.

of starlight, an important advance that required a lot of work (Hauser and Dwek 2001). The point at the left edge of the graph shows a measured upper bound on the background radiation at about 1-m wavelength. The energy in the radio background contributed by the measured counts of radio sources was known; I added that to the version of the figure I showed in my second colloquium.

I think the upper limit at microwave wavelengths in Figure 4.21 refers to a Bell Laboratories paper we were discussing in the Gravity Group, Hogg and Semplak (1961). We interpreted it as giving an upper bound of about 15 K. If so, I made a mistake in the figure: the wavelength is 5 cm, not 3 cm. I don't know why we overlooked the better reference, Ohm (1961). And we had not yet noticed – and pointed out to Bob Dicke – that he had placed a bound $T < 20$ K at 1-cm wavelength (Dicke *et al.* 1946). It is at these microwave wavelengths that fossil thermal radiation might appear: not so hot as to have an unacceptably large energy density for the relativistic big bang cosmology nor so cool as to be unobservable.

I later learned that Doroshkevich and Novikov (1964) had made a similar study of the cosmic radiation energy density. Their version of the CMBR spectrum is shown in Figure 4.5. The focus of their analysis was the mean energy density as a function of wavelength from the accumulated amounts of starlight and radio radiation produced by the galaxies. But they remark that the "Gamow theory" would produce a thermal spectrum at microwave wavelengths, and they refer to the paper by Ohm (1961) on the Bell Laboratories communications experiments discussed in Chapter 3 and by Hogg, Penzias, and Wilson in this chapter. It is a better bound than the one I showed.

The right-hand graph in Figure 4.21 shows my computation of the mass fraction X left in hydrogen at the end of the big bang thermonuclear reactions discussed in Chapter 3. Almost all of the rest of the baryons, with mass fraction $Y = 1 - X$, would be in helium. I did not compute the production of heavier elements, but felt it would be small. The arrow on the left is an estimate of X in the interstellar plasma, and the line on the right is my guess of a reasonable lower bound on X in the earliest generations of stars. The horizontal axes show the present radiation temperature computed for two possible values of the present mean mass density (in baryons; I wasn't thinking about nonbaryonic dark matter). According to my notes for the Wesleyan talk I pointed out that an interesting value of Y in a universe with a hot big bang could be associated with microwave background radiation that would be warm enough to be detectable. (Doroshkevich and Novikov 1964 also very clearly made that point.) I don't remember whether I mentioned the Princeton experiment aimed at its possible detection.

We were of course thinking about how we might interpret the experiment if it were found that there is undetectably little microwave radiation. According to the assumptions used in the figure the low temperature would imply an unacceptably low value of X (a large helium abundance). But there are ways out; I mentioned two at Wesleyan (according to my notes, which were my security blanket in those days). The first is that the big bang may have been cold. That can be reconciled with large X by postulating that there are enough neutrinos to prevent formation of neutrons. This is what is meant by the comment about leptons in the figure. I noticed only later that Zel'dovich (1962, 1965) had independently argued for a cold big bang. His reasoning, discussed on page 35, led him to the proposal that the early universe contained equal numbers of baryons, electrons, and neutrinos, that is, the lepton number is equal to twice the baryon number. That is all the leptons needed to eliminate helium production in a cold big bang.

The second way out I mentioned is that the universe is not even approximately homogeneous and isotropic: maybe there was no big bang. The homogeneity assumption is well supported by the observations now, but the evidence was sparse then.

In the paper Dicke and Peebles (1965) (which was submitted before we knew about Penzias and Wilson, but it has a comment added in proof) we mention a third possibility, that general relativity theory is not valid.

It is after all an enormous extrapolation of the theory from the meager tests we had then to its application on the scales of length and time of the expanding universe. At the time, Bob Dicke was very interested in the idea discussed on page 37. Perhaps the strength of the gravitational interaction was large in the early stages of expansion of the universe, maybe comparable to the strength of the electromagnetic interaction. Maybe the gravitational interaction is weak now because it has been decreasing for a long time. In this picture the rate of expansion and cooling of the early universe could have been rapid enough to have prevented significant element building.

We should have mentioned yet another possibility: in the steady state cosmology the continual creation of matter could have included helium or microwave radiation in large or small amounts: it's a free assumption. As I said earlier, my recollection is that at that time neither of us found the steady state philosophy interesting.

My notes for the Wesleyan talk suggest I had nothing useful to say about the astronomical determinations of the cosmic helium abundance Y. But at Princeton we were learning that Y is larger than seems likely to be accounted for by production in stars, and maybe is in line with a hot big bang. We mention that in Dicke and Peebles (1965), with a reference we had just learned to Hoyle and Tayler (1964), who knew a lot more about the astronomy than we did.

Bob and I did not refer to Osterbrock and Rogerson (1961). I now believe they were the first to present evidence for a critical result: the older stars that contain fewer heavy elements appear to contain about the same helium mass fraction as younger stars. They also point out the possibly very significant implication, that "the build-up of elements to helium can be understood without difficulty on the explosive formation picture (Gamow 1949)." I only remembered when writing this essay that I had referred to Osterbrock and Rogerson in a study of the structure of the planet Jupiter (Peebles 1964). I concluded that a small fraction of Jupiter's mass is in heavy elements, most of which had settled to a central core, while most of the mass outside the core is a mixture of hydrogen and helium. The mass fraction in helium that fit Jupiter's mass, radius, and rotational flattening is roughly consistent with the Osterbrock and Rogerson estimates, and with what Martin Schwarzschild told me about the composition of the Sun. The remark by Osterbrock and Rogerson about the explosive formation picture would not have meant much to me when I was making models of Jupiter: I knew close to nothing about cosmology. When I started thinking about a hot big bang I should have remembered the evidence for large Y in old stars. Bob Dicke liked to say that "we get too soon old and too late smart."

My second colloquium on cosmology was at the Applied Physics Laboratory at the Johns Hopkins University in Maryland, on February 19, 1965. I don't know why I was invited; maybe it had something to do with the fact that Alpher and Herman were at the Applied Physics Laboratory when they were developing the physics of element production in a hot big bang. But I learned about that connection much later.

In this colloquium I presented updated versions of the helium production calculation and the cosmic radiation spectrum. I had added to the latter a bound on the energy in microwave radiation from the absence of a discernible effect of its drag on energetic cosmic ray protons. That pretty convincingly ruled out the idea that the mass of the universe might be dominated by thermal radiation. But it did leave room for an interesting fossil thermal background.

I had asked David Wilkinson whether it would be appropriate to mention the Roll–Wilkinson experiment. You want people to know about your work, but only when it is unlikely someone else might be inspired to do it first. Dave assured me that no one could catch up with them at that point, so I mentioned the experiment. Ken Turner, a friend since our graduate student days in the Gravity Group, attended the talk. He told Bernie Burke about it. Burke brought the news to Arno Penzias and Bob Wilson. They were at the Bell Laboratories in Holmdel, NJ, not far from Princeton. They did not have to catch up: they had already done the experiment. Arno telephoned the news to Bob Dicke.

What was our reaction to the telephone call? I remember relief and excitement: they showed us that there actually is something to be measured, always a very good thing. That overwhelmed any chagrin over priority, and to me it still does, with one exception. The Nobel Prize rightly went to Penzias and Wilson: they made very sure of the reality of an unexpected result, and they made sure the world knew about it. But the Nobel committee should have included Dicke.

When and how did I learn that my first computations of light element formation largely repeated earlier work? My records reveal a few data points. I submitted a paper on my calculations to the journal *Physical Review*. The referee recommended rejection, saying that my calculations had already been done, and by whom. I revised and resubmitted several times. I have a draft dated January 1965 that has a reference to Alpher and Herman (1953), but I don't know whether this draft was the first to recognize that I was repeating old analyses. I have a copy of a letter I wrote to Hoyle and Tayler on February 1, 1965, acknowledging their prior work. I have a copy of my letter to *Physical Review* dated June 23, 1965, in which

I at last withdrew the paper. By then I had faced up to the fact that to make a meaningful contribution I would have to do a distinctly better computation.

Fred Hoyle also saw the need for a better computation, and he, Willy Fowler, and Bob Wagoner got to work. I met Bob Wagoner at a conference in Miami in December 1965. We exchanged ideas but not techniques of computation. I devised fixes for the numerically unstable reaction equations that work but likely would be close to incomprehensible to anyone else. Wagoner used the more familiar (to Fowler and Hoyle) techniques from the analyses of nuclear burning in stars, and he considered a larger set of nuclear reactions. I believe their computer code evolved into some that are used today. But the important elements of our results, in Peebles (1966) and Wagoner, Fowler and Hoyle (1967), agree.

Our ignorance in 1964 about the literature of this subject is legendary in the cosmology community, and legends beguile. I see the effect in Bob Dicke's comment (unpublished, dated 1975):

There is one unfortunate and embarrassing aspect of our work on the fire-ball radiation. We failed to make an adequate literature search and missed the more important papers of Gamow, Alpher and Herman. I must take the major blame for this, for the others in our group were too young to know these old papers. In ancient times I had heard Gamow talk at Princeton but I had remembered his model universe as cold and initially filled only with neutrons.

I think Bob apologized too much. I have the greater share of blame for poor homework: Bob was careful to stand back and let younger people in his group get on with research on their own. Our paper Dicke and Peebles (1965) did not give proper references to earlier work on the hot big bang, but we remedied that pretty quickly. I believe the citations are normal and proper in Dicke *et al.* (1965), the paper that offers the hot big bang interpretation of the Penzias and Wilson (1965a) detection. Because I have on occasion encountered the myth that our paper did not refer to earlier work I list our relevant references: Alpher, Bethe and Gamow (1948), Alpher, Follin and Herman (1953), and Hoyle and Tayler (1964). I don't remember whether the absence of a reference to the prediction of the CMBR in the present universe by Alpher and Herman (1948) signifies more than lack of careful reading. And the list of references is brief, but then this is a brief paper. In the late 1960s Dave Wilkinson and I systematically advertised the history of ideas in our lectures at conferences and colloquia; one sees an example in Figure 3.1 at the start of Chapter 3. And I think there is a full and accurate account of the history in Chapters V and VIII of *Physical Cosmology* (Peebles 1971).

Bob hated sloppy physics, a term he used on occasion to express strong disapproval. I don't remember his ever applying those feared words to me, though I do remember clear reprimands for less than careful physics. My homework in 1964 could be termed "sloppy," but I don't remember Bob or anyone else in the group chiding me about it then or later. We were caught up in the excitement of exploring rich and sparsely worked ground.

The other rich slice of physics I started pursuing in 1964 is the effect of the CMBR on the gravitational growth of small initial departures from an exactly homogeneous mass distribution into the present strong clustering of mass on the scale of galaxies. Here again Gamow (1948a) got there first. He pointed out that the matter temperature and density in the early universe determine the pressure, and the pressure sets the size of the smallest cloud of matter that gravity can cause to break away from the general expansion. This is the analog in cosmology of the Jeans criterion for the balance of gravitational attraction and the pressure gradient force of repulsion of a cloud of matter. Gamow also argued that the gravitational instability to the growth of mass clustering commences when the mass density in matter becomes larger than that in radiation. He was right, though his argument is not what we use today. A brilliant physicist can do that.

I was able to add something new. I found that when the universe was young and hot enough to ionize the baryons the drag of the radiation on the plasma is strong enough to prevent the gravitational formation of a non-relativistic cloud of baryons. That situation changes when the temperature dropped to about $3000 \, \mathrm{K}$, the plasma combined to largely atomic hydrogen and helium, and matter and radiation abruptly decoupled. I published the idea in Peebles (1965) (with, I am relieved to see, appropriate reference to Gamow 1948a on the Jeans length).

The departures from an exactly homogeneous mass distribution must disturb the spectrum and spatial distribution of the radiation. Here are my recollections of how this was worked out in the 1960s.

At the January 1967 Texas Symposium on Relativistic Astrophysics in New York City I presented a more detailed analysis of the behavior of the matter-radiation fluid prior to decoupling in general relativity theory. I took account of the effective viscosity from the diffusion of radiation through the plasma, and analyzed how the dissipation suppresses small-scale density fluctuations that act as pressure waves. I presented my paper on these considerations for publication in the conference proceedings, but because of turmoil at the publisher the proceedings never appeared in print. Richard Michie (1967) independently worked out main elements of this physics, but illness prevented this work from getting past the preprint stage. Joe Silk also independently worked it out, and he published (Silk 1967, 1968b), so the effect is properly termed "Silk damping."

Sachs and Wolfe (1967) derived the gravitational perturbation that, in general relativity theory, dominates the large-scale disturbance to the space distribution of fossil radiation from the big bang. Chapter 5 traces how that deeply influential relation was detected a quarter of a century later.

The acoustic (sound wave) oscillation of the matter-radiation fluid can leave characteristic patterns – that much later actually were measured – in the distributions of matter and radiation. In his contribution Sunyaev describes how he and Zel'dovich (1970c) derived analytic approximations to the analysis of this effect. I didn't know what they were doing until quite a while later; communications with people in the USSR were slow. But as it happened I took the next step (while enjoying sabbatical leave and the hospitality of Caltech in 1968–1969) of working out the radiative transfer analysis that is needed for a more complete computation of the residual patterns in the distributions of matter and radiation. Jer Yu, who had been my first graduate student, joined me in the numerical solutions (Peebles and Yu 1970). Our paper shows the power spectrum representation of the relict acoustic oscillations in the mass distribution, along with a pretty awkward way to represent the acoustic oscillations in the radiation distribution. Apart from that – and the dark sector, and the unsophisticated numerical methods – this is close to the standard physics used in the analyses of the measurements of the variation of the CMBR temperature across the sky, including the wonderfully precise WMAP data discussed in the next chapter.[8]

The CMBR also is disturbed by its interaction with plasma that in the present-day universe is hotter than the radiation: scattering by the hot free electrons pushes the spectrum of the radiation down from blackbody at long wavelengths and up at the short-wavelength end. The plasma in clusters of galaxies is particularly hot and dense. It produces an observable disturbance to the CMBR that has become a useful diagnostic of how cosmic structure formed. Ray Weymann was among the first to analyze this important effect in a hot big bang cosmology. He writes:

I then became interested in understanding the coupling (and subsequent decoupling) of the matter and radiation and came to the realization that the Compton

---

[8] Another piece of the physics is the combination of the primeval plasma into mostly neutral atomic hydrogen and helium, with small but important amounts of free electrons and molecular hydrogen. The computations are presented in Peebles (1968) and Zel'dovich, Kurt and Sunyaev (1968). Rashid Sunyaev explains the computation and how I learned about their independent work. There are quite a few other examples of this parallel development of ideas in the 1960s. I think it's not surprising: once people became conscious of the radiation these became pretty obvious things to work out.

interaction was the dominant interaction mechanism. To derive the frequency-dependent interaction and its diffusion approximation, one does need special relativity, and I was helped by a paper and correspondence with Willard Chappell in Boulder, Colorado, who helped me over an obstacle. The resulting paper was published in *Physics of Fluids* (Weymann 1965). Not very long after that I applied that diffusion equation to study the temperature history of the matter and radiation, and that involved studying the recombination era. I wrote up two papers, and I believe one was published (after a struggle with the referee; Weymann 1966) but the other only appeared as a Steward Observatory preprint. One of these papers calculated departures from the Planck function that would result under various (and I later realized mostly unrealistic) heating mechanisms.

Shortly after this I received a letter from Zel'dovich who pointed out that the diffusion equation I had derived had already been derived by Kompaneets, though I was totally unaware of it, as it was in a Soviet journal. About then the Zel'dovich and Sunyaev (1969) paper came out. If you read that paper you will see that my paper was referenced fairly extensively by them, but my paper had two serious defects: I did not derive the analytic expression which they did, but relied only on numerical work, and I only applied the work to the cosmic expansion and not to finite clouds of electrons.

My only regret in all this is that since I did all that work at Arizona, there was at that time no other theoretician there to talk to and so I was too isolated and the work I did there suffered from that.

I had been thinking about this effect too, and had worked out the theory by 1970, but that was after Sunyaev and Zel'dovich. I at first didn't much care that I was scooped because I didn't think the effect would be large enough to be observable. Dave Wilkinson straightened me out on that, though a convincing detection did take a lot of work.

These analyses assumed the microwave radiation really is a fossil from the very early universe. An alternative that had to be considered in the 1960s was that the CMBR was produced by sources in the universe as it is now or at modest redshifts. Galaxies are sources of optical and radio radiation; might they also produce the microwave background? This local source model was discussed by Sciama (1966), Gold and Pacini (1968), Wolfe and Burbidge (1969), and Pariĭskiĭ (1968). It was a serious possibility in the 1960s that demanded tests: measurements of the spectrum and angular distribution of the radiation. If a fossil from the hot big bang the radiation spectrum ought to be close to blackbody. That spectrum would not likely be produced by microwave sources at low redshift, and at adequate angular resolution the radiation would break up into the individual sources. Layzer (1968), and later Hoyle, Burbidge and Narlikar (1993), postulated that absorption and reradiation by dust relaxed the CMBR spectrum toward blackbody, and smoothed out the radiation. The picture didn't seem promising, though, because we knew distant radio-emitting galaxies are observed at CMBR wavelengths, with no indication of absorption.

My notes for a colloquium on March 17, 1966, at the University of Toronto show significant advances toward the measurement of the spectrum. In addition to the Penzias and Wilson detection at wavelength $\lambda = 7.4$ cm I could show the very recently published Roll and Wilkinson (1966) measurement at 3.2 cm and the CN temperature measurement (as in equation 3.13) at 2.6 mm by Field, Herbig and Hitchcock (1966). The fit to a thermal spectrum certainly looked promising. And there was another data point, the consistency with the helium abundance. By this time I was arguing that we had a significant case for the hot big bang model, and a serious challenge therefore for the steady state cosmology.

I can recall early reactions by a few others to our proposed interpretation of the CMBR. In July 1965 Bob Dicke and I attended a conference on general relativity and gravitation at Imperial College, London. The microwave radiation was not on the program but there were informal discussions. We met Fred Hoyle; I remember the talk as friendly, but, for whatever reason, short. I don't remember meeting Igor Novikov, but he recalls (p. 99) bringing the news of the microwave radiation back to Zel'dovich.

Bob Dicke showed us a letter he received from Zel'dovich. In this letter, dated September 15, 1965, Zel'dovich writes

I am not more so cock-sure in my colduniverse hypothesis: It was based on the assumption that the initial helium content is much smaller than 35% by weight. Now I understand better the difficulty of helium determination. You draw some conclusions from the observed helium content 25%. Are you sure it is not 35% or 15%?

It seems to me very desirable to measure the Planck spectrum corresponding to $3-4°$K at its maximum, at the wave-length ~ 1 mm, although it is a difficult task.

Undoubtedly your work will raise the interest to all sides of the problem and I sincerely congratulate you and your team on a success.

Zel'dovich went on to argue that Dicke's oscillating universe picture is "untenable as a consequence of unlimited growth of entropy." We knew the argument, but I think I recall that we were not so sure that entropy need be conserved in the bounce. This was before the full development of the cosmological singularity theorems Ellis discusses (commencing on p. 379), but we were aware of the general idea. I remember Bob saying, in effect, that general relativity predicts that a collapsing universe develops a singularity but it doesn't say whether the singularity applies to the whole universe or just a bit of it, maybe leaving a few black holes, while the rest of the universe expands again. What would happen to the accumulation of black holes as well as entropy over many cycles? I was inclined to work on something else.

In his letter of reply to Zel'dovich, dated October 5, 1965, Bob suggests that "the helium content of the proto-galaxy could very well have been zero."

That was inspired by his fascination with the possibility that the strength of the gravitational interaction decreases as the universe expands. If so it would make the rate of expansion of the early universe much larger than in the standard model. If the expansion were fast enough there would be negligible light element production at high redshift. I don't remember whether I told Bob about the Osterbrock and Rogerson (1961) evidence for large helium in relatively old stars.

At the time of this letter there was in the literature a direct measurement of the microwave radiation at just one wavelength, and one indirect measurement from the interstellar molecule CN. Zel'dovich seemed to be ready to accept that the spectrum likely will prove to be blackbody, but quite a few others reminded me that that is a considerable extrapolation from the measurement of one (or two if you trust CN) point on the spectrum. I remember a conversation with Phil Morrison in a noisy room. He said, in effect, measure the energy of the sound in this room and convert it to an effective temperature. You'll get an absurd value. He bet one guinea that the same is true of the microwave radiation, that measurements at other wavelengths would not follow the thermal spectrum. I think it was at the 1967 Texas Symposium that he agreed that he had likely guessed wrong and paid me one pound and one shilling. I met Ralph Alpher at that meeting, for the only time, but our conversation was short, I think because I was rushing to catch the train home.

By the end of 1966 Howell and Shakeshaft (1966) and Penzias and Wilson (1967) had added a data point at 21-cm wavelength to the measurements at 7 cm, 3 cm and 2.6 mm. The spectrum up to the expected peak looked encouragingly close to blackbody. Not long after this measurement, in a letter dated December 21, 1966, Dennis Sciama wrote to Bob Dicke,

As you may have heard I have recanted from the steady state theory, and have taken such a liberal dose of sackcloth and ashes that I am now more orthodox than the orthodox (though I don't suppose this phase will last long). Anyway you can tell Peebles that I now nearly believe that the excess background has a black body spectrum. I hope to see him and you in New York so that I can capitulate in person.

Sciama's new phase did last: he continued to work on the relativistic big bang cosmology, with particular attention to clues to the physics of the dark matter (Sciama 2001).

By 1970 three groups had attempted to measure the CMBR energy spectrum at wavelengths near 1 mm, where the spectrum is expected to break away from the power-law form that applies at longer wavelengths. As Zel'dovich had remarked, and Harwit (p. 329) and Weiss (p. 342) describe, that "is a difficult task." From 1970 to 1990 a series of experiments indicated that the CMBR spectrum significantly differs from blackbody near and shortward of the blackbody peak. The beautiful experiments by Mather *et al.* (1990) and Gush, Halpern and Wishnow (1990) at last showed that the spectrum is wonderfully close to thermal.

I have no complaints about the two-decade-long apparent anomaly in the spectrum – we were seeing first-rate science in progress – but it did confuse the subject and it led me to think about other things. That mainly was the statistical analyses of the clustering and the dynamical analyses of the motion of matter on large scales. At the time that was a better subject for me to work on anyway. It is the sort of thing I like doing, the field was ripe for exploration, and it grew into a component of the second critical test of the cosmological interpretation of the CMBR, the signature in its variation across the sky of its interaction with the growing inhomogeneity in the mass distribution.

I can date my work on measures of the cosmic clustering of matter to the March 1966 colloquium in Toronto. Sidney van den Bergh asked me how I could be sure the universe really is close to homogeneous in the large-scale average. I offered as evidence the CMBR, which we already knew is quite smooth, consistent with a near uniform large-scale mass distribution. The argument is pretty indirect, of course. Sidney countered that George Abell's map of the distribution of rich clusters of galaxies (Abell 1958) does not look very smooth. I said it doesn't look all that rough, considering the sparse sampling. I think I can remember Sidney's words, "you could check that." I worked out a method of checking it on the flight back home, and Jer Yu improved and applied it in his PhD thesis (Yu 1968; Yu and Peebles 1969).

The obvious measures for this project are second moments: the two-point correlation function and its transform, the power spectrum. My initial choice of the latter was influenced by Bob's preference and by the arguments in *The Measurement of Power Spectra* (Blackman and Tukey 1958). That proved to be right for measurements of the large-scale distributions of matter and radiation, but I learned that correlation functions are better suited to measures of the nonlinear clustering of matter on relatively small scales.

I continued the analysis of statistical measures of the distributions of extragalactic objects – n-point correlation functions and their transforms – and of the dynamical evolution that might produce the observed clustering, for more than a decade. There was a positive reason: this was rich fallow ground to explore. And there was a negative one: I mentioned the spectrum anomaly that beclouded my thoughts about the CMBR. Though I like to work alone, I needed help in this data analysis and interpretation, and it appeared. Along with Jer Yu, I am deeply grateful (though it may not have always been apparent at the time) for collaborations on these statistical analyses with Martin Clutton-Brock, Marc Davis, Jim Fry, Margaret Geller, Ed Groth, Mike Hauser, Dan Hawley, Bernard Jones, Diego Lambas, Mike Seldner, Bernie Siebers, and Raymond Soneira. All were volunteers. Young people somehow tend to sense when and where things of possible interest are happening.

In 1969 I gave a graduate course at Princeton on current topics of research in cosmology. John Wheeler insisted that I turn the course into a book, and he sat in the back of the room and took notes until I agreed. That so unnerved me that I wrote *Physical Cosmology* (Peebles 1971). By then I understood that cosmology is a real physical science that offers fascinating issues of theory and observation. It was a science with a limited empirical basis, to be sure. A measure of that is that I could present a reasonably complete survey of the science (apart from the subtleties of the astronomical observations that the title was meant to indicate I would not attempt to address) in just 280 pages. I marshaled evidence for the homogeneity assumption – the cosmological principle – and concluded that the case was encouraging but not definitive. A decade later the case was much stronger, but resistance to the assumption died out more slowly, a not unusual phenomenon. The last section in the chapter on the Primeval Fireball – the name John Wheeler had suggested for the CMBR – has the title *Is this the Primeval Fireball?* My answer was cautious, largely because of the apparent anomaly in the measurements of the spectrum at wavelengths near 1 mm. The case for the fossil interpretation of the CMBR is close to compelling now: we have a vastly improved spectrum measurement, and detailed evidence that the radiation has the predicted disturbances caused by its interaction with the mass distribution at decoupling and along the line of sight. And we have the elegant concordance of the theory and observations of helium and deuterium. But all that is the subject of Chapter 5.

I close with some thoughts inspired by reading the exchange of letters between Dennis Sciama and Bob Dicke. In his reply, dated December 30, 1966, Bob writes

I was very happy to learn that you have abandoned steady state theory, but I do not recommend that you take too orthodox a position. A number of peculiar things are showing up that favor the scalar-tensor theory and I had hoped that you would be one of the few people who might be convinced when the observations were good enough to warrant it. Another reason for being unorthodox is that it's fun.

This is very characteristic of the central lesson Bob gave me by word and example: don't take received wisdom too seriously, but you had better take the science very seriously. Starting in Peebles (1984, 1986) I enjoyed better than a decade of fun pointing out the observational challenges to the then orthodox adoption of the Einstein–de Sitter cosmology (without Einstein's cosmological constant, and negligible space curvature). I spent about as much time constructing alternatives to the CDM model (with its set of assumptions about the initial conditions for structure formation). I meant this model (Peebles 1982) to serve as a simple example showing why the improving limits on the anisotropy of the CMBR were not necessarily inconsistent with the idea that galaxies and clusters of galaxies grew by gravity out of small primeval departures from homogeneity. As the model became popular I became nervous, because it was easy to think of alternatives that could equally well fit the still very loose observational constraints. Here again the consensus developed before the evidence warranted it, but in this case the orthodox view proved to be on the right track. I folded at Peebles (1999) when the advances in the tests had become so rapid that my alternatives were being ruled out as quickly as I could produce them.

The observational basis for cosmology now is far better than anything I would have imagined in the 1960s, and the case for the hot big bang far more compelling. With Dennis Sciama I have become "more orthodox than the orthodox." But with Bob Dicke I doubt that we now know all the physics relevant for the observational analysis of the evolution of the universe and its contents from high redshift (let us say from light element production, $z \sim 10^{10}$) to the present. We are attempting to draw spectacularly large conclusions from what still is an exceedingly limited collection of data. I expect this active field of research will continue to fascinate the next few generations.

4.7.2 David T. Wilkinson: Measuring the cosmic microwave background radiation

Dave Wilkinson's leadership in the exploration of the CMBR, through his own research and the education of other key players, continued from the identification of this radiation to his central role in a last great experiment, the Wilkinson Microwave Anisotropy Probe.[9]

---

[9] David Wilkinson was one of the group who planned this book. He did not live to write a contribution (d. September 5, 2002), but Dave's voice comes through in this transcript of an interview conducted by Michael D. Lemonick on July 25, 2002, and recorded by The Educational Technologies Center, Princeton University. The comments we added are in square parentheses and footnotes.

DW: My name is David Wilkinson. I'm a professor in the Physics Department at Princeton. I work in cosmology and astrophysics; I do experiments. I came to Princeton in 1963, was lucky enough to find a hot research topic and rode that to tenure, so I've been here ever since.

Q: What do you think led to your being a scientist, and in particular a physicist?

DW: I became a physicist because of a course in engineering I took in college, called "Cement." And I couldn't imagine taking a whole course in cement.[10] I enjoyed my freshman physics course, so I decided I would become a physicist and not a cement engineer. That really was the reason. Plus I really liked physics.

Q: Where did you go to college?

DW: The University of Michigan. I went to school at the University of Michigan.

Q: And where did you grow up?

DW: I grew up about 30 miles west of Ann Arbor in a little town called Michigan Center.

Q: Were either of your parents scientists?

DW: No. My father didn't graduate from high school. My mother worked her way through teachers college at Kalamazoo and ended up teaching math. So I think I got some of her genes for the math and science side. But I got the practical genes from my dad; that's why I'm an experimentalist. He could build anything and fix anything.

Q: So you majored in physics in Ann Arbor. Was there any particular area of physics that you specialized in?

DW: No, not as an undergraduate. As a graduate student, first of all I got a degree in nuclear engineering because that was the hot topic at the time and one could walk out with a Masters and get a fantastic salary of $10,000 a year. But I soon decided I didn't want to build reactors and I went into more of a particle physics mode. I did my PhD measuring how strong a magnet the electron is. It had little to do with cosmology but it was a lot of fun. And I had a great thesis advisor [Dick Crane].

Q: How did you come to Princeton?

---

[10] On other occasions Dave mentioned steam tables.

DW: Fortunately my PhD thesis turned out to be pretty important. Bob Dicke here at Princeton, people at Columbia, Harvard, and Yale, had all tried to do this experiment, and Dick Crane and I did it better. So the old boys network went to work and I got my choice of where I wanted to go. Things were a lot different in those days. And I decided I wanted to come here and work with Bob Dicke on gravitation.

Q: In what sense did you work on gravitation?

DW: When I first got to Princeton I worked on gravitation with Bob Dicke. He was doing ground-based experiments and had just started working on the [shape of the] Sun, and I was intrigued by that project. In the end, I didn't work on it but I realized that I had a real fundamental interest in astronomy. Then Bob suggested a project which involved building a small radio telescope and that just completely clicked with what I wanted to do.

Q: What were you going to do with this radio telescope?

DW: [About the time I came to Princeton] Bob Dicke independently had dreamed up the idea of a microwave background left over from a hot phase earlier in the universe. Not only had he gotten the idea that the universe was filled with this thermal radiation, perhaps, but he had invented in 1946 the instrument to [find it – the] so-called Dicke radiometer, which is famous in radio astronomy circles. So he sort of drew a picture on the blackboard and said, OK boys, go build this. So Peter Roll and I went off to build this little radio telescope that ended up on the top of Guyot Hall [pictured on pp. 214 and 223] on one of those turrets of the building.

Q: Did you at that time have any preference for any particular model in cosmology? Did you like the idea of a big bang?

DW: Cosmology was just completely in its infancy when I came to Princeton. There was still a huge debate going on whether it was big bang or a steady state universe. The steady state universe always looks like it does now, in the past and in the future. You have to play a few tricks with physics to do that, but philosophically it's very satisfying to think the universe will always look like this. And of course there was the big bang theory named by [Fred] Hoyle as a joke. It said that the universe started in a very hot condensed state and then expanded out and its still expanding. So the theories were so crude at that point. There was no data except that the universe was expanding. It was very hard to have any kind of an objective opinion. Of course if Dicke's idea worked out – incidentally this was an idea that had been well published by George Gamow and his colleagues twelve years earlier, but we did not know about it – if that idea worked out, that was very strong evidence for a big bang. There was no way that this heat radiation could be naturally produced in the steady state.

Q: So you went out to build this radio telescope. How was Jim Peebles involved in this project?

DW: We formed a little group based on Dicke's idea to explore it. Jim did the theory behind it. If there was a big bang would this radiation still look like heat radiation? Would it have the spectrum (the intensity *versus* wavelength) that one expects from heat radiation? Or would that have gotten distorted somehow between the big bang and now? That was the key calculation that had to be done. He also did a calculation on making elements in an early universe which also, unbeknown to us, had been done by Gamow's group. So Jim did the theory, Peter Roll and I built the instrument, and Bob was the great advisor.

Q: Even though you didn't have a personal opinion about which cosmological model was correct, did you have any sense that if you found this radiation it would be a very big deal?

DW: Yes. If we found this radiation it was certainly going to be a big deal because it would resolve this basic argument about big bang and steady state. Plus it would give us a tool for examining the physics in the very early universe before any stars or galaxies formed. And that was unprecedented: to be able to measure a probe that came right out of the big bang. There was a lot of anticipation. There wasn't a whole lot of hope.

Q: Why wasn't there hope?

DW: The idea that we might actually find this radiation seemed kind of remote to us. First of all, there was no other data to indicate that we were living in a big bang universe. It seemed rather fantastic that this remnant would be around and nobody would have discovered it before. It's not a weak phenomenon. But it turns out, the way radio astronomers do their work, they have much better sensitivity than they need but they can't detect this radiation because it's coming equally from all directions – almost. [Radio astronomers subtract the radiation received when the antenna beam is on an object from the radiation received when the beam is on apparently blank adjacent sky. The subtraction eliminates unwanted radiation originating in the instrument and atmosphere, but it also eliminates an isotropic sea of radiation.] So the more we thought about it and read papers in radio astronomy, the more we realized that, yes, this thing could be out there and nobody would have seen it. You need a very special type of radio telescope to do it.

Q: Tell us the story, the now famous story, of the day you were sitting in Bob Dicke's office having lunch and the phone rang.

DW: The group that was looking for this microwave background, which is what we call it, consisted of four people, Bob Dicke, the leader; Peter Roll and I, the experimentalists; and Jim Peebles, the theorist. Every Tuesday at lunch we would meet in Bob's office and discuss the progress and problems and so forth and try to figure out what we needed to do to get there. There were some very specialized pieces of equipment we had to build, and it wasn't obvious how to do these things. Well, one Tuesday we were sitting there and the phone rang. (That often happened; Dicke was a famous guy so people called him all the time.) He picked up the phone and we went on with our conversation as usual, and then we heard him say "horn antenna." Well, that was one of the very special things you needed to do this experiment. And then he said, "cold load" – cryogenic load – and that was the other thing you needed to do this experiment. So now we were pretty tuned in because at this stage we were about halfway through building the apparatus [with a horn and cold load]. We hadn't gotten it on the roof to observe yet. So we listened to the rest of the conversation, which didn't go on more than 5 minutes, and Dicke hung up the phone and he said – I'll never forget his words – "Well boys we've been scooped." He immediately, in 5 minutes' discussion with Arno Penzias, realized that they [Arno Penzias and Bob Wilson] had been looking at this microwave background, this heat from the big bang, for a year trying to figure what was wrong with their instrument. And to their great credit Arno Penzias and Bob Wilson stuck to it. Often experimentalists will sort of write [things like] that off and say, OK, well here's a little effect that I don't want to deal with, there's probably no important science in here, it's just some quirk in my apparatus. And they overlook it and go on and do their measurements. Well Penzias and Wilson didn't do that. They stuck in there; improved their measurements. That's why Dicke was convinced so quickly, because they had their ducks lined up. They could answer all of our questions. So we went up and visited [Bell Labs] about a week later, looked at their data, looked at their apparatus, and it was obvious that they were seeing what we were looking for.

Q: When you found out that you had been scooped, did you stop working on your experiment?

DW: Oh no! Just because they had found the radiation we didn't stop. In fact we sped up because our apparatus was designed to measure a different wavelength than Penzias and Wilson had used. And this was the crucial test of the idea. No one would believe that what they were seeing was heat from the big bang without measuring this spectrum that I talked about, intensity *versus* wavelength. [Thermal radiation] has a very special shape. So we charged ahead in order to try and verify this spectrum at a different wavelength than they had. The discovery papers [on the Bell Labs detection and the Princeton interpretation were submitted] on my birthday in May and our [measurement] paper [was submitted the following January]. So we were about six months behind them.

Q: What was the reaction of the astronomical community to these papers? Was the big bang accepted pretty much immediately?

DW: The astronomers did not like it much, and the physicists didn't like it much – for completely different reasons. The physicists didn't understand any cosmology at that time. It's completely different now; a lot of physicists work in cosmology. At that time, Dicke's group and a few others were the only ones working in cosmology. So there was no way that they could evaluate the science. And certainly with one measurement at one wavelength everybody was skeptical, including us. We stuck our necks out and published [a paper (Dicke *et al.* 1965) interpreting the Penzias and Wilson (1965a) result] saying we think this is heat from the big bang. That was pretty roundly laughed at. Even after we got our data point I got a lot of questions at meetings and got grilled. But gradually people started accepting it. These big paradigm shifts in science are always hard to swallow because, whether you like it or not, you're in one camp or another. Certainly the steady staters did not like this at all. The big bangers had a little trouble with it, because why did it take till 1964 to discover this stuff? Radio astronomy had been around for 15 years. So there was a lot of sort of detailed knowledge that needed to be accumulated before you really realized that these measurements probably indicated discovery of heat from the big bang.

Q: Once the idea of the big bang started to be accepted and people really did accept that this was radiation left over from the fireball, what did you decide to do next? Did you have any thought of leaving cosmology and doing some other experiments?

DW: As the idea was gradually accepted that this radiation really was from the big bang, more and more people started coming into the field and making measurements, of all different sorts. All of which agreed with predictions of the big bang theory. I saw this as a wonderful opportunity to do some groundbreaking research, because here was a brand new phenomenon coming from the very early universe, something we never had before – not even come close to it. This radiation dates from when the universe was only about 300,000 years old, and that's in a 14 billion year old universe. So this stuff came right from the beginning and it looked like we probably had an opportunity to do some really fundamental measurements of the early universe. I would have been crazy to get out of the field at that point. There were just too many opportunities.

Q: What was your next series of experiments? What did you decide to do next?

DW: The next thing we did, after the initial verification that the spectrum looked OK, was to ask ourselves, is this stuff really coming from everywhere in the universe? That was a crucial test. If this was some kind of new radiation from our own galaxy then it would be concentrated in the Milky Way. If it was truly a universal phenomenon, then it should be coming the same from all directions. Bruce Partridge, who was here at the time, and I modified the original apparatus [that had been used to measure the intensity] to scan the sky and look for little wiggles [in the intensity]. This is the so-called anisotropy in the radiation, which is a big industry these days. Well, we didn't have very much sensitivity. Radiometers these days are a million times more sensitive than the thing we had. We set this thing up on top of Guyot – again. Part of the experiment was to try and switch the beam a lot. So we had a big reflector that would come up in front of the antenna and deflect the beam up to [the North Pole, and then the reflector] would go down, the beam would go off to the equatorial plane. Well, we didn't quite apply enough oil to this thing so it started squeaking and the undergraduates were really annoyed by this thing because it went on all the time so it was squeaking away at night. So somehow those guys scaled the wall of Guyot and went up there and dismantled our reflector. This is one of those funny stories about Princeton undergraduates and what they'll do to do something different. Anyway that experiment went on for a year. All the data came off on chart recorders with pen and ink. Bruce and I would come in every day and spend about two hours reading those charts by eye, writing down long columns of numbers because at this point computers weren't around. There was no way to record the results digitally. And after a year we concluded that yes, this radiation was very [close to] isotropic, better than a tenth of a percent. And again it fit the prediction of the big bang theory. That was really a part-time thing, to carry us over to when we could build new technology. Meantime, while we were taking this anisotropy data, we were building much more sensitive receivers.

Q: When did you start using the more sensitive receivers?

DW: We started using the more sensitive receivers in the late 1960s, early 1970s; took them to mountaintops because water vapor in the atmosphere bothered us so we wanted to minimize the amount of water overhead. Mountaintops seemed like a good idea. Turned out it wasn't because of all the turbulence going over the top of the mountains. So the next thing we did was put our radiometers in scientific balloons and fly them from Texas. I had a wonderful graduate student named Paul Henry (who's very active in the [Princeton University] graduate alumni association). He built a radiometer with my help, trotted off to Texas, attached this thing to one of these big balloons and sent it up to 90,000 feet. Very successful piece of work, pushed the limit on the fluctuations down quite a bit – we almost discovered what is called the dipole in the radiation. That is, half the sky is warmer than the other half because we're moving [through the radiation in the direction of the warmer half of the sky], so there is a Doppler shift that makes half the sky look warmer. Peebles had predicted this, predicted its magnitude. And if you look at Paul's data, he saw it, but not with enough conviction that we were willing to say we've discovered the dipole.

It's another curious story. Paul saw the dipole at about the right magnitude, but almost completely in the opposite direction from what we had predicted. To predict the direction you assumed that the center of our galaxy is fixed with respect to the radiation and that we are moving through the radiation because of rotation of the galaxy. So you know which direction we're moving and that should be the warmer direction in the sky. Well it turned out the warmer direction was the other way. There was a lot of head scratching about that. I spent several days in the library trying to convince myself the astronomers had the right sense of rotation of the galaxy, which they did. So the only interpretation was that the galaxy was actually moving very quickly in the opposite direction, and it's turned out that's the case. But it was one of those surprises in science, those things you don't expect that happen, and you do a lot of head scratching before you publish something like that.

Q: So you could have discovered the dipole and the bulk flow in the same experiment.

DW: Yeah. In this really crude apparatus that Paul and I built. It was a real Rube Goldberg by today's standards.

Q: Meanwhile, while you were doing these balloon experiments and looking for the dipole and maybe even seeing it, I understand that by the mid 1970s you also began talking to people about a satellite experiment. How did this COBE (Cosmic Background Explorer) satellite business start?

DW: It became clear that what one needed to look really carefully at this anisotropy and to look very carefully at the spectrum of the radiation to see if it fit this classical thermal spectrum – intensity *versus* wavelength – [was] to get into space. You had to get the atmosphere out of here. These are pretty delicate measurements, and the atmosphere causes all kinds of trouble, mainly from water vapor and oxygen emission. And it's clumpy in the atmosphere, and you see all these clumps go through; it makes your signal noisy.

Several of us who were active in the business – Rai Weiss, John Mather, Mike Hauser, and I'm sure I'm forgetting somebody[11] – got together at [the Goddard Institute for Space Studies] near Columbia University, and started talking about a satellite to do both of these jobs, to look at the spectrum and to look for the fluctuations, if there were any. That was a long haul. It took several years, of course, for NASA to go through its usual procurement procedures and start sending us money. It was a fairly complicated satellite, complicated orbit. So there was a lot of interaction with NASA engineers. The science team got pretty big and unwieldy as people wanted to join. Then Challenger [the Space Shuttle disaster] came along and put us back 5 years because COBE was supposed to be launched on the shuttle. Well that wasn't going to happen after Challenger because we needed a west coast launch and NASA cancelled its west coast launch facility after that. So the Goddard Space Flight Center engineers completely reconfigured the satellite.

Well, that's a little bit of a distortion. We always had in the back of our head that maybe we'd have to go on a Delta rocket, so it wasn't a complete coincidence that this thing could be modified to fit into a Delta. We didn't have to change any of the structural stuff. We did have to make some modifications in shielding and so forth.[12] So that took several years.

---

[11] Hauser's recollections of this meeting are presented on page 418. His records show that the meeting included John Mather, who initiated the meeting, Michael Hauser, Dirk Muehlner, Patrick Thaddeus, Rainer Weiss, Joe Binsack, and David Wilkinson.

[12] Michael Hauser describes the situation in more detail. "In the initial COBE studies, NASA required the designs to be compatible with either a Delta or Shuttle launch, but early in the design phase NASA decided on the Shuttle launch. The spacecraft and instruments were built and being tested when the Challenger disaster occurred in 1986. After some months of uncertainty, NASA decided to launch COBE on the last remaining Delta rocket. GSFC engineers had to re-design, build and test the spacecraft structure and Earth–Sun shield (deployable instead of fixed as in the Shuttle version design). The FIRAS and DIRBE instruments for

Finally COBE got launched in 1989. Early in 1990, only about [eight] weeks after launch, John Mather announced COBE's measurement of the spectrum and it was spectacular. When I first saw it in December, the hair literally stood up on the back of my neck. I can remember the feeling because my students and I had been plugging away at this curve one point at a time for 25 years, and here all of a sudden was the whole curve spelled out in great detail with tremendous accuracy. No question about it: we were looking at a spectacularly accurate thermal spectrum. When John announced this at the [American] Astronomical Society meeting on January [13, two] months after launch, he talked about the apparatus first and tried to convince people that the experiment was working. And of course people [thought that his] not showing us the results [meant] there must be something wrong. Then he threw down the spectrum and the whole audience stood up and applauded. It's very rare at a scientific meeting for that to happen. But there was expectation. People knew what COBE was going to try to do and didn't have much faith that it would happen, unless you knew a lot about the apparatus. Then all of a sudden there it was – clear as a bell. The universe's big bang. No doubt about it.

[Two] years later George Smoot from Berkeley announced the [COBE] results on the fluctuations, which are much smaller and harder to measure so it took years really to begin to see them in a convincing way; years of averaging, scanning, averaging. Again [the result, this time the detection of anisotropy] was quite a sensation. I'm not quite sure why but the popular press grabbed hold of it. A few scientists made some outrageous statements like looking at the face of God and all that kind of stuff. So it got hyped up to the point where it was ridiculous. I was almost embarrassed to go out and give a talk because of course our colleagues were saying, what are these guys doing? Why are they hyping this up so much?

Q: Was the discovery of anisotropy a significant result, an important result?

DW: It was very important to finally measure the level of the anisotropy and that's because the theorists had all kinds of reasons for believing it had to be there. The main reason is that unless you have some fluctuations in the density and temperature in the early universe, very tiny fluctuations – one part in a hundred thousand for the density fluctuations – you can't make galaxies and clusters of galaxies and all the magnificent structure that we see in the universe today. You can't start out with a perfectly smooth matter

measurements of the CMBR spectrum and the CIB required only minor modifications, but the DMR anisotropy instruments had to be re-designed to fit within the reduced volume of the Delta rocket shroud. In spite of these challenges, COBE was launched in 1989, the year in which COBE had been scheduled to be launched on the shuttle at the time of the Challenger loss."

distribution way back there and end up 14 billion years later with what we see in the sky. These were hard calculations. So first [the theorists] said, well, [the CMBR temperature fluctuations] will be a part in a thousand. Well we got to that level pretty fast with the measurements. Well it wasn't there. Well it will be a part in ten thousand. Well that didn't work out so well either when we got there with our measurements. So they were beginning to panic, literally panic. Because the whole standard cosmological model didn't make any sense if those fluctuations weren't there. So a lot of this hype [about the anisotropy detection] that went on came not from the COBE team but from the theorists who were just breathing a big sigh of relief that yes, OK, things make sense again. And [a temperature fluctuation of] a part in 10^5 is very well accepted now. These little fluctuations are only 30 microkelvin from place to place across the sky – not an easy measurement to make.

Q: So this satellite goes up, was very successful, and as always the question is now what? What's the next important thing to do? Tell us a little bit about your reluctance to work on another satellite and how you finally decided you had to do it anyway.

DW: Well, what to do next after COBE was a big question. Now, this field had attracted a lot of very good experimentalists around the world. So there was a lot of activity, from balloons, from the ground, from mountaintops, from everywhere you could go, still hampered, however, by not being able to measure the whole sky with a clear shot not looking through any atmosphere. So a lot of ingenious experiments went on during that period of time. We did some in our group here that were good experiments, from Saskatoon and from Chile and wherever we could go where we thought we could get a good quiet sky to look through. But again it became clear that a satellite was needed to really go after these wiggles in detail and to get the smaller ones. COBE had very big beams so the [measured] bumps on the sky were large. One needed to actually have a telescope on a satellite so you get narrow beams and get higher resolution. Everybody knew this, knew that that was the way to do it. But it turned out there wasn't really an appropriate mission defined by NASA to do this. It didn't need a $2 billion-grade observatory, you just didn't need that big a satellite. On the other hand it couldn't be done with a $30 million- – what's called a small explorer – satellite. So some of us were behind the scenes urging NASA to start a new program called the medium-sized satellites. This is another one of these things in science where you recognize that there's the need for [an agency to change] their plans a little bit; takes years. Also, I was not eager to get involved with another satellite project. I like the model of a couple of graduate students and a professor and some undergraduates building an apparatus, going off to Canada or Chile, and making measurements as a team and so forth; and not these huge enormous groups with literally hundreds of engineers, dozens of scientists. And COBE was not a good experience sociologically. The team mostly got along but when it didn't get along it was pretty painful. So I wasn't at all sure I wanted to do this again. Went out to Jet Propulsion Lab because they have a good satellite program out there. I asked them if they were interested, they said oh yeah. Came back; never heard from them for a year and a half. [At] Goddard Space Flight Center though, a friend of mine there named Chuck Bennett called me up out of the blue not knowing that I had gone out to JPL and said, don't you think there ought to be a satellite to measure this anisotropy? I said great, I think so. You want to do it? And he said yeah. I said OK: a couple of requirements. Faster and cheaper I like, [but] it won't be better because it's faster and cheaper: the science team has to be small and tight and everybody has to do something [on] the science team. One of the problems with COBE was that there were people on the science team who really weren't contributing and there was a lot of resentment around that. Chuck completely agreed because he had gone through the COBE experience as well. So that was the beginning of MAP, the microwave anisotropy probe [now WMAP]. JPL came in with a proposal with University of California collaborators, Caltech collaborators. MAP won the competition and we started building it. And that was a completely satisfactory, satisfying experience. It was a small group. Everybody worked. All the jobs were done where they should have been by the people who had the expertise. Very little was farmed out to industry where you have a lot of trouble with contracts, engineers that don't understand the science. So MAP was a very good experience.

Q: I understand the heart of the satellite was actually built right here at Princeton. Tell us a little bit about that.

DW: In the division of labor for MAP we all agreed that the expertise for building the instrument was here at Princeton. We had built a lot of instruments. We had a lot of experience with very high-tech microwave components. We had a good relationship with the National Radio [Astronomy] Observatory in Charlottesville, where there was an ingenious engineer named Marian Pospieszalski who knew how to build an amplifier that was absolutely key to the whole thing. There are 80 of these amplifiers in this satellite, all built by Marian's group. The whole mission couldn't have worked without it. But we had a good relationship with Marian and with the Observatory. So it really made sense to build the microwave instruments here. That's where the expertise was; it just made sense. On the other hand, the satellite and all the associated electronics and support equipment obviously should have been built at Goddard because that's where they have the expertise. That's the way we divided the labor and it worked out very well. We built the instruments here, shipped them down to Goddard, they integrated them into the spacecraft. We did a lot of the testing here before we shipped them, and we did spacecraft-level testing down there. The project went very well.

Q: Tell us the story of the contract.

DW: Working with NASA is sometimes kind of frustrating because the scientists and engineers there usually don't get involved with contracting and the financial reporting and all of that stuff. And there are a lot of requirements. So when the Princeton contract to Goddard to build these instruments [arrived] it was about an inch and half thick and I obviously wasn't going to read this thing – I was too busy trying to build instruments. I flipped through it and it was all boilerplate. Just stuff they pulled out of the files. It was appropriate maybe to Alcoa Aluminum or somebody like that but it certainly wasn't appropriate to Princeton University. At that point NASA was not collaborating so much with small university groups. Large labs like the Johns Hopkins Applied Physics Lab, places like that, yes. But for small groups like ours the interface was very rough. The financial reporting business I understood: we had to comply. So we actually had to hire a person to do that, Susan Dawson. She was great; she took all that off my table. I sent the contract back and said send me a three-page contract and I'll sign it. But forget about all this boilerplate. And they did, I think primarily thanks to Chuck Bennett, who really wanted this project to go. He was able to convince the contract people that we were not some big aerospace company. That was one of the rough spots but that got resolved very quickly.

Q: Where physically were these instruments built?

DW: The instruments were built in Jadwin Hall in one of our labs in the gravitation group there. Most of the work was done by Norm Jarosik [a senior research staff member at Princeton] and Michele Limon [a postdoctoral fellow at Princeton]. We built a lot of specialized equipment here at Princeton in the machine shops. Having a first-rate machine shop was just essential. We would not have gotten that $8 million contract without a good machine shop. That's the sort of thing that people don't often think about.

I once alarmed the Dean by saying, I don't care what you do with the library but don't take away our machine shop. That shook him up a little bit. The facilities, the people that you have to back you up, and the machines, are really important. NASA actually bought us a couple of very nice numerical machines in order to do this project.

4.7.3 Peter G. Roll: Recollections of the second measurement of the CMBR at Princeton University in 1965

Peter Roll is retired, after 25 years as a university administrator of technology. He is currently working on the development of a community web portal for the retirement community in which he and his wife live, near Austin, Texas.

My perspective on the 1965 discovery of the CMBR is quite different from that of other contributors to these essays. Dave Wilkinson and I had our first measurements of the CMBR in the summer of 1965. We satisfied ourselves and our colleagues – Bob Dicke and Jim Peebles – that they were valid. Shortly after this, I left Princeton to join the staff of the Commission on College Physics in Ann Arbor. One thing led to another in my career, and by 1971 I had gone into academic administration full time at the University of Minnesota. Research became, for the remainder of my life, a spectator sport in which I played a support role in a variety of administrative ways. I've remained an active spectator, keeping up with scientific press reports on developments in which most other authors of these essays were directly involved.

Dave Wilkinson and I began work on the Princeton measurement of CMBR in 1964 – I had finished work on the Eötvös–Dicke experiment the previous year and we had written it up for publication (Roll, Krotkov and Dicke 1964), and Dave Wilkinson had recently joined Bob Dicke's research group. Dicke set both the theoretical context for our work – looking for remnant radiation from the big bang – and the experimental approach – using the Dicke radiometer he had invented in 1946 at the MIT Radiation Laboratory. Jim Peebles was doing the theoretical calculations, keeping us informed of how they were related to our experimental work. Dave and I were experimental physicists with no previous experience in radio astronomy and little experience working with microwave electronics and liquid helium. But we learned, and we designed and tested the equipment, with encouragement from Dicke and other groups in the Palmer Physical Laboratory. The work progressed well, and by February 1965 we expected to get data that summer.

In his last interview, Dave Wilkinson described the telephone call from Arno Penzias to Bob Dicke during one of our weekly lunch meetings. Dave's description is exactly as I remember it, with the exception of the length of the call. Dave described it as short, about 5 minutes, while I remember it as long, about 30–40 minutes. Visits were exchanged with Penzias and Bob Wilson at the Bell Laboratories Holmdel site and Princeton, and we all knew that the Princeton group was going to be number 2 rather than number 1 on the discovery.

What did we discuss during these exchange visits? Two things have stuck in my mind, though I can't trust my recall too far after 40 years. The first is that Dave and I quizzed Penzias and Wilson on the details of their equipment – how they had dealt with the many difficult problems to eliminate sources of systematic errors. They satisfied us that they had done this properly, and we shared information on what we were doing about these same problems. We learned about pigeon droppings in the horn-reflector antenna that Penzias and Wilson were using. Our equipment at Princeton was smaller for the shorter wavelength (our 3 cm versus their 7.3 cm) and we could more easily cover it when not in use. Even though the birds themselves were plentiful on our Guyot Hall observing tower (Figure 4.22), heat radiation from pigeon droppings in the antenna was not a significant issue for us.

The second detail we asked about was what Penzias and Wilson were looking for when they started their measurements. My recall is that they

Fig. 4.22. The first Princeton CMBR experiment, on Guyot Hall. Peter Roll's outline appears behind the instrument; David Wilkinson is holding the screwdriver. Photo: Robert Matthews.

started out to make an absolute measurement of the radio flux from the Andromeda galaxy, for which they would need an accurate measurement of any background flux from the sky around Andromeda. I suspect my recall of this detail may be, at best, a little oversimplified and incomplete.

The story of how the two research groups learned of each other is told well enough in the 1978 Public Broadcasting System Nova program, *Whispers from Space*. Jim Peebles gave a talk on his work at the Johns Hopkins Applied Physics Laboratory in early February 1965, with Dicke thinking that we were far enough along with our apparatus to finish and get data before anyone else could do so – because it would take them longer than that to build the equipment. Professor Bernard Burke, a radio astronomer at MIT, either attended the lecture or heard about it – he knew a little about what Penzias and Wilson were up to and suggested to them that they should contact Dicke. What Bob Dicke quickly recognized was that, if someone already had the apparatus and had started or completed the measurement, they would beat us.

We had been pretty sure that, if what we were looking for turned out to be cosmological, it would be an important scientific discovery. Realizing we would be number 2 created, I think, a certain amount of "awkward-ness" among those of us at Princeton. It was a disappointment, of course. I hope I speak for Dave Wilkinson and Jim Peebles in saying that we all felt more disappointment for Bob Dicke than for ourselves – Bob had been so close to a big one more times than most scientists. Dicke, I suspect, felt more disappointment for Wilkinson and me than for himself.

We understood that these things happen, and that being number 2 was still very important. At that time, the explanation was by no means certain, and none of us could be dead sure that there wasn't something wrong with the measurements. Penzias and Wilson were initially stymied by what they had, and we hadn't yet gotten data to examine. Number 2 would be important to confirm the result and get a second point on the spectrum – if it wasn't thermal, then our explanation could not be correct. At Princeton, we all got back to work, knowing that we were on to something important.

Shortly after the exchange of visits between the Princeton and Bell Labs groups, Bob Dicke informed us that each group would publish a letter, to appear back-to-back in the July 1965 issue of *The Astrophysical Journal*. The first letter would be by Penzias and Wilson (1965a) announcing the discovery, followed by a second from the Princeton group interpreting the result as remnant thermal radiation from the big bang (Dicke *et al.* 1965). We got to work on our letter, based largely on the work of Peebles and Dicke, but including a description of the work Wilkinson and I had begun.

Over the years, the "awkwardness" associated with the discovery has been, for me, explaining the situation to others. Broadcast of the PBS Nova program in 1978, followed shortly by the Nobel Prize award in the fall of that year, was the first time this work had been in the public eye enough to trigger these questions. Did I – or the Princeton group – feel in some way "cheated by circumstances?" I certainly didn't feel that way. All of us with the Princeton group at the time were sorry that Bob Dicke had not been included. I think we all understood also, in different ways, why the Nobel Committee did not do it that way. A 25-year history of research findings and conjecture preceded the discovery – it is summarized well by Peebles and Partridge, and documented in detail by many of the other essays in this collection. If nothing else, this left a confused situation for the Nobel Committee, and one about which there are differences of opinion. The discovery of the CMBR certainly deserved recognition. The Nobel Committee made a good choice, and it may have been the only one they could make.

My family and I first watched the PBS Nova program as a rerun in January 1979, after its first broadcast a few months earlier and after the Nobel Prize had been announced. We watched it, in fact, in an empty hospital room that nurses had set up for us across the hall from where my wife was recovering from surgery. Our children ranged from 9th grade to college junior in age, and I had previously told them the story and explained why Dave Wilkinson and I should not have received or shared the Nobel Prize, I thought they understood. But when Penzias and Wilson appeared on screen showing their apparatus, the kids began exclaiming, "Is that them, dad – are those the guys that won the prize? ...Boo! Hiss! Boo! You guys took the prize away from our dad!" So I explained it to them again. Their outburst on that occasion was a somewhat more candid and immature way of expressing what many others have asked. With a few more years behind them, I know they understand now. Soon I will show the tape and explain it to two very bright grandchildren.

I've told the story of this discovery many times, including to classes on cosmology I've given several times in the Senior University of Georgetown. (I rely on the 1978 Nova video tape to tell the part about Princeton and Bell Labs, however, and I leave the rest to questions.) This Senior University is an "institution" formed by several fellow residents of the "active adult community" in which my wife and I have lived for the past ten years. Its 600-odd students are almost all nonscientists – bright, mature adults with a lot of experience and accomplishments in their lives. They are fascinated not only by the story of how our universe began, but also by how and why scientists do this kind of work and arrive at some really strange conclusions – conclusions that are supported by a web of evidence from many different fields of research. Despite the fact that they are supposed to be objective, each scientist experiences the story personally and tells it differently. In this regard, science is no different than any other area of human endeavor.

The inside story of the discovery of CMBR, and the understanding of our universe to which it led over the past 40 years, is a magnificent example of the scientific method – messy, as it *really* is:

Looking back from 1965: Early research in the creation of heavy elements, intergalactic molecular spectra, the radiometer developed in radar research at MIT – a quarter century of missed hints and clues, almost but not quite pieced together more than once – finally pieced together, and two groups coming up with results at nearly the same time.

Looking forward from 1965: A discovery that was initially controversial has been so well documented with a variety of measuring techniques, and new and much more sensitive detectors used at high altitudes, from satellites, and over a wide range of wavelengths. The CMBR was first examined and thought of as uniform in all directions. It became possible to measure the direction and speed of our Galaxy's motion through absolute space by looking at a small asymmetry in the intensity of CMBR – hotter in one direction than in the opposite direction. Theories emerged on the earliest history of the universe, including Guth's strange superexpansion in the first instant of the big bang; and on how and when stars and galaxies began to form. These theories had to be consistent with one another, or they wouldn't be accepted. It became possible to calculate the distribution of tiny fluctuations in the CMBR and to measure these fluctuations from the WMAP satellite, distinguishing between some valid and invalid theoretical concepts and establishing numerical values for some of the important properties of the universe in which we live.

What a different understanding of the universe this is now, compared to the time when the steady state and big bang theories were actively contending with one another! During these past 40 years, many other concepts and variations were tried and found wanting, either because of theoretical inconsistencies or observations that did not support them.

The story of this one discovery has several of hallmarks of the scientific method, in addition to the messiness mentioned above:

Hypothesis about what you are looking for. How do you expect the research to turn out? Whether this concept is well founded or speculative is beside the point. Dicke's real contribution to the original CMBR work was just that – it was his idea to look for red-shifted thermal radiation from the big bang. A corollary to this principle, however, is to be skeptical and challenge your own conclusions, especially if they support your biases. In drafting the second of the back-to-back letters to *The Astrophysical Journal* in the spring of 1965, Dicke incorporated a statement that the CMBR detection was evidence for a closed universe that would one day contract back on itself – a concept tied to his work on the Brans–Dicke scalar theory of gravitation. In a chance meeting of the two of us, I argued that he should remove this statement, leaving intact the discussion of ramifications of the CMBR for open and closed models of the universe (flat was thought to be very improbable at the time). I was quite uncomfortable disagreeing with a person for whom I had the utmost respect. Neither of us could have guessed that results, 40 years later from the WMAP satellite, would show the distribution of tiny fluctuations in the CMBR and confirm a flat universe so convincingly.

Careful documentation. When Dave Wilkinson and I completed our first measurements, we took time and care to document, in a 1967 article in the *Annals of Physics*, what we did and how we did it, in complete detail (Roll and Wilkinson 1967). I had done this earlier with the results of the Eötvös–Dicke experiment, because the validity and limits set by a null result are the important part of the experiment. We did likewise with our CMBR measurement, because, at the time, it was controversial and not at all accepted that it was a thermal spectrum; if there was anything wrong with our methods and analysis, we wanted others to be able to find it. In hindsight, this was completely unnecessary. There have been so many measurements by so many techniques confirming the blackbody properties of the CMBR and more, that the specifics of how we did the second measurement have become almost irrelevant. Nevertheless, it was important to both of us at the time to complete the job properly.

Persistence. There are three examples of this among the people I worked closely with at Princeton. The first is Bob Dicke, who devoted the last half of his professional life to gravitation and cosmology – devising conceptual/theoretical models, experiments, and observations to understand better the nature of this basic law of physics and the physical nature of our universe.

The second is Dave Wilkinson, who spent his entire career after graduate school following the trail of the cosmic microwave background, eventually into space. Dave's scientific legacy is his two decades of work on satellite observations of the CMBR, culminating in the Wilkinson Microwave

Anisotropy Probe satellite. Results are still coming out of data from WMAP, as recently as two weeks before I write these words. All of us who knew him, even from way back, grieve that he is not still among us to witness the results of his dedication.

The third is Jim Peebles, who started as Bob Dicke's student and stuck with his study of galaxies, cosmology, and related matters from a more theoretical perspective – but still related closely to observations and measurements. I am not as familiar with the details of Jim's work since I left the field, but I hear about it often enough to know that he has been at it consistently and persistently for 40 years.

I'm quite sure that all three of my former colleagues have contributed as much, to science and society, by the students they trained and mentored as by the research they have pursued. Some of them have become successful scientists in their own right – others have gone off into other fields, as I did, and contributed in other ways.

An autobiographical appendix: notes on what drove me to physics, and then to leave for a different career. I came to physics from a family with no particular interests or talents in things scientific. From an early age, I had a knack and interest in things mechanical and quantitative. I entered Yale as an undergraduate with many interests. Before my senior year, I had taken no physics at Yale beyond a noncalculus introductory course. (I did sit through several graduate courses at Heidelberg during my junior year on an exchange scholarship.) My first job out of Yale (nuclear reactor design at Westinghouse) and my graduate work in experimental nuclear physics back at Yale were both interesting and rewarding. But I've also played the French horn all my life. My motivation for physics was at least partly to understand the physics of that treacherous instrument, so that I might improve my skills as a performer. This, however, was not a fashionable area of physics research, and gravitation and cosmology turned out to be far more interesting.

Finally, when I became a full-time administrator at the University of Minnesota in the 1970s, I was able to continue teaching a course in Musical Acoustics and engage in a little research and dissertation supervision with the Departments of Music and Music Education. I learned from the late Arthur Benade (Case Western Reserve) that the basic physics of the French horn and other brass instruments is governed by the Webster horn equation (Bell Labs, ca. 1916), which is none other than the Schrödinger equation with a transformation of variables. And I did learn how to play the horn better because of this work in the 1970s.

In 1965 I left Princeton for a year with the Commission on College Physics in Ann Arbor. A major activity that year was a report on "Computers in Physics Education" – the first ever report on the role of computers in higher education. When I joined the Physics faculty at the University of Minnesota in 1966, I was quickly identified as "... an expert on computers in education ..." By 1971 I was serving on so many committees, doing interesting work for the University and the state, that I moved full time into academic administration, with a portfolio including computers; radio, television, and audio-visual services; and library technology. In 1984, I moved to Northwestern University as Vice President for Information Technology, leaving behind my vestigial teaching and research in musical acoustics. From there I moved in 1992 to Executive Director of netILLINOIS, a nonprofit internet service provider mostly for Illinois educational institutions in the early days of the Internet. In 1995, I retired and moved with my wife to a new Sun City development in Georgetown, Texas.

In hindsight, it turns out that the theme in my life since 1971 has been networking and communities, rather than physics. This began with my appointment to a Cable Television Advisory Committee of the Metropolitan Council of the Twin Cities in late 1971, where the theme was cable TV as a community service network. Through most of the 1970s and 1980s, I was a board member of EDUCOM, an organization that pioneered networking to support academic communities and introduced higher education to the Internet. At Northwestern, I set the stage for a proper networked campus, though it did not get far off the ground during my tenure there. As I approached retirement, it was clear that the Internet was the platform for the "community network" that so many of the activist younger generation were promoting in the 1970s in Minnesota. And so I moved to Sun City, Texas, with an interest in seeing how the Internet might become a community network as it matured. And this is a work in progress. We started with a Computer Club that now has 2000 members (out of 7800 residents) and are finally in the process of implementing a community web portal, which will be our community network.

Throughout this 35-year period, the scientific and engineering research communities have been the creators of the platform for community networks of all kinds – ARPANet, BitNet, Usenet, TCP/IP, and all the others. These networks migrated into the larger society, finally, after 1989, when Tim Berners-Lee developed the World Wide Web at CERN, and in 1993 when Larry Smarr, an astrophysicist and Director of the National Center for Supercomputing Applications at the University of Illinois Urbana-Champaign, fathered the first web browser, Mosaic. The Internet as we know it today was catapulted into society and the economy by the particle physics and astrophysics research communities, as a tool that has improved scientific communication and made progress in science faster, more efficient, and more accessible. It has transformed not only research, but also society and the economy. Even retirement communities such as the one in which we now live.

One of the issues which interests many of our fellow retirees is why the US taxpayer should fund research in basic science. Cosmology really doesn't have that much impact on everyday life. I conclude my Senior University classes in Cosmology with this question: What is the return on this investment in basic research? The answer to this is now unbelievably easy. The economic impact of the Internet is the return on investment in particle physics and astrophysics research for the last n years – you pick the number of years, and the dollars work out just fine.

But this economic impact is all an accident – it's not why any of us do or have done research in things like the CMBR – those reasons are much more personal and complex.

4.7.4 R. Bruce Partridge: Early days of the primeval fireball

Bruce Partridge is a cosmologist turned radio astronomer who has taught at Haverford College for 38 years. He spent five years, 1965–1970, in the fabled Gravity Group at Princeton working on the "Primeval Fireball" (the CMBR) and primeval galaxies. He also served six years as the Education Officer of the American Astronomical Society, is president-elect of the Astronomical Society of the Pacific, and even survived eight years as an academic administrator at Haverford.

I will start as I propose to continue, in a quite personal and even anecdotal tone. I'll begin with my interest in astronomy, awakened in my teen years by building two reflecting telescopes with my father, and end with studies of the spectrum and anisotropy of the cosmic microwave radiation begun at Princeton in 1965.

In my college years, I bounced back and forth between history, physics, and astronomy. In retrospect, I can see that these were pointing toward my eventual fascination with the evolution of the universe and how we can determine it. Physics ended up as my major, but I got some grounding in astronomy as an undergraduate. In the early 1960s, Princeton University was just developing an undergraduate astronomy track. To gain admission, one had to take an elementary astronomy course designed primarily for the dimmest of undergraduates. We used a text coauthored by the professor, a text that mentioned the word "universe" only twice, both times misidentifying it with the Milky Way Galaxy. Fortunately, my subsequent courses were with George Field, a master teacher as well as a visionary astronomer. In 1960 I took from him a course that dealt in part with cosmology; that section of the course was based on Hermann Bondi's (1960a) thin book, *Cosmology*. Bondi's book was a fair representation of the state of cosmology at the time: attention was focused on cosmological models and possible observational tests of them. The largest scientific question in the field was whether the steady state model fit the (meager) data better than what we now call "big bang models."

It is perhaps a mark of how small a dent cosmology made on me that I elected to do research with George Field in the areas of interstellar grains and radio astronomy instead. But my main focus in my last year at Princeton and thereafter at Oxford was in quantum physics (my Oxford DPhil was on optical pumping in helium gas). Nevertheless, fascination with large-scale questions in astronomy was ticking away in the background. I recall attending, in 1964, a meeting of the Royal Astronomical Society (RAS) to hear about the newly discovered phenomenon of quasars. It was at that meeting, incidentally, that I first encountered Dennis Sciama, and noted both his wonderful ability to explain scientific principles clearly and his collegial treatment of a very young Stephen Hawking.

So, when it came time to apply for postdoctoral positions, I looked to groups in both Britain and the USA that were bringing techniques of physics to bear on astronomical or cosmological questions. My Princeton background led me to send an application to Bob Dicke. Bob's invitation to join the fabled "Gravity Group" was the crucial event in my scientific career.

As a 25-year-old with a scant knowledge of cosmology, I walked into Bob Dicke's office in the late summer of 1965. I knew of Bob's ongoing work on the Eötvös experiment, but his enthusiasm in 1965 was more firmly directed toward either explorations of Solar oblateness (as a test of relativity and the scalar-tensor variant) or the newly discovered microwave background radiation. Generous as always, he offered me a free choice, and then took me to see the two experimental setups. We went first to the Solar oblateness experiment, housed in a small wooden hut down by the Princeton Observatory. The hut was crowded with complicated electronics, many of them lock-in amplifiers, a Dicke invention I came to love and rely on. But the assembly of electronics was rather daunting. In contrast, the microwave background apparatus looked comfortingly simpler and even familiar – I had used microwave techniques in my thesis research. And I thought Dave Wilkinson would be a fine person to work with. Boy, was I right!

With great good fortune, I chose as my first effort in the Gravity Group to work with Dave on designing and running what became the first specifically planned CMBR anisotropy experiment. The way that experiment was planned and carried out provides some useful lessons on how one should – and should not – design an experiment.

Dave and his colleague Peter Roll (whose contribution commences on page 213) had a year or so earlier designed an experiment to detect the radiation left over from the big bang. This instrument, shown in Figure 4.23, was specifically designed to make an absolute measurement of temperature or intensity of the CMBR.

To measure or put limits on the anisotropy of the radiation requires a quite different approach. On the one hand, anisotropy measurements are easier, since they can be made comparatively (is this part of the sky hotter than that part?). On the other hand, Dave and I recognized that to be meaningful, such an experiment needed to be much more sensitive, and to produce temperature measurements accurate to a few parts in 1000. Penzias and Wilson (1965a) in their discovery paper had already noted that the "excess noise" they picked up is approximately isotropic, with any variations in intensity below about 10%. We aimed to improve this limit by nearly two orders of magnitude. The plan was to scan a circle in the sky at constant

Fig. 4.23. The former pigeon coop atop the Geology Building that housed the Roll–Wilkinson (1966) CMBR spectrum measurement and the 1965–1967 Princeton "isotropometer." Photo: Robert Matthews.

declination over a long enough period so that any diurnal variations would cancel out. A dipole distribution in the CMBR temperature would then produce a 24-h variation (in sidereal time), and a quadrupole distribution would produce a variation at 12-h period.

As anyone who has lived in New Jersey knows, however, the atmosphere over Princeton is not exactly stable. To cancel out the atmosphere to first order, we needed to make calibration observations of a stable, unmoving region of the sky through a comparable air mass. We thus elected to switch the beam (observing direction) between the north celestial pole (the fixed point) and a point an equal angular distance away from the zenith to the south. We thus ended up scanning a circle at declination $\delta = -8°$. There were two levels of beam switching. First, we switched at about 1000 Hz back and forth between our main horn antenna and a much smaller antenna pointed toward the zenith. As a further control, we switched the beam of the primary antenna itself every few minutes by raising a reflecting sheet to divert the beam to the north celestial pole. This was the Princeton "isotropometer" housed in an unused pigeon coop on a tower of Guyot Hall (Figures 4.23 and 4.24).

The kilohertz signal was phase-sensitively detected, and plotted out using a pen and ink chart recorder. (Mentioning a pen and ink chart recorder to scientists today must be the functional equivalent of telling my children that I walked 3 miles each day to catch the school bus. Both are true.) Dave and

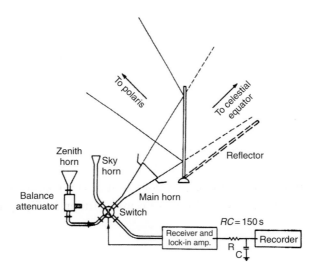

Fig. 4.24. Schematic of the "isotropometer," showing the moving reflector used to zero the instrument (Partridge and Wilkinson 1967). ©1967 American Physical Society.

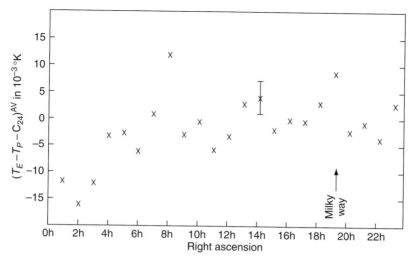

Fig. 4.25. Results of the scan of a circle at declination $\delta = -8°$ in the experiment at Princeton University. The fractional temperature fluctuations are $\delta T/T \sim 3 \times 10^{-3}$ (Wilkinson and Partridge 1967). ©1969 Nature Publishing Group.

I and a handful of undergraduate students working with us then read the output of the chart recorder by hand to determine the differences between the declination $\delta = -8°$ circle and our constant calibration point, the north celestial pole. We ran this experiment for substantially more than a year to help average out diurnal effects. Some of those results appear in Figure 4.25.

It soon became clear that atmospheric noise was completely dominating the signal. By late 1966 we were planning improvements. It would have helped, for instance, if we had been able to switch the main beam more rapidly, but we were aware that the ferrite devices used for switching are themselves a source of noise and potential systematic error, a problem later encountered in another anisotropy experiment by Dave Wilkinson and Paul Henry (Henry 1971). So we took another approach to doing a better experiment, trying to find a place where the atmosphere is more benign. We probably should have leapt immediately to the conclusion that we needed to get above the atmosphere altogether, as Dave later did in his pioneering balloon experiments, and as George Smoot and his colleagues later did with their U-2 experiments (Smoot, Gorenstein and Muller 1977). But we were a frugal pair, so we decided instead to find the place in the United States with the least cloud cover. Dave discovered that that is southwestern Arizona, and I found out that there is an Army base at Yuma, smack in the middle of this relatively cloudless zone. Through my father's Army connections, I got us permission to move an improved isotropy-measuring device to the Army's Yuma Proving Ground.

We faced some constraints in designing the equipment. Both of us were busy teaching and could not spend much time in Yuma. So we needed to design a fully automated station that would take data and record it and that needed no daily maintenance. The equipment was designed with the main horn antenna pointed down, to prevent the collection of dust, rain, dead moths, etc. We also designed the equipment to scan two circles in the sky as well as the constant reference point, the north celestial pole (the reference horn also pointed there). Finally, we also took much greater care to prevent radiation from the ground entering the main antenna through its side lobes – see the ground screens identified in Figure 4.26. I took charge of designing the structure to support the main antenna, as well as the rotating beam-switching device, a tilted, elliptical mirror. I recall bringing my designs to Bob Dicke, who took a brief look at them and said, "Well, it is certainly sturdy." By that he meant that I had overdesigned the strength of the contraption by several orders of magnitude – I suspect it was at least as "sturdy" as the Army's top line tank!

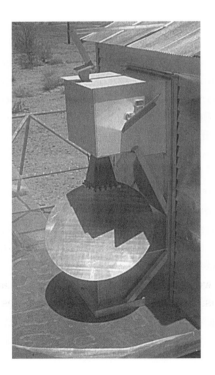

Fig. 4.26. A refined experiment to look for anisotropy in the CMBR. Note the inverted horn and the use of ground screens to minimize stray radiation from the ground (and the sturdy construction).

Now, if you're designing a remote experiment, you need to have it in a place where casual hikers or hunters are not likely to poke around in it. The management at the Army's Yuma Proving Ground suggested that we use a securely fenced area at the outer edge of the base. It was securely fenced because it was the site at which the Army tested the integrity of nerve gas shells. There were racks and racks of nerve gas shells of various sorts lying about in the desert, left out to see whether or when they would leak. Needless to say, the area was both securely fenced and patrolled.

So, in the summer of 1967 we packed the monstrosity I had designed plus some additional equipment (see below) into a large U-Haul truck, and set out for the west. Dave used the trip as a family vacation; I got to drive the U-Haul. When we arrived at the Yuma base, Dave was appointed a temporary captain, and I got to be a lieutenant. The U-Haul was costing us a fortune to drive back and forth to our remote site, so we bought an item designated as a "personal transport device" to the NSF, otherwise known as a moped, for me to commute to the instrument.

We soon had the equipment up and running. Since useful computers were still a ways in the future, the basic control mechanism for the experiment was derived from a rewired digital clock, and the data were printed out on a line printer (whose values still needed to be recorded and sorted by hand).

While we were installing the Yuma apparatus, Dave and I were finishing up two papers on the results of the first anisotropy experiment at Princeton. One of those was written in a crummy motel room in Yuma using the only available horizontal surface, the top of a beer cooler. I remember sitting on the dirty green shag carpet, drinking Coors, and the excitement of reaching millikelvin levels in anisotropy.

We also had some time to learn about the nerve gas from soldiers on duty near our site. They explained the use of gas masks – we were issued with them in case the nerve gas really did leak – and told us that the first way to detect leaks of nerve gas was to check out the rabbits. Near each stack of nerve gas shells there was a hutch containing several standard laboratory rabbits. These were placed there since rabbits are highly sensitive, it appears, to nerve gas. So the first alarm for leakages was increased rabbit mortality. Well, in our brief time in Yuma, the rabbits started to die. There was considerable consternation, not least on our part, until a wise veterinarian pointed out that all the rabbits had been bought at the same time, and all of them appeared, entirely naturally, to be reaching the rabbit equivalent of three score and ten. Since we had escaped the nerve gas, Dave and I joined the soldiers' favorite game of sitting in a cargo container while someone set off a military-strength tear gas grenade. The game was to see who could last the longest before bolting for fresh air. We were young then.

As the Yuma experiment came on line, Dave and Bob Stokes left for the second main leg of the summer's work, a refined, multi-frequency measurement of the spectrum of the CMBR. I later joined them, traveling through the desert on my trusty "personal transportation device."

The idea here was to measure the temperature of the CMBR at three different frequencies – later extended to four by Paul Boynton – using very similar apparatus, so that the temperature measurements could be securely intercompared. In particular, the hope was to show that the spectrum of the radiation we were studying is not an exact Rayleigh–Jeans form, with energy density that varies with frequency ν as $u_\nu \propto \nu^2$, but instead shows some curvature as the peak of a blackbody spectrum at temperature about 3 K is approached.

So we designed radiometers having similar beam sizes, all able to couple to a common calibration cold load. To prevent systematic errors, we designed the main horn antenna to look downward at an angle, making it easy to couple to a tilted dewar containing the cold load without moving the apparatus (Figure 4.27; Stokes, Partridge and Wilkinson 1967). Thus, to deflect the beam to the zenith in order to measure the CMBR, we needed to use an oversize reflector. We also arranged the reflector to be movable, so that the main beam could be cast through different zenith angles, enabling us to measure the atmospheric emission with the same equipment used for the absolute temperature measurements. One of the three radiometers used in the 1967 campaign is shown in Figure 4.27, along with the experimental setup.

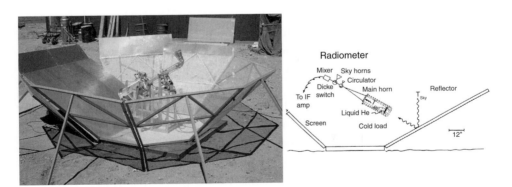

Fig. 4.27. Photo of one of the three radiometers used on White Mountain, California, to measure the spectrum of the CMBR (Stokes, Partridge and Wilkinson 1967). The horn antenna is coupled to a large-diameter cold load. On the right is a schematic of the radiometers (Wilkinson 1967). ©1967 American Physical Society.

It is worth mentioning the care we took to avoid systematic error. Dave Wilkinson, as all who knew him will attest, was extraordinarily careful about finding and eliminating, or at least modeling, sources of systematic error. We took great precautions, for instance, to control emission from the ground leaking into the side lobes of the antennas we used. We were conscious that emission from the walls of the calibration cold load could present a problem, and for that reason we expanded the beam and used a large "over-moded" cold load immersed in liquid helium. I have already mentioned quasisimultaneous measurements of the atmospheric emission. And we also took account of the possible emissivity of the reflecting surface.

The result of this work was to produce temperature measurements at three wavelengths with substantially smaller error bars than previous workers had been able to obtain. The error bars were small enough to show rather convincingly that the spectrum of the CMBR does indeed begin to turn over at high frequencies, as expected for a 3 K blackbody (Figure 4.28). And the final temperature we derived from combining observations at the three frequencies gave a value $T = 2.68^{+0.09}_{-0.14}$ K, in remarkably good agreement with the COBE satellite results that came along nearly two decades later (Stokes, Partridge and Wilkinson 1967; Wilkinson, 1967).

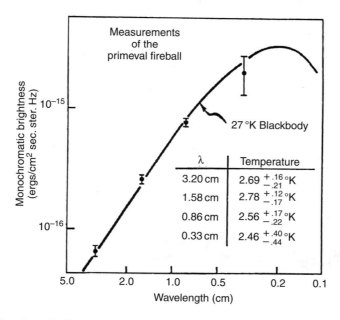

Fig. 4.28. Results of the Princeton measurements on White Mountain, showing a departure from the Rayleigh–Jeans law $u_\nu \propto \nu^2$ (Partridge 1969). The crucial 0.33-cm measurement was made by Boynton, Stokes and Wilkinson (p. 312). ©1969 American Scientist.

The spectral observations were carried out at the highest place in the United States with electrical power, the White Mountain Research Station maintained by the University of California. Not surprisingly, other groups had figured out that this was an excellent place from which to observe the microwave background. When we arrived, we discovered Bernie Burke and his colleagues busy assembling apparatus that looked an awful lot like that shown in Figure 4.27 (Ewing, Burke and Staelin 1967). Our group and his agreed to work entirely independently, so as not to influence one another's results. Yet another group, Welch *et al.* (1967), also recognized the value of high-altitude observations. However, they encountered problems with the design of their cold load calibrator, and perhaps as a consequence came up with too low a value for the CMBR temperature. On page 295 Welch describes how that turned out.

Even for these measurements, there was a constant struggle against the atmosphere. Uncertainties in the amount of emission from the atmosphere, particularly from water vapor, dominated the error budget. Throughout the experiment, we were worried about possible frequency-dependent systematic errors that could bias our results. I suspect it was at this stage that Dave came to recognize the value of balloon experiments, and even more of a satellite experiment to get out of the atmosphere altogether. Nevertheless, working with Bob Stokes and Paul Boynton, Dave went on to do one more ground-based temperature measurement in these years, the measurement carried out at 0.33-cm wavelength at the High Altitude Observatory in the Colorado Rocky Mountains (Boynton, Stokes and Wilkinson 1968). They found $T = 2.46^{+0.40}_{-0.44}$ K, which within the errors is consistent with the modern value. And, as a footnote, I went on to join an Italian–Berkeley–Haverford team that returned to White Mountain 15 years later to make refined spectral measurements at five wavelengths, 0.33–12 cm (Smoot *et al.* 1985).

What were we trying to accomplish with these early experiments? With the wisdom of hindsight, it is clear that we were beginning the process of mining the CMBR for cosmological clues. But in the years 1965–1968, the full value of spectral and anisotropy measurements was far from appreciated. The beautiful and influential theoretical work on the power spectrum of CMBR fluctuations, for instance, lay years in the future. So what were we really trying to accomplish?

First and foremost, we were trying to establish that the microwave radiation detected by Penzias and Wilson (1965a) is indeed cosmic, and not coming from more local sources in the Solar System, the Milky Way, or galaxies or from some other class of extragalactic objects. In the mid-1960s there were plenty of skeptics, and numerous noncosmological explanations of the "excess noise" reported by Penzias and Wilson. We recognized that strong proof of cosmic origin lay in two fundamental tests: the blackbody shape of the spectrum of the radiation and its isotropy on both large and small scales.

Electromagnetic radiation pervades the universe. At radio wavelengths it is dominated by the emission from galaxies and quasars. Could the "excess noise" detected by Penzias and Wilson simply be the high-frequency tail of this background? The spectrum holds the key to the answer. Emission from radio galaxies is typically dominated by the synchrotron process, producing a power-law spectrum $u_\nu \propto \nu^{-\alpha}$ with α generally in the range 0.5–1.0. This is very different from a thermal spectrum where $u_\nu \propto \nu^2$ at long wavelengths. Another possibility is "free-free" emission from a thin plasma with non-relativistic electrons, which typically produces a power-law spectrum with $\alpha \simeq 0.1$. Such a spectrum, too, is easy to distinguish from the truly thermal or blackbody spectrum expected from radiation left over from a hot, dense state of the early universe.

More difficult to distinguish from a true blackbody spectrum is *gray-body* – emission from an optically thin but higher temperature source. At wavelength $\lambda \gg 0.3/T$ cm, with temperature T measured in kelvin, gray-body emission can have the same ν^2 dependence as true blackbody emission, but the spectrum peaks at shorter wavelengths. To confirm true blackbody emission at $T \simeq 3\,\mathrm{K}$, we needed both to confirm the ν^2 dependence at long wavelength and find evidence for the peak expected near 0.1-cm wavelength.

It is worth repeating how unlikely it is to find a purely thermal spectrum in the cosmic setting, where densities tend to be very low. Only if the universe were many orders of magnitude denser than it is now, could true thermal equilibrium have been established. If the microwave background radiation truly does have a thermal spectrum, it not only establishes the cosmic origin, it also shows that the early properties of the universe were radically different from those prevailing today.

By 1967 we had the answer: we were seeing curvature in the spectrum consistent with a peak at a wavelength of about 1 mm (Stokes, Partridge and Wilkinson 1967; Wilkinson 1967).

Isotropy is the second test. An observer not moving with respect to the comoving coordinates of the universe discussed in Chapter 2 (p. 9) would see that radiation left over from the big bang appears isotropic (apart from the disturbances caused by the departure from a smooth mass distribution).[13] On the other hand, a Solar System origin would be expected to

---

[13] Parenthetically, this would not be the case if the universe itself were expanding in an anisotropic way. Ellis (p. 380) notes that this idea was of considerable interest in the 1960s.

produce intensity variations tied to coordinates fixed with respect to the direction to the Sun in the sky. Sources in our Galaxy would presumably produce radiation that peaks in the direction of the galactic plane, in an anisotropic distribution akin to the concentration of bright stars in the plane of the Milky Way. Such a distribution would introduce a large dipole moment, and particularly a quadrupole moment, into the distribution of CMBR intensities. If the radiation were somehow produced by a myriad of extragalactic radio sources, as suggested for instance by Wolfe and Burbidge (1969), it would be "grainy" on a small scale. More precise limits on anisotropy on both large and small scales would, we hoped, kill off these noncosmological explanations. This hope motivated our work, and we soon showed (Partridge and Wilkinson 1967; Smith and Partridge 1970) that the radiation is indeed highly isotropic on both large and small angular scales.

Nor were challenges to the cosmic origin of the CMBR mounted solely by inventive theorists. At least one experimental result, the pioneering rocket measurement of Shivanandan, Houck and Harwit (1968; see Harwit's piece here) seemed to favor a graybody spectrum. The results naturally raised doubts about the cosmic origin of the microwave background.

All of these results, attacks on the very notion of the "primeval fireball," were very much on our minds as we mounted the experiments described above and wrote up our results.

One anecdote encapsulates the skeptical air of the times. In 1969, as I recall, I gave a talk on our Yuma experiment at a meeting at Caltech. In the question period, a formally dressed, middle-aged man in the back asked, in effect, "Given that you see no change in emission as the sky passes overhead each day, how do you know your equipment is even switched on?" Fortunately – since the questioner was Charles Townes – I gave an appropriate answer, describing in detail the care we took to calibrate the instrument.

So I would say that in the 1960s, we were on a mission to convince the skeptics, an attitude that strongly colors an early review of the primeval fireball published in the spring 1969 issue of the American Scientist (Partridge 1969). I would suggest, however, that there was another influence at work. We were, after all, working for Bob Dicke, acknowledged as the master of the beautiful null experiment. These are experiments designed to test the absence of some physical effect by establishing more and more stringent upper limits on the magnitude of the effect. Dicke's ultrasensitive version of the Eötvös experiment, for instance, showed that there are no differences in the way gravity acts on different chemical elements to a level of roughly one part in 10^{11}. I will speak only for myself here, but part of the motivation driving me was to do better and better null experiments on the CMBR, in particular to establish lower and lower upper limits on possible fluctuations in the CMBR (see Partridge 2004). In other words, I am confessing to having been driven less by theoretical concerns or predictions than by an experimenter's lust to do the best possible experiment, and to cover as much parameter space – in this case sensitivity and angular scale – as possible. As Dave and I were planning better experiments to push down limits on the amplitude of the temperature fluctuations on degree scales, I was also thinking about ways to limit anisotropies on small angular scales. This required the use of larger aperture devices, since the angular scale goes approximately inversely as the diameter. Others – Ned Conklin (1969), Eugene Epstein (1967) and Penzias, Schraml and Wilson (1969) – had already set upper limits on arcminute-scale fluctuations; Paul Boynton and I realized we could reach both smaller angular scales and higher sensitivity using a 36-ft telescope operated by the NRAO. Our results turned out to be only mildly interesting, and I mention them simply because they reflect at least one person's motivation in these early years – to set the best possible limits on fluctuations at all angular scales.

That I was not alone in this aim is reflected in the way in which anisotropy measurements were presented in these early years, and for at least a decade afterward (Figure 4.29). What is shown is basically a plot of upper limits on

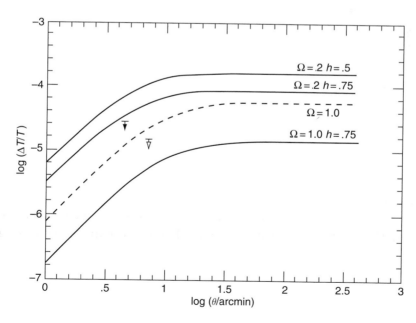

Fig. 4.29. An early (and poor) way of representing upper limits on anisotropy in the CMBR on various angular scales.

the fractional temperature fluctuations $\delta T/T$ across the sky, with little reference to any underlying theory of what the angular spectrum of anisotropies might be (though we did know that the overall amplitude would be affected by the mass density fluctuations). Also reflecting the focus on upper limits is the fact that a paper I wrote with Italian colleagues in the early 1980s was initially rejected solely on the grounds that the upper limit we established was not as low as the upper limit somebody else had established, despite the fact that we were working at degree angular scales and the "better" experiment was at arcminute scales.

Is it possible that the drive to set lower and lower limits on CMBR anisotropies has a contemporary analog in the drive to determine the cosmic equation of state parameter w that is supposed to describe the evolution of the dark energy density? Does the (expensive) effort to improve limits on this effect parallel our efforts 40 years ago to improve limits on the CMBR temperature anisotropy? Indeed, the same question could be asked about plans to measure another cosmic parameter, the ratio, r, of tensor to scalar anisotropy perturbations. Clever scientists are designing better and better methods of refining measurements of r and w, but without much theoretical guidance (especially on r).

The importance of the role in theory brings me to the crucial contributions made by our colleagues in the Soviet Union, well described in this volume by many of the key players. As Malcolm Longair points out, few Soviet scientists, and certainly none of the stature of Yakov Zel'dovich, could travel abroad in the 1960s. So some of us went to the Soviet Union: Malcolm for a long and fruitful stay and I for a much shorter pair of visits in 1968 and 1969. The summer of 1968 was a difficult one to be traveling behind the Iron Curtain: the Soviet Union had just cracked down on their Czech satellite. Those of us from the US were initially advised by the State Department not to go to the Soviet Union, but after a brief wait in Britain, several of us decided to make the trip, albeit with some trepidation.

During the wait in Britain, I gave an impromptu talk on the Princeton CMBR results at Cambridge. In the audience was Fred Hoyle and several other Steady Staters who, to put it mildly, viewed our results with skepticism. At one point in the middle of my talk, one of Fred's colleagues marched to the front of the hall and took over the viewgraph to quibble with the statistics I was presenting. I eventually wrested back the viewgraph and continued, somewhat shakenly, to give my talk. Given my deep admiration for Fred Hoyle (eight years earlier I had hoped to work with him on a PhD but ended up at Oxford instead), I remained sad that he could never accept the cosmic origin of the microwave background. As late as 1988, he was still holding out, and we crossed gentlemanly swords at a meeting in Bologna over the importance of spectral measurements in establishing a truly cosmic origin of the 3-K background.

Back to 1968. The US Embassy in Britain finally gave us a green light and a group of US and European astronomers flew to Leningrad on the way to Moscow and eventually to a scientific meeting in Soviet Georgia (Figure 4.30). We could not have been greeted more warmly and more positively by our Soviet colleagues. Within hours of arriving in Moscow, I was taken to Yakov Zel'dovich's apartment. The big room in the apartment was equipped with a conference-size wooden table and at least one blackboard. He and his younger colleagues – Andrei Doroshkevich, Igor Novikov, and Rashid Sunyaev among them – made me feel welcome, fed me cucumbers and tomatoes as well as fizzy pear juice and endless cups of tea. And then they subjected me to the most detailed, searching, and exhilarating questioning I can remember. They wanted to know every single detail of the US work on the CMBR. On the experimental side, I believe I met their expectations. I think I did rather less well in explaining the many theoretical advances which colleagues like Jim Peebles were working on. After squeezing me dry, Zel'dovich and his colleagues explained to me some of the ideas then being

Fig. 4.30. Bruce Partridge, Rashid Sunyaev, Vladimir Dashevskii and Malcolm Longair, left to right, at an International Conference on Gravitation and the Theory of Relativity, Tbilisi, USSR, September 1968.

pursued in their research group. It was, in effect, a mini-seminar on modern cosmology, delivered with Zel'dovich's characteristic clarity and enthusiasm. Here too, I could have done a better job as a "channel" between Zel'dovich's brilliant group and scientists in the US working on similar problems. As the years have gone on, the details have grown a bit fuzzy in my mind, but my reverence for Yakov Borisovich has not changed a whit. As striking as his command of physics was the broad reach of his interests. In another encounter, a decade later, my wife and he talked about Freud; as a professional psychologist she was struck by the range and depth of his questions about an author who, to put it mildly, was not exactly favored in the Russia of the times.

I met Zel'dovich for a fourth and, sadly, final time in the summer of 1981, thanks to an invitation from Rashid Sunyaev to visit Russian radio astronomy facilities, including RATAN, where Sergey Trushkin gave us a memorable tour. We spent a few days with Yakov Borisovich in Yalta, where he was spending a summer vacation by the sea. On this occasion, he did his best to explain to me the new ideas bubbling up from Andrei Linde and others on inflation. Much of this was done on the beach, with Zel'dovich drawing diagrams in the sand. I wish I could claim that I had fully understood what he was saying, and that I had brought these ideas back to the US with the same precision and clarity that he used. I failed Zel'dovich and his colleagues pretty miserably. At one point, he became exasperated with me when I did not understand how a vacuum – false or not – could do anything of interest in cosmology. I was not keeping up and he grew frustrated. So he gave up and we went swimming. He attacked the waves with the same enthusiasm he did everything else, and I crept a little bit back into his good graces by keeping up with him, in swimming at least. And he did arrange to get me out of the clutches of the Soviet police in Yalta after I had made a mistake by photographing the wrong building at the wrong time ... but that's another story!

Finally, I would like to explore a phase change that occurred in the field in the early 1970s. It may surprise some of you, who see the CMBR as a giant growth industry, to learn that by the end of the 1960s interest in it was waning, or at least changing. That is reflected in the rate of publication of papers dealing with the CMBR shown in Figure 4.31, taken from the "history" chapter in my book, 3 K: *The Cosmic Microwave Background Radiation* (Partridge 1995). Before exploring some reasons for the temporary dwindling interest in the CMBR, let me make a much more positive point: this is the time when many other groups here, in Europe and in Russia began to take an interest in improved CMBR experiments.

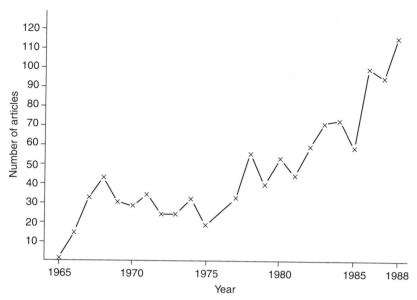

Fig. 4.31. The number of CMBR-related papers published each year (Partridge 1995). Note the lack of activity in the years 1969–1977.

The emergence and success of these groups is linked to one of the reasons I see for the cooling of ardor at Princeton. By the late 1960s we had, in effect, done all the easy experiments. The experiments Dave and I did were, as I liked to say, "one-Cadillac scale," costing of the order of $10,000 to 20,000 each. They were constructed almost entirely from commercially available components. To improve these experiments, new detectors and optics were needed, as were instruments designed specifically for the detection of CMBR anisotropies or the precision measurement of the CMBR spectrum. In addition to new technology, better observing strategies were needed (recall my remarks above on problems created by the atmosphere). The same strictures applied to the use of existing radio telescopes: observers had pushed them to their technological limits as well.

New groups brought new techniques and technologies to bear. I want to mention specifically the introduction of bolometric detectors into the field, and to praise the foresight of people like Paul Richards and Francesco Melchiorri, and of Rai Weiss, who has written for this volume. Francesco was a pioneer in the field, who unfortunately passed away in 2005.

So there was a pause while new technologies and techniques were brought to bear. Along with new groups joining the field, Dave Wilkinson wisely moved in the direction of balloon experiments. I got interested in the use of radio-frequency interferometry to probe yet smaller angular scales. The introduction or exploration of these new techniques took time, and that is in part responsible for the drop in activity in the CMBR field in the early 1970s.

Another factor, at least in the case of Princeton's Gravity Group, was the explosion of other interesting things to do in astrophysics, ranging from pulsar timing to searches for "primeval galaxies." The experimentalists of the Gravity Group found lots of other intriguing things to do while we waited to sort out new CMBR technologies and techniques. Dave, for instance, began to explore limits on extragalactic optical backgrounds and oversaw Marc Davis's pioneering search for primeval galaxies. I mounted a separate search for primeval galaxies, and got interested in observational tests of the Wheeler–Feynman (1945) absorber theory (Partridge 1973) and searches for bursts of radio-frequency emission. Both Dave and I, joined by Ed Groth and Paul Boynton, spent a lot of time from the spring of 1969 on making precision timing measurements of the optical pulses of the Crab Nebula Pulsar. In an ironic twist, we felt we had discovered evidence that the Crab Nebula Pulsar is slowing down due to the loss of energy by gravitational radiation; it turned out that nature had thrown us a curve ball in the form of a glitch in the pulsar period. But another, cleaner pulsar system would reveal energy loss by gravitational radiation and win the Nobel Prize for Russell Hulse and Joe Taylor.

I will end by floating an idea that may be strongly colored by retrospective wisdom. Could the lull in CMBR activities have been in part influenced by the fact that we were beginning to pay some attention to theoretical predictions as to the properties of CMBR anisotropies and spectral distortions? That is, instead of blindly trying to set better and better limits on both anisotropy and spectral distortion at a range of wavelengths and scales, were we, I wonder, beginning to recognize (a) how hard it would be to see meaningful spectral distortions and (b) that the amplitude of anisotropies would in general be very small except on certain angular scales? Frankly, my recollection of my mood in the late 1960s and early 1970s is now a little too hazy for me to say for sure. What I can say is that the five years, 1965 to 1970, were not only the years that truly established physical cosmology, but were a hell of a lot of fun!

4.8 Developments at Cambridge

4.8.1 Malcolm S. Longair: Cambridge cosmology in the 1960s

Malcolm Longair carried out his postgraduate studies at the Mullard Radio Astronomy Observatory of the Cavendish Laboratory, University of Cambridge from 1963 to 1967. He spent the academic year 1968–1969 as a

Royal Society Exchange Fellow to Moscow. Subsequently, he was Astronomer Royal for Scotland and Director of the Royal Observatory Edinburgh from 1980 to 1991. He returned to Cambridge in 1991 as Jacksonian Professor of Natural Philosophy and from 1997 to 2005 was Head of the Cavendish Laboratory.

Cosmology in Cambridge in the 1950s and early 1960s was dominated by the controversy between Martin Ryle and Fred Hoyle over the interpretation of the number counts of extragalactic radio sources. I was only involved in this rather acrimonious debate toward the end of the most controversial period, which most people would concede ended with the establishment of the thermal nature of the CMBR and its remarkable isotropy.

I have given many more details and full references to the most significant events in the development of steady state cosmology and the controversy over the number counts of radio sources in my book, *The Cosmic Century: A History of Astrophysics and Cosmology* (Longair 2006).[14] An interesting feature of the development of steady state cosmology was that most of the proponents were UK cosmologists and this at least partly accounts for the fact that steady state cosmology was taken much more seriously in the UK than in the USA.

The facts of the controversy and its resolution are now well known and will not be repeated here. What has been less appreciated are the very different scientific agendas which Ryle and Hoyle pursued. Both were larger than life personalities who had honed their research skills under periods of extremely high pressure during World War II. Hoyle was a brilliant astrophysicist whose extensive imagination led to some of his most remarkable achievements; for example, the prediction of the triple-α resonance, as well as to the controversies which dogged his career. Hoyle's biography is the story of a disadvantaged Yorkshire schoolboy with a very strong independent streak from the very beginning, who, by sheer ability and hard work, attained the premier position in British academic astronomy, the Plumian Professorship of Astronomy at the University of Cambridge.

Although trained as a physicist, Ryle was primarily an electrical engineer with a genius for making complex radio-receiving systems operate. In particular, he understood how aperture synthesis techniques could be used to provide both angular resolution and high sensitivity for radio astronomical observations. The principles of aperture synthesis were understood by 1954, but putting them into practice was a virtuoso technical achievement which

---

[14] Helge Kragh's (1966) *Cosmology and Controversy: The Historical Development of Two Theories of the Universe* can also be thoroughly recommended.

involved a considerable team of researchers and support staff under Ryle's charismatic leadership.

Ryle had remarkable physical intuition but little time for complex mathematical arguments or the intricacies of theory. Early in the history of the development of radio astronomy, profound personal animosities developed between a number of the participants in the debates. The excess of faint radio sources was certainly exaggerated in the 2C survey of 1954 and this experience made Ryle reluctant to release his results until he was sure they were fully understood and as secure as they could be. I was often present when he lamented the fact that what had taken him years to get right through the dedicated efforts of his team could be dismissed in a stroke by a theorist.

These events had positive and negative impacts upon the work of the Cambridge radio astronomy group. The positive side was that new astrophysical and cosmological opportunities had been opened up. In 1956, the radio observatory moved to a disused wartime Air Ministry bomb store at Lord's Bridge and, in acknowledgment of a grant of £100,000 from the electronics company Mullard Ltd, the Mullard Radio Astronomy Observatory was opened in 1957. The technical successes of the interferometry programs led to construction of the first fully steerable aperture synthesis radio telescope system, the One-Mile Telescope completed in 1965, and then to the 5-km Telescope completed in 1972. These were undoubtedly the most powerful radio telescope systems in terms of angular resolution and sensitivity for a number of years.

The negative side was that the group became very much more defensive in its interaction with outside groups. Great care was taken to ensure the reliability and completeness of subsequent catalogs and radio maps before they were made available outside the group. The accusation of secrecy recurred with Jocelyn Bell and Antony Hewish's discovery of the radio pulsars in 1967. This was such a surprising and important discovery that every effort had to be made to ensure that the data were absolutely secure before the data were made public. Even those of us in the next door offices did not know what was going on until Hewish gave a colloquium in the week of publication of the discovery paper in *Nature*.

The contrasting approaches of the experimental and theoretical astrophysicists were exacerbated by the split which occurred in the early 1960s between the Cavendish Laboratory and the newly formed Department of Applied Mathematics and Theoretical Physics (DAMTP). Hoyle was a member of DAMTP, as was Dennis Sciama, who was also a strong proponent of steady state cosmology. The differing points of view were institutionalized and observational and theoretical cosmologists pursued their research programs independently.

The resolution of the controversy over the radio source counts only came with the construction of the next generation of radio telescopes which had higher angular resolution and hence were less sensitive to the effects of source confusion. The radio source counts derived from the 4C catalogs showed a clear excess over the expectations of Euclidean world models. The optical identification programs led to the discovery of quasars in the early 1960s. By the mid-1960s, the evidence was compelling that there was indeed an excess of extragalactic radio sources at large redshifts and this was at variance with the expectations of steady state cosmology.

It was a great sadness that relations between Hoyle and Ryle were so soured by the controversy over the radio source counts. In early 1967, Peter Scheuer and I attempted a reconciliation between them. We believed the new data were very compelling that the excess of faint sources was real and that there was no need to prolong hostilities. The four of us got together in Hoyle's newly founded Institute of Theoretical Astronomy in Cambridge to try to effect a reconciliation. We talked for about 45 minutes but, sadly, there was no longer any common ground. Hoyle and Ryle simply repeated their entrenched views. It was one of the saddest events of my career.

In fact, during the 1960s, the differences in interpretation of the source count data were confined to a relatively small, but prominent, number of individuals. The great strength of Cambridge astrophysics expanded enormously through the 1960s. The foundation of Hoyle's Institute of Theoretical Astronomy in the mid-1960s provided positions for a very powerful group of post-doctoral fellows including Peter Strittmatter, John Faulkner, and Peter Eggleton as well as regular summer visitors including Geoff and Margaret Burbidge, William Fowler, and the next generation of their research associates including Don Clayton, Bob Wagoner, Wal Sargent, and their colleagues. A good appreciation of success of this venture is included in *The Scientific Legacy of Fred Hoyle* (Gough 2005) in which Hoyle's wide range of scientific interests are reviewed mostly by his collaborators. At the same time, in DAMTP, Dennis Sciama took on a succession of brilliant research students who worked on quite different problems in general relativity, relativistic astrophysics, and cosmology. Among these, George Ellis completed his PhD in 1964, Stephen Hawking in 1966, Brandon Carter in 1967, Martin Rees in 1967 and Malcolm McCallum in 1971. The intellectual vitality of Cambridge astrophysics and cosmology was remarkable.

Despite the differences between a few of the most senior astronomers, the staff members, postdocs, and graduate students of the various departments maintained good working relationships and, to encourage interactions, the Astronomers' Lunch Club met every week. This was open to all astronomers working in Cambridge and helped to maintain intellectual contacts between the groups. These were lively occasions and for many years I was secretary and treasurer of the Club which met regularly at Clare Hall.

In 1964, while I was completing my first year of research at Cambridge, I attended a course of lectures given by Hoyle on the problems of extragalactic research. It was a rather remarkable audience and included many who would become future leaders of astrophysical research – Martin Rees, Roger Tayler, Peter Strittmatter, John Faulkner, Russell Cannon, Bob Stobie, and many others. Hoyle would arrive with, at best, a scrap of paper with some notes and expound an area of current research. One week, the topic was the problem of the cosmic helium abundance.

At that time, helium was one of the more difficult elements to observe astronomically because its high excitation potential meant that it could only be observed in very hot stars. Already by 1961, Donald Osterbrock and John Rogerson had shown that the abundance of helium seemed to be remarkably uniform wherever it could be observed and corresponded to about 25% by mass. In Fred's lecture, he discussed the recent preprint by O'Dell, Peimbert and Kinman (1964) concerning the helium abundance in a planetary nebula in the old globular cluster M15. Despite the fact that the heavy elements were deficient relative to their cosmic abundances, the helium abundance was still about 25%.

Hoyle reviewed the evidence on the cosmic helium abundance and then described the work of Gamow, Alpher, Herman, and Follin concerning the problems of synthesizing the heavy elements in the early phases of the big bang. Although helium is synthesized in the central regions of stars during their long phases of evolution on the main sequence, it is most unlikely that this process could have created as much helium as 25% by mass of the baryonic matter in the universe.

By 1964, it was possible to carry out primordial nucleosynthesis calculations more accurately. At that time, Roger Tayler had just returned to Cambridge and was present in the audience. Hoyle and Tayler realized that they could undertake much more precise calculations and, in the next week, they and Hoyle's research student John Faulkner worked out the details of the formation of helium in the early phases of the big bang. The audience had the privilege of being present as a key piece of modern astrophysics was created in real time in a graduate lecture course.

John Faulkner told me about the remarkable events which took place over that weekend when they were analyzing the results of his computations.

Hoyle became very excited indeed about the results of the computations because he thought the models were predicting a very high helium abundance, which exceeded what was observed – he thought he had found a reason to reject the big bang scenario. In fact, however, the results settled down to about 25%, in remarkable agreement with observation and essentially independent of the overall baryonic matter density in the universe. Hoyle and Tayler's paper was published in *Nature* (Hoyle and Tayler 1964).

One consequence of the big bang model which Hoyle and Tayler did not mention explicitly in their paper was that the cooled remnant of the thermal radiation present during the very hot early phases should be detectable at centimeter and millimeter wavelengths. According to Roger Tayler, he had included this result in his draft of their paper, but it did not appear in the published version. Alpher and Herman's prediction had been more or less forgotten when Gamow's theory of primordial nucleosynthesis had failed to account for the creation of the chemical elements.

The very next year, the CMBR was discovered by Arnold Penzias and Robert Wilson (Penzias and Wilson 1965a), more or less by accident. There is no need to take this story further since many of the key players give a blow-by-blow account of this crucial discovery, the number of earlier "near-misses" and the remarkable story of the wealth of cosmological information which the CMBR contains.

Wagoner, Fowler and Hoyle (1967) repeated the analysis carried out by Hoyle and Tayler, but now using all the available cross sections for many more nuclear interactions between light nuclei and with the knowledge that the CMBR has radiation temperature 2.7 K. When Hoyle first presented these results in Cambridge, many of us were surprised that it seemed as though he had been converted to the big bang picture. As in the paper itself, however, equal weight was given to the idea that these computations could also be applied at very much higher baryonic densities to very massive stars, which collapsed and "bounced". The nucleosynthesis of the expansion phase was exactly the same as a universe of very high baryonic mass density. The densities were so high that heavy elements could be synthesized in these stars. The subsequent paper by Wagoner (1973) concentrated upon the primordial synthesis of the light elements.

There is a delightful sequel to the story of Hoyle's approach to steady state cosmology. He always affirmed that the idea of the continuous creation of matter was the most important aspect of the theory. On the occasion of his 80th birthday in 1995, I invited him to lecture to the Cavendish Physical Society. He was delighted to accept this invitation because he had given his first lecture on steady state cosmology to the Cavendish Physical Society in 1948. Hoyle remarked wryly that, in his development of the theory of continuous creation, his only mistake had been to call his creation field C rather than ψ. The exponential expansion of the early universe according to the inflationary picture involves a scalar field ψ which performs exactly the same function as Hoyle's C field, or Lemaître's cosmological constant Λ. Then, Hoyle would have been the originator of the inflationary picture of the early universe.

4.8.2 John Faulkner: The day Fred Hoyle thought he had disproved the big bang theory

John Faulkner is Professor Emeritus of Astronomy and Astrophysics at the University of California, Santa Cruz. He was one of Fred Hoyle's many graduate students from 1960 to 1964, subsequently spent two stimulating years as a postdoctoral fellow in William A. Fowler's Kellogg Laboratory at CalTech, and in 1966 returned to Hoyle's fledgling Institute of Theoretical Astronomy, Cambridge. In 1969, after receiving an offer he couldn't refuse, Faulkner moved to UC Santa Cruz.

On a Sunday evening in late February or early March 1964, I witnessed a remarkable sight at Fred Hoyle's home in Clarkson Close, Cambridge. I had just reported to him the results of computing the helium content of the universe from a presumed hot big bang origin. Those computations were performed according to a prescription Fred had lain out in full in a lecture only the previous morning.

For a wide range of initial conditions, the results showed what we at least regarded as a "high" helium content. (That characterization has to be understood in the context and against the particular prejudices of those days.) Fred reacted ecstatically to the news I had brought him. He got up from his armchair in the study alcove where he invariably read, wrote and talked with students, and walked triumphantly about the adjacent living room, shaking his fist in the air and declaring, "We've disproved the big bang!"

Within a short time (see Narlikar's contribution), Fred came to realize instead that he might well have found some of the best supporting evidence then possible for the big bang (short of observing the CMBR itself). Here I try to describe what led up to Hoyle's remarkable moment of mistaken euphoria, and explain it in its historical context.

I became Fred Hoyle's graduate student after obtaining a distinction in Part III of the Cambridge Mathematical Tripos. The grueling Part III year was the traditional hurdle for those wishing to be admitted as graduate students the next year. Those allowed to take it (only those students with first or high upper second class degrees in Part II of the Tripos) were neither fish nor fowl – no longer undergraduates but not yet admitted as graduate students. We all knew that at least our Cambridge research futures hung on how we did in the six end-of-year exams. A distinction (a "star") would guarantee success, a mere pass would lead to uncertainty. (We were however assured that as already highly selected graduates, we could all simply walk into an appropriate graduate program at any other university fortunate enough to have us apply!)

After the exam results were announced, those interested in pursuing research on the applied side were asked to meet in the sacred small lecture room on the first floor (second floor US) of the Arts School, where, despite its deceptive name, physics and astrophysics luminaries like Dirac, Hoyle, Mestel, and later Sciama gave their high-powered Part III lectures. G. K. Batchelor, the chairman of the recently established DAMTP, read out some pertinent sentences: "Professor Hoyle would like to see the following students tomorrow morning at his home ... At 9 o'clock, J.V. Narlikar, at 10 o'clock, J. Faulkner, at 11 o'clock, S. M. Chitre." Thus did we learn the order in which we had finished in the exams.

When I arrived at 10 o'clock, Fred congratulated me on my Part III performance, and then showed me a handwritten list of potential research areas, inviting me to make my preference known. I pointed to Relativity and Cosmology. "Oh, that's a pity," he said, "because Narlikar just chose that, and I would rather not have two students in any one area. Is there anything else that takes your fancy?" "Well," I replied, "I've also found myself quite interested this past year in Stellar Structure and Evolution." "Oh, very good," said Fred. "There's still a lot to be done in that area. I think you'll find it a good choice."

So that's how I became a stellar theorist. When I look back now on how things developed for me in the next decade or so, I believe that this rather brief and informal research selection process did result in a choice of field that best suited my capabilities.

Jayant Narlikar and I had often studied in the same library areas during our Part III year. We were on pleasant nodding terms, but it was not until we found ourselves both graduate students of Fred's that we became very good friends. Jayant knew of my originally expressed interest in relativity and cosmology. Early in 1961 he learned of a summer school on "Evidence for Gravitational Theories," to be held in Varenna on the shores of Lake Como, and asked if I would be interested in attending it with him. Lake Como? Italy? I jumped at the chance.

That summer school was memorable for several reasons. Most importantly, I saved Barbara Hoyle from drowning, and was by that singular event admitted to the Hoyle family's inner circle. I briefly mentioned this in my after-dinner speech at Cardiff's memorial meeting for Fred Hoyle (Faulkner 2003). The circumstances are described more completely in Simon Mitton's (2005) biography of Fred.

There I also met, for the first time, relativists of world class like Bondi, Dicke, Schild, and Weber. While I enjoyed their lectures immensely, I also learned a lot in informal conversations. I shall have to recount Alfred Schild's amazing wartime saga elsewhere; it involves his mother in England receiving the first postcard he was allowed to send her, as an alien internee in Canada, a year after she thought that he had died – he had been declared officially lost to a bombing at sea, en route from England.

One more amusing memory of these noted relativists stands out. A group of eight or ten lecturers and students had taken a ferry across the lake for dinner. Walking back along the far shore road as darkness fell, we saw that there were clearly some planets visible in the sky. An argument broke out among the senior relativists as to which planets they were. I knew what they were, and said so, but they were all skeptical. In an attempt to resolve the matter, they fed some lire into a scenery-viewing telescope mounted on a lakeshore post, and attempted to elevate it so that they could look at those planets. I regret that I didn't own a camera in those impecunious days – a photograph of some of the world's leading relativists kneeling, in full observing mode, would have been priceless.

Apart from a death-defying ride at more than 100 mph through lakeside galleria in my hotel room-mate R. U. Sexl's convertible Porsche (in which he claimed he had beaten the known driving speed record from Vienna to Como), one other memory remains quite vivid. In one of the cosmology lectures, a question arose about the consequences of a certain possible observational selection effect for the deduced acceleration parameter of the universe, q_0. Suppose that as the redshift z became larger, one increasingly failed to see galaxies occupying the less bright tail of the intrinsic luminosity distribution. Then in practice, instead of viewing an average and unvarying "intrinsic standard candle" among the galaxies one actually could observe, one's assumed standard candle would in fact become intrinsically more luminous with z. What would be the effect of such a selection bias with increasing

z on the deduced q_0? After some hand-waving discussion, all the big shots agreed on the answer.

Something bothered me about their conclusion. That evening, while Sexl was out impressing the local young women, I went through an analysis that assumed a simple dependence of L_{gal} on z. I found that in their hand-waving discussion, the pundits had only considered something that was but part of the full effect. When properly evaluated, the conclusion was reversed! Sexl chortled at my result when I showed it to him later that night. I gave Fred my analysis after breakfast the next morning. He immediately saw the point and accepted what I had done. He made an announcement before the lectures proper started that day, saying (in paraphrase) "We all thought yesterday that the effect would be such and such, but my student John Faulkner (whose main research area is stellar evolution) has shown that it's just the opposite, and here's why ...." In the break several people came up to me and ruefully patted me on the back.

Fred suggested that I write up my analysis as a letter to *The Observatory* Magazine. Back in Cambridge I literally did that, in careful fountain pen handwriting. This was in the days BP – before the photocopier existed. I naïvely gave Fred that sole copy of what I had written, asking for his opinion before I sent it off. (I was only a first year student; I had never published anything.) First weeks, then months went by. I eventually asked him what he had thought of it. He said it had had been fine, and asked whether I had submitted it. When I told him he had the only copy, he was sure he must have given it back to me. The missing account never did show up, however, and I was so far into my stellar work by then, I was loth to take time to reconstruct the now fading argument. Some time later, a slightly extended version of what I had done was published by someone else – and in *The Observatory*, to rub salt into the wound. Thus my first independent piece of research had all come to naught. I have sometimes found myself wondering whether Fred's reluctance to have two research students working in the same general area was somehow subconsciously involved in his misplacing or losing that manuscript of mine. (Shortly after this book was sent to press, my long lost manuscript and the cover letter that had accompanied it were both found, after 47 years, among Fred Hoyle's preserved papers in the library of St John's College, Cambridge.)

What I'm now going to write may surprise those who know of my long distaste for computing and distinct preference for analytical work. Nevertheless, I fairly rapidly became one of Fred's two best computing students, the other being Sverre Aarseth. Fred had pioneered the automatic computation of stellar evolution, first coding in inscrutable machine line instructions with

Brian Haselgrove on the original Cambridge Maths Lab Edsac I computer. Several months into my first year as his student, Fred returned from an extended visit to the US with an IBM program now coded in FAP, essentially mnemonically written IBM machine instructions. He had developed a severe distrust of systems programmers, claiming that every time he returned with a working code to any machine, they had made it harder to run. He used specially-written FAP subroutines for all his needed mathematical functions. His printing subroutine came from someone he trusted at Westinghouse; its main snag was that all minus signs were printed as ampersands! We all became quite accustomed to reading them that way. The Westinghouse program also had very limited output formatting. I came across a Fortran manual, realized the potential advantages of using it, taught myself Fortran from that manual, and then bided my time.

In those days, for Fred's group computing meant commuting. (Fred had quarreled with Maurice Wilkes, director of the Cambridge Maths Lab, and forbad us from having anything to do with it. I nevertheless first learned Edsac II machine code surreptitiously, and later Edsac Autocode, which I still claim was one of the most practical programming languages in which to become quickly proficient and productive.) So instead of computing in Cambridge, Fred had us do our work on what was then the first IBM 7090 in Europe. The problem was that it was in Wardour Street, off Oxford Street in London. That meant a draining commute from Cambridge several days a week. We started with revised FAP programs. However, the day came when Fred said he desperately needed both some fairly standard and more complicated functions and integrals to be evaluated for a variety of input values. I coded the whole thing up in Fortran, and went up with it to London the next day. Hard to believe, but it worked first time! I returned with a sheaf of beautifully formatted results. When I showed them to Fred the next day, he first admired the clear formatting, then pointed at the now visible minus signs. "What are these?" he asked. "They're minus signs," I replied. He looked at me, blinking. "Well, this is obviously the way to go, isn't it?" And the instruction went out to the other students to learn and use Fortran from now on.

There were various other ways in which Fred realized that I was at that time unusually good at computing, which has a bearing on this unfolding story. Basically, he learned that he could trust what I did.

I am now approaching the main topic of this contribution, the computation of the helium content of the universe. But before I get to that, I also need to lay out what the understanding of the helium content in various objects was when the 1960s began. That requires going back even a bit earlier. (I have found that many of today's cosmologists have very little idea of this history; a certain local physicist is a case in point.) Much of what I shall summarize is laid out more completely in my contribution to *The Scientific Legacy of Fred Hoyle* (Gough 2005).

The brief story is that until the end of World War II, astronomers thought that stars contained either very little, or no more than 35% or so hydrogen content (X) by mass. In 1946, Fred Hoyle swung the hydrogen pendulum over to 99% or more. This was before the steady state theory was a glimmer in its originators' eyes.

Hoyle's very interesting and significant paper (Hoyle 1946) changed forever the former notion that less than half of stellar material was hydrogen, and that the "metal content" Z of many stars exceeded 10%. This paper does not now receive anywhere near the attention it deserves. At the time Fred wrote it, many people, in particular those generally interested in stellar structure, still believed that the initial hydrogen content of the Sun was quite low – between about 0.35 and 0.50 by mass fraction. Schwarzschild (1946) had published a model of the Sun just a little earlier; it still had $X = 0.47$, and a huge amount of metals $(Z = 0.12)$. *That* was the standard picture at that time. In his own 1946 paper Fred noted that there were growing indications in other astronomical areas – in studies of both the interstellar medium and stellar atmospheres – that the previously reigning view could not be right. He then pointed out, in paraphrase, "You get a much better fit of model main sequences to the observations if instead of something like 35% hydrogen and huge amounts of metals, you assume that the stars generally contain less than about 1% by mass in the form of heavy elements and a really substantial amount of hydrogen."

As one sees by looking at his figures, the fits for low metals and a value for μ (the mean molecular weight) between 0.5 and 1, and indeed closer to 0.5, pass through or closer to the general run of most of the points than those of the earlier presumed compositions. Thus Fred was the first stellar theorist to champion what with some modification is essentially the present picture – metals of no more than a few percent, the rest predominantly hydrogen – quite a revolutionary point to be making at that time, and one in which he perhaps went to extremes. (Indeed, in the summary of his paper he stated, "The data suggest that at the time of condensation of the stars at least 99 per cent by mass must be in the form of hydrogen." He would employ such a figure for Population II stars as late as 1959, and urge it on his students in the early 1960s; see below.)

That particular year, 1946, was of course two years before the emergence of the steady state theory. In that theory, it was philosophically satisfying and convenient to have all the newly created material be hydrogen, that is of the simplest conceivable atomic form. I'm inclined to think that Fred's desire to have a unified picture, together with his argument from stellar structure (as just presented), really led him to largely favor a high hydrogen content (and indeed, a *very* high hydrogen content) in the oldest stars in particular, for quite some time. He still favored that point of view when I became his student in 1960.

In two stellar evolution papers (Haselgrove and Hoyle 1959; Hoyle 1959a) Fred still used and promoted additional arguments for preferring a hydrogen content of 99% for the oldest Population II stars. With a badly needed revision in the CNO energy generation rate, Haselgrove and Hoyle laid the foundation for an oft-quoted result in Fred's immediately following paper (Hoyle 1959a). From his computations, and a rather involved argument invoking the brightness of RR Lyrae stars, he first announced to the world a result that he knew would be considered startling at the time: that "the age of the Galaxy must be in excess of 10^{10} years." Fred's paper has often been quoted for this eye-catching result (it is printed in italics), but no one who does so ever seems to note (or even notice?) that such a large value for the age of the galaxy (again, "large" at that time) was only obtained for input parameters that would raise eyebrows today: $X = 0.99$, helium mass fraction $Y = 0.009$, and currently evolving masses quite a bit above a solar mass ($M > 1.3 M_\odot$). For $X = 0.75$, much closer to today's understanding of the hydrogen content of Population II stars, he obtained only $t \simeq 4.8 \times 10^9$ years, a value his additional RR Lyrae argument led him to reject. I've always felt it was a little naïve or disingenuous for later authors to quote the much higher age with great approval, without even the tiniest *caveat* paying attention to Fred's choice of input parameters.

In the end, of course, Fred was right about the age of the galaxy, but not for the range of parameters or the argument he had used. To go further here is beyond the scope of this contribution.

In lectures and conversation from 1960 onward, the idea we stellar students received from Fred was "Of course, the helium content of the oldest stars is very low." (When we undertook extensive main sequence calculations in 1962, he suggested we calculate our Population II models for $X = 0.985$.) I did not once hear such ideas questioned. Indeed, I began to notice a curious thing. Roger Tayler had returned to Cambridge in 1961, originally to act as an intermediate mentor for Fred's many students. He started giving a number of voluntary but meticulously prepared lectures on a wide variety of topics. (The reputation he so gained was partly how he became a regular DAMTP faculty member.) Yet, even when the course was on "The

Abundances of the Elements," the topic of the helium content of the oldest stars seemed to be skirted, in much the manner that Fred would skirt it. It was always explained that helium absorption lines can only be seen in very hot stars, and that all such presumably pristine main sequence stars of Population II had regrettably evolved away to other parts of the color–magnitude (or HR) diagram. By the time any subsequent descendants were hot enough to display helium lines again, one could not be sure whether or not they were (indeed they all probably had been) contaminated by the products of nuclear burning welling up from their deep interiors. Thus was argued away the very possibility that their surface abundances reflected their original compositions in any direct way.

Yet, ironically, Geoff and Margaret Burbidge had already been responsible for planting a questioning seed in my mind at the outset of my research studies.

Fred was away in the US for the first few months of my contemporaries' research careers. Before leaving, he suggested some useful reading, certainly more than enough to keep me fully occupied. In addition to Burbidge, Burbidge, Fowler and Hoyle (1957), of course, he suggested long articles from the classic Volume 51 of the *Handbuch der Physik* (Arp 1958; Burbidge and Burbidge 1958). Chip Arp's article first introduced me to his wonderful comparative work on globular clusters, and the correlation between their metal contents and the form of their horizontal branches in the HR diagram. The article by the Burbidges was dense with useful information. Within it I found an intriguing argument spelled out in only a few sentences. They pointed out that if the luminosity of the Milky Way Galaxy had been essentially constant during its presumed lifetime, then attributing that luminosity to hydrogen burning would have resulted in an average enrichment of the galactic helium content by at most 1 or perhaps 2%. That simple but profound argument had a strong impact upon me. In another paper (Burbidge 1958), Geoff remarked that this approach would be capable of producing the observed Population I helium contents, only if, for example, the Galaxy had been 100 times brighter in the first tenth of its lifetime. (But even he seemed to find that particular sleight of hand unconvincing.)

The Burbidges hadn't taken the argument to the point that it necessarily implied a much higher primeval helium content in the oldest stars. Nevertheless, I was left with that thought planted in my mind by their original argument. Perhaps there had been a fairly "high" helium content initially, one that had only been modestly altered subsequently. That thought was ever present in my mind even though I don't recall ever voicing it to Fred.

I have to say, looking back, that we in Cambridge seem to have remained remarkably ignorant of the impact of the work on helium abundances by Osterbrock and Rogerson (1961), among others. It is particularly surprising as Fred certainly made known to us his appreciation of Don's work in explaining the radii of low-mass main sequence stars. I simply cannot explain it, apart from the fact that we appear to have been in our own insular bubble at that time. (The self-satisfaction that was part of the Cambridge scene is well known.) As Osterbrock reported in his contribution (and as he discussed with me shortly before his untimely death), this failure to take note of what was being found out about helium at that time in other areas of research was apparently common to most stellar structure investigators, including the two men Don seemed to admire more than all others, Schwarzschild as well as Hoyle.

I now finally come to the computation I performed.

In the Lent (winter) term of 1964, Fred Hoyle gave a remarkable series of lectures solely intended for a graduate audience, on Extragalactic Astronomy and Cosmology. These nonexaminable lectures took place in the usual Part III venue, from 11 am to noon on a Tu.–Th.–Sat. schedule. Several faculty (including Tayler and Lynden-Bell) and most astronomical postdocs, as well as then current graduate students, attended that course. I was known for taking rather verbatim notes; the following summary is taken from my notebook. [Here and in what follows I shall enclose clarifying insertions or essentially editorial comments in square braces.]

Narlikar presented a couple of early lectures when Fred was unavoidably absent, but the bulk of the course consisted of Fred apparently working his way through problems that were then of particular interest to him. My notes say that Fred described these as "conversation classes," rather than lectures. The course began by discussing quasars, with much quoting of Maarten Schmidt's work. It went on to radio galaxies, massive objects, cosmic rays, etc.

Cosmology proper entered at Lecture 15. Half way through it, Fred mentioned "Gamow's addition" of nuclear reactions to a hot beginning, and said (because Gamow started with pure neutrons), "You're balanced on a razor edge in this problem." [He meant that with a pure neutron beginning, the density at a given starting temperature could be neither too low nor too large.] He then stated "The empirical evidence is that He/H by mass is $\sim 1/2$." [I found myself wondering where that came from, since it seemed at odds with what he had promoted in Population II stars. He must have been meaning, in Population I.] Near the end of the lecture, without explicitly giving his own view, he remarked that "Schmidt has a thing about

He^4/H by number being about 0.13 in almost all things ever examined. For example, the number (whatever it may be) is constant over the face of Andromeda. Galaxies would have to be considerably brighter than they are to give this ratio." [Aha! That otherwise gnomic remark indicates that he was thinking of the problem of enriching an originally low universal helium content.] He finished the lecture with a remark that when he went through the corresponding analysis to Gamow's but with no assumption about density, he actually found that he wound up with a number ratio of about 0.13, independently of any density consideration. [At that stage, he really didn't indicate what the differences might be in the other physical assumptions.] He ended with two questions. "Are stars only negligible changers?" "Is the composition largely determined by, say, massive star nuclear reactions?"

At the start of Lecture 16, Fred declared that he now had a pedagogical dilemma. He had done something more carefully than the previous time, with startling results! But first he needed to complete something else he had left hanging. He switched gears abruptly at the end of the lecture, asking out of the blue if anyone could tell him the form of the cross section for positron plus neutron going to proton plus antineutrino. He remarked that he could get it by detailed balancing for the (electron, proton) cross section from a preprint on that by Bahcall, but he had concerns about two factors in it before deducing what he wanted.

The key lecture, number 17, took place on a Saturday morning. Fred now explained that because of the presence of electron–positron and various neutrino–antineutrino pairs the analysis would differ from Gamow's attack. He reduced the problem to a fairly simple differential equation for $n/(n+p)$. [Here n and p are the neutron and proton number densities. He was taking account of the reactions in equation (3.9) on page 32 that determine the abundance of neutrons that could end up in helium.] He concluded that the "critical quantity" $n/(n+p)$ would "freeze out" at a value of about 0.135. He said he had calculated this for one chosen starting condition, using Simpson's rule and counting squares on graph paper. However, it clearly needed to be done for a number of starting conditions, to check both the value he had obtained and the expected insensitivity to initial conditions above a certain initial starting temperature.

He was looking at me as he said this, and I took it as an open invitation to do just that. As had happened before, I thought that he had no other way of quickly obtaining detailed desired results for varied initial conditions. Indeed, I was then working on an analogous but rather more complicated computing problem with him and Narlikar, which resulted in a paper submitted soon afterward, in May (Faulkner, Hoyle and Narlikar 1964). [I had almost finished my thesis work, and was employed that year as a research assistant to Fred on a DSIR grant.] I do not recall Fred saying at that time that he was working with anyone else on the helium problem; up to that point, there is no mention of Roger Tayler in my notes. Fred then remarked in his lecture that the odd factors that still concerned him could result in a final value for $n/(n+p)$ of about 0.15 so that one could end up with about 30% helium, and "this is a pretty good number." [In this context the final neutron fraction $f = n/(n+p)$ at freeze-out is a much more convenient variable than the number density ratio $N = $ He/H, the beloved variable of interstellar medium or planetary nebula investigators. Under the assumption of negligible heavy element concentrations, and assuming each surviving neutron combines with a proton to ultimately produce helium, Y becomes simply $2f$. So values of f in the range 0.135–0.15 imply $Y = 0.27$–0.30. That simple doubling appears to be what he consistently used in this context.] He concluded by saying "If we didn't know about massive objects, we would unavoidably come to the conclusion that we live in a radiation universe." [In retrospect, this struck me as an attempt to "save the phenomenon" – the phenomenon being his preference for virtually no helium to begin with. How could he explain away observed "high helium" at apparently an expected Big Bang value?!]

My reaction to the end of the lecture was to have a quick lunch and then head over to the Maths Lab. One could run one's own programs there. The equations to be integrated were relatively simple, and I was sure that with luck I could complete the needed work later that same day, or on Sunday if need be. I wrote the necessary autocode program and punched it onto the input paper tape, which I unfortunately tore in my excitement. After necessary repairs and backtracking, more paper tapes bearing the results spewed out of Edsac and I hurried over to a reader to interpret them. The results were a tribute to Fred's simple methods. For the input values from which Fred had obtained $n/(n+p) \simeq 0.135$, I found 0.1337! I rather self-consciously inscribed "Helium content of the Universe" on my copies of the tapes, waved them triumphantly at some grad students I knew, and headed home. [Those tapes followed me on several moves until I lost track of them. They could still be somewhere, very brittle by now, in old packing cases in my garage.]

At that time Fred, who had fallen out severely with G. K. Batchelor, was rarely to be seen in the DAMTP. [He tended to just give his lectures in the Arts School near the old Cavendish Laboratory, perhaps drop in afterward to the DAMTP in the far corner of the Cavendish to see people fleetingly and pick up any mail, and then go home.] So, I made the usual arrangement by phone to see him at his home. The allotted time was shortly before dinner on Sunday. I arrived by bicycle as predinner martinis were being served. I gave him a copy of my program and of the results. His reaction, described at the start of this contribution, surprised and indeed disturbed me. Did he really think that there was somehow reliable evidence for old stars having very little helium? What could it be? [I had concluded that any other astronomers who might still think this had been bowled over by the power of Fred's own personality – even if they were otherwise philosophically opposed to the steady state theory – but I didn't dare say so! As several people could confirm, I've expressed this viewpoint to confidants over the years.] I came out of this brief reverie to hear Fred now saying something about the lowest values being most relevant as a test, and that O'Dell had observed number ratios $N = $ He/H as low as about 0.09. [In fact, that lowest result was indicated to be extremely uncertain, though it would have implied a value for f of about 0.13, or of Y about 0.26. Most of what O'Dell considered his more reliable results were substantially larger than this.] I felt confused and unsure of my grounds for feeling that what I had witnessed was a little bizarre. I knew that O'Dell had written a paper the previous year about helium abundances in rather bright field planetaries (O'Dell 1963), but those results were not necessarily relevant. There was nothing in those studies that would necessarily link them to an old population. However, I thought I had seen a very recently arrived preprint by a collaboration including O'Dell; that was probably more relevant. I left thinking that I needed to find that preprint.

I found it the next morning in the small DAMTP library. It was a very recently submitted paper, and had probably been received only in the last week or so, which meant that Fred might not have seen it yet. [The published paper – O'Dell, Peimbert and Kinman (1964) – shows that it was submitted to *The Astrophysical Journal* on February 10, 1964.] The authors reported what as far as I knew was a truly remarkable and ground-breaking analysis, namely that of the first planetary to be observed in a truly classical globular cluster. They had found that the planetary nebula K648 in M15 had $N = $ He/H (by number) $= 0.18 \pm 0.03$. Yet oxygen was low relative to solar values by a factor of about 60! Here then was an incredibly "high" helium value (corresponding at face value to $Y \simeq 0.41 \pm 0.04$) coupled with a typically low "metal" content characteristic of the oldest Population II stars. While it could still be argued that this truly high helium represented some kind of nuclear "dredge-up," was it likely that the oxygen content could remain apparently so pristine? The authors thought not, and I tended to agree with them, although one could still argue that the case was entirely circumstantial.

I became further alarmed at the thought that Fred might be going out on a limb of his own making. Yet who was I to make this point? I decided to talk with Roger Tayler, whose office was just down the corridor from mine. (Though by then a regular faculty member, Roger continued to act as an intermediary with Fred.) Roger understood my concerns, and said he would speak with Fred about it; he told me he had already been discussing some aspects of the physics with Fred.

The next morning, at the start of Lecture 18, Fred reported that I had done the computations for a slew of initial starting temperatures, saying "John Faulkner's computations gave 0.1337 where I had 0.135!" [Malcolm Longair told me at Cambridge's memorial meeting for Fred that this announcement had had a strong effect on him.] Fred went on to discuss possible changes in some of the factors or coefficients, clarified the contribution to helium enrichment that stars could make from their characteristic lifetimes and luminosities, and then turned to massive stars. He reinterpreted time-reversed cosmological results in terms of collapsing massive star matter densities rather than expanding radiation-dominated energy densities. His reconstructed table from this exercise still included my computed value of 0.1337 for $n/(n+p)$, in a table that now had more variations in it. [How he had obtained those results was a mystery to me. They were all given to four significant figures. Surely these weren't from his square-counting method? Perhaps they were from a massive star mass-scaling of my original result, that was neither obvious to me then, nor now.]

He proceeded to do further "tidying up," as he called it, in the 19th and last lecture. This was very hard to follow, as changes in coefficients and consequences were flung about at dizzying speed. In the cosmological case he finally examined what limitations there might be for the assumptions to be consistent for a given temperature, concluding that at $T = 3 \times 10^9$ K for example, the matter density needed to be between 10^{-3} and $10^3 \, \mathrm{g \, cm}^{-3}$. The former limit came from the presumed current mean universal density with "an upper limit of T now to about 3 degrees." [! Fred had often mentioned in lectures that several current mean energy densities were effectively equivalent to the energy density in such an ambient temperature.]

Turning to massive objects, for which he declared "the situation is much simpler," he then made an intriguing statement that finally confirmed part of Roger Tayler's involvement in the problem: "Roger Tayler has pointed out that some of the approximations are not valid down at $T_9 \simeq 0.4$." [Here $T = 10^9 T_9$ K.] Fred concluded they were still good at $T_9 \simeq 0.6$, but that the radiation density factor would be off by a factor of about 2 at $T_9 \simeq 0.3$. Going back to helium abundances in stars again, he quoted Eggen as claiming that in the Hyades, the ratio of helium to hydrogen by mass was as high as 3/2. [I put a disbelieving "!" after this in my notes. As I wrote my thesis up, I showed that this conclusion and several others like it were based in part on far too naïve interpretations of simplistic power-law homology results. However, I also realized that self-consistency meant that the Hyades stars were more distant than previously thought. Although that also had cosmological implications – the Hyades providing the first rung on the cosmic distance ladder – it's too convoluted a story to pursue here.] Fred also remarked "One really ought to get an unevolved subdwarf to see if it has far more, or far less helium than it should have ... but it's difficult to see what to do."

I heard no more about the progress of Hoyle and Tayler's work until their paper was published (Hoyle and Tayler 1964), as I was moving to Caltech that September. Looking at it more carefully than I was able to then, I find traces still remaining of what I took for a long time to be Fred's main motivation – to test and (he seems to have hoped) disprove the big bang picture. While that may be an overly strong conclusion, I can find no other way to explain his mistaken, ecstatic (and revealing?) moment of triumph when I showed him my computed results. Also, there are several numbers in it that I just don't recognize. Those include the leading coefficient in the factor determining the rate of change of the neutron fraction, the numerical value of the result attributed to me (although it could be at a different temperature cutoff point) and, correspondingly, the final fraction attributed to helium production in massive objects. [Both such quantities are quite a bit larger than Fred mentioned in his lectures, according to my notes.]

Until recently, I was under the delusion that Hoyle and Tayler thanked me in their paper for doing the computation I performed. I've been quite wrong about this. Instead, they simply wrote, "Mr. J. Faulkner has solved the equation for several starting temperatures."

Finally, the most peculiar oddity and indeed lacuna is this – what I believe to be the key preprint that brought down the low helium house of cards (O'Dell, Peimbert and Kinman 1964) is not referenced! Instead, the result from that paper is attributed (but only implicitly) to O'Dell's sole-author paper (O'Dell 1963), in which it does not appear. The dilemma Fred found himself in is contained and highlighted in two successive sentences in Hoyle and Tayler (1964) that assert "...low values of He/H are of more interest in relation to the original composition of the Galaxy than high values. However, O'Dell's high value of 0.18 ± 0.03 for the planetary nebula M 15 [meaning: in M 15] is of special interest because O'Dell also finds a low value for the ratio O/H ...."

For me, related scientific consequences followed from this 1964 brush with the big bang. I had become a primarily stellar theorist with the inside knowledge that the big bang would necessarily produce a "high" initial helium content in the oldest stars. I resolved to allow for the possibility of both high and low helium when I tackled the outstanding problem of understanding horizontal branch (HB) stars in globular clusters. (My results were ultimately published in Faulkner 1966.) The "high helium" models produced a far better fit to Arp's observations. In an immediately following paper (Faulkner and Iben 1966) Icko Iben and I showed that for Population II stars, "high helium" evolutionary tracks off the main sequence were much steeper than those for models containing low helium. This resolved yet another long-standing puzzle in stellar evolution. When a careful comparison of the best observed local subdwarfs with the Hyades main sequence also provided evidence for an initially high helium content (Faulkner 1967), it essentially completed the case for high helium from classical stellar structure alone.

My experiences in 1964, exciting though they briefly were, also helped to determine what I would not do in the near future. When I arrived at the Kellogg Laboratory in September with my first (and then only) "high helium" HB model almost burning a hole in my pocket, I rather brashly, reluctantly but firmly declined Willy Fowler's generous offer to become Bob Wagoner. As it turned out, of course, Bob Wagoner himself was far more suited to play that role.

4.8.3 Robert V. Wagoner: An initial impact of the CMBR on nucleosynthesis in big and little bangs

Bob Wagoner is Professor of Physics, Emeritus, Stanford University. A continuing research interest is the physics of compact objects, including their roles as sources of gravitational radiation detectable by LIGO and other facilities.

Timing may not be everything, but it certainly can help. In 1960, when I was a mechanical engineering undergraduate at Cornell, I attended the Messenger Lectures of Fred Hoyle on cosmology. That experience, and books such as Dennis Sciama's (1959) *The Unity of the Universe*, opened my mind. I received my PhD in physics at Stanford in 1965. My thesis was on general relativity, although I spent part of one summer working (amid many spiders) on Ron Bracewell's radio telescopes. I was on my way to a research fellowship at Caltech just as the discovery of Penzias and Wilson (1965a) was announced. Soon after my arrival, Willy Fowler invited me to join him and Fred in an exploration of the consequences of this sea of photons, using nuclear astrophysics as a cosmological probe.

Details of my view of the development of primeval nucleosynthesis through 1973 can be found in a review (Wagoner 1990), where references that I have omitted here can be found. In keeping with the scope of this volume, my focus here will be mainly on the 1960s.

However, I begin by mentioning the first prediction of a cosmic radiation temperature (5 K) by Ralph Alpher and Robert Herman (1948), based on their work with George Gamow on what is now called big bang nucleosynthesis. It may not be well known that they neglected the overwhelming influence of neutrinos in establishing the neutron–proton ratio (and thus the synthesis of heavier nuclei), so that the approximate agreement with the eventual observation was fortuitous. Hayashi (1950) provided the correct interaction rates, and Alpher, Follin and Herman (1953) provided the first complete description of the standard model of the evolving major constituents (but no baryons except protons and neutrons) of the early universe.

It is somewhat of a mystery why this knowledge was not employed to recalculate the abundances until Zel'dovich (1963a) and Hoyle and Tayler (1964) considered the production of the key nucleus, helium (^{4}He). Fermi and Turkevich had developed a nuclear reaction network just before 1950. Zel'dovich concluded that a reasonable production of helium required that the present temperature of the fossil radiation is high (20 K), apparently because he believed indications of a low observed primordial abundance of helium (see the contribution by Novikov on page 99). This led him to (temporarily) abandon the big bang model.

Hoyle and Tayler provided more details of their (approximate but realistic) calculation, showing that the neutron–proton ratio when the weak interactions "froze out" essentially determined the abundance of helium, which was weakly dependent on the photon–baryon ratio. They noted, however, that conditions within exploding supermassive stars could be similar to that in the early universe (but with fewer photons per baryon). They also noted that the observed energy density of starlight only required the production of 10% of the observed amount of helium. Thus they concluded "that most, if not all, of the material of our everyday world has been 'cooked' to a temperature in excess of 10^{10} K." It also may not be widely known that they were the first to note that the number of types of neutrinos affects the expansion rate and thus the abundance of helium.

I was very fortunate to be a postdoctoral fellow (1965–1968) in Caltech's Kellogg Lab when it was a major hotbed of theoretical astrophysics. The emerging revelations of the nature of quasars only added to the excitement produced by the realizations of the consequences of the cosmic microwave radiation. The enthusiasm of Willy Fowler for many aspects of science and life (parties, etc.) infected everyone.

In our collaboration, Fred's point of view was of course influenced by his continuing belief in the steady state universe and thus the production of helium and other light elements within exploding supermassive stars (which we dubbed "little bangs"), complementing the ordinary stellar production (Burbidge *et al.* 1957; Cameron 1957) of the heavier elements. However, he was also impressed by the fact that if the helium was produced mainly by ordinary stars and their resulting luminosity was somehow universally thermalized at a time close to the present epoch, the radiation temperature would be close to 3 K.

The most critical element in my computer code was the nuclear reaction data provided by Willy and his group and many other nuclear physicists. Of course, we also had to extrapolate or otherwise estimate the rates of a few reactions that had not been measured at the relevant effective energies (usually 0.1–0.5 MeV, except for neutrons).

I presented our first results at the April 1966 annual meeting of the National Academy of Sciences (Wagoner, Fowler and Hoyle 1966). The calculation involved 40 nuclei and 79 nuclear and weak reactions. At about the same time, Jim Peebles published his calculation of the abundances of helium and deuterium within both the standard model and universes with different expansion rates (Peebles 1966). The accuracy of his predictions of the abundance of ^{2}H and ^{3}He was reduced by the limited number of nuclear reactions included.

Our results were published the following year (Wagoner, Fowler and Hoyle 1967). As indicated above, we considered a large range of the baryon–photon ratio, corresponding to big and little bangs. Our major conclusion was that reasonable agreement with observed abundances of ^{2}H, ^{3}He, ^{4}He, and ^{7}Li could be achieved if the universal baryon (matter) density was about 2×10^{-31} g cm^{-3} (a factor of 2 less than the presently accepted value from WMAP and other data sets; Spergel *et al.* 2007). However, the abundance data that was available was from within the Solar System (Earth, Sun, and meteorites), so we did not know how relevant it was. On the other hand, the predicted abundances have stood the test of time. We also explored the effects of inhomogeneity and neutrino degeneracy (large lepton–photon ratios). Within little bangs (larger baryon–photon ratios), carbon and heavier nuclei were produced, but the abundances did not closely resemble those observed unless the bounce occurred at a temperature of about 10^9 K. Fred believed that this could happen in the first generation of (supermassive) stars (usually termed "Pop III stars," after the classification of stellar populations into the younger Pop I and the older Pop II).

My summer of 1967 (and 1971) at Fred's new Institute of Theoretical Astronomy at Cambridge was very memorable. Willy, Don Clayton, and I occupied the first "office" in the hut in the sheep pasture behind the present Institute. The only building housed the IBM 360-44 computer, which I had to myself a large part of the time to tune my nucleosynthesis code. Many discussions with Fred focused on the properties of supermassive stars (sometimes over martinis while watching cricket), and with Willy and the Burbidges on abundance issues.

My involvement in nuclear astrophysics essentially ended a decade later. Exploration of other big bang models (Wagoner 1967, 1973), including the results of Peebles (1966) and those within anisotropic universes (Hawking and Tayler 1966; Thorne 1967), revealed to me that in general, only three factors affected the abundances produced. They were

(i) The number of baryons per photon.
(ii) The expansion rate, dependent upon the theory of gravity, anisotropy, and other forms of mass–energy density (other neutrino types, gravitational radiation, magnetic fields, etc.).
(iii) The neutron–proton ratio, dependent upon the lepton (neutrino) number per photon and the neutrino phase-space distribution (if the expansion was anisotropic).

The agreement of the abundance of ^{4}He with that produced within the standard model, and the detection of interstellar deuterium (Rogerson and York 1973) then strongly supported the conclusion that the density of ordinary matter was far short of that required for a flat universe. It was very gratifying that the early universe produced precisely those nuclei that stars or cosmic ray spallation could not.

The power of this deep probe of the early universe is based upon the fact that its physics is known, from the heroic efforts of many nuclear physicists (Fowler, Caughlan and Zimmerman 1967, 1975) and the discovery and subsequent measurements of the blackbody flux of cosmic microwave radiation.

4.8.4 Martin Rees: Cosmology and relativistic astrophysics in Cambridge

Martin Rees is Professor of Cosmology and Astrophysics and Master of Trinity College at the University of Cambridge.

When I enrolled as a Cambridge University graduate student in October 1964, after undergraduate work in mathematics, I had no particular research project in view, and minimal confidence that I had made the right choice – indeed I seriously thought of switching to economics. But I ended up with few regrets, because of two bits of excellent luck which I couldn't initially foresee.

First, I was assigned as one of Dennis Sciama's supervisees. I already knew of Sciama through his splendid lecture course on relativity, and had read his book *The Unity of the Universe* (Sciama 1959). He had charisma; he inspired his research group with his infectious enthusiasm; he followed developments in theory and observation along a broad front; and he was a fine judge of where the scientific opportunities lay. When I joined this privileged group, George Ellis had completed his PhD, and was starting a postdoc; Stephen Hawking was still a graduate student, two years ahead of me; my closest contemporaries in the group were Brandon Carter, Bill Saslaw, and John Stewart. Within a few months I felt I had made a fortunate choice.

But there was a second piece of luck: the mid-1960s were years of ferment in observational and theoretical cosmology. The discovery of the CMBR was of course the preeminent event, but these years also saw the emergence of "relativistic astrophysics:" the first high-redshift quasars, the discovery of neutron stars, and the first results from space astronomy (especially X-ray astronomy).

Dennis Sciama was "plugged in" to all these developments. He encouraged his students to interact and to learn from each other. He eagerly shared new preprints (and correspondence, news of conferences, and so forth) with his students and postdocs, and with other colleagues such as Roger Tayler. (For instance, I learned during coffee-time sessions about Hoyle and Tayler's work on helium formation, and the parallel work of Peebles. Also about the debate with the Moscow relativists about the nature of singularities.)

In the late 1940s, Fred Hoyle, Thomas Gold, and Hermann Bondi – then all in Cambridge – had proposed the steady state theory, according to which the universe, although expanding, had existed in the same state from everlasting to everlasting: as galaxies moved away from each other owing to the expansion, new atoms were continually created, and new galaxies formed in the gaps. This theory never acquired much resonance in the USA (and still less in the Soviet Union). But its three advocates were vocal and articulate people; and in the UK, especially in Cambridge, the theory was widely publicized and discussed. And it was indeed a beautiful concept. Sciama himself espoused it, and indeed described himself as its most fervent advocate apart from its three inventors.

The steady state theory was (rightly) touted as being a good theory because it was vulnerable to disproof. It made definite predictions that everything was the same, everywhere and at all times (that is, at all redshifts). Therefore if things were different in the past from now, that was evidence against it. Even if there were evolutionary changes, optical astronomers in the 1950s were unable to detect objects at sufficiently large redshifts (and look-back times) for such changes to show up. However, radio astronomers realized that some of the discrete sources detected in their surveys were "exploding galaxies" too far away to be detected optically. Although the redshifts of individual objects were unknown, it was possible to draw inferences from the relative numbers of apparently strong and apparently weak sources (since the latter would, statistically at least, be at greater distances). In particular, the number of sources brighter than flux density S would scale as the $-3/2$ power of S in a Euclidean universe (as discussed on page 55), and when expansion and redshift were taken into account, the $\log N - \log S$ plot in a steady state universe would be flatter than the Euclidean slope. The first credible evidence against a steady state came from Martin Ryle's radio astronomy group in Cambridge (based in the Cavendish Laboratory), and from the Australian group headed by Bernie Mills. The slope (at least at the bright, high-S, end) was steeper than $-3/2$. Such a steep slope was incompatible with steady state. Ryle interpreted it (correctly as we now recognize) by postulating that we lived in an evolving universe where galaxies in the past (when young) were more prone to indulge in the "explosive" behavior that rendered them strong radio emitters.

For me, coming fresh to the subject in around 1964, the skepticism that greeted Ryle's evidence was perplexing. Ryle's claims – indeed everything he had claimed from 1958 onward – seemed compelling to me (and have indeed been vindicated by later developments). But I later realized that the skepticism of the "steady statesmen" was not simply irrational obstinacy. Some of Ryle's previous data, in particular the earlier 2C survey, had turned out to be unreliable, owing to "confusion" caused by inadequate angular resolution. Moreover, he had initially vehemently opposed the suggestion that the so-called "radio stars" – discrete radio sources with no obvious optical counterpart – were actually distant galaxies. To add even more irony, it was actually Thomas Gold who first made that suggestion – which of course became the cornerstone of Ryle's later argument in favor of an evolving universe. This "baggage" dating back to the early 1950s perhaps helps to explain why the steady statesmen held out against the evidence of the source counts. There was also, it has to be said, a personal antipathy between Hoyle and Ryle – two outstanding scientists of very different style.

Sciama took Ryle's data seriously, but when I joined his group in 1964 he was still clinging to the steady state theory. He conjectured that many of the unidentified sources were nearby. The apparent steepness of the $\log N$–$\log S$ relation could then (he argued) reflect nothing more fundamental than a local deficit. But when the sources were revealed to have high redshifts, he abandoned this model (and never went along the route of saying that redshifts were noncosmological). The clinching evidence that led Dennis Sciama to abandon the steady state was a very simple analysis that he and I did together on the redshift distribution of quasars (Sciama and Rees 1966). By 1966, more than 20 radio sources in the 3C catalog had been identified with quasars with known redshifts (extending up to $z = 2.01$ for 3C9). We applied to this small sample a crude version of the "luminosity/volume" or V/V_m test developed by Rowan–Robinson (1968) and by Schmidt (1968). If the universe were in a steady state, the quasars of the highest intrinsic luminosity should have been uniformly distributed in comoving volume. But when we split them into redshift bins, each bin corresponding to a shell containing the same comoving volume as the others, the quasars were concentrated in the high redshift bins. This evidence suggested that quasars were more common (or more luminous) in the past – just as Ryle had argued was the case for radio sources.

In a big bang model, the redshift distribution of quasars tells us little about the geometry of the universe, but something about the astrophysical evolution of galaxies – indeed I still work on the implications of such data for galaxy formation, reionization of the intergalactic medium, and cosmic structure formation. The detection of the CMBR of course offered far stronger evidence for an evolving universe than the radio source counts. Attempts to attribute the CMBR in a steady state model to a population of discrete sources were even more contrived than those required to reconcile the theory with radio source counts and quasar data. The attraction of the steady state model was that everything of cosmic importance must be happening somewhere now, and therefore must in principle be accessible to observations. The theory's advocates believed – as was reasonable in the 1950s – that in a big bang model crucial processes would be inaccessible. But it has turned out that we can indeed observe "fossils" of the formative early eras of cosmic history soon after the big bang. The CMBR itself, of course, is one such relic; so also are cosmic helium and deuterium, and the fluctuations in the CMBR. So Sciama's disappointment was short-lived and he became quickly reconciled to the big bang – indeed he espoused it with the enthusiasm of the newly converted.

In parallel with these observation-led advances, the 1960s saw a renaissance in general relativity – a subject which had for several decades been rather sterile, and sidelined from the mainstream of physics. The impetus came from Roger Penrose. In my first year as a graduate student, I heard Penrose speak in Cambridge about his concept of a "trapped surface." I understood little of it, but was nonetheless fascinated. Roger Penrose is the kind of person who, even if you don't understand (or don't believe) what he's saying, gives the impression that an unusually insightful brain is at work. His thinking is not merely much deeper than most of us can manage – it is of a very special geometrical nature. Sciama was quick to seize on the importance of Penrose's new concepts. (Indeed it was he who had persuaded Penrose, whose PhD was in pure mathematics, to shift his interests to relativity.) Sciama encouraged some of his students to attend a lecture series that Penrose was giving in London. The most important outcome was Stephen Hawking's subsequent collaboration with Penrose, which led to the singularity theorems for gravitational collapse. The main import of Penrose's work for cosmology – as described in the article by George Ellis on page 379 – was an adaptation of these arguments to show that there must have been a "singularity" in the past of our universe, even if it was irregular at early times.

There was, at that time, a substantial research effort (spearheaded by George Ellis and a series of collaborators) aimed at investigating and classifying the various classes of homogeneous but anisotropic cosmological models. This was an interesting exercise in its own right. However, a special motivation came from Charlie Misner, who spent the academic year 1966–1967 on sabbatical in Cambridge. It was from Misner that we learned about the so-called "horizon problem," that causal contact becomes worse in the early phases of a Friedmann (decelerating) universe, rendering it a mystery that the present universe seemed so uniform and synchronized. Misner noted that causal contact would have been better if the early expansion had been anisotropic – best of all in the "mixmaster" model where there was an alternation in the axes of fast and slow expansion. The aim of the "Misner program" was to show that a universe could have started off (and homogenized) via a mixmaster phase, but that the initial anisotropies would have been erased, either dynamically or via neutrino viscosity. This program failed – and until the invention of the "inflationary" universe, more than a decade later, most of us probably thought that an explanation of global homogeneity would have to await a quantum-level understanding of the singularity. It was coincidental that the theoretical advances in relativity, instigated by the new "global methods" that Penrose pioneered, happened concurrently with the discovery of the CMBR.

It was a further coincidence that, during the 1960s, objects were discovered where general relativity was crucial, rather than a trivial refinement of Newtonian gravity – discoveries that stimulated the new research area of "relativistic astrophysics." The discovery of quasars (and, later, of neutron stars) indicated that objects probably existed in which the crucial features of Einstein's theory would have to be taken into account. Black holes of course are the most remarkable prediction of Einstein's theory. The Schwarzschild solution, discovered in 1916, represents the simplest black hole. They were speculated about in a rather half-hearted way by astronomers and cosmologists in the 1930s to 1950s. But the term "black hole" was not used until 1968, when it was coined by John Wheeler, and it was only in the late 1960s that theorists really clarified the nature of black holes.

A more general solution, discovered in 1963 by Roy Kerr (1963), was believed to be a description of a collapsed spinning object. The biggest breakthrough actually came from the work of Israel, Carter, Hawking, and others. They showed that Kerr's solution was generic, in the sense that any black hole would end up being described by this particular solution of Einstein's equations. Any gravitational collapse leads, after the emission of gravitational waves, to a black hole described exactly by two numbers, its mass and its spin. So black holes proved to be just as standardized as an elementary particle.

The number of people involved in these theoretical developments was even smaller than the experimental and observational community – indeed most relativists were associated with one of three "schools," those centered in Princeton, Cambridge, and Moscow. Communications were far less immediate than today (especially, of course, between East and West in the Cold War era). However the interactions that occurred were almost invariably cooperative and friendly. My own work was mainly on astrophysics and on galaxy formation: for this work, the new paradigm of the hot big bang was the essential backdrop, rather than being at the focus. My aim was to understand how galaxies produced so much radio power, how they became quasars, etc. It was already fairly clear that the power generation involved gravity, although, despite early advocacy by Salpeter, Zel'dovich, and Novikov and (especially) Lynden-Bell, it wasn't as clear as it would become in the 1970s that a single huge black hole was implicated.

I continued to be uneasy, until the early 1970s, about the apparent coincidence between the energy in the CMBR and the energy that could be supplied by astrophysical sources (via hydrogen burning or via gravitational collapse), but this proved of course a blind alley and distraction.

I can lay claim to two minor positive contributions directly related to the CMBR. One (Rees and Sciama 1968) concerned what is now sometimes called the "Rees–Sciama effect" – the perturbation in the CMBR due to a transparent gravitational potential well along the line of sight (for example, a cluster or supercluster of galaxies). In the linear regime, this is subsumed in what is normally called the "integrated Sachs–Wolfe effect" – it is non-zero except (to first order) in the Einstein–de Sitter universe. However there is a distinctive effect due to virialized clusters. Had Sciama and I known then the actual amplitude and scale of clustering, we would not have felt it worthwhile to explore these higher-order effects. But at that time there was no way of ruling out large-amplitude density fluctuations on gigaparsec scales (indeed there were early – and in retrospect misleading – indications of such clustering from the distribution of quasars over the sky). This effect has only recently been detected. My second contribution (Rees 1968) addressed the possible polarization of the CMBR. The simplest illustrative examples of this effect arose in anisotropic but homogeneous models (though the effect was obviously present in more general models). This work stimulated an early search by Nanos (1974, 1979), but it was more than 35 years before polarization was actually detected.

In CMBR studies, a consensus has generally quickly developed whenever there has been an advance – this is in contrast to (for instance) the prolonged debate and perplexity about the physics of AGNs and quasars. This is because the CMBR data, though challenging to obtain, are "cleaner," and the relevant fluctuations are in the linear regime. Successive developments – the CDM paradigm, the CMBR fluctuation spectrum, and so forth – have led to a well-established set of cosmological parameters. It has been a privilege to have followed a subject where progress has been sustained so consistently for 40 years, and to have known many of the scientists to whom these historic advances are owed.

4.9 Critical reactions to the hot big bang interpretation

4.9.1 Geoffrey R. Burbidge and Jayant V. Narlikar: Some comments on the early history of the CMBR

Geoffrey Burbidge is Professor of Physics at the University of California, San Diego. He served for six years as Director of the Kitt Peak National Observatory. His latest major award, jointly with Margaret Burbidge, is the Gold Medal of the Royal Astronomical Society. Jayant Narlikar served as

Founder Director of the Inter-University Centre for Astronomy and Astro-physics in Pune, India, until his retirement in 2003. He is now Emeritus Professor at IUCAA. Among his current interests is exobiology.

Both of us were asked to describe our views of the ways we first approached this topic. We have decided to combine our contributions but present them separately because we came to the basic ideas from different directions. Geoffrey Burbidge became interested in the CMBR from his early association with the fundamental problem of the origin of the chemical elements. Jayant Narlikar had been interested in alternative cosmologies and was, therefore, concerned with the problem of how the CMBR could be produced without a hot big bang. Each of us has given a "first person" account. As we had the benefit of close interaction with Fred Hoyle we have folded in his views also wherever necessary.

The Approach Taken by Geoffrey Burbidge. My first interest in this area came during the period 1955 to 1957 when Margaret Burbidge, Fred Hoyle, Willy Fowler, and I were solving in detail the problems of the origin of the elements (Burbidge *et al.* 1957).

I realized that the large abundance of helium in stars ($M_{\mathrm{He}}/M_{\mathrm{Baryon}} \equiv Y \cong 0.24$) meant that there must be a very special place, or an era, when there had been a great deal of hydrogen burning. At that time, the value of the Hubble constant was thought to be $180\,\mathrm{km\,sec^{-1}\,Mpc^{-1}}$ (Humason, Mayall and Sandage 1956), so that the Hubble time was $H_0^{-1} \simeq 6 \times 10^9$ years. Taking the luminosity of the Milky Way to be about $10^{44}\,\mathrm{erg\,sec^{-1}}$, this meant that over 6×10^9 years the total mass of helium that was produced by hydrogen burning would be far less than 24% of the total mass of helium $\simeq 2.5 \times 10^{10}$ solar masses.

I did not realize at the time that my argument was very similar to that which had been made by Alpher, Bethe and Gamow (1948) a decade earlier. At the time of the first calculation by Gamow, Alpher and Herman, Hubble and Humason (1931) had given a value of $H_0 = 550\,\mathrm{km\,sec^{-1}\,Mpc^{-1}}$, so that $H_0^{-1} \simeq 2 \times 10^9$ years and the discrepancy between the observed abundance of helium and the amount which could be attributed to hydrogen burning in stars was even larger. However, in contrast to me, Gamow and his colleagues had discussed the basic physics of the big bang and concluded that helium could only have been made in the early universe. Up until then it had been assumed that in Friedmann models, in the beginning the rest mass energy is much greater that the radiation energy. The immediate effect of the change to a radiation-dominated universe was to require that the scale factor of the universe $a(t)$ is proportional to $t^{1/2}$. Omitting electron–positron pairs, the radiation temperature T is inversely proportional to a. Thus the radiation temperature T is proportional to $t^{-1/2}$. With radiation alone and no neutrinos $T_9 = 15.2 \times t^{-1/2}$ where T_9 is measured in units of 10^9 K and t in seconds. However, the numerical coefficient 15.2 is modified by the presence of electron–positron pairs and by neutrinos. For temperatures high enough for the electrons and positrons to be relativistic, and for two mass-less neutrino types, the numerical coefficient is changed from 15.2 to 10.4. So long as the energy in the early universe is dominated by radiation the equation above holds.

But the next step in the discussion was completely *ad hoc*. The mass density of stable nonrelativistic particles, explicitly neutrons and protons, decreases with the expansion of the universe at a rate proportional to a^{-3}, i.e. as $t^{-3/2}$. Calling this density ρ_b, Alpher and Herman (1948) took $\rho_b = 1.70 \times 10^{-2} t^{-3/2}\,\mathrm{g\,cm^{-3}}$ with the coefficient 1.70×10^{-2} being the *ad hoc* step. There is *nothing* in the theory which fixes this value. It is a free choice, chosen to make things right, in this context to obtain the calculated value of the helium abundance Y to agree with observation. Thus, while the big bang theory can explain the microwave background, it tells us nothing about the helium abundance unless we *choose* a numerical value which enables us to do this.

This is fine if you come to the problem of the helium with a belief in the big bang. And this is what most contributors to this book have done. But I came to the problem with no cosmological beliefs.

In the 1950s a debate was going on between the majority of cosmologists, who believed in a beginning, and a few, particularly Hoyle, Bondi, and Gold, who had developed an alternative, the steady state cosmology (Bondi and Gold 1948; Hoyle 1948). By the late 1950s, standing on the sidelines in Cambridge, I realized how unpopular the steady state theory was, since at the time there was a very unpleasant dispute going on between Ryle and his group on one side, and Fred Hoyle. In the early 1960s, Hoyle and Narlikar (1961) gave an alternative interpretation of the radio source counts to show them as consistent with the steady state theory, whereas Ryle insisted these provided strong evidence against the steady state.

Returning to my own work on the origin of helium, I made a calculation assuming that all of the baryonic matter of the universe with a density $\rho_b = 3 \times 10^{-31}\,\mathrm{g\,cm^{-3}}$ had the same helium abundance. I then showed that if it were produced by hydrogen burning the energy density must amount to $\approx 4.5 \times 10^{-13}\,\mathrm{erg\,cm^{-3}}$ (Burbidge 1958; see also Bondi, Gold and Hoyle 1955).

In my paper I offered several possible scenarios for the production of helium. It could have been produced in the early universe if there was one;

it could be due to higher luminous phases in galaxies for periods during their lifetimes; or I speculated it was possible that we were overestimating the real cosmic abundance of helium because the ratio of helium to hydrogen was much smaller in the low-mass stars which make up a large part of the total mass, than it is in the hot stars and nebulae in which the abundances can be determined spectroscopically.

The key point that I missed, as did Bondi, Gold and Hoyle (1955), who had made a similar calculation in 1955, arguing that the energy must have come from red giants (in 1958 I had missed the Bondi, Gold and Hoyle paper), was that the energy density corresponding to the production by hydrogen burning when the energy was degraded to blackbody form would give a blackbody temperature of 2.75 K!

If these results had been publicized, they might have been seen as predictions based on observed quantities of what the temperature of the blackbody radiation would turn out to be, if it were detected. But of course this never happened.

As he told me many times later, Fred Hoyle had realized all along that the hydrogen burning in stars was a possible source of the helium and that it would lead to a powerful background radiation field. Much later he and I took very seriously the fact that the CMBR energy density is so close to what the prediction from the hydrogen burning origin would give, and concluded that all of the light isotopes D, ^{3}He, ^{4}He, and ^{7}Li also have a stellar origin. In other words *all* of the isotopes in the periodic table are due to stars. Our paper on this topic was rejected by *Physics Review Letters*, obviously because very convinced big bang advocates refereed it. However, it was finally published in 1998 in the *Astrophysical Journal Letters* (Burbidge and Hoyle 1998).

A key point that most physicists were unaware of throughout the 1950s, 1960s, and 1970s, and in particular the large number of those who believe in the standard model still appear to be unaware of it, is that in 1941 A. McKellar at the Dominion Astrophysical Observatory in Victoria made an estimate of the radiation field in which the interstellar molecules CN and CN$^+$ are bathed, and stated that if this was blackbody the radiation temperature is $1.8\,\mathrm{K} < T < 3.4\,\mathrm{K}$. The exact quote from his paper (McKellar 1941) is as follows:

Dr. Adams has kindly communicated to the writer his estimate of the relative intensity, in the spectrum of ζ *Ophiuchi*, of the $\lambda 3874.62$, $R(0)$ interstellar line of the $\lambda 3883$ CN band and the $\lambda 3874.00$, $R(1)$ line, as 5 to 1. $B_0 J''(J'' + 1) + \ldots$ has the values 0 and $3.78\,\mathrm{cm}^{-1}$ for the 0 and 1 rotational states and for the two lines

$R(0)$ and $R(1)$ the value of the intensity factor i are, respectively 2 and 4. Thus from (3) we find, for the region of space where the CN absorption takes place, the "rotational" temperature,

$$T = 2{.}3\ K.$$

If the estimate of the intensity of $R(0)/R(1)$ were off by 100 percent, this value of the "rotational" temperature would not be changed greatly, $R(0)/R(1) = 2.5$ giving $T = 3{.}4\,\mathrm{K}$ and $R(0)/R(1) = 10$ giving $T = 1{.}8\,\mathrm{K}$.

Had this been generally known in the 1950s, and been put together with the result quoted earlier, the history of what most people want to believe about the CMBR and its origin might be different.

At the time in the early 1960s when Fred Hoyle and George Gamow were debating cosmology, Fred was aware of this result, and used it when Gamow would argue that the temperature was likely to be much higher. I first learned of this result from Fred in that period.

My view of the subsequent history (*as I saw it*) is as follows. In the early 1960s Robert Dicke and J. Peebles reworked the ideas of Gamow, Alpher, and Herman. Since Dicke was a superb experimentalist, he proposed that an attempt be made to detect the radiation. This is what he and David Wilkinson set out to do. But, of course, before they achieved any result there was the serendipitous discovery by Penzias and Wilson (1965a).

But throughout the 1960s the ideas emanating from Princeton and also from Moscow from Zel'dovich's group led almost everyone to believe that the radiation could only be a remnant of a big bang and would be of blackbody form.[15] It would be proof that the steady state theory was wrong. With the Penzias and Wilson discovery, while there was still no proof that it was blackbody, it was thought that the verdict was in.

Even Fred Hoyle began to doubt the correctness of the steady state cosmology, and in his address to the British Association in September 1965 he came as close as he ever did to concluding that the steady state would not work. Starting at that time, he began to discuss a modification of the steady state which in the 1990s, with J. V. Narlikar and me, was turned into the QSSC – an oscillating model still over the long term a steady state universe (Hoyle, Burbidge and Narlikar 1993).

[15] It was in this period that my view that cosmological ideas are driven as much by the views of leading scientists as by actual observations was strengthened. I was present at meetings where early rocket observations were reported which did not confirm the blackbody idea. Those were immediately severely criticized by leading theorists who did not understand the experimental details but were absolutely convinced that the blackbody nature must be correct. They eventually turned out to be right, but their prejudice was obvious.

Jayant V. Narlikar's View. I recall that one day in 1964, Fred Hoyle walked into his office in the DAMTP in a rather disturbed mood. He confided: "I believe, I have found the strongest proof for the big bang." With his previous encounters with Martin Ryle and his colleagues in the Cavendish, I wondered if there was some new evidence from radio astronomy that had unsettled Fred. "No," he added, "my own calculations suggest that helium was mostly made not in stars but in a high temperature epoch in the past. I find that if the density–temperature relationship is properly adjusted one can get almost 25% helium."

For someone who had worked long and hard on stellar nucleosynthesis to demonstrate that most of the chemical elements were made in stars, this finding had come as a shock, even though it was he himself who had done the calculation. His work with Roger Tayler was subsequently published in *Nature* (Hoyle and Tayler 1964) and quickly became a much-cited paper . . . probably it was the only paper Fred wrote with conclusions close to favoring the big bang scenario. Nevertheless, he left an alternative possibility open, namely the existence of supermassive objects that allow stellar nucleosynthesis to generate adequate helium. This possibility is also discussed briefly in the classic paper on nucleosynthesis by Wagoner, Fowler and Hoyle (1967).

Even so, Fred did not relate the 1964 finding with the possible existence of relic radiation. The result struck him as very important only in 1965 after the discovery of the radiation by Penzias and Wilson (1965a). Although the blackbody nature of the radiation had not been established in 1965, its finding together with helium abundance apparently had the effect of convincing him of the existence of a high temperature phase early in the universe.

It was against this background that he delivered his oft-quoted speech to the British Association for the Advancement of Science (Hoyle 1965) in which he came close to supporting the big bang cosmology at the expense of his own steady state theory. One popular magazine in the USA likened this reaction to the problematic situation of Lyndon Johnson abandoning his membership of the Democratic Party to join the Republicans!

I had worked with Fred on many aspects of the steady state theory, and felt that Fred had "given in" too soon. Dennis Sciama, another strong adherent of the steady state idea, also felt the same, although within a couple of years he changed over to the big bang point of view. In the meantime, Fred had second thoughts on the matter. Both he and I, along with Chandra Wickramasinghe, felt that alternative explanations of the radiation background should be looked for. The reasons were mainly as follows:

(i) There are radiation backgrounds at various other wavebands and these are mostly traced to astrophysical sources. Can the microwave background be shown to originate from astrophysical sources radiating mainly in infrared and microwaves?

(ii) Following a more general line of argument, there are galactic and extragalactic astrophysical processes with energy densities comparable to the newly discovered microwave background (CMBR), for example cosmic rays, magnetic fields, and galactic starlight. So to ascribe a relic interpretation to the CMBR gives an unexplained coincidence of energy density.

(iii) The fact that if all helium in the universe were made in stars the resulting energy density would be comparable to that of the microwave background which has already been highlighted in this paper suggested a nonrelic interpretation.

I will discuss these possibilities briefly from a modern standpoint.

It was shown by Wolfe and Burbidge (1969) that the multiple source hypothesis would generate a microwave background that was too inhomogeneous for agreement with the preliminary limits on anisotropy. The only way to escape from this conclusion was that the sources were far more numerous than galaxies and typically weaker than galaxies. Such a population was considered rather unlikely and has not been found.

The search for an astrophysical process to generate the CMBR in the Milky Way Galaxy or in clusters of galaxies led Hoyle and Wickramasinghe to various scenarios involving interstellar dust: dust that could convert starlight or other energy into a thermalized form with the energy density found in the CMBR. Narlikar, Edmunds and Wickramasinghe (1976) wrote a paper suggesting how this could happen using dust grains in the form of whiskers. The scenario was plausible but it was not clear that it would meet the various observational constraints that were being placed on the properties of the CMBR.

The idea of Narlikar, Edmunds and Wickramasinghe (1976) could be applied to a situation in which it was assumed that there had been a lot more starlight initially because of greater stellar activity, which led to most of it being thermalized by whiskers. This idea, however, ran into problems with the original formulation of the steady state theory, which would not allow any epoch-dependent process. Nevertheless, Hoyle and Wickramasinghe persisted with the efforts to study the thermalization process in detail.

Eventually the process was shown to work, not in the original steady state cosmology but in its variant, the *Quasi-steady state cosmology*. This cosmology was proposed by Hoyle, Burbidge and Narlikar (1993) and it envisages a long-term steady state universe with short-term oscillations. The e-folding time of the long-term steady state is around 1000 Gyr, whereas the period of a typical oscillation is around 50 Gyr. We refer the reader to the details given in Hoyle, Burbidge and Narlikar (2000) and to later references (Narlikar *et al.* 2003). So far this alternative is able to achieve the following:

(i) Explain the CMBR as a relic of stars burnt out in the previous oscillations with the present temperature of 2.7 K related to stellar activity at present observed in the universe. See Hoyle, Burbidge and Narlikar (1994) for details.

(ii) A Planckian spectrum at all wavelengths except possibly at wavelengths longer than 20 cm. (There the galactic noise anyway dwarfs the cosmological effect.)

(iii) An angular power spectrum that explains the main peak at around $l = 200$, as arising from typical clusters at the last minimum scale epoch (Narlikar *et al.* 2003).

(iv) The dust density required for thermalization being consistent with that needed for dimming distant supernovae.

(v) A weak polarization on the scale of clusters arising from magnetic alignment of whiskers scattering the radiation.

(vi) Independent evidence for the existence of whisker dust from various astrophysical scenarios.

Fred Hoyle firmly believed that an alternative interpretation of the CMBR along the above lines would turn out to be closer to reality than the standard interpretation. What were the attitudes of the other two coauthors of the steady state theory? I never had the chance to discuss the CMBR with Tommy Gold. By 1965 he had already moved away from cosmology and I do not think he worried too much about the issue. Hermann Bondi had likewise developed other interests. However, I had met him on several occasions. Once in an interview on the All India Radio, Pune, during the 1990s I had asked him what he felt about the steady state theory in the light of the observations of the CMBR, especially by COBE. He replied that to him the steady state theory had been attractive from the Popperian point of view: it made definite statements which could be checked against observations. That the CMBR spectrum had turned out to be so close to the Planckian was, in his opinion, a very difficult observation for the steady state theory to explain. So he had felt that the theory was no longer viable. Like most cosmologists he had been unaware of the above work on alternative cosmology, but seemed pleased that perhaps such an explanation of the origin of the CMBR might succeed.

Going back to 1965, one can say today that while the big bang scenario has been taken a good bit forward in the last four decades, the alternative explanation has also made considerable progress and deserves to be critically examined side by side with the standard explanation.

4.9.2 David Layzer: My reaction to the discovery of the CMBR

David Layzer is the Donald H. Menzel Professor of Astrophysics Emeritus at Harvard University. He is the author of two books, Constructing the Universe *and* Cosmogenesis, *and was an associate editor of the* Annual Reviews of Astronomy and Astrophysics *for 30 years.*

Cosmology became a science in the 1920s. During that decade Hubble's observational program with the 60- and 100-inch telescopes on Mt. Wilson supplied compelling evidence for the hypothesis that guided his program and was its central finding: that the observable universe is a fair sample of the universe as a whole. Friedmann's (1922) theory of a uniform, unbounded fluid, based on Einstein's theory of gravitation in its original form, predicted that such a fluid cannot be static but must expand from an initial singular state in the finite past. And to round off the decade, measurements of the redshifts of faint distant galaxies by Hubble and Humason showed that the system of galaxies was in fact expanding in the way predicted by Friedmann's theory. The next major advance in observational cosmology was the discovery of the CMBR by Penzias and Wilson (1965a).

Not everyone was surprised. George Gamow had suggested that heavy atomic nuclei were formed by successive neutron captures in an early hot universe. Using measured neutron-capture cross sections he and his colleagues deduced the temperatures that would have had to prevail when the expanding universe was dense enough for successive neutron captures to produce (approximately) the observed relative abundances of heavy nuclei. On this basis they predicted that the radiation field, eventually decoupling from the matter, would retain its thermal character and would now have a temperature of about $10\,\mathrm{K}$. (Of course, as we now know, this prediction rested on a false premise. The heavy nuclei were formed in the cores of massive stars, not in a hot, dense cosmic medium.)

Others were surprised. The steady state cosmology, put forward by Hermann Bondi and Thomas Gold (1948) to explain a discrepancy between the estimated age of the universe (based on measurements of Hubble's constant) and the estimated age of the Earth, was still popular, especially among British cosmologists. In Sweden, Bertil Laurent and Oskar Klein had suggested that the universe is finite and bounded, an expanding island floating in empty space. These cosmological models became instant casualties of Penzias and Wilson's discovery. A thermal radiation field with a temperature of 3 K couldn't be formed in either of them.

Proponents of an initially cold Friedmann universe were also surprised. Lifshitz's (1946) theory of the growth of density fluctuations in a Friedmann universe had shown that thermal fluctuations in a uniform gaseous medium were many orders of magnitude too small to evolve into self-gravitating systems. To overcome this difficulty Zel'dovich (1962) suggested that an initially cold cosmic medium would solidify when its density reached approximately one tenth the density of water. Then, as it continued to expand, it would break up into solid chunks large enough to cohere under their internal gravitational attraction.

The path that led me to Zel'dovich's hypothesis was different. In 1951 I was a postdoctoral fellow in Ann Arbor, working on problems in atomic physics, when I came across a copy of Otto Struve's (1950) book *Stellar Evolution*. I was especially intrigued by Struve's account of binary stars and theories of their origin. Though half the stars in our neighborhood belong to binary or triple systems, neither of the two main hypotheses for the formation of binaries – the fission hypothesis and the capture hypothesis – could account for this fact. It occurred to me that if stars had formed in close proximity to one another – if the cosmic medium had once been a uniform distribution of strongly interacting protostars – then, as the medium continued to expand, most of the protostars would have ended up in small groups, the most stable of which would be binaries.

This thought immediately suggested to me that all self-gravitating systems might have been formed in this way, as clusters of smaller systems. The earliest stage in this process of hierarchical gravitational clustering would have been the formation of the smallest objects held together by their own gravity rather than by chemical cohesion. Clusters of these objects would evolve into planetary systems, clusters of these evolving systems would come together in larger self-gravitating clusters, and so on, up to galaxies, clusters of galaxies, and clusters of galaxy clusters. I wrote a short paper (Layzer 1954) in which I argued on the basis of this picture that the Solar System could have evolved from a cluster of marginally self-gravitating chunks of matter. I argued that this picture could explain why satellite systems like those of Jupiter and Saturn mimic the Solar System.

But it was just a picture, not a theory. Atomic physics was still the focus of my research. I hadn't studied general relativity nor read Lifshitz's (1946) seminal paper. I knew that the universe was expanding, and I assumed (correctly but for no good reason, then) that self-gravitating systems were not expanding with it. And that was the extent of my knowledge. So I began to study general relativity, with a view to acquiring more insight into the interplay between the disruptive tendency of the cosmic expansion and the tendency of overdense regions to contract.

Zel'dovich, in his 1962 paper, had used a theory of the growth of cracks in a stressed solid to estimate the sizes of the primordial fragments. His aim was to show that random (square root of N) fluctuations in a uniform distribution of these fragments would be large enough to evolve into self-gravitating systems. My approach centered on energetic considerations. Its aim was to understand not just how an initially uniform cosmic medium could ever become unstable against the growth of density fluctuations but to understand how it could become and remain unstable against the growth of density fluctuations on progressively larger scales. I reasoned that because the gravitational interaction has no inherent scale, gravitational clustering would have to be a self-similar process. Thus a log–log plot of (primordial) binding energy per unit mass against cluster mass would have to be a straight line, extending from the smallest self-gravitating systems to clusters of galaxy clusters. Observational evidence supported this conclusion; and the predicted slope of the relationship (based on a theory developed in Layzer 1968 and 1975, my 1968 Brandeis lectures in Layzer 1971, and my book *Cosmogenesis*, Layzer 1990) agreed with the observed slope. Moreover, the theory predicted a coincidence first pointed out, I believe, by Fred Hoyle (1953): the gravitational binding energy per unit mass of our own planetary system (and, presumably, others as well) is approximately equal to the chemical cohesion energy per unit mass of a typical solid (and of solid hydrogen).

By 1965 most of this work had been done, though not all of it had been published. So I greeted Penzias and Wilson's announcement with mixed feelings. Like most people who had opinions on such matters, I found the experimental findings and their interpretation convincing. Also like most people, I recognized that they would have momentous consequences for cosmology. At the same time, I felt pretty confident that the picture of hierarchical gravitational clustering was essentially correct. So I had to face the question: Can the existence of a thermal radiation background with a temperature of 3 K be reconciled with the picture of gravitational clustering in a cold universe?

If, as most people assumed, the background radiation was the remnant of a primordial fireball, its almost precisely thermal character would be easy to understand. On the other hand, if it was created by the burning of hydrogen into helium later in the history of the universe, two conditions would have to be met. The universe had to have been opaque to the background radiation (at the temperature it had then). And the mass density of hydrogen converted into radiation had to be less than the closure density. These conditions work in opposite directions. The farther we go back in time, the easier it is to construct conditions under which the universe will be opaque to radiation at the appropriate temperature. But because the energy per unit mass of the radiation field diminishes like the reciprocal of the cosmic scale factor, the second condition puts a lower limit on the epoch at which the radiation could have been created. Could both conditions be met?

A quick and dirty calculation suggested that they might be – though it would be a tight squeeze. So there seemed to be no reason to abandon the scenario of gravitational clustering in a cold universe – at least not yet. But to survive, the scenario needed to pass more stringent tests.

In the cold universe, as in standard hot models, helium is formed during an early era of nucleogenesis. Following a preprint by Jim Peebles, Michele Kaufman (1970) studied under what conditions this could be done in an initially cold universe. Her results were promising, but left unanswered a key question: Would helium created in an early cold universe be subsequently transformed into still heavier elements? Subsequently, Anthony Aguirre (1999) devised reasonable cosmological models that are cold enough to solidify at the appropriate time but warm enough to prevent helium from being consumed in the production of heavier nuclei.

Can the background radiation be adequately thermalized in an initially cold universe? The most recent calculations, again by Aguirre (2000), indicate that the answer is yes.

An attractive feature of the cold universe scenario is that it requires a large fraction of the (ordinary) matter in the universe to be nonluminous. For in the cold universe, the background radiation is produced by an early generation of massive (and supermassive) stars, whose ejecta supply both the dust that thermalizes the radiation and the nonluminous matter that makes itself known through its gravitational effects. This is attractive because it makes the existence of dark matter/missing mass a necessary feature of the universe, required by the production of the background radiation. And it makes two testable predictions. It predicts that the dark matter is ordinary matter and it predicts a small range of possible values for the ratio between dark matter and bright matter.

Recent observations of the microwave background and of the redshifts of distant galaxies seem hard to reconcile with the cold universe scenario. On the other hand, the standard hot scenario still lacks a compelling account of the origin of self-gravitating systems in the expanding universe. Whatever our views on the issue of hot versus cold – unlike most of my colleagues I remain an agnostic – we can all agree that Penzias and Wilson's discovery has changed not just the face but the character of theoretical and observational cosmology.

4.9.3 Michele Kaufman: Not the correct explanation for the CMBR

Michele Kaufman is a scientist in the Ohio State University departments of Physics and Astronomy. Her current research uses the Very Large Array of radio telescopes, the Hubble Space Telescope, and the Spitzer Space Telescope.

When I was an undergraduate, I heard Dr Tommy Gold say in a public lecture that the density and temperature of intergalactic gas were uncertain by factors of 10^{12}. Later, as a graduate student at Harvard in 1964, I started research under David Layzer's supervision by calculating the expected radio-to-microwave background radiation produced by a combination of emission from discrete extragalactic radio sources and intergalactic free-free emission. The goal was to try to place limits on the amount of intergalactic ionized hydrogen. I included the effect of self-absorption. An earlier paper on the intergalactic free-free spectrum by Field and Henry (1964) had omitted self-absorption.

Before the Penzias and Wilson (1965a) result was widely announced, Arno Penzias visited Princeton, MIT, and Harvard, and at Harvard, he was directed to talk with me. Thus I learned that Penzias and Wilson had measured the background radiation at 4.08 GHz. This provided my model with an important constraint on the values of the intergalactic electron temperature and density, and in the summer of 1965 I published a paper in *Nature* on this with the conclusion that intergalactic free-free emission could account for the background measured by Penzias and Wilson (Kaufman 1965). This paper attracted some attention as the then only published alternative to fossil thermal radiation from a hot big bang. After the microwave background was measured at other frequencies, it was clear that intergalactic free-free emission was not the correct explanation for the CMBR. Reviews of the CMBR continued to reference my 1965 paper as a suggestion that did not pan out.

I later switched research areas from cosmology to galaxies, especially individual spiral galaxies. My research in the past 25 years has included detailed studies of spiral tracers in the grand-design spiral M81 and detailed multi-wavelength studies of galaxy pairs involved in grazing, prograde encounters (with Debra and Bruce Elmegreen). Our HST image of NGC 2207/IC 2163, part of the latter study, has appeared everywhere in the national news media, including the front page of *The New York Times* as well as scholarly journals (Elmegreen *et al.* 2006).

4.10 Measuring the CMBR energy spectrum

4.10.1 Jasper V. Wall: The CMB – how to observe and not see

Jasper Wall served as Director of the Royal Greenwich Observatory and of the Isaac Newton Group of Telescopes, La Palma. He is now Visiting Professor, University of Oxford, and Adjunct Professor, University of British Columbia.

In 1965 Donald Chu, Allan Yen and I made extensive sky brightness measurements at 320 and 707 MHz. Comparison told us that something was wrong with the zero point, wrong by the same few degrees at each antenna and at each frequency. Here is the story.

Engineering was in my blood, via father and grandfather. I grew up in the Ottawa Valley, in a happy and stimulating household in which the mantra was "This works so well we must take it apart to see why." Clocks, toasters, cars, plumbing, house electrics, lawn mowers, washing machines, hi-fi; nothing was safe from my Dad and his two young sons. Inevitably it was off to do Engineering at Queen's University, from where I graduated in 1963. But well before 1963 I had found the conventional branches of engineering to be less interesting than I had wished. I headed off into Engineering Physics, great training for applied research postgrad studies. But in what? I had spent a couple of summers at the National Research Council in Ottawa, working in the radio astronomy group. It seemed to me at the time that astronomy was perhaps of passing interest and might offer decent engineering challenges. The astronomy got me in the end, but the engineering background paid rich dividends at various times in my later professional life. The immediate challenge was radio astronomy instrumentation, which I set out to do in a Master's degree program in the Department of Electrical Engineering at the University of Toronto, starting autumn 1963.

My joint supervisors were Donald MacRae, Professor and Head of the Department of Astronomy, and the brilliant and enigmatic J. L. (Allan) Yen, Professor of Electrical Engineering, theorist, instrumentalist, expert on Toronto Chinese cuisine (chopsticks were an early part of my graduate education) and a man who required almost no sleep. I saw both my supervisors but rarely, and then only when I was in trouble with them, this more frequently than was comfortable. I learned through the standard apprenticeship system, the senior grad students mentoring the new student intake. I learned most from Ernie Seaquist, who was well into his PhD program in the Astronomy Department. He was patient and generous to me with time precious for his own extensive radio astronomy program, and by example he taught me far more than just radio astronomy.

My project was to measure absolute temperatures of the galactic background at 320 MHz, using the pyramidal horn already installed at the David Dunlap Observatory (DDO), Richmond Hill, 19 miles north of Toronto. The horn itself (Figure 4.32) was in relatively good shape, needing some cleaning to remove certain avian deposits of the sort that Penzias and Wilson (1965a) encountered in their researches. The challenge as I mapped it out was (a) to build a reasonably low-noise amplifier and Dicke-switching receiver and (b) to design and build a reference cold load for the switching system, one with absolute temperature known to specified accuracy. The measurements were then simple drift scans, with the horn turned to the north celestial pole at periodic intervals for a reference level. This level would be calibrated by replacing the horn input with the reference cold load input. There were impedance-matching subtleties involved, as long-serving radio astronomers will recognize.

First task – to build a new receiver at 320 MHz. Field effect transistors, FETs, had just become available, actually working at this high a frequency! Low noise as well! But they cost real money, all of $34 each. In a rare interview with Allan, I got the money and the transistor. Next day I blew it up. (In retrospect I begin to understand the supervisor problem.) I managed to extract funds for a second one, and, after walking around it for an afternoon, made a decision on how to handle it which helped me the rest of my life. It's just another transistor! Handle with ordinary care – otherwise I couldn't see how I would get anywhere. It worked. I applied the lesson later when dealing with original astronomical plates. Treat them as you treat glass, with respect, but without awe. More tense and more "careful" $\equiv$ greater risk and less research.

The second FET ran throughout the project. The new receiver was built with help of George Watson, a solitary soul working out at Richmond Hill:

Fig. 4.32. The pyramidal horn antenna, aperture 3.7 by 2.8 m, used at 320 MHz for my galactic background temperature measurements.

a craftsman, a perfectionist, and a delight, whose stories, unrepeatable and certainly unprintable, enlivened many of my days and nights in the little frozen cabin at Richmond Hill, while adding a certain breadth to my graduate education. More supervisor trouble ensued when in the course of transporting a frequency generator to the cabin (they weighed about 150 kg in those days), I settled the old radio astronomy station wagon axle-deep into the Observatory grounds in soft spring mud.

The cold load was a real challenge. Nobody really knew how to proceed, and the one I fashioned was the best technical achievement of my MSc. It did work well, and I was confident of its noise temperature – but note that it was a liquid nitrogen cold load, at about 80 K. This was close to the mean galactic brightness temperatures; but of course a long way away from CMBR values.

I heard/read of the CMBR as my observations progressed. Reaction (a): nothing to do with me; I'm a galactic (semi-) astronomer, working at too low a frequency and too high a mean brightness. Reaction (b), with minimal cosmic consciousness and from a radio astronomy point of view: surprise, Ryle was right after all – but a singular beginning? Steady state was conceptually much easier to handle.

And following this two minutes of deep thought, back to reality – the horn antenna had half-power beamwidths of $19.0° \times 22.5°$. Absolute temperature mapping requires correction for the response in side- and back lobes, of course. Thus I built a scaled version of the horn, complete with supporting structure, smaller by a factor of 9 and operating at 2.88 GHz. I mounted this on the antenna range turntable on the roof of the Electrical Engineering building, with a distant horn-reflector plus S-band generator to provide the signal. The main-beam and first side-lobe patterns agreed remarkably well with the main-beam measurements of the main horn using drift scans of the Sun, a point source (only 30 arcmin in size!) to the fat beam of the horn. The side and back lobes enabled me to estimate the spillover radiation.

There were many delays, including my MSc course load and stormy winter weather. Measurements began in February 1965 and continued to June; I covered the hottest part of the sky but by June (Figure 4.33), interference

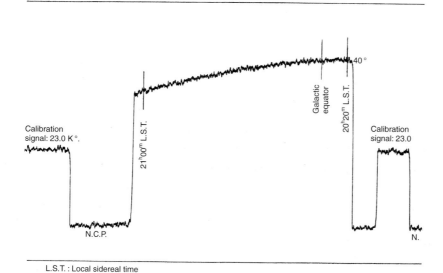

L.S.T. : Local sidereal time

Fig. 4.33. A chunk of drift scan, this one at declination $\delta = 40°$ complete with periodic visits to the North Celestial Pole and calibration-signal injections.

from the USAF Buffalo base essentially halted the observations. I could not finish the cold (galactic anticenter) parts, another sore point between me and supervisors. My MSc thesis, complete with the iterative calculations to remove side- and back-lobe responses, was completed in October 1965. In parallel Donald Chu ran a sister set of measurements at 707 MHz, using a 2.5-m precision horn-reflector at the Algonquin Radio Observatory of the National Research Council of Canada. The techniques he used followed mine precisely, including construction of a scaled model of the horn-reflector. His measurements and mine were to be used to calibrate in absolute terms higher-resolution galactic plane surveys at DDO with a new 10-m paraboloid reflector (for which I did commissioning and feed design.) These together with polarization measurements which Ernie Seaquist was working on were to provide comprehensive data on the Milky Way emission. This grander scheme never happened.

In November I set off for Australia, where I had been offered a scholarship at the Australian National University to do a PhD in a collaborative radio-optical program between Mount Stromlo Observatory and the Australian National Radio Astronomy Observatory at Parkes. John Bolton was to be my supervisor. My seduction by astronomy was complete. Engineering cropped up later in my life in building CCD systems, commissioning telescopes etc.; but it was astronomy now where my commitment lay.

Donald Chu, finishing the same patch of sky I had done, likewise left for different things, a proper job in his case with the then largest computer company.

In the excitement of starting a new life in a country where snow drifts across the telescopes were no longer a problem, the brightness temperature measurements were temporarily laid aside.

The rest of the story has a certain inevitability about it. Donald Chu had made some tentative comparisons of his data with mine; he found unsatisfactory answers. We knew roughly what the emission spectrum of the galactic background was – this synchrotron emission continuum from long-blown supernovae had a brightness spectral index of about -0.5 to -0.7 (Yates and Wielebinski 1967). Comparison of the 320- and 707-MHz results at independent map points by Donald and myself yielded a spectral index of -0.3, far too flat. Trying to reach indices in the "recognized" range meant zero-point errors outside our estimates. In 1965 we had left it at this: we had both moved on.

In 1967 or 1968, as cosmological consciousness dawned, I realized what had happened. Subtracting $3\,\mathrm{K}$ from both of our sets of measurements yielded spectral indices in agreement with the "known" results (Figure 4.34).

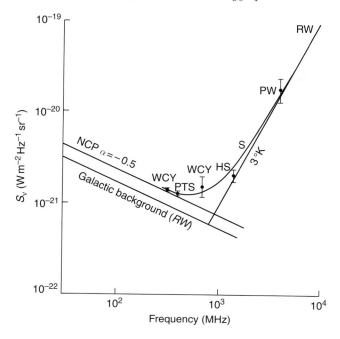

Fig. 4.34. The surface brightness measurements, circa 1969, from Wall, Chu and Yen (1970). PTS: Pauliny–Toth and Shakeshaft (1962); PW: Penzias and Wilson (1965a); HS: Howell and Shakeshaft (1966); RW: Roll and Wilkinson (1966); WCY: Wall, Chu and Yen (1970). ©1970 CSIRO Publishing.

I collected the data together, redigitized it, and finally wrote up the experiments (Wall, Chu and Yen 1970). There was no great urgency at this stage.

In retrospect a dedicated CMBR measurement would have been simple. We had only to cover the colder parts of the sky, put our two sets of measurements together with a prior on the galactic emission spectral index, and a measurement of the excess radiation was there. We were a bit late in the time frame – but if we had got on with it in the first years of our MSc degrees rather than spending them wading through forgotten courses on plasma physics, the result would have been waiting for us.

The most astonishing aspect to me in hindsight was just how easy it would have been to make the measurement successfully, using the horns we already had, and a financial outlay of almost nothing.

I blame VLBI (partially). If Allan Yen had not become preoccupied with this (Broten *et al.* 1967) I know his razor-sharp mind would have seen the possibility; he read everything and was on top of everything. I know that excess radiation was in his mind – although he never mentioned CMBR or excess radiation to me, his annoyance when I had been unable to finish measuring colder parts of the sky convinced me of this. This too came in retrospect.

The CMBR subsequently played little part in my career of observational cosmology. I stuck to AGNs and their spatial distribution, together with schemes of (unified) beaming models. Most of this was with radio-selected samples. There were perhaps just three points of contact:

(i) In carrying out the (1984 version) deepest survey at 5 GHz with the VLA, Ed Fomalont, Ken Kellermann and I put limits on CMBR fluctuations in the range of an arcminute and a bit less (Fomalont, Kellerman and Wall 1984). These were the best upper limits at the time; but they were far from real detections at these angular scales, as we now know. Perhaps our main contribution was to determine how to minimize cross-talk between the antennas, a help to subsequent experiments. Even so, the VLA for all its power was never the instrument for CMBR fluctuations.

(ii) The standard model has the CMBR dipole, 1 part in 1000, explained as the Earth moving at $370 \, \text{km s}^{-1}$ relative to the rest frame, with apparent temperature brighter in the direction of motion. The predicted motion should be visible in the number counts of distant objects, their combined surface brightness enhanced in the direction of motion of the Earth. There are serious difficulties in looking for this dipole in discrete objects: how distant, how to select, how to perform widescale calibration; what to do about obscuration, how to get beyond the cluster-dominated epoch. A uniform all-sky survey of radio sources offers hope, however, as Ellis and Baldwin (1984) pointed out. After completion of the superb NRAO VLA Sky Survey (NVSS; Condon *et al.* 1998), that hope could be really entertained. It took much work to understand the systematics of the survey, and much work to remove the nearby objects from it – but in the end Chris Blake and I succeeded in observing the dipole (Blake and Wall 2002), agreeing in magnitude and direction with Earth motion as implied by the CMBR (Figure 4.35). This remains the only detection of the velocity dipole in discrete galaxies, objects formed long after the epoch at redshift $z \sim 1100$ corresponding to the last scattering surface from which we see the CMBR. The mean redshift of our radio galaxies is about unity. The universe is therefore showing large-scale homogeneity at this epoch, and further analyses coupled with new deep and wide sky surveys can refine this result. Although few doubt the interpretation of the dipole in the CMBR, the detection in real

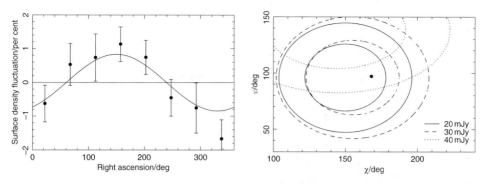

Fig. 4.35. Left: measured amplitudes of the deviation from mean surface density for NVSS sources, as a function of right ascension. (Note that the direction of the CMBR dipole lies – accidentally – close to the Celestial Equator.) The predicted amplitude is shown as the solid line. Right: error circles (1σ, 2σ) representing the direction of the NVSS dipole for samples selected at different flux-density levels. The point denotes the direction of the CMBR dipole.

objects represents one of the tests the CMBR needs to pass if it is truly a relic of the big bang (Ellis 2002).

(iii) With superb results from WMAP (Bennett *et al.* 2003), and with the Planck mission on the horizon, we would like some reassurance that the fluctuations we see in the CMBR are not contaminated by extreme inverted-spectrum populations of radio-millimeter sources. To this end, with Rick Perley, Robert Laing, Joe Silk, and Angela Taylor, I recently proposed a 43-GHz VLA survey of some 2 square degrees of the northern sky to search for such a population. This is the highest frequency search for extragalactic radio sources – and it found very few (Wall *et al.* 2006). We conclude that at small angular scales and the high frequencies of the measurements of the power spectrum of the angular fluctuations of the CMBR, there is little to fear from discrete radio source contamination.

I offer some conclusions.

(i) The CMBR was there all the time in our 1965 data; and we could have done the measurements earlier with specific attention to detecting it as a part of our absolute flux measurements. It would have come in somewhere between 3 K and 5 K at a guess. I think it's a stretch to say that we would have believed it on its own; our frequencies were a little low. But had there had been contact with cosmologists such as between Penzias and Wilson (1965a) and Dicke *et al.* (1965), then it might have been different. Too if my cosmic consciousness had not dawned so slowly, it might have been different.

(ii) I cannot have any regrets. My MSc project was superb for starting research in observational astronomy. How better to learn everything about the basics of radio astronomy? Every aspect in the process was revealed to me in glaring detail, all the pitfalls, noise, bandwidth, line-loss, mismatch, spillover, ground radiation, antenna patterns, conversion of antenna temperature to brightness temperature.... It was baptism by fire, and I did love it, I think. It is next to impossible for a student nowadays to learn about instrumentation in depth at any wavelength, and I grudge a big vote of thanks to my supervisors Donald MacRae and Allan Yen for so comprehensively dropping me into it.

(iii) It is possible to observe and not see. After all, Donald Chu and I were only a couple of engineers playing around with horn antennas...

4.10.2 John R. Shakeshaft: Early CMBR observations at the Mullard Radio Astronomy Observatory

John Shakeshaft is an Emeritus Fellow at St Catharine's College, Cambridge. He served for many years as Editor of Monthly Notices of the Royal Astronomical Society.

At the time of publication of the Penzias and Wilson (1965a) paper, I was a member of staff in the Radio Astronomy Group of the Cavendish Laboratory, the physics department of the University of Cambridge, having been an undergraduate and graduate student at Cambridge, the latter under the inspiring supervision of Martin Ryle. I had had an interest in cosmology and measurements of cosmic radio radiation for over ten years. Indeed my first published scientific paper, in 1954, had the title *The Isotropic Component of Cosmic Radio-Frequency Radiation,* although I advise readers not to bother to search it out. At that date, low-noise receivers for the microwave range had not yet been developed, so interest was concentrated at lower frequencies. Westerhout and Oort (1951) had shown that the survey of galactic radiation at 100 MHz (or Mc/sec as we called it) by Bolton and Westfold (1951) could be explained by assuming that most of the radiation came from "radio stars" distributed through the Galaxy in the same way as the common Population II stars of types G and K, although it was necessary to add in an isotropic component besides. They suggested three possible explanations for this extra component but found none to be satisfactory. Subsequent to their paper, extragalactic sources much more intense than normal galaxies had been identified, such as the so-called "colliding galaxies" Cygnus A, and I attempted an estimate of the integrated contribution due to these. Interestingly perhaps, in view of later controversies about the number counts of radio sources and their cosmological significance, I concluded that the isotropic component could be accounted for by standard relativistic cosmological models but not by the steady state theory. Shortly after publication, however, the general realization that galactic radio emission is largely due to synchrotron radiation from cosmic ray electrons in the interstellar magnetic field vitiated both the Westerhout and Oort model and my conclusion from it.

Toward the end of that decade I began work, with graduate student Ivan Pauliny-Toth, on a survey of the background radiation at 404 MHz ($\lambda = 74$ cm). This was intended as part of a study over a range of frequencies to determine the spectrum of the galactic radiation as a function of direction, which could provide information on the dependence of cosmic ray electron density and interstellar magnetic fields on position within the Galaxy. It was therefore important, if brightness temperatures at different frequencies were to be compared, for these temperatures to be absolute values rather than merely relative values in different directions. We used an 8-m diameter dish on an alt-azimuth mount (it was in fact a German radar dish "liberated" after World War II by Martin Ryle, and one of the two dishes that Graham Smith used as an interferometer to determine accurate positions of the sources Cygnus A and Cas A, enabling their optical identifications by Walter Baade and Rudolph Minkowski), an electron beam parametric amplifier and a Dicke-type radiometer with a liquid nitrogen reference source. The published survey (Pauliny-Toth and Shakeshaft 1962) was used later both at Cambridge and elsewhere to correct CMBR measurements for the contribution due to galactic radiation. The experience gained on determining losses in the antenna and connections, and ground radiation in the far-out side lobes, was also helpful in measurements of the CMBR a few years later.

The news in 1965 of the serendipitous result from Holmdel of 3.5-K CMBR at 4080 MHz (the second such major astronomical discovery from there, the first being Karl Jansky's accidental finding of the galactic radio emission in 1931) was received at the Mullard Observatory with great interest but no real surprise, since our work on radio source surveys over the previous 12 years had left us convinced that the universe was evolving and not in a steady state; the discovery of radiation from a big bang therefore fitted naturally with these ideas. Although the data from the 2C survey had been overinterpreted in terms of actual sources, the ingenious $P(D)$ probability analysis by Peter Scheuer (1957) of the deflections D of the interferometer records themselves, without the identification of individual sources, showed conclusively (to us, at least) that the slope of the radio source counts $N(\geq S)$ was proportional to $S^{-1.8}$, significantly steeper than the $N(\geq S)$ proportional to $S^{-1.5}$ expected for a uniform Euclidean model, and even more so than the values expected for Friedmann and steady state models. By 1965, the increase in the numbers of actual identifications of distant radio galaxies and quasars had confirmed the excess of sources at large redshifts, and subsequent studies have shown that Scheuer's result for the slope was indeed correct. One of the merits of the steady state theory was said to have been that it gave specific predictions, unlike the Friedmann models, but its proponents seemed very reluctant to accept that these predictions were in conflict with the observations.

I realized that we were in a position fairly easily to check the Holmdel temperature value at a different frequency, namely 1407 MHz ($\lambda = 21.3$ cm), which would help to determine whether this component of radiation had a thermal spectrum as predicted. With the aid of graduate student Tim Howell, a copper horn of beamsize $13° \times 15°$ was set up inside the 8-m dish mentioned above, itself surrounded by a wire mesh screen, 30 m square, lifted at the edges, so the horn was doubly screened from ground radiation. The horn was connected to a Dicke radiometer, with a termination in liquid helium as the reference source. This consisted of a metal film resistor at the end of a 75-cm length of low-loss coaxial line. The temperature distribution along the line was measured and the effective noise temperature at the upper end calculated to be $T_L = 5.9 \pm 0.2$ K. When the leads from the horn and cold load were interchanged in the circuit, the alteration of receiver output gave a direct measure of the temperature difference with no contribution from any asymmetry of the switch and leads. The receiver output was calibrated by measuring the differences between terminations immersed in water at various temperatures.

Observations were made at night with the horn directed toward the zenith (declination $\delta = 52°$) at a right ascension such that the galactic radio emission was at a minimum. A temperature difference $T_H - T_L = 0.9 \pm 0.1$ K was found, implying that $T_H = 6.8 \pm 0.3$ K, representing the combined contributions from (a) galactic radiation and the CMBR, (b) atmospheric emission, (c) ground radiation, and (d) losses in the horn and waveguide-coaxial connection. For (b) we took the value of 2.2 K, derived by Dave Hogg for a wavelength of 20.7 cm (see below), and assumed an error of ± 0.2 K. For (c), we measured the polar diagram of the horn and estimated a value less than $0.1\,K$, and for (d) we calculated a contribution of $1.3 \pm 0.2\,K$. The sum of (b), (c), and (d) was $3.5 \pm 0.3\,K$, leading to a value of $3.3 \pm 0.5\,K$ for the minimum background brightness temperature. The galactic contribution to this was found by convolving the reception pattern of the horn with the brightness temperatures measured in the survey at $74\,cm$ mentioned earlier and scaling the result to $20.7\,cm$ by assuming T to be proportional to $\lambda^{2.7}$.[16] The result was $0.5 \pm 0.2\,K$, leaving a CMBR value of $2.8 \pm 0.6\,K$ (Howell and Shakeshaft 1966), which turns out to be gratifyingly – if fortuitously – close to the currently accepted value of $2.725\,K$.

The atmospheric absorption for frequencies up to $8\,GHz$ is due predominantly to nonresonant absorption by molecular oxygen, and Hogg (1959) had calculated values from $400\,MHz$ up, on the assumption of a line-broadening constant of $0.75\,GHz$ per atmosphere. Our search of the literature for experimental measurements of the absorption by observations of the extinction of extraterrestrial sources as a function of zenith angle had revealed a relatively wide scatter of values, with some in very poor agreement with Hogg's predictions. To throw more light on this problem, we carried out measurements of our own at 408 and $1407\,MHz$. The interpretation of these involved consideration of the change of apparent angular size of the source in question due to differential refraction in the atmosphere. After applying the necessary corrections, our results fitted well with Hogg's curve, but we then realized that some of the earlier workers had either not applied the refraction correction or applied it with the wrong sign. Judicious reworking of the earlier results, where necessary, then produced a satisfactory agreement between theory and experiment (Howell and Shakeshaft 1967a).

On completion of the initial measurement of the CMBR at $1407\,MHz$, we tried to check whether this component of the cosmic radiation could be detected at the lower frequencies of 408 and $610\,MHz$, although the dominance of galactic radiation in this range would cause increased uncertainties. Studies by Peter Scheuer (1975)[17] and by Ray Weymann (1966) had suggested that deviations from a blackbody spectrum might be present at low frequencies. We used optimal scaled horns with beamwidths of

---

[16] To my embarrassment, Jim Peebles, in reviewing this piece, has noticed that the actual wavelength corresponding to a frequency of $1407\,MHz$ is of course $21.3\,cm$ rather than $20.7\,cm$, as appeared in the original paper and as I unthinkingly copied above. He is the first person in the last 40 years to have pointed out this blunder to me. At this late date I do not have the original working material available and so cannot determine whether the quoted temperature of 2.8 ± 0.6 K might require modification. Any such change would only be in the second decimal place, already omitted due to the size of the error.

[17] Presented in an article written in 1965 for *Galaxies and the Universe*, and eventually published in revised form in 1975.

15°, screened from ground radiation by wire mesh, and Dicke radiometers with liquid helium reference sources as before. After applying corrections for other contributions as at 1407 MHz, the effective brightness temperatures from the region of the celestial North Pole were 24.3 ± 0.9 K at 408 MHz and 10.4 ± 0.7 K at 610 MHz, the ratio of these, 2.3 ± 0.2, being significantly less than the ratio of 3.1 ± 0.1 expected for the galactic contribution with a temperature spectral index of -2.8 ($T \propto \lambda^{2.8}$). This implied that there was indeed an extra component of radiation characterized by a temperature close to independent of wavelength, that is, a blackbody spectrum. Further analysis indicated that, if the spectrum of this component were blackbody, the excess temperature would be 3.7 ± 1.2 K (Howell and Shakeshaft 1967b). Unfortunately, the error in this value was such that no new upper limit could be put on the epoch of ionization of the intergalactic gas.

This work concluded for over 20 years observational studies at the Mullard Radio Astronomy Observatory of the CMBR, since other groups much better equipped for work at high frequencies had vigorously entered the field, but they were subsequently taken up again with the building by Paul Scott and others of the Cosmic Anisotropy Telescope (CAT), a three-element interferometer which, in 1996, was the first telescope to detect structure in the CMBR on angular scales smaller than the main peak in the angular spectrum (Scott *et al.* 1996). This was followed by the Very Small Array (VSA), now observing from Tenerife the anisotropies on angular scales between 15 arcmin and 2°, and the Arcminute MicroKelvin Imager (AMI) to study the Sunyaev–Zel'dovich effect in high-redshift clusters and proto-clusters of galaxies. In addition to this observational work, there has been theoretical modeling of background fluctuations, and the Cambridge Planck Analysis Centre has been set up in preparation for the launch in 2009 of the European Planck Surveyor satellite.

Other authors in this volume have noted that, if the attention of observationalists had been drawn to the matter, the CMBR could perhaps have been detected (or recognized as such) years earlier than in fact it was. It is, for example, unfortunate that in neither of the two editions (1952 and 1960a) of his influential textbook *Cosmology* did Hermann Bondi refer to the possibility, nor did Fred Hoyle (1959b) in his paper *The Relation of Radio Astronomy to Cosmology* at the *Paris Symposium on Radio Astronomy*. We must hope that sufficient of the astronomical literature is now available on the World Wide Web for rapid searches which could prevent oversights of this kind in the future.

4.10.3 William "Jack" Welch: Experiments with the CMBR

Jack Welch retired from teaching Astronomy and Electrical Engineering at UC Berkeley in 2005 but continues as the Alberts Professor in the Search for Extraterrestrial Intelligence. He was Director of the Radio Astronomy Laboratory at Berkeley from 1972 to 1996 during which time the BIMA Millimeter Telescope Array was built and operated. He continues his research in the interstellar medium and star formation and is currently working on completion of the Allen Telescope Array.

My introduction to the question of the absolute radio brightness of the sky came from a talk that I heard at a meeting of the IEEE Antennas and Propagation group held in Palo Alto in 1961 or 1962. The Speaker was R. W. De Grasse, one of the team of engineers at the Bell Telephone Laboratories that had developed a communication system for the Echo project. He described the horn-reflector antenna and maser receiver amplifier that had been built at Crawford Hill in New Jersey. As a young radio engineer just beginning work in radio astronomy at Berkeley, I was enormously impressed with the quality of the instrumental work and the care taken with the system noise measurements. I remember him saying that they assumed the sky background temperature to be zero but were uncertain about an excess of a couple of degrees or so in their summary of system noise contributions. The excess was thought to be pick-up in the antenna side lobes (Ohm 1961). At the time, I had no idea what to expect for the background.

A few years later, I read the letter in *The Astrophysical Journal* by Penzias and Wilson (1965a) describing their beautiful background measurements with that same antenna. Using a new receiver at 4.08 GHz with a new reference load (Penzias 1965), they were able to report with certainty an excess of about 3.5 K that had to be ascribed to the cosmic background. The companion paper by the Princeton group (Dicke *et al.* 1965) with the plausible interpretation that the radiation was the blackbody radiation remnant of an earlier stage of an expanding universe was very exciting. George Field, who had recently joined the Berkeley Astronomy Department, was very taken with the new finding and realized that earlier observations of the excitation of interstellar CN (Herzberg 1950) might be consistent with the new radio observations. The excitation of the first rotational level of the CN line corresponded to background radiation at a wavelength of 2.6 mm, suggesting that the excess radiation was that of a blackbody in agreement with the Princeton group interpretation. At the time, our group was developing receiving equipment at wavelengths near 1.0 cm for radio astronomy and studies of atmospheric emission with a small antenna. George urged us to attempt a measurement of the CMBR to help determine its spectrum at the shorter wavelengths.

We decided to take a detour from our other program to study the background at a wavelength of 1.5 cm. An important piece of information about the universe had been found. We might be able to add to that, and it would be an interesting instrumental challenge. At 1.5-cm wavelength, the background emission from the atmosphere is rather high at sea-level sites, and we planned an observation from the High Altitude Barcroft Laboratory of the University of California White Mountain Research Station. Sam Silver, the Director of the Space Sciences Lab at the University of California, had outfitted a trailer for remote observations, and we were able to take it to the Barcroft Laboratory for our observations. The atmospheric emission brightness is typically only 3–4 K at the 12,400-ft altitude of the Barcroft Laboratory. Our technique was conventional. We used a Dicke radiometer that compared the brightness of the sky as detected by a standard gain horn and associated receiver with that of blackbody loads at known temperatures, and we made tipping measurements to extrapolate the brightness to zero air mass. One difference in our system was that we used a load at the temperature of liquid nitrogen as our low-temperature reference rather than a liquid helium load. We felt that we could characterize it well and it would be easier to manage at the remote site than a liquid helium load such as those used by the other groups. As a check, we measured a liquid helium load in the lab at Berkeley with our system and found the correct temperature. We spent the summer of 1966 making background observations at the high-altitude site.

Our reported result, 2.0 ± 0.8 K, was disappointing (Welch *et al.* 1967). The final uncertainty was large. The reproducibility of individual measurements was limited by the scatter in the measurements of the liquid nitrogen load brightness. Because of the greater temperature difference between the sky brightness and that of liquid nitrogen, the extrapolated results were subject to greater random errors. In addition, our mean value was low in comparison with the results of the other measurements available at the time of our publication. The average, particularly including the first radio detections (Penzias and Wilson 1965a; Roll and Wilkinson 1966) and the temperatures derived from CN measurements (Field and Hitchcock 1966; Thaddeus and Clauser 1966) were pointing to a blackbody temperature of 3.0 K or even higher, outside our error limit. As the more accurate measurements, shortly thereafter from the Wilkinson group (Stokes, Partridge and Wilkinson 1967) and others, and finally from the COBE satellite (Mather *et al.* 1990) came in, we were somewhat relieved that the limit of our error just included the final blackbody temperature, 2.725 K.

A year or so after our publication, I was reexamining the characterization of the pyramidal horn for some other calibrations that we were planning and discovered that I had made a mistake in the model tipping curve that we had used for the background measurements. Correcting for that properly, we would have had 2.3 ± 0.8 K for our result, a little closer to the final accurate temperature. Since that miscalculation was small compared to our random errors, we did not think it appropriate to publish it. In retrospect, I realize that was a mistake. The systematic error is, of course, different from the random errors, and it should have been reported.

We subsequently returned to our original program of getting a short wavelength telescope running for other astronomical observations, particularly for studies of Solar System objects and the interstellar medium. There we had some nice results with the first discoveries of polyatomic molecules in the interstellar medium revealing the molecular clouds where stars are born (Cheung *et al.* 1968, 1969). Then we proceeded to develop interferometry at short wavelengths for interstellar medium and star formation studies as well as for other fields.

Our most recent encounter with CMBR studies occurred when we discovered that we were making some accurate ground-based flux measurements of Jupiter at the same time that they were being made by the WMAP satellite in the course of its calibration (Page *et al.* 2003a). We had just completed our study when the WMAP results were announced. Our measurement was made at a wavelength of 1.05 cm (Gibson, Welch and de Pater 2005), in between the two longest WMAP receiver bands and close to the center of the Jovian ammonia inversion absorption band. Our accuracy for the Jovian flux was about 1.5% and it fell nicely between the Jovian fluxes of the two adjacent WMAP observations that had comparable accuracies. I think everyone was pleased with the good agreement between these independent calibrations of Jupiter. Our result enabled us to get a fairly accurate measure of the upper Jovian atmospheric ammonia abundance. Absolute calibration to 1–2% accuracy was essential for getting a good Jovian atmospheric model, and the WMAP results helped with that as well.

Some of the best memories from the earlier period were of discussions with Dave Wilkinson, an experimentalist of extraordinary capability.

4.10.4 Kazimir S. Stankevich: Investigation of the background radiation in the early years of its discovery

Kazimir Stankevich is Honored Worker of Science and the head of the astrophysics section of the Radiophysical Research Institute, Nizhny Novgorod.

The second Russian research center for radio astronomy (the first was in Moscow, in the Physical Institute of the Academy of Sciences, FIAN) was formed in the State University in Gorky (now Nizhny Novgorod) in the 1950s. Experimental investigations, which started in 1947, developed in the radiophysics department of the Institute of Physics and Technology under the guidance of Professor G. S. Gorelik. The head of the radio astronomy section in the Laboratory of Oscillation in the Physical Institute of the Academy of Sciences, Professor S. E. Haykin, contributed a lot to the formation of experimental radio astronomy. He got the section of radiophysics to take part in a program of research into radio-wave propagation in the atmosphere using the Sun, the Moon and other cosmic objects as sources of radiation. At the same time measurements of radiation fluxes, angular sizes and coordinates of the sources of radiation were carried out. It was in that section where the radio astronomical research group (under the guidance of V. S. Troitsky) arose; several radio telescopes were created, and the research for the program was successfully conducted. The influence of Professor V. L. Ginzburg (the head of the subfaculty of radio-wave propagation at the university at that time, later a Nobel Prize winner) on the development of radio astronomy was tremendous. Students and graduate students fell under his influence and were carried away with radio astronomy. His lectures on radio astronomy and the origin of cosmic rays became the introduction to the most important problems of this new science and promoted formation of lines of investigation.

After my graduation in 1957 from the graduate school, subfaculty of radio-wave propagation, I was assigned to the section of microwave radio astronomy headed by V. S. Troitsky in the newly formed radiophysical research institute attached to the State University. The section of Vsevolod Sergeyevich always gave a lot of attention to the development of new methods of investigation. In the 1960s, on Troitsky's initiative, we began to elaborate the technique for precise measurements of weak radiation noise intensity from cosmic sources. The method developed by V. S. Troitsky, V. D. Krotikov, K. S. Stankevich, and N. M. Tsetlin uses temperature calibration of the antenna by means of blackbody radio-wave radiation in the Fraunhofer zone of an antenna.[18] By means of this method flux densities

---

[18] The calibrating disk, or "artificial Moon," typically had diameter close to 1 m, and was placed on a cliff far enough away that the angular size of the disk at the telescope was smaller than the

were measured with an uncertainty of 2–3%. This method was called "artificial Moon," since Vsevolod Sergeyevich took an active part in the study of radio-wave radiation from the Moon and pioneered the use of this method applied to this object. Precision measurements of the Moon's disk temperature in a wide range of frequencies were necessary to him for remote probing of its surface layers. A model of the Moon's surface layers was created and the properties of the lunar material were predicted based on these results. I applied this technique for calibration of fluxes in the spectra of strong, discrete sources for the purpose of increasing the precision of primary standards for the absolute radio astronomical scale of flux densities from cosmic objects.

In 1960 there took place an important event which influenced my research interests a lot. Iosif Samuilovich Shklovsky (1960) published his theory of a long-term decrease of the flux density from a supernova remnant. He showed that the difference could be seen when the fluxes are measured a few years apart. Before that nobody could presume that cosmic sources were able to vary their power sufficiently fast. Evolution was expected to be possible but at insignificantly low rates. Hence, if this effect were seen, a new branch of radio astronomy would appear, the study of evolution of the radiation from supernova remnants and of the energy processes inside them. Precision measurements of fluxes and spectra of radio-wave sources were required to detect and study the evolution of young remnants of supernovae. Based on the results of observations during 1961 to 1964 we detected the decrease of flux density from the supernova Cas A at several wavelengths in the centimeter range (Lastochkin and Stankevich 1964). We were among the first groups to confirm Shklovsky's theory.

There were plenty of discoveries in radio astronomy in the 1960s. Quasars, the strongest sources of extragalactic radio-wave radiation, were detected in 1960, and by 1963 they were shown to be quasi-stellar objects. At that time the first catalogs of discrete radio sources and their spectra were published. Other notable discoveries include the emission line of hydroxyl, OH, at $\lambda = 18\,\text{cm}$, astrophysical OH masers, the radio recombination lines of atomic hydrogen, and the variability of radio-wave radiation from quasars and radio galaxies. The discovery of the extragalactic microwave background radiation in 1965 was an outstanding event in radio astronomy. It was detected by means of techniques and equipment used in radio astronomy and hence

---

angular resolution of the telescope, but large enough to be an appreciable source of radiation. The disk was close to a perfect absorber at the wavelength of observation, making it a source of blackbody radiation at the disk temperature.

there appeared a new object in the realm of radio astronomy – the whole universe!

Before that discovery none of my colleagues in FIAN, Kharkiv, or NIRFI (the Radiophysical Research Institute in Nizhny Novgorod) or other observatories was engaged in cosmology; everybody was carried away with the achievements of radio astronomy. Sometimes we heard about cosmological problems during the All-USSR conferences. I recall a dispute over a singularity and the search for solutions which exclude it. Certainly there was a search for objects with large redshift within the scope of radio and optical astronomy, but it was nothing but accumulation of data. We still were far from cosmological generalizations. Deep observations of the sky and counting of radio-wave sources started some later, after 1965. From our point of view the radiation at 3 K was the first invasion of cosmology into the realm of radio astronomy (or vice versa). After the discovery of radiation at 3 K cosmology became an experimental science: the model of the hot universe was supported, and besides it was the second fact in favor of evolution of the universe, now with a hot commencement. By the way, the singularity which scared everybody so much has moved elegantly to the realm of physics of extreme states of matter at high pressure and energy density. Cosmology became an interesting and attractive science, and we, radio astronomers, could be of use in the study of it. Several years later as a professor in the chair of radio astronomy and radio-wave propagation I was giving lectures on cosmology. And I am deeply convinced that cosmology should be taught to students specializing in any branch of physics.

I learned about the discovery of the excess isotropic cosmic radiation at wavelength 7.35 cm (Penzias and Wilson 1965a) and its interpretation as relic radiation of the hot universe (Dicke *et al.* 1965) published in the summer of 1965 from my Moscow colleagues, since foreign journals were available in Moscow several months earlier than at the periphery. I recall that I read the papers in October or November in the library of the Physical Institute of the Academy of Sciences.

At the time the radiation at 3 K was discovered I was already experienced in precision measurements of the absolute values of flux densities from discrete sources by the method of the "artificial Moon," which calibrated the antenna temperature by thermal radiation of a discoid blackbody in the beam (far-field) region of the radiation pattern. We used this concept to measure the background radiation temperature. Calibration of the antenna temperature during reception of the background radiation was performed by means of thermal radiation from two reference absorbers. They were kept at different temperatures: that of liquid nitrogen and of the surroundings.

The antenna radiation pattern was shielded from the ground. Our research group (the engineers, V. P. Lastochkin and V. A. Torkhov, and me, the senior researcher) possessed all the necessary tools and materials to measure the temperature of background radiation at the wavelength of 3.2 cm. The whole year of 1966 was devoted to construction of the reference radiators and improvement of the measurement technique. So it was not until the winter of 1966–1967 that we were able to perform the measurements on the roof of the radiophysical institute. The air temperature was −25 °C. That was believed to be the best most stable weather conditions.

The background radiation temperature we measured was 2.2 ± 0.3 K at 3.2 cm. On March 4, 1967, we submitted our paper to the journal *Radiofizika* (Stankevich, Lastochkin and Torkhov 1967). By that time a measurement at this wavelength was already published by Roll and Wilkinson (1966). We were not pioneers, but we achieved our goal – we developed an alternative technique of absolute measurements of background radiation. It seemed to be of great value since this method could be applied in a wide range of wavelengths, from millimeters to decimeters, which was important for detailed study of the background radiation spectrum. The main task at that time was to prove that the spectrum of this radiation at 3 K is consistent with the spectrum of a blackbody of the same temperature. From this point of view the millimeter range of wavelengths was of first-rate interest.

Nobody but A. E. Salomonovich, the head of the laboratory for millimeter radio astronomy at the Physical Institute, could possess a sufficiently sensitive and stable radiometer for the 8-mm wavelength range. He was the first in the Soviet Union to construct radiometers of that kind. Alexander Efimovich agreed to execute a joint measurement of the 3-K radiation temperature using the technique we developed for a similar measurement at wavelength 3.2 cm. He expected that the radiometer for a measurement at 8.2 mm, which was constructed by his order in the machine shop of the Institute of Radio-Engineering and Electronics of the Academy of Sciences, should be finished by June of that year (1967). The radiometer was developed by engineer V. I. Puzanov. I had to elaborate the technique and produce the reference blackbodies, including the one cooled down to the temperature of liquid nitrogen. So, by spring (April) of 1967 the list of authors was formed and the time of joint measurements, June, was clarified.

Radio radiation from the cloudless atmosphere is quite changeable and in summer is 5 to 6 times greater than the background radiation at 3 K. This is why to separate the contribution from the atmosphere one had to determine the brightness temperature of the atmosphere during the experiment. This goal was achieved by measuring the absolute value of the radiation intensity received at two angles to the horizon, 90° and 30°. For this purpose the reference radiators were constructed in such a way that one could tilt them to use the same cooled radiator at the two angles. This is why the volume of the cavity had noticeably increased and we needed about 50 l of liquid nitrogen per filling. For the whole experiment we would need not less than two cubic meters of liquid nitrogen. It was impossible to get such an amount of liquid nitrogen in Gorky. A. E. Salomonovich proposed that the experiment be performed at the radio astronomy station of FIAN in Pushchino. Alexander Efimovich organized a meeting of the radio astronomers at FIAN in Pushchino before the experiment. I presented the measurement technique and detailed the plan of the experiment. Our colleagues L. I. Matveenko and R. L. Sorochenko supported us and took part in the organization of observations later on. Our proposals were approved. In the upshot the experimental setup was installed at the radio astronomy station and liquid nitrogen was delivered from Moscow in a vacuum flask container once a week, supplied by the cryogenic laboratory of FIAN.

The experiment to measure the background radiation temperature at the wavelength of 8.2 mm started at the end of June and was successfully completed by early August. The temperature was found to be equal to 2.9 ± 0.7 K, and the Planckian character of the background radiation was confirmed as far as 8-mm wavelength. The paper *Measurements of the Temperature of the Primordial Background Radiation at 8.2-mm Wavelength* was submitted on August 17, 1967 (Puzanov, Salomonovich and Stankevich 1967). At that time there were no published papers on measurements of the CMBR in the millimeter range. There was a publication of a measurement at wavelength 1.5 cm (Welch *et al.* 1967), where the temperature of the background radiation was reported to be 2.0 ± 0.8 K, so there was some confusion about the spectrum of the radiation. Our paper improved the situation and favored the Planckian character of the CBR.

Our published papers attracted the attention of the well-known physicists Ya. B. Zel'dovich and R. A. Sunyaev to our research. At that time they were engaged in the study of the evolution of matter and radiation in the hot model of the universe. They inferred that the spectrum of the background radiation might possess some peculiarities. Heating of the primordial plasma in the universe in a stage of expansion before recombination of hydrogen could result in deviations in the Rayleigh–Jeans part of the background radiation spectrum. Another issue in cosmology was the density of intergalactic plasma. It could produce detectable radio radiation in the spectrum of the CMBR. One had to search for signs of it at long wavelengths, $\lambda \sim 50$ cm. By the end of 1967 the only known measurements of the background radiation temperature in the decimeter range were for the wavelengths of 20.7 cm, with the result $T_{CMBR} = 2.8 \pm 0.6$ K (Howell and Shakeshaft 1966), and 21.2 cm, where Penzias and Wilson (1967) reported $T_{CMBR} = 3.2 \pm 1.0$ K. Of course it was not enough to determine the peculiarities of the spectrum. The study of the background radiation implied intimate knowledge of the characteristics (temperature and spectral index) of nonthermal emission: continuous radiation from the Galaxy and unresolved extragalactic sources of radiation. Those quantities can be determined by simultaneous absolute measurements at three different wavelengths. My young colleague S. A. Pelyushenko and I performed such measurements according to the technique described above for wavelengths of 15 cm, 20.9 cm and 30 cm at the testing area of the institute in Zhimenki in the summer 1968. The paper (Pelyushenko and Stankevich 1969) was submitted on July 12, 1968. We determined that the temperature of the extragalactic background radiation does not appreciably vary with wavelength in the range 15–30 cm, and that the temperature equals 2.5 ± 0.6 K. This showed that radiation from intergalactic gas does not manifest itself in this wavelength range.

Weymann (1966) pointed out that radiation by intergalactic plasma could produce a 10–20% increase over the thermal spectrum at $\lambda \sim 30$ cm, with a larger increase at longer wavelengths. Howell and Shakeshaft (1967b) found $T_{CMBR} = 3.7 \pm 1.2$ K for wavelengths of 49.3 cm and 73.5 cm and reported a high upper limit, 2 K, for possible deviation from the background radiation spectrum at 73.5 cm. This is why additional measurements were necessary to find at least a tighter upper limit on possible deviations.

Limitations due to mechanical problems excluded the use of horn antennas and reference blackbodies cooled down with liquid nitrogen for measurements at wavelengths above 30–40 cm. That led me to propose using the Moon as a screen from cosmic radio emission. The studies by V. S. Troitsky showed that its disk temperature is constant and well known in the decimeter wavelength range, and polarization effects are negligible for wavelengths longer than 3 cm. In other words, the Moon could be used as an intensity standard. The experiment had to be performed for two wavelengths by means of antennas with high angular resolution. At that time there was a lack of large parabolic antennas in the Soviet Union. I. S. Shklovsky solved this problem. He asked J. G. Bolton to invite me to perform measurements with the Parkes 210-ft radio telescope. My visit to Sydney, Australia, at the Division of Radiophysics CSIRO came about with financial support from the School of General Studies, Australian National University, in September 1968. With support of a grant from the School of Electrical Engineering,

University of Sydney, Richard Wielebinski and his postgraduate student
W. E. Wilson took part in the project to accomplish work on absolute calibrations. Richard was engaged in the study of the spectrum and polarization
of galactic radio emission. He was also interested in absolute calibrations of
nonthermal radiation. Preparation for observations with the radio telescope
took quite a while. The radio telescope was equipped with a radiometer system for 635 MHz. It was necessary to create a supernumerary channel for
408 MHz using a radiometer constructed by Richard Wielebinski, and construct and install an integrated feed assembly for simultaneous reception at
two wavelengths. We performed rather complicated absolute measurements
of the cosmic background in February and May 1969. We made use of the
Parkes 210-ft radio telescope at wavelengths of 47.3 cm and 73.5 cm, and
a large horn antenna was used at 73.5 cm to determine the absolute temperature of the sky in the calibration point at high galactic latitude. The
temperature of the CMBR for these wavelengths was found to be 3.0 ± 0.5 K
(Stankevich, Wielebinski and Wilson 1970), which is in good agreement
with the average value 2.7 K from observations in centimeter and millimeter
ranges. It followed that radiation from intergalactic plasma did not appreciably affect the background radiation spectrum in the decimeter wavelength
range.

From the combined analysis of the Parkes results and the data for absolute
measurements at wavelengths of 15 cm, 20.9 cm, and 30 cm it was found that
the CMBR spectrum in the Rayleigh–Jeans range agrees with the radiation
spectrum of a blackbody of 2.7 K temperature, and no deviations from this
spectrum as a result of emission by cosmic plasma at any stage of evolution
of the universe were revealed. I consider this conclusion to be the main
contribution from the work we accomplished.

4.10.5 Paul Boynton: Testing the fireball hypothesis

*Paul Boynton, Professor of Physics and Astronomy at the University of
Washington, was in his youth a dedicated member of the Princeton Gravity
Group from 1967 to 1970.*

In his essay, *Six Cautionary Tales for Scientists*, Freeman Dyson com-
pellingly warns against idolatry of "Big Science" and the unacceptable cost
of failing to nurture the nimble spirit of exploration inherent in "small
science" (Dyson 1992).

During the last few decades, large institutionalized scientific projects have
sometimes played a productive role in extending our grasp of the natural

order, but big science did not bring about the discovery of the CMBR. I believe that one could not find more compelling support for the value of quick, modest, "table-top" approaches to research than the essays collected in this volume. These accounts portray a vitally diverse community of experimentalists rapidly and resourcefully responding to a new landscape of phenomena to be observed and new hypotheses to be tested – while in constant conversation with their theoretician colleagues. This process was clearly a triumph of small physics.

Whether an experimentalist is drawn toward a career in big or small physics may be a matter of circumstance, but for some only small is beautiful. My path to the Princeton Gravity Group, where small physics was doctrine, led directly through the valley of the shadow of Big Physics.

A few miles east of the University proper, lies Princeton's James Forrestal Campus. When I was a physics graduate student, it was the site of two major research facilities: the Plasma Physics Laboratory (PPL, where the C-Stellarator was then under construction) and the Princeton-Pennsylvania Accelerator. Both would become familiar haunts.

By the time I arrived at Princeton in the fall of 1962, I was thoroughly pumped up to join the quest for controlled fusion at PPL. My undergrad senior thesis on an obscure plasma instability led to working in Jim Drummond's Plasma Physics Group at the Boeing Scientific Research Laboratories in my hometown, Seattle, during the year following graduation. In fact, this millennial dream of realizing a virtually limitless source of pollution-free energy was largely my motivation for applying to grad school, and only to Princeton.

There were several hints that PPL was truly Big Physics: (1) hundreds of scientists, engineers, and technicians were focused on the project, (2) there were many layers of administration, and (3) the lab was the terminus of two major "pipelines." Through one, invisibly flowed millions in federal funds. The other was conspicuously part of the landscape. From the eastern horizon of the Forrestal campus, marching toward the lab across the wooded New Jersey countryside was a procession of steel high-tension towers carrying 110 kV power lines that could easily meet the needs of a small city, but these lines ended abruptly at the PPL substation. Clearly, something big was underway, and I was excited to be part of it.

Within a year, however, I began to see that my research advisor and his colleagues were thinking in terms of decades to achieve controlled fusion, with commercially marketable electrical power still farther away. Not much has changed since then, for even now the best hope for meeting such a goal is project ITER, which is still years away from initial tests.

As twenty-somethings generally consider themselves both invulnerable and everlasting, it was not the abstract specter of passing decades that deflected me to another field of study. Something more immediate and visceral tarnished the luster of my vision of contributing to a new age of progress made possible by such a clever manipulation of the natural world. At PPL I had found myself in the midst of a host of earnest physicists and engineers laboring toward this common goal, yet this was just one of several such labs around the world. The prospect of being a soldier in a vast army of similarly engaged minds, all driven by committee-defined strategies and milestones, did not appeal to me. Near the end of my second year, I jumped ship and opted for a thesis topic in experimental high-energy physics.

This may seem a ludicrous choice considering the current size of typical high-energy collaborations, but this was 1964 and our PPA group was only three: Tom Devlin, my thesis advisor, Julie Solomon, and I. While writing my thesis, attending American Physical Society meetings, and visiting other laboratories, I could see that high-energy experiments were becoming more ambitious and complex. My next moment of awareness regarding the real world of Big Science occurred on a visit to Brookhaven National Laboratory, where I encountered an experiment instrumented with rows and rows of rack-mounted electronics all networked with a PDP-8 computer (or perhaps several) for data acquisition and control – and yes, tended by a vast army of grad students, postdocs, and diligent assistant professors. Once again I got the message, and this time I knew which way to turn. The choice was nearly as elementary as that between sex and death.

During my student years at Princeton, I had been fascinated by the elegant simplicity of Professor Robert Dicke's fundamental physics experiments: measurement of the gravitational redshift of solar lines, a sophisticated revisiting of the Eötvös experiment, the implication of solar oblateness for testing general relativity theory, and probing the relation between active and passive gravitational mass. All were the subject of seminars or colloquia I attended. In each, profoundly basic hypotheses were being tested, yet none was more than a tabletop experiment – deceptively simple in construction, but combining sophisticated application of symmetry principles with electronic and mechanical tricks that yielded not only superprecise measurements, but also immunity to subtle systematic errors. Many were configured as null experiments with attendant advantages. At each turn in considering experimental design it seemed Dicke's group miraculously converted daunting challenges into distinct advantages. In conception and execution each project was impressively clever. Intimacy seems a curious label here, but a kind of intimacy between researcher and instrumentation is possible in this realm of small physics, a relationship unbuffered by committees, collaboration meetings, project management, or mega-budget politics. Small was beautiful, and it really appealed to me.

In early spring 1965, the Physics Department was abuzz with the revelation that members of Dicke's Gravity Group had been developing an apparatus to detect the remnant radiation from a putative hot big bang origin of the universe, and that researchers at Bell Labs in northern Jersey may have serendipitously planted their flag first. Nevertheless, in my mind Dicke had done it again – another simple-in-concept, yet dazzling tabletop experiment – and this time with cosmological implications. Although once more intrigued, my thesis project required me to leave for a year to set up a beam line at the Berkeley 184-inch cyclotron so that our high-energy group could attempt to measure cross sections for the excitation of isobaric analog states through pion-double-charge-exchange interactions with various complex nuclei. Perhaps another time...

The lingering appeal of the Gravity Group's activities weathered my preoccupation with thesis research and I joined the Gravity Group as a postdoc in the early summer of 1967. Professor Dicke had a spacious office on the upper floor of Princeton's historic Palmer Lab, where, in the vaguely medieval ambience of the basement, Dave Wilkinson, Jim Peebles, Mark Goldenberg, Bruce Partridge, and I, along with grad students, Bob Stokes, Paul Henry, Karl Davis, and Ed Groth, shared a large, windowless laboratory area. Rather cluttered and somewhat dimly lit, the room featured doorways along one wall that led to seldom-used nooks with black walls and ceilings for light-sensitive experiments. These darkened recesses lent a convincing touch to what might have been termed a "modern Gothic" workplace. Nonetheless, to me (and I believe to all of us) it was as comfortable and inviting as home.

Dedicated in 1908, Palmer Lab was built during Woodrow Wilson's time as president of Princeton University. Compared to the elegant, expansive grace of Jadwin Hall, built in the late 1960s to replace it, Palmer Lab's charm pales; but fares better when contrasted with the quarters it replaced, the John C. Green School of Science.

As a young faculty member in 1906, Owen Richardson (Nobel Laureate 1928) describes Princeton's 19th century physics facility in a memoir:

I remember getting quite a shock when I was first introduced to the part where I was expected to set up a research laboratory. This was a kind of dark basement, ventilated by a hole in the wall, apparently accidental in origin, and inhabited by an impressive colony of hoptoads, which enjoyed the use of a swimming pool in one corner. However, with the help of the Clerk of Works, these visitors and their amenities were got rid of and a lot of good work was done in it. Looking back on those days, I think they were in many ways the most satisfactory of my life.

Although there were no toads in our space, I do identify with Richardson's sentiment. I could not (and still can't) imagine a more exciting place, shared with these splendid colleagues amid the exhilarating conversation and consuming activity of engaging small science that loomed so very large.

In the more rarefied atmosphere of the upper floors of Palmer, I recall late-afternoon and evening informal sessions when Dicke, Peebles, Wheeler, and others would enthusiastically explore the romance of the "primeval fireball" picture – tracing out the manifold implications of a hot big bang. Everything was beginning to fit together: helium and deuterium synthesis, the CMBR, structure formation, primeval galaxies. This was a picture that had come alive – not just because of the discovery of the CMBR by Penzias and Wilson (1965a), or even the construction of a detailed expansion scenario by Peebles and Dicke, which had presaged that discovery (Dicke *et al.* 1965). In no small measure, the energy driving this effort emerged from the Gravity Group's firm conviction that remnant radiation from a hot big bang, if it existed, could be detected. Dave Wilkinson and Peter Roll acted on that confidence by building hardware and devising observing strategies based on microwave radiometer techniques Dicke had developed decades earlier at the MIT Radiation Lab. Jim Peebles, equally convinced, gave regular lectures on "physical cosmology" where he distributed weekly Ditto-machine notes bearing a striking resemblance to the monograph he later published under the same title (but not in that now-unfamiliar purple ink). This was the quintessential tabletop experiment, and in later years Dicke would sometimes muse: "What we need is another really good idea like the Fireball Radiation."

It was, in fact, a great idea. Developed independently at Princeton in 1964, but anticipated by Gamow and his colleagues Alpher and Herman nearly two decades earlier (Alpher and Herman 1948; Gamow 1948a,b). Unlike Dicke, Gamow did not grasp the directly observable, unique consequence of such a sweeping vision of cosmic evolution. Even in his earlier letter (Gamow 1946) motivating consideration of element production via cold neutron coagulation, he begins with a sketchy glimpse of the big bang, and clearly invokes initially hot matter subsequently cooled by expansion, suggesting an associated radiation component without explicit mention (until the 1948 publications). This letter coincidentally appears in the same volume of *Physical Review* as the 1946 paper describing Dicke's microwave measurement of an upper limit of $20\,\mathrm{K}$ to the temperature of (presciently labeled) diffuse cosmic matter (Dicke *et al.* 1946). Even so, Dicke remained unaware of Gamow's work for nearly 20 years. This curious disconnection is discussed elsewhere along with the observation that by 1948 technological developments would have allowed Dicke to carry out the CMBR detection program he later initiated in 1964 (Boynton 2005).

The stimulating synergism experienced within the Princeton group of true believers spread through the larger community of cosmologists at a more measured pace, and frequently met with the skepticism of proper scientific conservatism. At an American Astronomical Society meeting in Philadelphia later that year, I witnessed a particularly memorable instance of our confident enthusiasm regarding these new ideas encountering a dubious scientific establishment. On a gray, wet day in late fall the entire Gravity Group, Dicke included, boarded a chartered bus to mount a major presence at sessions devoted to the current status of various CMBR/big bang issues.

I vividly recall Jim Peebles giving a characteristically animated presentation regarding some aspect of big bang cosmology, waving the chalk about as his tall frame enthusiastically strode back and forth in front of the board while the Princeton contingent nodded and smiled encouragingly. Just as Jim was pulling together several points to form a particularly insightful synthesis, Professor George McVittie, eminent elder astrophysicist who was seated directly in front of me, could contain himself no longer. In a clearly exasperated tone he interrupted Jim exclaiming: "One can make any point at all with a little slap-dash arithmetic!" Jim turned with a flourish and grinning broadly replied, "My arithmetic may seem slapdash, but I can assure you it is impeccable."

McVittie recoiled only by silently sinking a bit in his chair as Peebles moved confidently on without pausing. Startled by this exchange, my full attention was riveted on McVittie's next move. There was nary a twitch, but in the intensity of the moment I was certain I saw a phantom curl of smoke rising above his shining, pulsing pate. It was an extraordinary rejoinder and our hero never broke stride. Forty years later, upon my inquiry regarding this encounter, Jim surprised me by producing a copy of McVittie's hand-written apology penned that same evening.

The group's continuing sense of mission sprang largely from the fact that in late 1966 only two radiometric observations of the microwave background were in hand, both made from New Jersey: the Bell Labs detection and antenna temperature measurement at $\lambda = 7.35$ cm, and the Princeton follow-up at $\lambda = 3.2$ cm. These data were consistent with a λ^{-2} power-law brightness spectrum, and therefore suggestive of the long-wavelength, Rayleigh–Jeans tail of thermal radiation – with implications of considerable interest to the Gravity Group. Of course, this could indicate either thermal-equilibrium (blackbody) emission, or a dilute, higher-temperature (gray-body) source. Wilkinson, now joined by Bruce Partridge and graduate student Robert Stokes, laid plans to examine the spectrum in more detail.

During the summer of 1967, Wilkinson, Partridge, and Stokes embarked on their seminal expedition to a high-altitude research station on White Mountain Peak above Owens Valley, California. They toted redesigned, refined, microwave radiometers operating at wavelengths $\lambda = 3.2$ cm, 1.58 cm, and 8.56 mm, to better establish the spectral distribution of the CMBR by observing at three wavelengths with three independent instruments of the same design, calibration method and observing technique.

These wavelengths largely avoid thermal emission from broad atmospheric O_2 and H_2O molecular resonance lines that constitute a major radiation background when attempting ground-based measurements. In addition to evading these lines, by observing from an altitude of 12,000 ft, above much of the integrated atmospheric water vapor density profile, the most problematic aspect of this local microwave background should be reduced: a random, time-variable contribution to the radiometer signal from molecular H_2O due to its inhomogeneous distribution coupled with atmospheric currents, which we commonly referred to as "atmospheric noise."

This expedition was the Gravity Group's second-generation refinement of their resolute goal to test the primeval fireball hypothesis. Theoreticians in the community at large had come up with various nonthermal, even non-cosmological mechanisms to explain this large, isotropic, apparently cosmic background, but none of these alternative processes filled the universe with blackbody radiation as would be expected of a hot big bang fireball. Bruce Partridge sets out an excellent, comprehensive account of both theoretical and experimental activities associated with the discovery and early history of the CMBR in his monograph 3K: *The Cosmic Microwave Background Radiation* (Partridge 1995). See also the superb contributions to this volume by Jim Peebles and Bruce Partridge.

Although the basic goal that summer focused on better establishing the brightness spectrum of the microwave background over a limited range of wavelengths, the real prize would be to distinguish a generic λ^{-2} power law tentatively indicated by that pair of earlier measurements, from the unique behavior of the Planck law: that the radiation brightness begins to drop below the power-law extrapolation of the Rayleigh–Jeans tail as one proceeds to shorter wavelengths, as is illustrated in Figure 4.36. At that time, detecting this blackbody signature was the obvious, feasible test of the cosmological origin of the CMBR, other than more precisely measuring the background

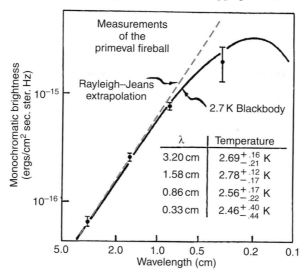

Fig. 4.36. A 2.7-K brightness spectrum compared to the extrapolated Rayleigh–Jeans power law (determined by the $\lambda = 7.35$-cm Bell Labs results and 3.20-cm point shown here) and superposed on White Mountain and Climax measurements. The $\simeq 30\%$ departure of the Planck curve from the power law at $\lambda = 3.3$ mm is easily visualized, but not so easily measured. As discussed in the text, lines of constant antenna temperature lie parallel to the Rayleigh–Jeans power law in this plot (adapted from Partridge 1969; ©1969 American Scientist).

isotropy (which Wilkinson and Partridge were actively pursuing). Peebles had already shown that throughout the evolution of an expanding, hot, homogeneous, isotropic, model universe up to the decoupling era, the thermal equilibration time was much shorter than the expansion time, therefore necessitating a thermal equilibrium post-decoupling radiation remnant as well. Moreover, that first pair of New Jersey brightness measurements was quantitatively consistent with thermal equilibrium radiation specifically at a temperature of 3 K. So if truly blackbody in nature, the peak of the associated spectrum would lie near wavelength $\lambda = 1$ mm, and consequently the distinct shape of the Planck spectrum might even be observable in the shorter wavelength measurements planned for White Mountain.

The introductory paragraphs of the paper in *Physical Review Letters* reporting results of the White Mountain expedition (Stokes, Partridge and Wilkinson 1967) emphasize the importance of this test: "To be convinced that one is seeing true blackbody radiation, and not that of a hot gray body, it is necessary to go to shorter wavelengths and look for the curvature in the spectrum due to quantum effects." For 3-K radiation the Planck function lies below the extrapolated long-wavelength power law by 15% at $\lambda = 1.58$ cm, and nearly 30% at 8.56 mm. Was detection of such departures within the reach of these planned ground-based measurements?

To appreciate quantitatively the challenge posed by this test, and to set out terminology for the following discussion, a brief technical digression seems appropriate, but the reader may skip the next paragraph to find a brief summary in the one after without losing the thread of this tale.

Although the spectrum of the radiation brightness for the blackbody and graybody options under consideration here are properly represented in Figure 4.36, these quantities are not directly observable with a Dicke radiometer. By design the output of our radiometer is proportional to the microwave power coupled into the radiometer by its antenna, and in a pre-selected, narrow, frequency band $\Delta\nu$ (centered on $\nu = c/\lambda$). For isotropic, thermal-equilibrium radiation, this narrowband power is proportional to the product $\lambda^2 B_\nu \Delta\nu$, where B_ν is monochromatic brightness (the Planck function) and the factor λ^2 enters because the effective solid angle of the antenna pattern is determined by diffraction. Moreover, defining $x = hc/\lambda k\tau$, where τ is the thermodynamic temperature of the blackbody cavity, the radiometer output is seen to be proportional to $\lambda^2 B_\nu/2k = \tau[x/(e^x - 1)]$, which has the units of temperature. Thus, when properly calibrated with a blackbody of known τ, the radiometer output directly indicates a "temperature" that is proportional to the microwave power detected by the radiometer, and this observable is referred to as antenna temperature. In the Rayleigh–Jeans regime ($\lambda \gg hc/k$, i.e., for small x) antenna temperature is manifestly constant (independent from λ) and equal to the thermodynamic temperature, τ. As λ decreases, the antenna temperature drops below τ just as the Planck function drops below the Rayleigh–Jeans, power-law extrapolation of the Planck function in Figure 4.36; that is, $T_\text{Antenna} = \lambda^2 B_\nu/2k = \tau[x/(e^x - 1)]$. From these definitions, it follows that $T_\text{Antenna}/\tau \simeq 0.7$ for $\tau = 3$-K blackbody radiation at $\lambda = 8.56$ mm as stated earlier. Consequently, this 30% differential reduction in monochromatic brightness of the blackbody relative to the graybody case is seen in practice to be measured by the radiometer as an absolute reduction of T_Antenna below the longer-wavelength value τ.

Given that the Dicke radiometer output is calibrated to indicate antenna temperature, T_Antenna, and if the CMBR is blackbody radiation with $\tau \simeq 3$ K, testing the graybody null hypothesis against the blackbody alternative, amounts to determining if there is a statistically significant reduction in the antenna temperature measured at $\lambda = 8.56$ mm compared to the mean antenna temperature ($T_\text{Antenna} \simeq \tau$) associated with the longer wavelength (Rayleigh–Jeans) measurements at $\lambda = 3.2$ cm and 7.35 cm.

A discriminating hypothesis test would not be possible at $\lambda = 8.56\,\mathrm{mm}$ with the relatively large water-vapor fluctuations (hence antenna temperature fluctuations) that plague sea-level New Jersey observations. Would rising to an altitude of more than two miles, thereby reducing the vertical column density of water vapor and the rms fluctuation thereof, more than compensate for the increased, intrinsic strength of water vapor emission resulting from observing at such a short wavelength? A definitive answer to this question followed from analysis of the White Mountain data.

Atmospheric emission noise was not the only concern. The 1965 CMBR observations from the roof of Guyot Hall on the Princeton campus (Roll and Wilkinson 1966, 1967) indicated that the error budget for antenna temperature measurements made under the expected improvement in atmospheric conditions at White Mountain would probably be dominated by the uncertainty in instrumental systematic effects, not random atmospheric fluctuations. The mission was doomed unless radiometer design and calibration techniques were significantly improved to reduce the magnitude of these biases, and more precise empirical methods devised to establish quantitatively the associated uncertainties therein.

The White Mountain data subsequently validated both this concern and the effectiveness of the remedies. At wavelengths $\lambda = 3.2\,\mathrm{cm}$ and $1.58\,\mathrm{cm}$ the observed statistical fluctuations from water vapor emission were significantly diminished at higher altitude as expected. Combined with the hard-won reduction in the measurement uncertainty of instrumental systematics, the resulting overall improvement in measurement precision, even at $\lambda = 8.56\,\mathrm{mm}$, was gratifying. The time and ingenuity devoted to reducing systematic error associated with these new radiometers had been crucial to achieving the comparatively precise antenna temperatures reported at all three wavelengths, and the overall estimate of the CMBR temperature was tightened to roughly $\tau_{\mathrm{CMBR}} = 2.7 \pm 0.1\,\mathrm{K}$. Moreover, these successes reduced measurement uncertainty to the point that a significant reduction in antenna temperature at $\lambda = 8.56\,\mathrm{mm}$ was observed relative to the average of the new, more precise values now established at $\lambda = 3.2\,\mathrm{cm}$ and $1.58\,\mathrm{cm}$.

A standard test of the graybody null hypothesis can be phrased roughly as this question: given the observed values and their uncertainties, what is the probability that the value of the antenna temperature measured at $\lambda = 8.56\,\mathrm{mm}$ is *not* statistically distinguished from the measurements made at substantially longer wavelengths? If that probability is small enough, the graybody hypothesis should be rejected. More specifically, we want to conduct a one-sided test, asking for the probability that the short-wavelength value does *not* exceed the long-wavelength values in order to reflect the asymmetric aspect of our alternate hypothesis. In this latter formulation, calculating from the radiometric data, the probability that the graybody null hypothesis might be true was about $1/160 = 0.0062$, which can be expressed as a confidence of 99.4% $(1 - 0.0062 = 0.9938)$.[19] By most standards this is shy of grounds for definitive rejection, but stronger than expected given the apparent measurement difficulties that had been faced. This fortunate outcome was noted in a summary remark in the paper in *Physical Review Letters* reporting the White Mountain results (Stokes, Partridge and Wilkinson 1967), "We believe this is good evidence for spectrum curvature, and argues strongly against a hot gray-body source."

There were, however, three subsequent indications that a rejection of the graybody hypothesis with confidence of 99.4% was not considered strong enough. First was the decision to proceed with the tentative plan to conduct another high-altitude measurement at an even shorter wavelength, $\lambda = 3.3\,\mathrm{mm}$, where the antenna temperature of a 2.7 K blackbody is only 1.1 K, or 1.6 K less than a putative graybody source consistent with the longer wavelength measurements. The second is found in the introduction to our paper in *Physical Review Letters* reporting the $\lambda = 3.3\,\mathrm{mm}$ result (Boynton, Stokes and Wilkinson 1968): "Recent microwave radiometer measurements at wavelengths of 3.2 cm, 1.58 cm, and 8.56 mm have indicated a blackbody of temperature 2.7 °K, but we have not yet demonstrated conclusively the spectral curvature expected from quantum effects." Dave Wilkinson's presentation to the 1979 Copenhagen Symposium, *The Universe at Large Redshifts*, holds the third (Wilkinson 1980). Regarding the 1967 and 1968 series of ground-based measurements, he writes: "So, the main goal of these experiments was achieved; a gray-body spectrum was ruled out with 5-σ significance." In this publication he cites two papers in support of this conclusion, both reporting results at $\lambda = 3.3\,\mathrm{mm}$. One is the 1968 observation by the Gravity Group (Boynton, Stokes and Wilkinson 1968). The other, a more precise, corroborating experiment by a group from the Aerospace Corporation (Millea *et al.*[20] 1971). The Princeton measurement alone rejected the graybody hypothesis with 99.998% confidence.[21] Comparison of a 2.7-K thermal spectrum and its extrapolated Rayleigh–Jeans power law is shown in Figure 4.36, along with the results of the Climax and White Mountain brightness measurements and CMBR

---

[19] This corresponds to rejection at a significance level of 2.5σ – the equivalent-single-tail-Gaussian-critical point for a t-test with $t = 2.67$ on 26 degrees of freedom.
[20] The same people had generously provided us with the E-Band diode mounts in the fall of 1967.
[21] This corresponds to rejection at a significance level of 4.2σ – t-test with $t = 5.11$ on 25 degrees of freedom.

thermodynamic temperatures. (Partridge presents another discussion of this set of measurements; see Figure 4.28.)

So, if a much stronger test of the "fireball hypothesis" were possible at this shorter wavelength, why wasn't a 3.3-mm radiometer included in preparations for the 1967 White Mountain project?

There were at least two reasons. First, a credible atmospheric emission model with some empirical foundation was used to estimate the magnitude of this challenge. The model indicated that for the same atmospheric conditions (same observing altitude, temperature, water vapor column density, water vapor inhomogeneity, and wind profile) not only would the water vapor emission be five times stronger at $\lambda = 3.3\,\mathrm{mm}$ than at $8.56\,\mathrm{mm}$, but the absolute antenna temperature fluctuations would also be five times larger (for the same antenna beamwidth). This strong dependence of emission on wavelength is primarily due to the broad tail of the very strong H_2O atmospheric line at $\lambda \simeq 0.6\,\mathrm{mm}$. Even though the antenna temperature difference between the graybody and blackbody hypotheses is 2.3 times larger at $\lambda = 3.3\,\mathrm{mm}$, this contrast would be swamped by the still larger increase in water vapor emission fluctuations. Second, at that time there were no reliable, commercially available mixer diodes operating at frequency $90\,\mathrm{GHz}$ ($\lambda = 3.3\,\mathrm{mm}$), a critical component of the Dicke superheterodyne radiometer. Moreover, completing a mixer-diode research and development program before the summer of 1967 was out of the question, and the expected increase in atmospheric emission noise would seem to make this effort a fool's errand in any case.

As it turned out, after analyzing the White Mountain $\lambda = 8.56\,\mathrm{mm}$ data, and thereby better understanding the magnitude of water-vapor fluctuations, prospects for a $\lambda = 3\,\mathrm{mm}$ attempt appeared somewhat more grim. But David Wilkinson had a trick in mind. He determinedly scheduled a second expedition – this time to the National Center for Atmospheric Research (NCAR) High Altitude Observatory at Climax, Colorado in the midst of the Colorado Rockies, a mile higher than Denver ... and in the dead of winter. In fact, he chose the dates according to the historical record to maximize the chance that we would encounter ground-level temperatures of $-20\,^{\circ}\mathrm{C}$ or below. His idea was to exploit the precipitation of atmospheric water vapor as ice crystals, precipitously reducing the partial pressure of molecular H_2O, and, therefore, the fluctuating emission background as well.

On the basis of this inspired strategy, Wilkinson, Stokes, and I made plans to develop our own mixer diodes, construct a 90-GHz Dicke radiometer, and assemble a reflector and ground shield to be deployed at Climax in March of 1968.

My part in this work brought me to frequent interaction with Bob Dicke – an experience by which I came to appreciate the true greatness of this man. While Wilkinson, Partridge and Stokes toiled on White Mountain, my task was to get up to speed regarding microwave design principles generally, Dicke radiometers in particular, and the handicaps posed at that time specifically by millimeter-wave mixer diodes and klystrons. For the basics I studied Dicke's ground-breaking radiometer-design publication from the 1940s (Dicke 1946b), and several of the MIT Radiation Lab volumes to which he had contributed substantially. Also, I frequently sat with the master himself, whom I found quite helpful and accessible – he relished talking about the "old days" at the Rad Lab. For me, these sessions could be daunting as well as enlightening. Not because of his manner, which was always relaxed and unassuming, but that the depth and range of his knowledge, experience, and analytical powers were sometimes overwhelming. In time, however, I came to admire Bob Dicke for reasons beyond his towering intellect. The following glimpses illustrate for me this man's grace and true virtue.

A few weeks before I joined the Gravity Group that first summer as a fresh-faced postdoc, I answered my home phone one afternoon and heard the greeting, "Hello, this is Mr. Dicke." At first, I didn't recognize who this might be. The only Mr. Dicke I knew was Arnold Dicke, one of my classmates, and this voice was nothing like Arnold's, nor would Arnold have identified himself to me as "Mr. Dicke." Milliseconds later I realized just who was at the other end of the line – a person whom I would have referred to with great deference as Professor Dicke. He congratulated me on my National Science Foundation fellowship award, said that he looked forward to talking with me about choosing a project that would fit into one of the group's current research activities, and asked if I would please stop by his office the next day. Since that incident I have noted for special respect those people who intentionally do not use their titles.

Months later, while walking with Dicke late one afternoon along the basement hallway of Palmer Lab toward the Gravity Group warren, we encountered a longtime building custodian who walked with a pronounced limp. He had always struck me as a pretty rough character with an arresting similarity to my image of Stevenson's Long John Silver. I was somewhat surprised when my professorial companion interrupted our conversation to stop and shake the man's hand. Surprise turned to shock when Dicke greeted the custodian by referring to him as "Gimpy." Bob Dicke and Gimpy (obviously pleased and grinning) exchanged a few humorous jibes before moving on. They seemed an odd duo, but the easily displayed affection and regard for each other engraved on me another life lesson.

From the sidelines I observed many similar encounters. In my mind, Bob Dicke was a giant – deserving all the accolades and personal regard a world can bestow. Yet, in his actions and general disposition, he revealed his own view of himself: a man like any other, obligated to defer to greatness in mind or deed, wherever he found it. I am reminded of Jane Austen's line describing a man of considerable accomplishment and wherewithal who might have grounds for self-important vanity, but whose eventual internalization of egalitarian principles made him an endearing, even though towering, character – "Indeed, he has no improper pride."

As a neophyte, I had the good fortune to know several such men whom I came to view as shining archetypes. Each easily fulfilled that key requirement of greatness: never seeming to live inside it, preferring to occupy the adjacent, less populated ground of humility. In this context I am compelled to recognize another stimulating contributor to Gravity Group meetings, another giant and also someone I greatly admire, John Wheeler.

At that time John's office could be found in that rarified atmosphere of Palmer Lab's upper floor mentioned earlier, and to which I only occasionally ascended. However, while on leave for an academic quarter at the University of Washington in the 1970s, John occupied the office next to mine. Our nearly daily conversations let me know him more closely as an admirable colleague and valued friend. Curiously, just as with Dicke, this vignette also involves a custodian.

John's habit was to leave his office door open while working at his desk. He would close the door, though, when in conversation with a guest or colleague. While meeting with him on many occasions, there inevitably would be a knock at the door. Rather than shout out, "come in," he would jump up from his chair, stepping briskly to the door, always with a warm greeting for whomever awaited him. Like an unusually attentive ER doctor conducting triage, he would quickly answer a question or graciously request they return later. Whether custodian or colleague, John's spirit remained unchanged. The custodian, however, was invited to enter at once in order not to interfere with his duties.

Another exception was his lovely wife, Janette, whom he would invariably ask to be seated – a gentle signal for the current guest to leave, but still with a clear invitation to continue the conversation later. Whether Janette, who unfailingly received his immediate attention, or any of the rest of us, who would be welcomed at a later time, all knew they would receive John's undivided attention . . . until there might be a knock at the door. It was a mystery to me how he could be so selflessly accommodating. This was a man who carried in his shirt pocket five appointment books bound together with a rubber band – one book for each day of the week – so full was his calendar.

By the time Wilkinson, Partridge, and Stokes had returned from White Mountain, Dicke had introduced me to many subtleties of waveguide components. Particularly fascinating was the exploitation of various symmetries in the design of four-port devices, especially the "Magic Tee," and the attendant advantages of a balanced mixer – even though not solving the problem posed by having no known source of reliable mixer diodes.

The fabrication of microwave devices had been gradually improving since the push to develop 3-cm radar during World War II, but in the mid-1960s, 3-mm components were still near the development frontier. I heard from Bill Ernst at the Princeton Plasma Physics Laboratory, who used millimeter-wave instrumentation for plasma diagnostics, and from the microwave research group at the nearby RCA Laboratories, that mixer diodes were not the only weak link in available millimeter-wave hardware.

At that time klystrons were the only practical millimeter-wave local oscillator for superheterodyne receivers and were notoriously inefficient, ran hot, and were noisy due to amplitude fluctuations and mode shifting. This oscillator noise extended out to hundreds of MHz from the line center, and even with a balanced mixer could degrade receiver performance. Both groups suggested we could improve the klystron performance by immersing it in an oil bath. This turned out to be good advice, and we took the additional precaution of choosing the receiver IF band to extend from 0.5 GHz to 1.0 GHz, an order-of-magnitude increase in center frequency and bandwidth relative to the White Mountain radiometers.

For a balanced mixer to work well, a pair of diodes with well-matched characteristics are required; but we faced difficulty procuring or fabricating even a single device that behaved remotely like a diode when placed in a section of waveguide and driven at 90 GHz, let alone two similar ones. Dave Wilkinson kept making phone calls and found that engineers associated with the US Army's Communication and Electronics Command at Fort Monmouth, NJ, were working on just this problem and would meet with us. So Dave, Bob Stokes, and I drove off toward the Jersey shore one late-summer morning. We met with a few casually dressed technical people accompanied (or perhaps monitored) by a couple of officers in uniform. The conversation soon moved to details, and our questions became more penetrating. Because of security concerns the officers suggested it would be more productive for us to talk with certain people at the Aerospace Corporation who could sell us two GaAs point-contact mixer diodes in E-band (90-GHz) waveguide mounts. Although abruptly ending the conversation, this was a good tip.

These mounts (with tunable shorts) turned out to provide a good RF match between waveguide and diode. The resulting receiver performance seemed adequate for our application, but the diodes were mechanically fragile and extremely susceptible to static-discharge damage despite elaborate precautions. Ours survived as a pair less than a week in a relatively benign laboratory environment. Their delicate constitution did not bode well for out-of-doors service amid the rigors of mid-winter weather in the Colorado Rockies.

Desperation was mounting when our friend Bob Wilson called one day to suggest we contact his Bell Labs colleague at Crawford Hill, Dr Charles Burrus, who had been developing millimeter-wave mixer diodes in connection with in-house communication research. Charlie was a godsend – friendly, helpful, and seemingly unfettered by proprietary security concerns. Although our worst fears about having to learn to make our own diodes were confirmed, he explained the entire process in practical terms and sent us back to Princeton with an adequate supply of special materials to remake enough point-contact, Schottky-barrier diodes in our Aerospace Corporation mounts to replace years of failed junctions.

These materials consisted of a few feet of gold-copper wire, from which electrolytically to form sharpened "cat whiskers," and a few dozen tiny, 1-mm square, GaAs tiles to be indium-soldered to a post in the diode mount, then chemically etched immediately before attempting to make contact with a carefully inserted whisker.

I include this detail to provide the context for our concern that each time a diode failed at Climax, we would have to carry out this delicate process on the equivalent of a park bench. I say, *park bench* because the prescribed GaAs etchant was concentrated bromine – tricky to use without a fume hood. We knew our Climax accommodations would certainly not include access to laboratory facilities. Our plan was to conduct diode refurbishing in the great Colorado outdoors, with personnel properly upwind of any alpine breeze.

After much practice remaking diode junctions in Palmer Lab and testing the performance of each as a mixer, we came to the more sobering realization that recovering from a failure probably meant fabricating several new junctions to achieve a reasonable match to the still-functioning mate. We imagined setting up our microscope, chemicals, and etching equipment on top of a packing crate and manipulating tiny parts with gloveless fingers in subzero weather for hours at a time – a sufficiently grim prospect to give even eager postdocs a second thought.

About two months before we were to leave for Colorado, Charlie Burrus, already held by us in high regard, elevated himself to near god-like status by announcing in a phone call that he had, through micro-lithography and a host of other tricks, been able to deposit a fairly dense, uniform array of 2 µm gold dots on the surface of a pristine GaAs wafer before dicing it into tiles.

With this giant-step innovation, making a diode was now simply a matter of sharpening a whisker and inserting it into the diode mount to make contact with one or another of these gold dots. It might take several stabs to do so, but since Charlie had covered the GaAs surface between the dots with an insulating layer of SiO_2, achieving success was unmistakable. Moreover, diode junction performance was found quite uniform from dot to dot. Charlie's remarkable gift meant we would neither need to carry a bottle of bromine to Climax, nor freeze our fingers off. Maybe we could make all this work after all.

Our radiometer noise temperature was typically between 10^4 K and 2×10^4 K, which may seem high, but implies a one-sigma antenna temperature measurement uncertainty of only 5 mK (0.005 K) for a 10-minutes integration time, the smallest single contribution to the error budget for our CMBR observations.

Not surprisingly, the more practiced we became with diode fabrication and handling, the more robust they seemed to be. Although we went well prepared to rebuild diodes at Climax, the initial pair saw us through the entire set of observations.

Perhaps the most important lesson learned on White Mountain was that minimizing calibration errors and the uncertainty associated with various systematic effects could make the difference between success and failure, even if atmospheric fluctuations dominated the error budget. I'm sure this was why Dave made instrumenting the Climax cold load and refining the evaluation of calibration errors his preoccupation during that previous autumn, repeatedly testing and modifying both the hardware and the technique. In addition to work with mixer diodes, receiver fabrication, and testing, Bob and I had measured the side- and back-lobe response of the large conical horn antenna we fabricated by electro-forming copper on an aluminum mandrel, then passivated by gold plating to maintain a high-conductivity (low-emissivity) surface.

To prepare for Climax, we began packing our equipment in early March for airfreight shipment to Denver. The grant bought the whole team war-surplus parkas, but we purchased our own long underwear and suitable cold-weather boots. During the flight west out of Newark, there were moments when my excitement over the possibility of participating in a major scientific advance nearly fell under the wheels of my fear of falling short of that challenge.

As Dave made final arrangements for our travel, I suggested we stay the night in Denver at the Cosmopolitan because a close family friend, Robert Wilhelm, was the hotel manager. I was certain we would find ourselves well hosted, and during the flight I may have raised expectations a bit high. Carelessly, I had not alerted Mr Wilhelm's office until the day before our departure; and by not speaking with him directly was unaware that he had been, and still was, out of town as I strode into the hotel lobby that evening.

Upon checking in I learned of some difficulty regarding the reservation, and we found ourselves escorted to a rather small, disagreeable room looking out into an airshaft. The hotel provided two rollaway beds for Bob and me, leaving little maneuvering space in the already-cramped room. Dave would not let this episode rest, and for the next week or more ribbed me about our wonderful Denver accommodations as a result of my "special" connections.

Not highly motivated to remain in such tight quarters, Dave, a jazz lover, wanted to see what the Denver clubs had to offer. With Bob and me in tow, Dave lurched in and out of Denver's nightspots. In short order, it became clear that Denver was far from New York City in more ways than geographical. The best approximation to a jazz club he could find was actually a strip joint where the dancers lacked attire in the tradition of Kansas City – they'd gone about as fur as they could go. Bob and I flanked Dave at the bar, which also served as the stage, so drinks had to be passed between the dancers' feet. I can speak only for myself, but I'm sure my wide eyes betrayed me – revealing a pitiful attempt to feign ennui – while Dave sat snickering to his Svengali self upon realizing his corrupting influence on at least one of the less worldly fellows at his side. I'm also certain that Bob Dicke never heard of this delightfully tawdry incident.

The next day at the airport we loaded a rented truck with the crates containing our equipment and headed west and upward through deepening snow to the summit of Fremont pass. There, we found Climax, Colorado, at that time the site of NCAR's High Altitude Observatory, the world's largest molybdenum mine, and what had been (prior to 1965) the highest altitude, populated settlement in the USA.

We lived and worked in a simple cabin, set on a hillside with a spectacular view of Sheep Mountain from across a wide alpine valley. Giant columns of ice reached the ground from the roof on either side of the small porch, clearly marking the entrance to our quarters, and a thick snow pack blanketed the entire scene. Most of that first day was spent vigorously shoveling snow from the large concrete pad directly in front of the cabin, where we would set up the aluminum ground shield that would surround our precious radiometer. This wintry setting is shown in Figure 4.37 with the shield and instrument in

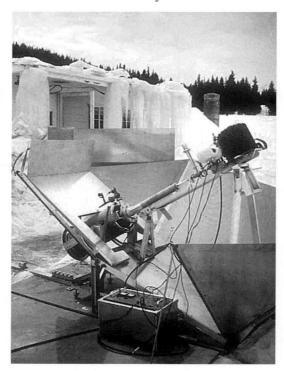

Fig. 4.37. The 3-mm radiometer operating near NCAR's High Altitude Observatory at Climax, Colorado. Note the huge icicles on our cabin in the background – this visit was in mid-winter to take advantage of a "freeze-dried" atmosphere. This photo shows the crucial step of calibrating the instrument by measuring the radiation temperature of a microwave absorber immersed in liquid helium contained in the wide-mouth dewar at the lower left protruding through the hole in the ground screen.

the foreground undergoing low-temperature thermal calibration by measuring the radiation temperature of a near-ideal microwave absorber immersed in liquid helium (Boynton, Stokes and Wilkinson 1968). There was also a high-temperature calibration point set by observing an ambient-temperature blackbody. The section of the ground-radiation shield on the camera side has been removed to provide a view of the liquid helium dewar coupled to the horn antenna of the radiometer.

The intense headaches that began for each of us that evening may have been unavoidable at such an altitude, but the day's exertion by normally sedentary workers had not helped. Even though the better part of a week was required to acclimate ourselves, during that time we managed to get things pretty well set up. Mild hypoxia not only slowed us down, but also led to a number of silly mistakes that often triggered spells of uncontrollable giggling – by Bob and me.

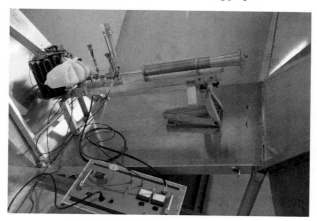

Fig. 4.38. This photo, again with the ground-radiation shield temporarily removed from the camera side, not only shows the instrument configuration for determining the atmospheric contribution to the sky temperature, but also renders the various elements of the radiometer more clearly visible: the main horn and the smaller, skyward-pointing, reference horn are seen directly connected to two ports of the microwave switch. A third port opens to an isolator, then on to the mixer diodes of this heterodyne receiver (which are wrapped in foam insulation and directly in front of the klystron oil bath at the far left). The fourth port is terminated with a smaller sky horn as is one port of the isolator. The ferrite switch alternately connects the main and reference horns to the receiver at a 1-kHz rate producing a modulated receiver output whose amplitude is proportional to the difference in microwave power received by these horns. The reference horn is pointed toward a fixed sky region, whereas the main horn beam scans through a range of zenith angles depending on the attitude of the hinged reflector panel to the right.

The age gaps among the three of us were not appreciable, but looking back I imagine that Dave saw this regression of our comportment as some cosmic test of his patience. Yet he managed to remain remarkably good-natured throughout our ten days at 11,000 ft.

The closer view of the instrument in Figure 4.38 shows the instrument configuration employed to determine the atmospheric contribution to the sky temperature. The radiometer remained stationary while the hinged panel on the right was placed in a sequence of five attitudes to reflect the horn-antenna beam to view the sky from zenith angles between 0° and 60°. The differential microwave power measurements made in this configuration allow isotropic radiation from above the atmosphere to be distinguished from atmospheric thermal emission (see Roll and Wilkinson 1967). For this purpose the reflector panel replaces the radiation shield with cutout to receive the calibration dewar as shown in Figure 4.37.

The best observing conditions (coldest, driest atmosphere) prevailed during the nighttime. After six days of preparation, we began the observing routine by transferring liquid helium into the cold load in early evening and conducting the interleaved sequence of atmospheric tips and calibrations (as outlined in Wilkinson 1967) until after midnight. This procedure was repeated over the four-day interval near the end of our stay.[22]

True to best practice, we did not analyze the data before returning to Princeton. Rather, we concentrated simply on following the data-collection routine we had carefully laid out and rehearsed in advance. Always there was the subtle, guiding hand of David who had worked the previous summer at an even higher altitude. From that experience he understood that "best practice" was not only a good idea, but was also forced upon us by the fact that even after nearly two weeks at altitude one cannot think carefully when deprived of oxygen (Figure 4.39). Despite this zombie-like

Fig. 4.39. Hypoxic zombies: David Wilkinson, Robert Stokes, and Paul Boynton.

[22] My recollections are admittedly idiosyncratic and myopically focused on the activities of the Princeton Gravity Group reaching to grab the brass ring by striving successfully to test the Fireball Hypothesis. Many other groups were carrying out and publishing CMBR observations prior to the completion of the Climax effort, and their contributions are acknowledged in Figure 2 of Boynton, Stokes and Wilkinson (1968). Of particular relevance to the story I tell in this essay is the elegant exploitation of low-lying rotational states in interstellar CN as a low-temperature cosmic thermometer operating at $\lambda = 2.64$ mm. This technique has an interesting history reviewed by Thaddeus (1972). The obvious question to ask here is how the papers by Field and Hitchcock (1966) and Thaddeus and Clauser (1966), which were published well before we laid plans for the short-millimeter-wave part of the spectrum, influenced the radiometric study carried out by the Gravity Group in 1968. Taken at face value and treated as upper limits (in view of possible excitation by other processes, which had not been fully evaluated at that time), these fairly tight CN constraints on the monochromatic brightness spectrum clearly favored the Fireball Hypothesis and were considered persuasive by some. I remember some concern regarding biases that could arise in the several inference steps undertaken to move from the observed optical transitions to quantification of excitation by millimeter-wave photons (the technique is rather indirect). This reservation, combined with questions about the local (rather than global) nature of the interstellar gas clouds studied – of which there were only two – kept us from averting our vision.

limitation, we were encouraged by the relatively modest atmospheric fluctuations we recorded. The air above us was certainly cold and apparently quite dry.

Most nights the temperature would drop below $-20\,°C$. Occasionally we would point a flashlight vertically upward. Although the beam was swallowed by the blackness of a moonless sky, one could visually observe the fulfillment of David's prophecy of a dehydrating atmosphere: tiny, rod-like, hexagonal-column ice crystals reflecting glints of light from minuscule facets while tumbling toward the ground – a fascinating phenomenon I had not seen before and not since.

After a day packing up our gear we returned to the airport. Arriving around noon, Bob and I delivered the crates to the airfreight platform and returned the truck. Dave smugly departed immediately on a flight to the east, thinking to leave Bob and me to suffer the privations of another night at the Cosmopolitan, Denver's finest.

Robert Wilhelm, a consummate hotelier, had long returned, and knew that Stokes and I had a reservation at his hotel that night. Upon asking to check in at the desk, we were instead escorted to the manager's wood-paneled office where we sat for some time sipping 50-year-old Tawny Port and chatting amiably about our Rocky Mountain adventure. Wilhelm apologized for not having properly lodged us previously when he was away and personally escorted to us to our rooms, accompanied by a cadre of bellmen. I say "our rooms" because we were lodged in a very large and quite elegantly decorated penthouse suite, replete with vases of fresh flowers and complimentary chocolates. Moreover, Robert and Irene Wilhelm, whom I have known and admired since my early childhood, invited us to a memorable haute cuisine dinner that evening.

David *never* admitted believing any of this. At some point, though, he had to sign off on the travel expenses for that trip. I've always wondered whether he noticed the absence of hotel and restaurant charges for that last night in the mile-high city where Bob and I snoozed in such splendor.

4.10.6 Robert A. Stokes: Early spectral measurements of the cosmic microwave background radiation

Robert Stokes is President and CEO of Versa Power Systems, a solid-oxide fuel cell development company in the Denver, Colorado area. After completion of his PhD at Princeton in 1968 he received an appointment as an assistant professor at the University of Kentucky where he continued work on the CMBR. Later he managed the engineering physics division at Battelle

Pacific Northwest National Laboratories, served as Deputy Director of the National Renewable Energy Laboratory, and was Senior Vice President at the Gas Technology Institute.

As an undergraduate student at the University of Kentucky in the 1960s, I was part of a generation with a growing interest in space science encouraged by the US educational system's response to the Soviet launch of the Sputnik satellite. At the end of my junior year I was selected to attend one of the first Goddard Institute summer study courses in space science at Columbia University organized by Robert Jastrow. After an intense summer of focusing on planetary astrophysics, our group was treated to a memorable tour of several US space science facilities, traveling aboard a chartered DC6 aircraft in August of 1963. The tour included visits to the NSF astronomical observatory at Kitt Peak, Arizona; the Marshall Space Flight Center in Huntsville, Alabama (the tour conducted by none other than Werner Von Braun); the NASA launch facility at Cape Canaveral; and NASA headquarters in Washington, DC. As a result, my interest in space science was greatly intensified, and when presented with the opportunity to attend graduate school at Princeton, a focus on space science was a foregone conclusion.

After a year of graduate study at Princeton and a couple of stimulating classes taught by John Wheeler, I managed to land a summer appointment working as a student research assistant for Bob Dicke and Mark Goldenberg, taking data on a special ground-based telescope designed to measure the Solar oblateness as a test of the predictions of the Brans–Dicke theory. That summer, Paul Henry, another graduate student, and I traded off making observations while observing practice by the Princeton hammer-throw athletes, hoping all the time that our apparatus would not be damaged by a mis-thrown 16-lb steel ball. By the end of the summer I had become a part of the graduate student cadre associated with the Gravity Group, led by Dicke and Wheeler and including their junior colleagues Peter Roll, Jim Peebles, Dave Wilkinson, Mark Goldenberg, Kip Thorne, and Bruce Partridge.

Jim Peebles had already begun some theoretical work on the implications of a hot fireball model for the early universe and the nature of any remnant radiation. Earlier modeling work by Alpher and Gamow (Alpher, Bethe and Gamow 1948) and measurements of atmospheric radiation by Dicke *et al.* (1946) had laid the groundwork for the research not pursued in earnest until the mid-1960s. But by 1965, Peter Roll and Dave Wilkinson had already begun operation of a 3-cm radiometer specifically designed to test the blackbody radiation hypothesis.

By the time the Princeton group had connected up with Arno Penzias and Bob Wilson at Bell Labs and published the famous 1965 companion papers (Dicke *et al.* 1965; Penzias and Wilson 1965a) on the CMBR, I had just completed my PhD qualifying exams and was looking for a dissertation topic. Roll and Wilkinson (1966, 1967) had sent their confirming radiometric measurements to *Physical Review Letters* in January of 1966 and there was great interest in investigating the spectral nature of the newly discovered isotropic microwave radiation.

Dave Wilkinson agreed to take me on as his first doctoral student at Princeton and set me to work building a 1.58-cm radiometer to make coordinated measurements with two radio telescopes he and Bruce Partridge were constructing for a second series of measurements at 3.2 cm and 8.56 mm. We had made arrangements to conduct several months of measurements at a high-altitude laboratory operated by the University of California at Berkeley in the White Mountains along the California–Nevada border during the summer of 1967 to establish a more precise temperature for the background radiation field.

The hypothesis that the microwave background radiation is, in fact, the primeval fireball rests heavily on the spectrum being that of a blackbody. Measurements completed prior to 1967 had all been consistent with a spectral index $\alpha = 2$ over a considerable wavelength range in the Rayleigh–Jeans region of a 3 K blackbody.

To be convinced one is seeing true blackbody radiation and not that of a hot graybody, it is necessary to go to short wavelengths and look for the curvature in the spectrum due to quantum statistical effects. In early 1967, Dave Wilkinson, Bruce Partridge, Paul Boynton, and I began a series of experiments aimed at refining the absolute radiometric techniques and extending the wavelength coverage to 3.3 mm, a wavelength sufficiently short to differentiate between a true blackbody and a hot graybody.

Four Dicke radiometers were constructed by the group using similar designs at wavelengths of 3.2 cm, 1.58 cm, 8.56 mm, and 3.3 mm and taken to mountaintop-observing sites to reduce atmospheric background. The 3.2-cm and 1.58-cm measurements were repeated to check earlier work and to provide an accurate determination of the spectral index. The 8.56-mm and 3.3-mm points were expected to show deviations from the frequency-squared dependence in the spectrum of a hot graybody, the deviations amounting to 20% at 8 mm and 300% at 3 mm.

A great deal of experience was gained from the Roll and Wilkinson radiometer that was operated atop a building on the Princeton University campus, and Dave Wilkinson in particular was able to build on his exacting electron $g-2$ PhD work at Michigan (Wilkinson 1962) to design an approach that dealt with systematic errors in the experiments. A number of other articles in this volume provide a good bit of detail and photos of the experimental apparatus used by the Princeton group (including the article by Bruce Partridge starting on p. 221), so I will not repeat the details here.

The 3.2-cm, 1.58-cm, and 8.56-mm experiments (Stokes, Partridge and Wilkinson 1967; Wilkinson 1967) were performed at an altitude of 12,470 ft at the Barcroft facility of the White Mountain Research Station, Bishop, California, during July and August of 1967. The 3.3-mm result (Boynton, Stokes and Wilkinson 1968) was obtained in March 1968 from an altitude of 11,300 ft at the NCAR High Altitude Observatory, Climax, Colorado. Figure 4.40 shows a photo of the author with the 3.3-mm radiometer at the Climax site.

Paul Boynton had completed a PhD in nuclear physics at Princeton (Boynton 1967) and decided to stay on as a postdoc in the Gravity Group. At about the time that Dave Wilkinson, Bruce Partridge, and I departed for Barcroft, California, via Yuma Arizona, Boynton started designing and procuring parts for the new 3.3-mm Dicke radiometer that was to be employed for the follow-up series of measurements.

Fig. 4.40. Robert Stokes with a 3.3-mm radiometer – Colorado 1968.

Whereas the microwave radiometer components for the longer-wavelength radiometers were mostly commercially available items, the microwave mixers for the 3.3-mm superheterodyne receivers were still very much a development-stage component in 1967. After my return from California, Boynton and I spent several months attempting to procure or develop an acceptable microwave mixer that would work at 90 GHz (3.3 mm). We visited several military development labs and received considerable help and loaned components from the Army lab staff in pursuit of a working radiometer in late 1967. At the beginning of 1968 we made the decision in consultation with Dave Wilkinson to give up on locating a reliable microwave mixer and put together a portable laboratory to transport to Colorado so that we could construct the detectors ourselves from GaAs wafers and gold-alloy sharpened cat whiskers. It seemed a lot like the early days of radio experimentation, but it worked! However, for the Colorado observations, we typically needed to change out the mixer once or twice during each of the all-night runs.

The choice of wavelengths for the radiometers was dictated by the location of atmospheric windows in the millimeter band. This absorption and subsequent reemission are the result of closely spaced pressure-broadened resonance lines that occur in the water molecule near 1.3- and 0.27-cm wavelength, and in the oxygen molecule near 0.5- and 0.26-cm wavelength. To further minimize atmospheric effects, measurements were performed at high altitudes during times of low absolute humidity. In addition to the usual problems with absolute measurements, the CMBR spectral measurements were made more difficult by the impossibility of modulating the signal due to its isotropy. Since the microwave background signal is the residue after one has accounted for everything else, control of the systematic effects and careful calibration were crucial.

The results of the four radiometer measurements made by the Princeton group using these techniques were all consistent with a 2.7 K blackbody. A graybody spectrum fitted to the measurements at 3.2 cm and 1.58 cm would have predicted a value five standard deviations above the result at 3.3 mm; thus these were the first direct radiometric measurements indicating spectral curvature.

Paul Boynton and I used an improved version of the original 3.3-mm radiometer carried to an altitude of 14.9 km in the NASA Ames Research Center Learjet to get the first direct radiometer measurement in which the atmospheric contribution was less than the cosmic background. The radiometer was not calibrated using a primary calibration source during the air-borne measurements. It was calibrated before and after flight. This experiment was the result of follow-on work by Paul Boynton and me after we left Princeton. Paul had taken an assistant professorship at the University of Washington, and I had an appointment as an assistant professor at the University of Kentucky. Much of the final preparations for the air-borne experiment were facilitated by my spending the summer of 1971 at Battelle Pacific Northwest National Laboratories in Washington State.

A typical air-borne experiment involved a flight to an altitude of 55,000 ft in order to perform the measurements above the tropopause of Earth's atmosphere. Even though the Learjet was pressurized, we were required to wear oxygen masks in case of a failure of the modified safety hatch that carried the radiometer antenna. On our last flight, after we announced to the ex-Navy pilot that the experimental results looked good, the pilot treated us to a perfect 1-g barrel roll without losing a drop of liquid helium from the calibration reference Dewar flask. Figure 4.41 is a photo of Paul Boynton and the pilot (left) just before takeoff on one of the flights.

By the early 1970s there was an excellent demonstration of the thermal spectrum as is beautifully illustrated in Figure 1 in Boynton and Stokes (1974). But bolometer measurements at submillimeter wavelengths indicated anomalies that many took seriously until COBE in 1990.

My subsequent career choices have taken me away from space science and cosmology to a focus on energy technology; however, I continue to follow developments in cosmology and space science as a highly interested individual.

Fig. 4.41. Paul Boynton (third from left) at NASA Ames in 1971.

4.10.7 Martin Harwit: An attempt at detecting the cosmic background radiation in the early 1960s

Martin Harwit is Professor Emeritus of Astronomy at Cornell University and former Director of the National Air and Space Museum. He is a mission scientist on the European Space Agency's Far-Infrared Submillimeter Telescope project, Herschel, of which the National Aeronautics and Space Administration is also a sponsor.

In 1963 I initiated an effort to look for the cosmic background radiation from space (Harwit 1964). The small research groups I started, first at the NRL in Washington, DC, and later at Cornell University, designed and constructed cryogenically cooled rocket telescopes to detect this radiation. My calculations showed that we would be hindered by zodiacal foreground radiation. Our telescopes eventually confirmed this, by detecting the zodiacal glow along with a number of strong diffuse galactic sources. We also obtained painfully false results on the submillimeter component of the 3-K microwave background radiation, due largely to emission by contamination carried aloft by our rockets (Shivanandan, Houck and Harwit 1968). In order to depict the many mishaps, missteps, and misconceptions that motivated me to initiate background observations in the early 1960s, I begin my account ten years earlier, when I was a graduate student.

In the spring of 1954 I found myself standing in the Physics Department office of Professor David M. Dennison at the University of Michigan in Ann Arbor. Dennison was an eminent molecular theorist. Sitting behind his desk, he was finding it difficult to tell me that I would not qualify for a PhD in physics because I really had no aptitude for science. Perhaps I should look at other occupations, because science clearly was not my métier.

Two months earlier, I had turned 23, and my career as a scientist already was reaching an unfortunate conclusion. I had come to Michigan to study chemical physics. Although my undergraduate major had been physics, I had taken an advanced chemical physics course before coming to Michigan and read through Linus Pauling's (1948) excellent book *The Nature of the Chemical Bond*, and Gerhard Herzberg's (1945, 1950) two books on molecular spectroscopy. The field looked genuinely exciting. At Michigan, I was assigned to carry out near-infrared spectroscopic work on peptide bonds in the laboratories of Professor G. B. B. M. Sutherland, who later became Director of the National Physical Laboratories in Britain. During my one-year apprenticeship, I learned a lot about infrared techniques but did not accomplish much. I was studying for my doctoral exams at the time, and neither the research nor the exams went well, which was why I was standing in Professor Dennison's office that day.

Having to leave the Physics Department and wanting some time to figure out what to do next, I stayed on in Ann Arbor that summer, and found a job in the laboratories of Professor Leslie Jones in the School of Engineering. He and his group were conducting upper atmosphere research with rocket-borne instrumentation. I did some optical design with the group and tackled whatever jobs needed doing. It was the first time I had real fun in science.

The war in Korea had not yet ended in 1954. In the fall of that year I received the then standard letter from my draft board, which began with the ominous words, "Greetings from the President of the United States," and explained how my "friends and neighbors" had selected me to serve in the United States Army. I was to report for my two-year stint of duty in January 1955.

Because I had earned an MA in Physics at Michigan by this time, the Army assigned me to the Chemical Corps at the Army Chemical Center in Edgewood, Maryland. This is where I began my real scientific training. Most of the civil servants in the Corps were chemical engineers and knew little about physics. But now, ten years after the end of World War II, the government was asking them to work on radioactive fallout, neutron doses from nuclear bomb bursts, and similar problems. I found that with a few visits to the base library, I could usually figure out what needed to be done, and although I was just an army private, I was given a lot of responsibility. Nevertheless, my civil service supervisor would send me to places like MIT or Woods Hole to verify with known experts that my calculations had been correct, and I enjoyed the opportunities offered by those visits.

In my second year in the Army, I was sent to Eniwetok and Bikini atolls in the Pacific for a few months to participate in what at the time was believed to be the first hydrogen bomb drop from an aircraft. We attempted to measure neutron doses at different distances from nuclear explosions, big and small. Some of them could vaporize an entire island in the atoll, others just left a small crater. To while away the time between work and snorkeling in the waters of the atolls, I had taken along a number of books, among them a popular astronomy book by Fred Hoyle. I no longer recall whether it was his *Frontiers in Astronomy* or *The Nature of the Universe*. Both books were out in paperback by that time, as were all the books I had taken along. Though Hoyle used no formulae and little technical language, I began to think that I would be able to do the calculations he was describing. It was quite fascinating.

At the end of my two years' service, I applied to graduate schools and was accepted by MIT on the strengths of what must have been great recommendations from Leslie Jones, my supervisor at the Army Chemical Center, and one of the MIT professors the Army had sent me to consult.

At MIT, there was no course requirement in one's major subject. But for a minor, a student was required to pass three advanced courses. I signed up for an astrophysics minor. It is hard to believe, today, but in 1957 MIT had no astrophysics curriculum. However, an exchange arrangement with Harvard permitted me to take three graduate courses there.

Tommy Gold had just been given a Harvard professorship. At the time, he was postulating that dust on the Moon would hop around in response to electrostatic bombardment from the Solar wind. Inspired by this, I begged and borrowed some equipment in the Research Laboratory of Electronics at MIT, learned how to blow glass so I could construct a vacuum tube for bombarding dust with electrons, and then saw the dust disperse when I turned on the electron beam. Tommy came down to MIT to see this late one evening. He was delighted and asked whether I might like to switch to astrophysics after receiving my PhD. I had done some calculations in one of his seminars, and he thought I should postdoc with Fred Hoyle. Of course, I was very pleased, though I still had my thesis work to complete.

I had come across the recently discovered Hanbury Brown–Twiss effect, and read the controversy surrounding it that aired in the journal *Nature* at the time. Edward M. Purcell's clean resolution of that controversy was particularly illuminating. I thought that the techniques developed for detecting the HB–T effect might provide a first opportunity to directly detect Bose–Einstein fluctuations in electromagnetic radiation from a source in thermal equilibrium. None of the experimentalists in the MIT Physics Department was particularly interested in my making these measurements, but Professor William P. Allis, a leading plasma theorist, said he would be willing to supervise the thesis if I could find the means to build the requisite apparatus.

The Naval Supersonic Wind Tunnel located on the MIT campus at the time was run by Professor John R. Markham of the MIT Aeronautical Engineering Department. One of the problems they were tackling was the detection of the hot exhausts of rockets and jet engines. This necessitated devices sensitive to the infrared radiation from these plumes. Improved sensitivity could be achieved by using not one detector, but two, and correlating their signals. This correlation technique was also needed for the Hanbury Brown–Twiss apparatus. The Aeronautical Engineering Department offered to buy as much of the requisite equipment as could be commercially obtained. They would use the apparatus during working hours, and I was free to use it for my thesis work at night. They generously also provided me with the assistantship I would need to finish my thesis work.

The fluctuations to be measured were minuscule, and detectors available at the time were still quite insensitive, but by April 1960 I had reasonably reliable results, and was finished with my thesis (Harwit 1960). Early in May, my wife Marianne and I embarked on the USS United States for me to spend a NATO-sponsored postdoctoral fellowship year with Fred Hoyle in Cambridge. Four years earlier, I had been inspired by his popular writing. Now, I would be working with him. I hardly believed my good fortune!

When we arrived in Cambridge, Fred was away on one of his prolonged visits to Caltech, and I had time to finish a paper I had begun while still at MIT. I had found a small error in a paper on galaxy formation in a steady state universe by Dennis Sciama. When I redid the calculation, it showed quite clearly that there was no way that a steady state universe could form galaxies at the replenishment rate required by the expansion of the universe, unless forces other than gravitation were at play.

I submitted the paper to the *Monthly Notices of the Royal Astronomical Society*, and some time later received an acceptance and an invitation to present the work at one of the monthly meetings in Burlington House (Harwit 1961). To my dismay, a week before my scheduled talk, I saw an announcement on one of the Cavendish Laboratory's bulletin boards that Hermann Bondi, one of the original creators of the steady state theory, was going to give a talk at King's College, London, on precisely the same topic of steady state galaxy formation the week after my talk at the RAS. I had heard that Bondi was a fierce debater. As secretary of the RAS he would undoubtedly be present at my talk. I expected a punishing onslaught, and at once began to prepare myself by reading everything Bondi had ever written on related subjects.

On the day of the meeting, I gave my talk, sparred with Bondi, but felt that I had acquitted myself reasonably. At the end of the session, I approached Bondi and introduced myself. At his suggestion we went to eat a hamburger and chat for a while before he had to take the train home to Sussex and I returned to Cambridge. He told me he had been the referee on my paper which had suggested some further work to him. Would I have time to come to King's College the following week to hear his talk? I was delighted, of course.

Munching on our hamburgers that evening after my RAS talk, I remember us talking about the future. I mentioned that on my return to the United States, I hoped to set up equipment to carry out infrared astronomical observations. Nobody was active in that area, and yet it seemed highly promising for astrochemical studies with infrared spectrometers.

After a great year in Cambridge, working with Fred Hoyle after his return from Caltech and writing a few papers with him, I returned to the US, to take up an NSF postdoctoral fellowship, this time at Cornell University where Tommy Gold had invited me to come. He had just moved to Cornell to start a powerful new department.

After my fellowship year, I accepted a one-year assistant professorship at Cornell, at the end of which I was free to take a leave of absence. I knew I wanted to carry out infrared astronomical observations and felt that ultimately infrared spectroscopy would offer great insights. But the Earth's atmosphere absorbs much of the infrared spectrum and, even worse, glows strongly in the infrared. To obtain a clear view of the sky in this wavelength band, I knew I would need to take telescopes above the atmosphere; moreover, these telescopes would have to be cooled to cryogenic temperatures. Otherwise the glow from the telescope would be far stronger than any celestial signal. At MIT I had built a sensitive cooled infrared apparatus. At Michigan, in Sutherland's laboratory, I had gained experience with spectroscopy, and in Leslie Jones's group, I had learned how to build apparatus carried aloft in rockets. All I needed to do was to put all this together.

At Tommy Gold's suggestion, I visited Herbert Friedman of the US NRL in Washington, early in 1963, to propose the possibility of starting an infrared astronomy program using rocket-borne telescopes. NRL had impressive credentials in ultraviolet and X-ray observations from rockets, but had not ventured into the infrared.

Friedman was very receptive. In a friendly meeting held in his offices, we agreed that I would come to work at NRL in the fall of 1963 and stay for a year, with fellowship support from the NSF. During this year, I would set up a group of NRL scientists and engineers to conduct a program in rocket infrared astronomy. At the end of the year, I would return to Cornell to set up a similar program there, and the two research groups established in this way would thereafter continue to compete in the newborn field.

During the summer of 1963, I sought to clarify the steps we would take. It was clear from the start that we needed to keep our efforts simple; the telescopes would have to be small. Our first efforts would have to be broadband photometry; spectroscopy would have to be delayed until we had more experience with the far simpler photometry. But even with these limitations, we thought we should be able to obtain reasonable measurements of large-scale features and an isotropic background. For background observations, a small telescope would suffice as long as it had a high throughput, i.e., it maximized the product of telescope aperture and angular beam dimension on the sky.

The background radiation I hoped we would observe was radiation I thought should have been emitted in the conversion of hydrogen into helium over the eons. Even though I had written two papers, while in Cambridge, to show the difficulties the steady state theory had in accounting for galaxy formation, I still thought that all the helium now observed must have been produced in stars. Like most astrophysicists at the time, I was unaware of the pioneering work of Ralph Alpher and Robert Herman (1948). Unfortunately, most of it had been largely ignored, forgotten, or discounted.

In 1963 the helium content of the universe was known to account for approximately one quarter of all the atomic mass in the universe. If the conversion of hydrogen into helium had all taken place in stars, then some of the energy liberated in the process should be observable in the infrared. I no longer recall why I thought the observation was feasible, but this was an easy calculation, and there were so many things like that "in the air" at the time. There just weren't very many astrophysicists then interested in cosmological questions, and many of these thoughts simply remained unpublished, though knowledgeable people were aware of them and exchanged ideas about them over tea or coffee. These were quick ideas that were not sufficiently substantive to warrant publication. They were somehow too obvious.

With thoughts about the accumulation of starlight in mind I presented a paper at a colloquium held at the University of Liège in late June 1963. In the proceedings of the conference I wrote (Harwit 1964)

(A)n interesting infrared observation concerns the frequently discussed suggestion that the overall cosmic background radiation might amount to as much as 3×10^{-11} watt/cm^2 in the infrared ...

To this I added a cautionary note.

(T)he cosmic flux could only be detected from the immediate vicinity of the Earth, if the radiation were concentrated in a very long wavelength spectral range where interplanetary dust grains are expected to emit inefficiently.

I showed that the thermal emission of the zodiacal (interplanetary) dust cloud would dominate the brightness of the infrared sky in the near- and mid-infrared part of the spectrum and wrote,

One now is in a position to discuss the detrimental effects that zodiacal dust reradiation will have on infrared astronomical observations ... (T)he nature of the most promising infrared observations is different from much of the work in the visible region. One often hopes to obtain information about diffuse sources of radiation, so that the zodiacal foreground glow may be an important hindrance ... At $42\,\mu$ this cloud would radiate of the order of 4×10^{-13} watt/cm^2-sterad-μ at large elongation angles within the plane of the ecliptic.

Even today, four decades later, the zodiacal glow remains an obstacle to determining the true extragalactic background in the near- and mid-infrared. We may ultimately have to rely on teraelectron volt (TeV) observations of distant active galactic nuclei to determine the rate at which this gamma radiation is destroyed through electron–positron pair formation, as it transits through the cosmic infrared background in extragalactic space.

The Cornell–NRL collaboration started in earnest in September 1963. A large number of technical problems had to be overcome in just 12 months if the work of the first year was to culminate in demonstrable success. NRL provided major resources to the effort. Joining me were scientists Douglas McNutt, Kandiah Shivanandan, and Blair Zajac, mechanical engineer Henry C. Kondracki, and electronic engineer John M. Reece.

Though the ultimate goal of the group was to construct telescopes cooled to liquid helium temperatures which would offer unencumbered observations across the entire spectral range from 1 μm out to several hundred microns, we quickly realized that the design of a liquid nitrogen cooled telescope would be considerably more simple. Such a telescope, though not as cold, would still make possible near-infrared observations of great sensitivity, since a telescope cooled to the temperature of liquid nitrogen, ∼80 K, would emit negligible thermal radiation at short wavelengths, and the near-infrared detectors in any case should operate optimally at this temperature. Once sufficient experience in the construction of these near-infrared telescopes was gained, we intended to quickly turn to the technically more difficult task of constructing liquid-helium-cooled telescopes that could be operated at temperatures of 4 K with the helium at atmospheric pressure, or ∼2 K if the helium was pumped down to very low pressure.

Many of the first launches were failures. Today, rocket launches have a better track record. But in the mid-1960s, failures of small sounding rockets to de-spin, pointing mechanisms to correctly orient the payload, delayed launches, and other problems often led to dismaying setbacks. Our efforts, like those of many others, were plagued by these difficulties.

I returned to Cornell University in the fall of 1964, whereupon Douglas McNutt took over the direction of the NRL group. We continued to collaborate on efforts that had been jointly started, but as these were completed, the two groups began to work independently and compete.

Shortly after my return to Ithaca, discussions with Dr. Nancy Roman, in charge of the astrophysics program at NASA resulted in grants to Cornell of an initial sum of $250,000 and annual budgets of $100,000, sufficient support to conduct a viable research program, initially with Aerobee 150, and later with the larger Aerobee 170 rockets. While NASA provided this initial outlay, we also obtained funding from the Air Force Cambridge Research Laboratories (AFCRL).

At Cornell I hired Henry C. Kondracki, who left NRL to move to Ithaca, New York, as full-time mechanical engineer. William Wernsing, an electrical engineer in Ithaca, also joined the group, as did Jim Dunston, a local jack-of-all-trades technician. James R. Houck, a graduate student just finishing a PhD at Cornell in solid state physics joined our small group after a couple of years. He was soon asked to join the Cornell faculty and the two of us established a long-lasting collaboration.

Constructing a liquid-helium-cooled telescope turned out to be a major engineering effort. A cryogenically cooled telescope had to be launched under vacuum. Otherwise, atmospheric gases would immediately condense on the optics. But vacuum vessels at that time tended to be constructed with thick steel walls making them far too massive to be launched on small rockets. A sufficiently light-weight design was needed. The thermal/mechanical design problem of constructing such a telescope, which could survive the vibrations and linear accelerations of launch and yet have minimal heat-conduction paths to the outer shell at room temperature, was difficult to solve.

Since the sensitivity of cryogenically cooled detectors in a cryogenically cooled telescope would be extremely high, observations were possible at high speeds. The bolometers favored by many ground-based observers were too slow to take advantage of this speed. Some of the photoconductors that had been developed for military purposes were far more promising. But it was soon apparent that the very low radiative background that a fully cooled telescope provided minimized the photon flux on these detectors, and correspondingly lowered the conductivity of the detector material. The detectors then attained extremely high resistances ranging up to $10^{11}\,\Omega$. Even small capacitive effects would then produce unacceptably slow response times. A major effort had to be undertaken to decrease response times and take full advantage of the detectors' potential sensitivity and speed.

The Earth's surface brightness in the infrared was expected to be nine orders of magnitude higher than the basic signals the detectors were able to detect from their $\sim 1°$ fields of view on the night sky. Extreme care had to be taken to baffle the telescope to eliminate any stray light from the Earth's limb that might be scattered or diffracted into the telescope.

It took us five years, and a succession of failures, before we were able to produce a successfully working liquid-helium-cooled astronomical telescope. Early designs incorporated a parabolic primary mirror with 18-cm aperture and focal ratio f/0.9. At altitude the entire telescope, except for the entrance aperture, was surrounded by liquid helium. We flew three different types of detectors on these flights – copper-doped germanium, gallium-doped germanium, and n-type indium-arsenide hot-electron bolometers – to cover progressively longer wavelengths between 5 μm and 1.6 mm (Harwit, Houck and Fuhrmann 1969).

We had, of course, been aware of the Penzias and Wilson (1965a) discovery of the microwave background radiation. Its stunning cosmological implications were widely discussed. This was truly exciting work and we were eager to check it out. But, it was not until 1968–1969 that our liquid-helium-cooled telescopes began to reliably work and we were able to attempt the detection of the expected submillimeter component of a background flux at ~3 K.

Even with a well-working telescope, we encountered difficulties in background radiation measurements. These produced a false signal with all the characteristics of an isotropic flux at the longest wavelengths, 400 μm to 1.3 mm. My colleagues and I initially reported these signals as possibly of cosmic origin (Shivanandan, Houck and Harwit 1968). But subsequent flights appeared to verify our findings (Pipher *et al.* 1971).

Part of the problem was due to rocket exhausts and other gaseous and particulate ejecta that accompanied the payload to great heights to form a diffuse, radiating cloud surrounding the telescope. More serious was diffracted radiation from the Earth's limb. Malcolm Savedoff of the University of Rochester first suggested to us that this was a potential problem. Although we had carefully checked to make sure that no significant amounts of scattered light could reach our detectors, we had no way of checking for diffracted submillimeter radiation. Our detectors worked only in a liquid-helium-cooled environment and we had no cryogenically cooled, evacuated test chamber to make the requisite tests. Jim Houck, however, redesigned the telescope's baffle structure for the next scheduled flight, and this effectively eliminated the false background signal we had observed.

The first successes of our rocket flights involved two quite different types of detections, and resulted from separate flights on December 2, 1970, and half a year later, on July 16, 1971. The first of these discovered and accurately measured the infrared radiation emitted by the circumsolar zodiacal dust cloud (Soifer, Houck and Harwit 1971). We detected radiation in three spectral ranges, at 5–6, 12–14 and 16–23 μm. At 70–130 μm we could initially place only an upper limit. More careful analysis provided a detection even at these long wavelengths (Pipher 1971). Both the three- and four-color photometry put the dust temperature at ∼280 K. My greatest surprise in these findings was that the dust radiated significantly more powerfully than I had predicted (Harwit 1964), indicating that the zodiacal dust grains were unexpectedly dark, scattering only a small fraction of the incident light, while absorbing and reemitting an appreciably larger portion. Thirty years later, I was pleased to see that the far more comprehensive COBE results of Kelsall *et al.* (1998) showed good agreement with the surface brightness of the zodiacal dust our rocket instrument had recorded.

The second discovery, made with the Cornell liquid-helium-cooled telescope on July 16, 1971, was the magnitude of the total infrared flux emanating from the galactic center and four other regions in central portions of the Milky Way at 5, 13, 20, and 100 µm (Houck *et al.* 1971). The 85–115 µm integrated flux over an area of $3° \times 2°$ around the galactic center was $7 \times 10^{-20} \, \mathrm{W \, m^{-2} \, Hz^{-1}}$, in excellent agreement with the balloon-borne result that had previously been obtained by Hoffmann, Frederick and Emery (1971). Excellent agreement for this wavelength range was also obtained for the galactic ionized hydrogen regions Messier 8 and NGC 6357. But the Cornell rocket flight also recorded the previously inaccessible flux from these three regions at 5–6, 12–14, and 16–23 µm (Soifer, Pipher and Houck 1972). Additional results cited by the same authors from an earlier flight provided the 100 µm flux for NGC 1499, a region previously unobserved at this wavelength.

More than a dozen years later, scans of the galactic center were also undertaken by the IRAS (Gautier *et al.* 1984). Though these authors did not compare their results to any previous work, the 100 µm maps of the galactic center published by the IRAS team gave peak fluxes which, within normal calibration uncertainties, were essentially identical to the Cornell rocket results published a dozen years earlier. Within such uncertainties, the 12 µm IRAS fluxes and the 12–14 µm Cornell detections also showed reasonable agreement.

Even though our paper on the background measurements at 400 µm to 1.3 mm was laced with cautionary comments, it gathered widespread attention. As we became convinced that the signals were actually due to contamination we withdrew the results but, for the next 30 years, I continued to feel badly about this mistake. Not until Jean-Loup Puget and his group derived the far-infrared flux from COBE scans did I begin to feel relieved (Puget *et al.* 1996). If the correct analysis of the various cosmic background components had taken more than another quarter of a century and ∼$500 million, roughly five hundred times more money than we had spent in the course of our entire rocket program, it was perhaps not so shameful to have been wrong. Sometimes it may be better to try difficult observations and fail, than not to try at all.

4.10.8 Judith L. Pipher: Being a young graduate student in interesting times – Ignoring the forest for the trees

Judith Pipher has been a member of the University of Rochester faculty since 1971, and is now Professor Emeritus of Physics and Astronomy. Among recent recognitions of her work is the University of Rochester's Susan B. Anthony Lifetime Achievement Award.

In the late 1960s, I was Martin Harwit's first graduate student working on a cryogenic rocket telescope experiment, and Tom Soifer was Jim Houck's first graduate student working on the same experiment. Jim had recently become a faculty member in the Cornell Astronomy Department, and he played a key role in directing graduate student activities. Tom and I were working on parallel developments – namely definition and construction of single-element detectors to work on the specific regions of the spectrum that were our mandate. Tom concentrated on the Ge:Cu photoconductors (sensitive from 5 to 23 μm) and I built a Ge:Ga detector for $\lambda < 120\,\mu$m, a GaAs detector for $\lambda \sim 150$–$200\,\mu$m, and an InSb Rollin detector (hot-electron bolometer) for 400–1300 μm. Each detector had its own tricky construction techniques, and etching these devices prior to soldering or fusing leads on them was a black art. Consequently, we made many poor detectors before succeeding. Commercial filters were unavailable to define my wavelength ranges, so I made so-called "Yoshinaga filters" by mixing and heating polyethylene and powders with appropriate restrahlen bands. Martin was on leave in Czechoslovakia the final year before flight, so Jim assumed responsibility for all aspects of the experiment, including keeping the technical areas entrusted to the graduate students moving forward. He definitely took exception to the dangerous aspects of my bonding leads to GaAs detectors in a hydrogen atmosphere, and the construction of "poison" filters, but really, we all had no choice. Our filters and detectors were characterized spectrally on the "gray pachyderm," a large Perkin Elmer spectrophotometer for the far infrared. Fortunately Jim was and is a very clever experimentalist, so he helped us adapt to this system. Ensuring paints were sufficiently black for these longer wavelengths was a problem – by analogy with acoustic "black" walls we devised a two-dimensional black paint for cryogenic use. Integrating spheres were unavailable for long-wave measurements of the paint properties, so we had to invent those also. Fun but odd tasks for aspiring astronomers.

All of these technical preparations were made in order to make measurements of the zodiacal background radiation, the galactic dust emission background radiation, and the CMBR on the short wavelength side of the expected 3 K blackbody spectrum, as well as far-infrared photometry of discrete sources. I was not then, nor am I now, a cosmologist. However, as described in Martin's piece on the rocket experiments, we attempted to measure the cosmic background intensity from 0.4 to 1 mm by comparing it against a radiator held at 4.2 K. It was a surprise to see a response indicating a warmer radiation field: in fact, as mentioned by Martin, the surprise occurred because we had inadequately measured the off-axis response of the telescope (we could not characterize the diffracted light lobes at submillimeter wavelengths). Jim addressed those by improving the baffling for the next rocket flights, and the "surprise radiation" disappeared. The whole affair was an interesting lesson in humility for a graduate student: paying attention to the many details of the detectors, filters, and calibration and not paying attention to the bigger picture had deleterious effects. A number of astronomical theories were advanced to explain these incorrect measurements: the other lesson this graduate student learned was that publishing a surprising experimental result too quickly is indeed a mistake.

4.10.9 Kandiah Shivanandan: The big bang, brighter than a thousand suns

Kandiah Shivanandan worked on infrared detector technology at the Naval Research Laboratory. Since retirement he has been active in giving lectures to young astronomers and to the general public. He is writing a book, Stars to Atoms to Cells.

I have taken the title from a passage from the Bhagavad Gita which has been translated as

If the radiance of a thousand suns
were to burst into the sky,
that would be like
the splendor of the Mighty One

When I was six years old in Malaysia, I used to look at the night skies and wonder whether there are human beings up there. In early 1950, as an undergraduate in physics at the University of Melbourne in Australia, I lived with Jesuits for a year, and we had discussions of religion and cosmology. I used to read extensively books by Einstein, Hoyle, and Gamow. Though my major was not in astrophysics or cosmology, I always took an interest in it during my undergraduate and graduate studies, attending seminars and visiting astronomical telescope sites, including the Parkes Radio Telescope where I analyzed data. During my graduate days at MIT I worked part time at American Science and Engineering with Riccardo Giacconi, Herbert Gursky, and Bruno Rossi on the first rocket-borne X-ray astronomy experiment. It detected an X-ray source in the Scorpius region of the sky (Giacconi *et al.* 1962).

In early 1960 I joined the NRL Astronomy Group under Herbert Friedman in Washington, DC. There UV and X-ray rocket-borne experiments for astronomy were being developed. Herb wanted to start an infrared program. Martin Harwit came to the NRL for a year as a visiting scientist, and he and I worked together to develop the first liquid-nitrogen-cooled rocket-borne sensor for near-infrared astronomical measurements.

With the discovery of the CMBR by Penzias and Wilson (1965a), Friedman suggested that we adapt our program to longer wavelengths, in the range of 10–1300 μm, which would include the predicted peak of the CMBR. We would use liquid-helium-cooled sensors. Harwit returned to Cornell and independently developed a similar program. There was an active technology transfer between the two groups. The semiconductor branch at the NRL had developed a sensitive 800–1300 μm detector cooled to 4.2 K and more adaptable for a rocket experiment than previous bolometers that had to operate at lower temperatures. Friedman suggested that I could use the project for my PhD thesis. Professor Clyde Cowan (with Frederick Reines, discoverer of the neutrino), who was at the Catholic University of America, and Friedman were my thesis advisors. Harwit also invited me to participate in his rocket experiments using the detector developed at the NRL, and I spent time at Cornell in integrating the detector with the system.

The NRL system flights had technology problems, but the Cornell flights were successful. Our preliminary observations indicated that the effective blackbody temperature of the radiation background at wavelengths in the range 0.4–1.3 mm is about 8 K (Shivanandan, Houck and Harwit 1968). That is larger than the effective blackbody temperature measured from the ground at longer wavelengths (and we now know that was caused by emission by local sources that entered our detector). But I was exhilarated. It was the first background radiation measurement done close to the CMBR peak.

4.10.10 Rainer Weiss: CMBR research at MIT shortly after the discovery – is there a blackbody peak?

Rai Weiss has been at the Massachusetts Institute of Technology since 1950, and is now Emeritus Professor of Physics. His recent research interest is the development of the gravitational wave observatory LIGO.

CMBR research at MIT began in Bernie Burke's radio astronomy group. Shortly after the first Princeton (Dicke *et al.* 1965; Roll and Wilkinson 1966) and Bell Laboratory (Penzias and Wilson 1965a) papers were published, Bernie's group made a measurement of the CMBR spectrum at 32.5 GHz (Ewing, Burke and Staelin 1967). Bernie suggested that my Gravitation Research group try to make a measurement of the spectrum near the blackbody peak.

At the time my group had a program sponsored by the Joint Services in the Research Laboratory of Electronics in studies of the consequences of a scalar component of the gravitational field. The idea had originated with Bob Dicke as a way of incorporating Mach's principle in relativistic gravitation. We had started an active program to see if G, the Newtonian gravitation constant, was changing by as much as one part in 10^{10} per year. Dicke considered this a possibility if there were a scalar field. Prior to this Dirac (1938) had hypothesized a similar change based on the dimensionless scaling of large numbers in nature. Our program consisted of measuring g, the gravitational field of the Earth at its surface, with a new type of absolute gravimeter, good enough to see changes of this magnitude over a year. The gravimeter involved measuring the electric field needed to support a plate against gravity by using the Stark effect in a molecular beam that passed between the plates. Associated with the g-measurement was a need to establish whether the shape of the Earth is significantly changing with time. To enable these measurements we had begun a program of absolute laser frequency stabilization, again using molecular beam techniques. The lasers would illuminate interferometers that measured the strain in local patches of Earth's surface as a means to establish if Earth's radius was changing. Altogether it was a somewhat fanciful program which luckily had several spin-offs which were successful while the main effort proved too difficult.

One of the spin-offs was the application of the frequency-stabilized lasers to other high-precision measurements. Gerald Blum and I (Blum and Weiss 1967) made a Michelson interferometer to repeat a measurement of the tired photon hypothesis that had been formulated as an alternative explanation of the cosmological redshift. The experiment had originally been carried out with Lee Grodzins (Weiss and Grodzins 1962) using the Mössbauer effect.

In that experiment, we passed 14-keV gamma rays through a heated tube emitting blackbody radiation at 1000 K to see if the blackbody radiation caused a frequency shift of the gamma rays.

The concept of this photon–photon scattering experiment came from Finlay-Freundlich's (1954) observation that bright stars seemed to have larger redshifts of their spectral lines than dimmer ones. From this somewhat half-baked observation, Finlay-Freundlich, in typical grandiose astrophysics style, had extended the idea to being an explanation of the cosmological redshift observed by Hubble. The ideas of the steady state cosmology and its "perfect" cosmological principle asserting homogeneity in time were so powerfully held at MIT at the time that any idea that would provide the redshift other than dynamically was most welcome. In this incarnation of the hypothesis, the photon–photon scattering was to have occurred between the ambient starlight and the light from the observed galaxy exhibiting the cosmological redshift.

We saw nothing in the first experiment. Then after the discovery of the 3-K CMBR, the notion became that the sea of microwave radiation would be the scatterers. That became the motivation for trying the experiment again, this time with microwave photons and visible light. That brought energies closer to those of the actual cosmological situation. Blum and I placed an X-band (3-cm wavelength) microwave cavity in one arm of the interferometer in which we turned the microwave power on and off at a frequency we would look for in the shift of the visible fringes. The interferometer itself was held on a fixed point on the fringe against slow drifts by a servo system with gain at low frequencies and none at the modulation frequency. At the frequency of modulation of the microwaves we were able to detect about 10^{-9} optical radians $Hz^{-1/2}$, limited by the quantum fluctuations in the phase of the laser light. Again even with a much higher sensitivity than the first experiment and with more relevant photon energies, nothing was seen. The experiment did lead to the idea of detecting gravitational waves by laser interferometry, as in the LIGO project, but that is another story.

At about the time of the photon-scattering experiment, I was asked to teach the graduate course in general relativity theory. As was typical in those more callous days, the teaching assignment was made several days before the beginning of the term. I may have been dumb enough in some earlier polling of the faculty to check off an interest in teaching such a course, but I never seriously meant it. Now, with this assignment and with a laboratory dedicated to the study of gravitation, it would seem inappropriate to say that I didn't really know any general relativity. I had taken a course given by Bob Dicke, with his particular take on GR, and also listened to an idiosyncratic version given by Eugene Wigner, during my postdoc stay at Princeton. But I did not know the subject, most critically I did not understand the tensor calculus and the Riemannian geometry. That was a remarkable term, where I would lock myself into a room with many references and try to understand what I was about to teach. Although I had control of the curriculum, and exploited teaching about the experiments and observations, there came a time when the mathematics had to be tackled. I found Brillouin's book on *Tensors in Elasticity and Relativity* (Brillouin 1964) a good place to learn.

The reason for telling this is that in the class was a very smart student, Dirk Muehlner (Figure 4.42), who had started at MIT in infrared solid state physics with Clive Perry. Dirk was toying with the idea of going into gravitation or astrophysics. When in the course it came time to explain a gravitational wave and how one might consider detecting one, I gave a homework problem: explore the idea of using laser interferometry as a means of measuring the geodesic deviation induced by gravitational waves. Is there any chance that laser interferometry might be sensitive enough to make a detection? Dirk got quite interested in the idea but then when the course got to using general relativity in cosmology that became even more fascinating for him. It was just about at this time that Bernie Burke made his suggestion to measure the spectrum of the CMBR near the peak. The suggestion came with some real help in that Alan Barrett, who was also one of the leaders of the MIT radio astronomy group, had run a ballooning program to measure atmospheric emission. There was ballooning experience in the group and also the offer of borrowing some critical equipment, in particular, a multi-channel portable instrumentation tape recorder.

Fig. 4.42. The MIT group, from the left Dirk Muehlner in the Palestine control room in 1971, Richard Benford in the MIT laboratory, and Rainer Weiss with the recovery crew in 1971.

Dirk and I decided that it would be interesting to measure the spectrum of the CMBR near the peak. Dirk's first research task was to look into the atmospheric absorption (and emission) at 30–300 GHz. This range of frequencies would attach to the existing measurements and would break new ground by observing at frequencies past the 3 K blackbody peak. Dirk's first findings from his library visits were that things looked pretty grim. It was clear that the three major atmospheric constituents that would cause trouble were molecular oxygen, water, and ozone. Oxygen has a strong magnetic electron and rotation spectrum in the 50 GHz region. Water lines are everywhere starting at frequencies above 20 GHz (the cause of the famous K-band fiasco from World War II) but becoming really awful above 90 GHz with a killer line at 420 GHz. Finally, ozone rotational lines are sprinkled throughout the spectral region but are weaker than the water lines because of the smaller electric dipole moment and larger partition fraction. Once seeing the line structure, we knew that a measurement near the peak required getting above the bulk of the atmosphere. A satellite measurement would have been ideal but clearly was not yet in the cards. Ballooning and suborbital rocketry were the only options. We chose ballooning in part because of Alan Barrett's offer but also because we felt that there was just not enough time or real estate in a sounding rocket to do the observation properly. The region around the 3 K peak could, we felt, still be done with some atmosphere in the line of sight.

To both of our amazement, Dirk found articles in the geophysics literature that flatly stated that the atmosphere was getting wetter as one got higher (Grantham *et al.* 1966; Mastenbrook 1966). There were papers that puzzled about this and were as baffled as we since such a possibility required a source outside the Earth or some complicated reservoir in the stratosphere. We got quite deep into this and began to look hard at the observations. When one said that the atmosphere is getting wetter what was meant is that even though the atmosphere was obeying its exponential dependence of the pressure (and density) with altitude, the fraction of the atmosphere that was water was growing as one got higher. The number of water molecules per unit volume as a function of height was not following the exponential decay with altitude. We smelled a rat.

The NRL was the principal source of the information from a program that had become routine (always dangerous, but excusable since the main purpose of the program was to search for radioactive fallout from USSR nuclear explosions). They were sampling the atmosphere as a function of altitude by placing an evacuated can in a rocket and then opening the can to the atmosphere while at altitude. After taking the sample the can was sealed and, once back on the ground, was shipped to a mass spectrometer. It was clear to us that the procedure was troublesome when the sample had a low pressure, for then the water adsorbed on the walls of the can during the initial evacuation played an ever-increasing role, until at the highest altitudes the adsorbed water constituted almost all the water in the sample. It was no miracle then that, as one got higher and higher in the atmosphere, the measurements implied that the water concentration was growing relative to the ever-reducing density of the atmosphere. We eventually went to visit the NRL and our suspicion was confirmed.

In planning our measurements Dirk and I made the assumption that the fractional water concentration remained constant once in the stratosphere. Even with this assumption, however, it was very clear to us from the absorption and emission calculations for the different lines of the atmospheric constituents that we would need to go to the highest altitudes attainable with balloons. We also realized that zenith angle scanning to correct for the atmospheric emission, the technique used at lower frequencies from the ground where the atmospheric lines are not saturated, would require a mixture of theory and observation. That is because some of the lines, in particular the water lines, are saturated even at the highest altitudes we could attain with the balloon. Armed with this primitive knowledge of the atmospheric constituents, and a beautiful atlas of the atmospheric line frequencies, strengths and line broadening parameters kept as the AFCRL Absorption Line Parameters Compilation, on tape, we made a strategy for the wavelength bands of the observations.

We also realized that measuring the CMBR at and above the peak, to really establish that the spectrum turned over, needed some profound changes from the way the previous ground-based measurements had been carried out. The most serious problem comes from the fact that there is a peak in the spectrum, which simply reduces the amount of power per frequency band. It arises from the vengeance of the quantum theory. The number of modes of the radiation field keeps growing with the square of the frequency. The photon occupation number per mode is kT at low frequencies, as demanded by equipartition in classical statistical physics. This gives the Rayleigh–Jeans part of the spectrum. But the mode occupation number becomes a dying exponential, with exponent $h\nu/kT$, as the frequency increases above the peak. Not that this is bad in its own right, for after all these are the ingredients that produce a peak in the spectrum. The trouble comes from the sources of radiation coming into the beam that are at higher temperatures than $3\,\mathrm{K}$. These warmer sources still are growing in strength unrelentingly with frequency, their blackbody peaks occurring at much higher frequencies. At frequencies below the blackbody peak the contributions of a 3 K and, say, a 300 K blackbody are in the ratio of the temperatures, not great but only a factor of a 100. At the frequency of the 3 K peak (180 GHz), the ratio of the contributions is about 600, while at a frequency of twice the peak (360 GHz) the ratio has become about 5000.

This dramatic ratio is compounded by the fact that the emissivity of metals and many dielectrics increases with frequency, and as mentioned before the atmospheric emission increases with increasing frequency as well. Another factor in the worsening situation is that scattering also grows as frequency in the ratio of the scale of the surface disturbances to the observation wavelength. The scattering from shield- and beam-forming optics becomes important because it can bring in radiation from hot surfaces outside the beam. The only physical optics phenomenon in favor of measurements at the higher frequencies is the diffraction that occurs where the beam size times the beam divergence angle is roughly equal to the wavelength. In particular, one can use smaller aperture optics at higher frequencies.

Given these unpleasant facts Dirk and I made several design decisions to assure as best as we knew that we had things under control. We decided to place the instrument in a large open-mouthed dewar to allow enough room for the beam to be formed by cryogenic optics and not brush against warmer edges. The radiometer itself was placed in a sealed copper can within the outer dewar. The outer dewar was the source of cooling and expendable helium (Figure 4.43). To assure that the critical beam-forming optics did not contribute to the radiation measured, we placed all of it in the can filled with liquid helium, which at altitude became superfluid. The superfluid is not only a good thermal conductor but also has a low dielectric constant and is a highly transparent medium. Although the superfluid has no viscosity, there is a critical velocity in the fluid at which vortices form and one begins to experience the transition to a fluid with micro-turbulence, and eventually, after cross-coupling of vortex lines, into a viscid classical fluid. The fluid dynamics set the highest modulation frequency of the mechanical chopper in the beam. Another aspect, not usually encountered in normal fluids, is the 15% change in the density of the fluid with temperature. The instrument can (Figure 4.44) is initially filled with normal liquid helium at 4.2 K at atmospheric pressure. In our design we let the helium pump down to pressure equilibrium with the ambient atmosphere as the balloon rose. At altitude 39 km the pressure is 2.5 mm and the temperature is 1.4 K, the helium has made the transition from normal to superfluid at the temperature of 2.1 K, the lambda point, and the fluid density has changed from 0.125 to 0.145 g cm^{-3}. To accommodate the density change and still assure

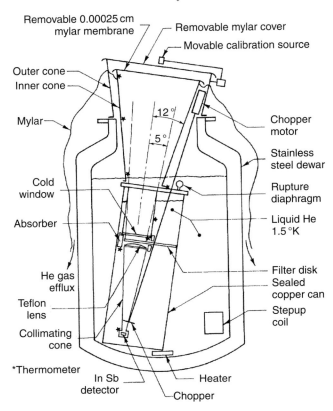

Removable 0.00025 cm mylar membrane

Removable mylar cover

Movable calibration source

Outer cone
Inner cone

Mylar

12°

5°

Chopper motor

Stainless steel dewar

Cold window

Rupture diaphragm

Absorber

Liquid He 1.5 °K

He gas efflux

Teflon lens

Collimating cone

Filter disk
Sealed copper can

Stepup coil

*Thermometer

In Sb detector

Heater

Chopper

Fig. 4.43. Schematic of the radiometer in the flight dewar. The initial radiometer and the second one, shown here, were similar. ©1973 American Physical Society.

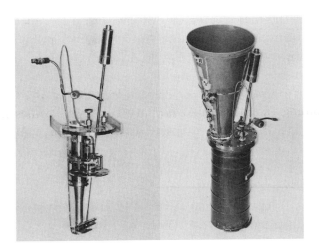

Fig. 4.44. Left: photograph of the radiometer showing the filter transport and chopper; right: the sealed copper can with the radiometer installed.

that the entrance window of the radiometer can would be in contact with the liquid helium, the window was recessed by 8 cm in the 50-cm can.

Aside from the cryogenic operation the radiometer was a standard system. A set of capacitive and inductive mesh filters defined the bandpass. The filters, made of silver patterns evaporated onto polyethylene, were multi-elements fused together thermally. The filters were particularly important in defining the high-frequency roll-off that supresses leakage of sunlight into the dewar. The roll-off was provided by a glass fiber-laced plastic window (Fluorogold) which also served as a cryogenic gasket material. The primary difference between our various flights was in the number and bandpass of the filters. The first flights had three filters while subsequent flights had six, one being a metallic reflector that completely blocked the incoming light. (However, this filter reflected the radiation generated by the detector, an unexpected radiative contribution to be discussed later.) The filters could be changed on command.

The collimator to fill the detector was a cone–lens combination which matched the beam hitting the detector area with a solid angle of close to π to the beam on the sky with an angle of 5°. The cone and lens were in the sealed can. At the bottom of the cone was a smooth disk Plexiglas chopper wheel with aluminum-evaporated sectors to bring the chopping frequency to 330 Hz, high enough to lie above the inevitable $1/f$ noise of the detector and preamplifier electronics. The chopper was driven by a long thin shaft with periodic bearings from a motor on the outside. The chopper could induce microphonics at the rotation frequency, but was not a serious source at the modulation frequency. Finally, the detector was a 5 mm on a side by 1 mm thick piece of indium antimonide which was immersed in the liquid helium. The detection mechanism was the small change in electron mobility in the semiconductor when the electrons absorbed the millimeter-wave radiation. The resistance of the material, measured continuously by a small bias current, became smaller as the electrons acquired more kinetic energy from the incoming radiation. The primary resistivity comes from electron coulomb scattering by ionic impurities in the material, where the deflection angle on scattering is reduced the greater the velocity of the electron. The time constant for equilibration of the electron gas to the temperature of the lattice, after it is excited by absorbing radiation, is approximately a microsecond. Although we understood the detection mechanism, we did not fully appreciate the systematic errors the radiation produced by the hot electron gas could make to our estimate of the incoming radiation. In principle, the calibration, if done with a truly nonreflecting load, would eliminate the problem. This is what we used at the end, but these calibrations were done on the ground and only relative gain-sensing calibrations were done in flight. The filter position containing the metal sheet helped in measuring this self-generated radiation by the detector.

The final element of the radiometer using the cryogenics was the amplifying system that converted the tiny voltages developed across the detector (fractions of nanovolts) to amplitudes more easily measured with room-temperature electronics. The low impedance of the InSb detector suggested that we use a passive step-up system to match the noise of JFET (Junction Field Effect Transistor) amplifiers operated at about 60 K, as cold as possible to reduce (Johnson) thermal noise but still high enough to avoid freeze-out of the charge carriers in the semiconductors. We employed a copper coil inductor and capacitor resonant circuit at the chopping frequency to produce the impedance transformation and accomplish the voltage step-up. The Q of the circuit was over 50, taking advantage of the reduction in resistance of copper at lower temperature. The coil, a possible source of noise, was surrounded by a superconducting lead shield and was potted with mineral oil to avoid relative motion between turns and the winding cores, which could be a source of microphonics.

The remaining parts of the instrumentation were more conventional but still fussy. Dirk and I had not previously constructed equipment that could survive the significant accelerations when landing, nor was the conventional lab practice to gain reliability really good enough. Systems had to work unattended and not fail at altitudes where convective cooling no longer applied since the atmosphere had such low density. We learned that on the scale of both costs and care a balloon payload was about 30 times more expensive and difficult than mounting something in the lab. Later I learned with the COBE satellite that carrying out an experiment in space was another factor of 100 times more costly and difficult. (I say this despite Werner von Braun's advertisement that space research would become as easy as ballooning. He actually said this at a committee meeting I attended in the mid-1970s while trying to convince the nation of the values of the space shuttle.)

A vignette of the first ballooning campaign I will never forget occurred because of a youthful and rash decision we made not more than several weeks before the package was to be shipped to the Balloon Base in Palestine, Texas. Both Dirk and I had become aware of the new integrated circuit operational amplifiers that had just come on the market. The chip was a 709 and looked like a little cockroach with eight legs. The chance to save enormously on the battery capacity and on the real estate required for the electronics was sufficiently seductive that we embarked on a solid three-day campaign to replace the discrete transistor electronics with these integrated circuits. After this almost complete overhaul of the electronics, the electronics boards passed our altitude tests and we decided, just to be sure, to make a final full system test with the radiometer at liquid helium temperature. To our horror the system worked but had a new and disturbing radiative offset. By luck we happened to have the FM radio in the lab on and noticed that the chopping frequency could be heard almost anywhere on the FM dial. We eventually traced the problem to powerful oscillations between 50 to several 100 MHz generated by the new integrated circuits. The circuits had been compensated with the filters recommended but were oscillating at frequencies past the ability of our instrumentation to detect them. The radiation was getting into the radiometer via the leads and then being detected by the InSb. In the nick of time another integrated circuit chip with internal compensation and, therefore, no high-frequency oscillation, the 741, became available. These operational amplifiers started a long and still active tradition of having the same functional connections on the pins to make these circuits readily interchangeable. Electronics has never been the same again.

Calibration of the instrument was accomplished by inserting a horn of known temperature and emissivity into the full beam of the radiometer. The voltage developed by the detector as a function of calibrator temperature was measured for each filter. The curves were consistent with integrals of the blackbodies of different temperatures over the instrument response. This was not important if all that was needed was a voltage associated with each temperature blackbody. It became important when trying to solve for the absolute atmospheric contributions from the elevation scanning when some lines were saturated while others remained unsaturated. One needed to know the detector volts generated per watt of incoming radiation at each wavelength, especially those at atmospheric lines. The absolute calibration was so fundamental to the measurement that we thought hard about performing it in flight but in the end found this so complicated and prone to failure that we resorted to a secondary calibration with a small blackbody source that could be brought into the beam by command. In the end such a calibration was only useful to measure an overall sensitivity and we had to trust the apparatus to not change transmission or spectral character. Later the COBE satellite, and Gush, Halpern and Wishnow (1990) in their rocket-borne observation, deployed in-flight absolute calibration.

The trickiest bit of the experiment design was at the interface between the cold world of the radiometer and the warm world we all live in. Indeed, it was the place that got us into trouble. Several functions needed to be satisfied. On the ground the atmosphere had to be separated from the cold part of the radiometer to prevent the catastrophe of air falling into the liquid helium-cooled surface and freezing there. At the same time, we wanted to observe the atmospheric emission as we gained altitude, in part to verify our model of the atmospheric radiation as well as to establish that the instrument was working. During the ascent from the pressure of 760 torr to the pressure of around 2.5 torr at float, the helium in the outer part of the dewar was pumped down from 4.2 K to 1.4 K. We needed to have an opening to the atmosphere for the pumping, but the pressure of helium had to be high enough to keep air from back diffusing into the dewar. Initially, we expected that we would need to heat the helium in the outer dewar to force enough evaporation, but it turned out there was enough efflux of gas due to the radiative load. The separation was done by tailoring a set of polyethylene bags. The bags allowed the helium to be pumped around the dewar circumference. At the top of the dewar, where the beam comes out to observe the sky, there were two hoops stretched with thin Mylar drum heads. The outer hoop sealed against a cowling which also clamped the outer bag. The outer hoop was intended to prevent debris from falling into the system. It was part of a helium-filled shield for the inner bag and hoop system. At altitude the outer hoop was removed, leaving only a shield cone coated with an insulator to reduce its emissivity at grazing angles. The second bag was attached to the outside of the shield cone. An inner hoop sealed the shield cone. The major part of the observations were to be made through the thin Mylar of the inner hoop, then at the end of the flight that hoop was removed to determine the radiative contribution of the hoop itself. Several of the early flights were made this way until in one flight a piece of ballast (lead shot) punctured the inner hoop Mylar sheet and we discovered that it was possible to observe at altitude without any cover. That is because the helium efflux from the dewar, if allowed to emerge through the shield cone, was sufficient to purge the air from the radiometer and keep the radiometer clean. Figure 4.45 shows a set of photographs taken at altitude of the entire sequence of hoop removals and dewar motions to allow the camera to look into the radiometer and observe if there had been air condensation. We observed only a small amount of deposition of nitrogen on the inner hoop after several hours at altitude. In fact to reduce the radiation from this film it was advantageous to remove the inner hoop earlier in the flight.

It turns out that in the initial flights the inner hoop was not far enough away from the main radiometer beam edge. The hoop caused sufficient scattering of radiation from a warmer shield to produce a radiative contribution which appeared to be maximum in one of the filter channels. That indicated

Fig. 4.45. A sequence of photographs taken during the June 1971 flight showing operation of the various covers and bags at altitude. In the top left panel both hoop covers are still in place. The secondary calibrator is visible. Note the debris on the outer hoop cover. In the top middle panel the outer hoop cover has been removed and now rests on the right. In the top right panel the dewar has been moved to bring the radiometer beam closer to the zenith. The inner hoop cover appears to be clean. In the bottom left panel the secondary calibrator has been brought into the beam. Now, about 4 1/2 hours after the outer hoop cover was removed, water frost is clearly visible on the inner hoop cover. The radiation from this water frost can be received by the instrument. In the bottom middle panel the inner hoop cover has been removed and the flow of helium gas is keeping air out of the cone. In later flights, the inner hoop was removed near the beginning of the time at altitude because we found that the helium efflux was sufficient to keep the radiometer input aperture clear of air and water. In the bottom right panel the zenith angle has been increased to make an atmospheric scan. The picture was taken to check whether the hoops flopped around – they didn't.

that there may be a distortion of the cosmic background spectrum at about 300 GHz. This was our principal blunder. In a way it was lucky that we had a free-fall with this instrument and that we had to rebuild. But I will say more about this later.

The very low levels of radiation that could be tolerated at altitude, and the desire to use zenith-scanning as a means of helping to remove the atmospheric contribution, necessitated special treatment in the ballooning art. A typical high-altitude balloon with payload capability of 500 kg is 100 m in diameter at altitude. In the beginning, we expected that a balloon packed in powder, to facilitate release from being folded in a box, would carry significant amounts of water into the stratosphere and we would have a local pollution of water lines worse than those in the atmosphere. We also wanted to make elevation scans as close to the zenith as possible. Both of these factors drove us to put about 700 m of nylon line between the payload and the bottom of the balloon. A flight train as long as this was impossible to launch because of low-level windshear near the Earth's surface. The strategy that the balloon base had developed with prior atmospheric constituent flights, which also needed to be a goodly distance from the balloon, was the use of a mechanical rope payout-reel that by air friction through propellers and capstan friction slowly released 700 m of nylon line when a command was given after the launch had safely lifted the payload off the ground. The reel enabled the launch of such a complex flight train but also increased the risk of flight failure. We did find the balloon to be a source of water vapor, and the reel was necessary to accomplish our measurement.

Our first flight made in beginners' innocence and with beginners' luck was a complete technical success, but it produced a cosmological mystery by giving a result consistent with the ground-based measurements of the CMBR temperature in the low-frequency channel, an excess in the middle channel that embraced the blackbody peak, and an interesting and useful upper limit in a band above the blackbody peak (Muehlner and Weiss 1970). This last result disagreed with large excesses measured by a rocket experiment of the Cornell and NRL groups (Shivanandan, Houck and Harwit 1968; Houck and Harwit 1969; Houck *et al.* 1972). There was no sensible way to reconcile the rocket excess with our measurements and it was clear that the rocket measurement was in error. Nevertheless, we too had an excess, smaller and now narrowed to a band between 180 and 360 GHz.

The result was so important that we decided to fly the same payload again with some small refurbishments, just enough to fix the damage caused on landing. The same flight techniques were used but this time things did not go so well. The apparatus worked but just as we reached altitude an errant command was given which terminated the flight by cutting the long line at the bottom of the balloon. Once we realized what had happened many commands (all unsuccessful) were given to unpack a parachute on the payload, but to no avail. The instrument package went into free-fall over an East Texas forest (Figure 4.46). It was not found for several weeks, until deer hunting season opened in early November. We had gone home before the package was found and Dirk went back to Texas to gather the pieces. It wasn't worth it: the package had not been designed to deal with an impact with the ground at 250 miles an hour. Not only were all our home-built cryogenics and electronics destroyed – worse still, the most expensive part of the payload, the instrumentation tape recorder which the MIT radio astronomy group had lent us, was in many pieces, a total loss. We did not have the funds to replace it. The reel, however, did return to the ground in one piece. It had a small parachute that was passively deployed in the

Fig. 4.46. A sketch made by Frank O'Brien of MIT, who I asked to join us for the second flight of the original radiometer. This was the flight that free-fell after an erroneous termination command was given just as we were reaching altitude. Frank, an old friend from Zacharias's atomic beam laboratory at MIT, was an engineer and machinist who taught us the ropes in experimental work. With the free-fall he unfortunately did not have much to do and made this drawing which captured a good bit of what was going on. In the end, as is elaborated in the text, the free-fall was a mischievous gift.

flight line and the long line, luckily, did not prevent the parachute from opening.

Much as the free-fall was a rude introduction to the hazards of ballooning, it was also a gift to keep us from confirming an erroneous result. In the process of designing, constructing, and testing a new radiometer and payload, we discovered that the excess in the midband of the first observation was from radiation scattered by the edges of the inner cone and hoop that made up the interface between the cold and warm world. When we were making the second payload we had more powerful and higher frequency sources to make careful beam maps into the side lobes. We found, not unexpectedly, that the beam profile was different for all spectral channels and, in fact, the spectral filter that was used in the mid channel caused side lobes with peaks at the angles associated with the edge of the cone and the hoop.

It took about a year to complete and test the new package. Most of the ideas for the radiometer were still valid and many had been proved to work in the first flight. We did increase the number of filter positions to allow better definition of the cutoffs in the blackbody integrals and, later in the flight series, to try to pin down more definitively the various atmospheric contributions. We made incremental improvements in the detection sensitivity and did a much better job of mapping the side lobes and shielding design of the instrument. Now burnt by experience, we horned in on the ballooning mechanics. We designed a more reliable reel. We took part in the design of the flight train, to avoid packed parachutes and any other command-driven procedure that affected the safety of the payload. A failure in the ballooning was now designed to result in a landing with a parachute irrespective of the failure of commands or noise triggering of the commands. In that year the NSF-sponsored NCAR Scientific Balloon Facility also upgraded its command and data communication electronics. This proved to be a significant step forward that substantially improved the reliability of the flights. In another big step forward Richard Benford, a technical jack of all trades who eventually became a full-fledged engineer, joined our little group. Now it was Dirk, Dick, and Rai who went to Texas to enjoy the high life.[23]

We made five successful flights with the new payload in two years (Figure 4.47). Two were made at stratospheric wind-turnaround in the spring and fall, when the upper-level winds change sign. This makes possible long flights, extending over a day or two, that remain in radio contact with Palestine, Texas. These are highly sought launch times, usually reserved for research that requires long exposures such as cosmic ray detection or measurement of periodic events in the atmosphere. Our helium consumption was large enough to limit useful observations to about 14 h, so we did not qualify for turnaround flights. Nevertheless, by being ready to go on 10 h notice we managed to get a turnaround flight as another group had to scrub their flight because of apparatus failure. The long durations allowed us to test for systematics and to carefully measure the contribution to the

---

[23] A comment on the location of the balloon base in Palestine, Texas. The balloon campaigns always lasted longer than we told our wives and girlfriends. Even though it rarely took longer than two weeks to get the instrument and flight apparatus together it usually required between five to six weeks to get a flight off and recovered. I used to ponder how this could happen. Weather reigns supreme in the ballooning business. The wind on the ground had to be less than 10 knots to be able to launch and for difficult flight trains the winds needed to be even smaller. The low-level winds, those at several hundred feet above the ground, could not be large, for otherwise there was the possibility of wind shear that could destroy a balloon. The favored launch conditions were not into clouds so that one could follow the progress of the launch. There were also conditions required in the recovery area. The lack of clouds was important there so that a chase plane could see where the payload would hit the ground after termination of the flight, and high winds and rain were not good for a successful termination. It turns out that East Texas, in such close proximity to the Gulf of Mexico which spawns much of the humid and thunderstorm-bearing weather in the Southern United States, was a particularly bad choice for a Scientific Balloon Facility. Not only was it difficult to achieve the needed benign weather conditions but also a good look at the map would indicate that in the summer when the stratospheric winds travel from east to west direct trajectories out of Palestine to the west all had the unfortunate property of intersecting the Mexican border. We had no reciprocity agreements with the Mexicans so that balloons that threatened to encroach on the Mexcian border had to be terminated. Locating the National Balloon Facility north of Palestine, say in Oklahoma, would have offered significantly higher probability of obtaining good launch conditions and much less probability of a threatened over-flight of Mexico. The trouble was that Oklahoma was not in Texas when the facility was inaugurated during the Lyndon Johnson years. One of Lyndon's gifts to ballooning was a significant number of separations and divorces.

Fig. 4.47. The launch layout for the first flight of the new payload. The package has been lifted by the launch vehicle ASCEND II. The payout reel with the 700 m of line is on the cart between the launch vehicle and the balloon. At the time when the picture was taken the top of the balloon was being filled with helium. The lower part of the balloon is lying on a tarpaulin between the cart and the clamp that is holding the balloon down. Although they are not visible, there are separate parachutes in the flight train for the reel and the payload. Once the balloon is released the launch crew member with his foot on the cart will follow the path of the balloon as it lifts the reel off the cart. ASCEND II will chase the balloon around the field until it is a little down wind of the vehicle and then release the payload. If the launch master has been successful, the payload will go forward and upward rather than backward and hit the vehicle. The launch ballet and the balloon dynamics are lovely to watch and much quieter than a spacecraft launch.

detected radiation by emission from water vapor generated by the balloon. We could also make zenith scans to help measure the atmospheric emission. Several of the flights used the largest and lightest balloons then being manufactured, and we were able to explore the change in the atmospheric lines with altitude near our flight altitude. This helped in validating our atmospheric emission models. The flights also differed in the extent and elegance of the ground shielding. That helped reduce the possibility of contamination by the radiation from the atmosphere and the ground that fell into the radiometer side lobes.

As we became more cavalier we began to improvise at the balloon base with new instrumentation that could help in understanding conditions at altitude. One of the most revealing innovations was the use of an old camera that was lying around. We fitted it with a heater and film advance motor to look at the top of the payload with flashbulb illumination. The sequence of pictures taken of the top of the system was timed by a Big Ben clock we had bought in the local drug store. We degreased it so it would work at the low temperatures at altitude. The sequence in Figure 4.45 shows all the hoop dynamics and the small amount of air frost deposited on the inner hoop cover. It also shows that there is no discernible deposition of the atmosphere in the radiometer mouth. This validated our experience that the helium efflux gas actually purged the air from entering the cold parts of the radiometer. One of the panels shows the secondary calibrator in position.

The results of these flights (Muehlner and Weiss 1973a) were published in *Physical Review* rather than *Physical Review Letters*. This was a significant mistake but we thought (at the time) it was justified because we felt that all aspects of the observation had to be explained and understood to make the measurement believable. *Physical Review Letters* offers rapid response but not much space, and we were not experienced enough to write our results in a convincing and short manner. To us the results were clear enough. The results of the initial flight were wrong: there is no excess near the peak. Our second set of measurements showed that there is a peak in the CMBR spectrum, the curve turning down at higher frequencies. There was no question that the Rayleigh–Jeans spectrum did not continue to high frequencies. We could reject the idea that the CMBR is produced by a diffuse dust shell around the Earth.[24]

There was still significant skepticism about our result from the people who had done the rocket observations. I remember well discussions with Martin Harwit about the atmospheric corrections that we had to make. Martin kept pointing out that our critical high-frequency data, which is

----

[24] Early on in thinking about the wisdom and importance of taking on the measurement of the spectrum of the cosmic background, I consulted Philip Morrison (Phil), who had recently come to MIT from Cornell to work with Jerrold Zacharias and Francis Friedman on developing new curricula for science education at all levels. I made an appointment with Phil to talk with him in his office and came at the designated time. Phil waved me in with his head in a book and asked me to explain my visit while he continued to read. I explained what I considered the importance of measuring the spectrum near the blackbody peak and also some of the difficulties we would experience in the measurement. I was not sure he was truly listening although it became clear that Phil was and that he was actually good at multiplexing his attention. His opinion was strongly expressed and very negative. He thought the CMBR was a mistake, that it was absurd to think that there could be a cosmic background that had equilibrated and that we in effect were embarking on a fools' errand. It was clear that the source of the radiation was something local, he thought possibly from a spherical shell of dust around the Earth. At the time he was still a strong proponent of the steady state theory of cosmology, which did not have an easy time with the CMBR. A few years later, after the results of the first balloon flights, I once again encountered Phil but by this time he had been through the conversion from skeptic to believer and he was just as strongly convinced that our result in the middle channel, showing an excess near the blackbody peak, was in error. So it goes with expert advice. Mind you, I have always been very fond of Phil and found him interesting to talk with and imaginative, it is just good that he was not in charge of funding or other serious matters. Several years later I helped him with demonstrations for a Nova program he called the "Whisper from Space." By that time we had made the series of flights with the new radiometer which gave results consistent with blackbody and we were good guys again. An enjoyable volume by which to experience Phil is a collection of essays and biographical sketches he wrote (Morrison 1995).

important in determining that there is a peak in the CMBR spectrum, was not all that different from the rocket results if one did not make the atmospheric correction. Furthermore, because this region of the spectrum is dominated by water lines, the zenith-scanning technique, so useful for measurements on the ground at longer wavelengths where the lines are not saturated, simply does not work. One had to solve for the radiation from a column of gas with Van Vleck–Weisskopf (almost Lorentizan) line profiles. The use of zenith-scanning measurements without additional calculation could give two limits. If the lines were fully saturated their radiative contribution would vary as the square root of the column density. An application of this case would over-correct the incoming radiation, leaving an apparently low CMBR contribution. In the other, unsaturated, limiting case the radiative contribution from the atmosphere would vary directly as the column density. Application of this case would undercorrect, leaving an apparently high CMBR contribution. One had to do it right by doing integrals. We did it right but the rocket people did not believe it. In part to make sure that stratospheric ozone was not playing tricks, we made a final flight with new filters to look specifically at more atmospheric lines (Muehlner and Weiss 1973b), which only confirmed our radiative modeling.

When Dirk and I wrote our long paper summarizing the results of the second radiometer flights, we presented the results associated with the radiation power measured through each of the spectral responses separately, both with the atmospheric corrections and without. These data could be related directly to the absolute calibrations made on the ground. We also attempted to get another handle on the incoming radiation spectrum by taking differences between the channels. This turned out to be troublesome because uncertainties in the atmospheric contributions contributed systematic errors, and differences increased the noise. We did not do the sensible thing of assuming a blackbody spectrum with an unknown temperature and making a χ^2 (least squares) fit to the temperature using the power measured in each spectral response. (Only after teaching experimental technique and data reduction in an undergraduate course did I become aware of the power of the χ^2 minimization).

Several years ago when Alan Guth was writing his book on the history of inflation (Guth 1997) he asked if those early balloon flights really had shown that the CMBR spectrum is blackbody. That led me to apply the χ^2 test to our data. The result is shown in Figure 4.48. The minimum of χ^2 is at 2.78 K. The value measured by COBE (Mather *et al.* 1990) and by Gush, Halpern and Wishnow (1990) is 2.726 K. At the time of

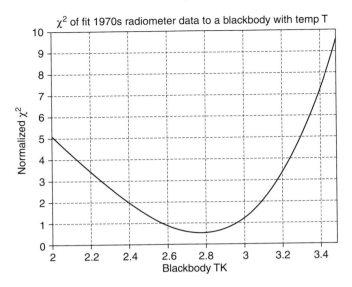

Fig. 4.48. Reanalysis of the data from all five flights made with the second radiometer. The calculated contributions from atmospheric radiation were used. The technique was to assume a blackbody spectrum with the temperature (horizontal axis) as the fit parameter in a χ^2 minimization. The value of χ^2 reduced to the number of degrees of freedom is the vertical axis. The smallest χ^2 is at a fit temperature of 2.78 K. This analysis was done in the mid-1990s to convince Alan Guth that we had measured the turnover in the spectrum. We should have analyzed the original data this way, but we didn't know enough.

our measurements the best estimate of the temperature was $2.72 \pm 0.08\,\text{K}$ (Peebles 1971, p. 141). Since our measurements probed shorter wavelengths this demonstrated that the CBMR spectrum is close to blackbody over the peak.

By the time we had finished writing up the results of the second radiometer spectrum measurements it was clear that the next step would be to use a Fourier transform spectrometer (FTS). This would help eliminate the atmospheric contributions since the water (but not the ozone) would show up as narrow intense lines that could easily be removed from the smoother spectrum of the CMBR. Two groups, one at Berkeley (Mather 1974) and another at Queen Mary College in the UK (Beckman *et al.* 1972) were already quite far advanced in preparing instruments for balloon flights. Herb Gush at the University of British Columbia in Vancouver was preparing a rocket instrument also using a FTS. Dirk and I went on to measure the anisotropy of the CMBR near the blackbody peak in a long series of flights that discovered a lot of galactic dust and not much else. But that is still another story.

4.11 Structure in the distributions of matter and radiation

4.11.1 Yu Jer-tsang: Clusters and superclusters of galaxies

Jer Yu is Chief Information Officer at the City University of Hong Kong.

I was a graduate student in the Physics Department of Princeton University from 1964 to 1969, and had the good fortune to work with Jim Peebles and David Wilkinson at the time of the discovery of the CMBR. I had the opportunity to observe at first hand the events that occurred during this exciting period, and to participate in some of the early work on the interpretation of this discovery.

I should begin my story with the decision I made in 1962 to go to the US to study. Although I was born in Shanghai, China, my family had moved to Hong Kong when I was seven. Thus, the first part of my education was mostly done in Hong Kong. In 1961, I was admitted by the University of Hong Kong into a program in pure mathematics. Back then, Hong Kong was nowhere near the international metropolis it is today. The university was fine, but it somehow lacked the excitement and diversity that I was hoping for. I was not sure what I really wanted to do with my life, but I was restless and yearned to go abroad to broaden my horizon. I began to make applications for admission to a number of US universities, from a list I compiled from the catalogs that I was able to find in the US Consulate. I was offered admission by several universities, and I chose the University of Michigan. UM has a long history of working with students from China, and enjoyed an excellent reputation in this part of the world.

I arrived in Ann Arbor as a transfer student in Engineering Physics in September 1962. Although I had started out not knowing anyone on campus, everyone that I met had been extremely kind and helpful. One of the persons I met was David Wilkinson, who was teaching a course I took. I cannot recall exactly how it happened, but soon I was working as a part-time assistant with Dave and his team in their experiment on nuclear magnetic resonance. This was my first ever paying job and, more importantly, the job gave me my first glimpse into what it was like doing research at the frontiers of science.

I completed my undergraduate degree at the University of Michigan and came to Princeton in September 1964. Unlike when I first arrived in the US two years earlier, this time there was someone I knew on campus before I came – David Wilkinson. Indeed, Dave's presence on the faculty of the Physics Department might have something to do with my being accepted by Princeton.

Every graduate student in Physics in Princeton was given a research studentship on admission, and was allowed to choose the area in which to work.

Naturally, I ended up working in the Gravity Group. There were a number of interesting projects going on to try to prove the theory of general relativity, one of which was Dave's experiment to detect the CMBR. Work had already begun to design and build a Dicke radiometer for this purpose. I was helping to assemble the equipment in the basement and to take measurements on the roof top. This was when I had my first lessons in cosmology. It was also when I met Jim Peebles.

Even as a graduate assistant, I could feel the excitement and intensity that were going on inside the group. It seemed like some new idea would be floated almost every other day on how the cosmic radiation could have an effect on our physical environment, and then plans were being put forward on how to observe such effects. News of the discovery of the radiation in 1965, if anything, had heightened the level of activities in the group.

In the meantime, I was really enjoying the good life of a Princeton graduate student. I had a reasonable stipend from my studentship that kept me free from any financial worries. The work in the Gravity Group was interesting and enlightening. Everyone was very willing to show me how things work and to teach me the theory behind it. The intellectual ambiance in the Physics Department was incomparable. I can recall the stimulating discussions in the weekly brown bag lunch seminar of the Gravity Group, the well presented departmental colloquia by invited speakers from around the world, the fascinating lectures by the Princeton professors in any number of the courses that I was free to take, or just the daily gathering of faculty and students for tea in the lounge in Palmer Laboratory. Added to this was the comradeship of all my friends in the Graduate College, coming from all over the world and working on so many different disciplines.

I passed my general examination, which was required of all physics graduate students, in the summer of 1966, and began to look for a topic for my thesis work. Since I knew that I did not have the knack for experimental work, I decided to do something theoretical. So one day, I walked up to Jim Peebles and asked him whether he would be my thesis advisor. Jim looked at me and said yes without the least hesitation. Thus, I had the honor of becoming Jim's first PhD student.

Jim began to talk to me about homogeneity and isotropy in the universe. He already had his theory of gravitational instability for the development of large-scale structures in the universe, and we began to look for observational evidence to support this theory. He gave me the two catalogs of clusters of galaxies, one compiled by George Abell (1958) and the other by Fritz Zwicky (Zwicky, Herzog and Wild, 1961–1968), and asked me to study the data to see whether there was any higher-level clustering of the clusters.

As a novice researcher, I began to read up feverishly all the papers on the subject that I could find, some relevant and some not so relevant. I was keeping extensive notes on everything that I had read, and spending a lot of time in the library. However, I was not making much progress with my thesis, until one day, in one of my regular meetings with Jim, he finally said to me something to this effect: "Jer, stop reading, start thinking." This one piece of advice had served me well for the rest of my life. You can always read about what other people think, but this is no substitute for your own thinking.

I had never done any observational astronomy myself and had not ever seen the galaxies and clusters (or the photographic plates on which they were recorded) in person. To me, the data were just x, y coordinates projected onto a spherical surface. I painstakingly transcribed the coordinates into 80-column punched cards so that I could feed them into what was, at that time, still a relatively new tool, a digital computer, for analysis. For the next six months, the half box of cards became one of my most precious possessions.

Princeton had one of the more advanced computers of the time (I believe it was an IBM 360/65) housed in the Engineering Quadrangle. I was making the trek between Palmer Lab and the Engineering Quad and submitting my deck of cards to the computer center almost on a daily basis. I was putting the data through all kinds of permutations to determine whether there were any significant patterns. On a good day, I could get two or three runs per day. Initially, I was hoping to find an exact mathematical formulation which would allow me to integrate over the data and come up with a definitive yes or no answer. Unfortunately I was not able to achieve this. In the end, I had to resort to using simulation to create a number of possible distributions, and then to compare the observed distribution against the simulated ones. In this way, I was able to draw some conclusions by inference (Yu 1968; Yu and Peebles 1969).

Another idea I had was to do a complete simulation of the evolution of the universe by following the development of some initial density fluctuations through the different epochs, and if everything works according to the theory, should finally be able to see large-scale objects emerge like bubbles in boiling water. Again, I was not able to completely solve the problem as I had formulated it, probably because the computers in those days were not powerful enough to do what I wanted to do.

I left Princeton in 1969, and after spending one year as a post-doctoral research fellow at the Goddard Institute for Space Studies in New York City, I returned to Hong Kong in 1970. The University of Hong Kong was just then getting its first mainframe computer, and had hired me because of my computer knowledge to manage this installation. Although I have not continued to work in the field of cosmology after I left the US, the experience that I had in Princeton was most memorable and rewarding. After all, there are not that many people in the world who can claim that he was there on the spot when the cosmic microwave radiation was discovered.

4.11.2 Rainer K. Sachs: The synergy of mathematics and physics

Ray Sachs is Professor Emeritus of Mathematics and Physics at the University of California, Berkeley. His analysis with Wolfe of the gravitational interaction of the CMBR with the departures from a homogeneous mass distribution is a central element in the cosmological tests. His current research interest is the mathematical modeling of radiobiological data.

In the 1960s, Artie Wolfe and I thought that inhomogeneities in the early universe sufficient to cause the presently observed lumpiness would lead to anisotropies in the observed temperature of the CMBR. Photons from different directions, coming from, and passing through, different time-dependent fluctuations would give information on the nature of the fluctuations. We used linearized perturbations of a spatially flat, general relativistic Robertson–Walker model to analyze density, velocity, vorticity, and gravitational wave fluctuations; their influences on the CMBR, regarded as consisting of test-photons, were worked out with general relativistic kinetic theory. We overestimated the size of the temperature anisotropies, but some of our ideas (Sachs and Wolfe 1967) were eventually supported by COBE and subsequent observations.

My most vivid recollection of the work concerns the interplay between mathematics and physics. Somehow, it seemed almost like an extension of the interplay between calculus and Newtonian dynamics in a freshman physics course.

The reader probably took such a course and may remember that it did not literally emphasize typical physical systems, whose methodical analysis usually requires additional background. Instead there were Newton's laws plus a magical zoo of idealized ropes, pulleys, weights, projectiles, billiard balls, levers, reaction forces, springs, pendulums, and (best of all) monkeys. Also, we learned how calculus really works, and what an immensely powerful unifying force it is; formal calculus proofs were blessedly absent; intuitive proofs were wonderfully present. The experiments we did in lab were boring to me but the *gedanken* experiments we analyzed for homework were endlessly fascinating. In optional reading Milne's (1935) beautiful little book on cosmology fit right in, despite (or perhaps because of) the fact that some of his main ideas do not actually work for the real world.

Tensor analysis is no less capable of unifying than freshman calculus, and the geometric approach to general relativity is even more elegant than three-dimensional Cartesian vector algebra. For me, our CMBR anisotropy work was, much as in a freshman physics course, mainly an exercise in applying mathematics to a highly idealized situation. Some examples may illustrate the flavor of the arguments.

1. One question concerned history-based versus law-based explanations in cosmology and in the rest of science. Biological evolution has a strong historical component, rather than being understandable in terms of equilibria or steady states resulting from general laws. We are what we are in good part because some 40 million years ago our lemur-like ancestors were what they were; to explain those proto-lemurs one must go still further back in time; etc. And in biological evolution we may (perhaps) be dealing with one unique process, so that our usual ideas of replicating experiments or assigning probabilities become obscure. Moreover, it seems pretty clear that chance played a significant role; Vice President Cheney is presumably not an inevitable consequence of basic natural laws.

Now is cosmology, contrasted for example with special relativistic quantum theory, also like that? My guess at the time was "yes." To explain lumps in the present universe we must go back to the lumps at the time when radiation and matter decoupled, which in turn can only be understood in terms of earlier lumps, and so on indefinitely. The modern idea, of including some initial conditions as part of basic physical law, could perhaps have saved me a lot of confusion but did not occur to me.

The two other atypical characteristics of biological evolution mentioned above also seemed to have probable counterparts in cosmology. Even more than in the case of biological evolution, the universe's history can probably best be understood as a unique process. Comparisons, probabilistic or otherwise, to possible alternate universes may be useful but probably have no fundamental significance.

Whether there is also anything accidental involved in this putatively unique history was puzzling, especially in those days, before inflation. The spacetime diagram in Figure 4.49 represents part of the history of a perfect-fluid Robertson–Walker general relativistic model. Causality is emphasized, that is, light-like geodesics are at 45° and only the conformal structure is considered. The world line represents part of the history of our Galaxy and here-now is shown as a dot. The main point is that in such a figure the big

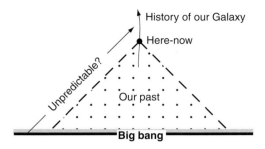

Fig. 4.49. A part of the history of the standard cosmology.

bang typically turns out to be a three-dimensional space-like surface, not a
point. Correspondingly there could be signals, as shown, coming from very
near the big bang, which have since been traveling toward us at the speed
of light and will eventually arrive with information (and other stuff) that is,
for us now, new and unpredictable.

So I felt at the time that the evolution of the universe, like biological
evolution, is a process where details of the past have dominant importance,
is a one-of-a-kind process, and is in some sense subject to blind chance. That
was worrisome, since most of the rest of physics is not like that at all. The
main properties of a hydrogen atom result from laws, not its own history;
there are many different hydrogen atoms; and blind chance plays no role in
their basic structure.

2. The whole concept of "now" was likewise worrisome. The dot in the figure
is indispensable for the discussion just given, and similar "now-dots" appear
to be needed in many other arguments. But the mathematical analysis,
which incorporates relativity of simultaneity, does not use any corresponding
preferred events. In our analyses, we tried to think through essentially this
point not only in considering what "blind chance" might mean but also when
we ran across the following question. Suppose the observed microwave temperature in a given direction is a bit larger than in other directions. Can this
be attributed to, for example, time-dependent density inhomogeneities, or
could it just mean that in the indicated direction we are somehow managing
to look a bit further out, and thus a bit further back in time to a hotter epoch
of the universe? Eventually we did find a consistent, coordinate-independent
answer that does not implicitly invoke absolute simultaneity.

Quite generally, "now-dots" like the one in the figure are really not legit-
imate in relativity. You only need to draw a Minkowski spacetime diagram
with a few time-like world lines in it to realize that no consistent, systematic
assignment of now-dots is possible without in effect imposing some version

of absolute simultaneity. Indeed a key part of the geometric approach to relativity is to consider a physical process as a unified history: past, present and future. Space-like slices through a history are at best conveniences and are often more misleading than useful. It is this four-dimensional aspect that accounts for the truly extraordinary simplicity (leaving quantum phenomena apart) of the actual relativistic universe compared to a hypothetical universe governed by vintage 1900 physics. Fundamentally the latter is much more complex because it is less unified logically (with space and time being distinct things instead of two different aspects of the same thing, energy and momentum conservation being separate laws, etc., etc.).

My guesses about "now" were (and are) the following: (a) For the reasons just given, "now" can't be allowed into a relativistic theory, and it is really highly embarrassing for such theories that all of us very strongly believe "now" is somehow very different from the rest of our history. (b) Perhaps when two people talk there are (especially if one of the two has extra frequent flier miles) small discrepancies between their perceptions of "now." It is tempting to blame this fact for the style of conversation one hears in a Berkeley cafe. But airplane speeds are not that large; the potential discrepancies generated are considerably less than a microsecond, and thus pretty harmless. The discrepancies could easily, in a highly interactive community like the human race, be subordinated to an implicit agreement on some kind of ad hoc consensus simultaneity. (c) In principle, however, there seems to be an important conflict between the impermissibility of "now-dots" and our overwhelming intuition that now is special. This conflict could be, as were Olbers' and Gibbs' paradoxes, an obscure signal of the need for some basic paradigm shift; if there is a simple resolution of the conflict I am not aware of it.

3. Some of the mathematical tools we used to analyze the behavior of photons were, truth to tell, motivated at least as much by formal analogies as by physics. I had learned from Jürgen Ehlers about the elegant way relativistic hydrodynamics treats fluid expansion, shear, and vorticity. The formal generalization to light beams was almost automatic; that approach eventually led to generalizations in terms of Liouville's theorem in an appropriate phase space, with which one can track the microwave photons as they come from and through distant matter inhomogeneities to us.

4. The main tool we used was likewise based in good part on a mathematical analogy – to first order time-dependent perturbation calculations in quantum theory. Specifically, we found that by linearizing the Einstein field equations around a Roberston–Walker perfect fluid spacetime (a method used earlier by Lifshitz 1946) one gets very instructive time-dependent

solutions, identifiable parts of which (essentially normal modes) correspond to gravitational waves, vorticity, or density and velocity fluctuations. The way in which each relevant physical aspect had just the appropriate mathematical counterpart, and vice-versa, seemed very satisfying. The perturbation solutions, not being restricted by the artificial symmetry assumptions essential to get fully nonlinear solutions explicitly, gave perspective on how inhomogeneities in the universe evolve in time and how they can influence anisotropies of the observed CMBR.

In summary, to me the most interesting aspect of possible CMBR anisotropies was the way the processes involved illustrated a synergy between mathematics and physics – to paraphrase Einstein, one of the most incomprehensible things about the universe is that mathematics can help us comprehend it. I was thus actually less interested in the universe than in the methods used to analyze it. In retrospect that seems odd, but that is what happened.

4.11.3 Arthur M. Wolfe: CMBR reminiscences

Art Wolfe holds the Chancellor's Associates Chair of Physics and is the director of the Center for Astrophysics and Space Sciences at the University of California, San Diego. His research interest is galaxy formation, with particular attention to gas-rich galaxies observed at high redshift.

I was a graduate student in physics at the University of Texas (Austin) when the CMBR was discovered in 1965. Although most cosmological models at that time were based on the assumptions of homogeneity and isotropy, there was little empirical support for either assumption. I was working with Ray Sachs, my PhD thesis advisor, on devising techniques useful for placing quantitative limits on departures from homogeneity and isotropy. After all, the universe of galaxies is observed to be quite lumpy, and it was unclear to us whether the global smoothness of the models was consistent with observations. Ray had been attending lectures by the astronomer G. de Vaucouleurs who emphasized the presence of superclusters of galaxies on length scales which, though not generally accepted at the time (they are of course accepted today), might have implications for the large-scale structure of the universe. In fact de Vaucouleurs always adhered to a model with zero mean density and structures on ever increasing scales; that is, a universe that is highly inhomogeneous on all scales.

I think it is important to emphasize that cosmology in 1965 was not an empirically-based branch of physics. Besides the Hubble expansion, we knew very little else about the universe. While the competing steady state cosmology was running into difficulties with the steep slope of the radio-source counts, Hoyle and collaborators were extremely resourceful in finding plausible scenarios to explain these data. In retrospect, the quasar redshift distribution indicated an evolving universe that was inconsistent with the steady state, but at the time the origin of the quasar redshifts was controversial, so this was not regarded as a definitive argument against the steady state model. As a result, the subject was in a state of flux with the big bang competing head-to-head with the steady state model. This was also reflected in the lack of good textbooks available for graduate students struggling to understand the field. Fortunately for me, I attended an excellent course on cosmology given by Engelbert Schücking. Unfortunately, most of the textbooks available at the time emphasized mathematical elegance at the expense of physics. A notable exception was the excellent monograph on cosmology by Bondi (1960a), one of the architects of the steady state cosmology.

Prior to the discovery of the CMBR, the only way to assess the large-scale structure of the universe was through observations of low-redshift galaxies. Ray Sachs and Jerry Kristian had computed a local power-series approach to this problem (Kristian and Sachs 1966). They showed how departures from pure expansion in the form of shear and vorticity could be inferred from measurements of galaxy shapes. But because this was a power-series expansion around here and now, their model was valid only for redshifts $z \ll 1$. The results of this exercise were illuminating in that Ray and Jerry found that the observations were consistent with a shear as high as 20% of the expansion rate, and with an even larger value for the vorticity. As a result, because observations of galaxies in the 1960s were not sensitive measures of large-scale kinematics, the data were consistent with significant deviations from the widely accepted idea of near pure expansion.

Detection of the CMBR by Penzias and Wilson (1965a) changed everything. Nobody I knew had been thinking about the CMBR even though it had been predicted earlier by Gamow and collaborators. I first heard about the Penzias and Wilson discovery at a 1965 seminar given in Austin by Nick Woolf. Because the CMBR was not mentioned in any of the text books or courses I took, I had to learn about an entirely new field from the bottom up. Fortunately for me, Ray handed me a paper copy of the 1965 preprint (electronic preprints did not yet exist!) article by Jim Peebles on the "primeval fireball" and its implications for galaxy formation. This preprint was really a blueprint for Jim's research for many decades to come. It contained terminology such as "the last scattering surface," "Thomson drag," etc. which, though familiar now, were revolutionary concepts in 1965. Jim's article was very different from most of the previous literature in cosmology. It was filled with physical rather than purely mathematical ideas. It introduced me to the concept that the CMBR was a truly global radiation field. I was astounded when I first realized that its mean temperature was an average over the present spatial particle horizon.

About this time Ray became interested in using the CMBR as a tool to study large-scale structure. The idea was to perturb Friedmann models to first order and see what effect gravitational perturbations had on the CMBR. The main reference in the study of linear perturbation theory was the classic paper by Lifshitz (1946). While Lifshitz solved the problem for the full suite of Friedmann models, we focused on the Einstein–de Sitter model: due to its mathematical simplicity the solutions to the perturbation equations could be expressed in terms of simple algebraic functions. This made it easier to compute light-like geodesics to first order. To calculate the effects of the perturbations on the CMBR temperature, we used the Liouville theorem for radiation to find that the present CMBR temperature in any direction is inversely proportional to $1+z$, where z is the redshift in the same direction. By computing the light-like tangent vectors we were able to find first-order corrections to temperature in terms of an integral along our past light cone over functions of the perturbed metric and its time derivatives. Later in my thesis I repeated this calculation by solving the collisionless Boltzmann equation for radiation to first order and obtained the same answer. Our solution for the temperature fluctuations included contributions from vector and tensor terms, which are physically related to vorticity and gravitational radiation. We focused instead on the scalar terms because they contained a first-order gravitational potential that was a solution to the Poisson equation with density perturbations as the source. As a result we derived an expression in which the temperature perturbation $\delta T/T$ is proportional to the density perturbation $\delta\rho/\rho$.

To estimate $\delta T/T$ we assumed $\delta\rho/\rho \approx 10\%$ on scales $d \sim 300$–$1000\,\mathrm{Mpc}$. In the 1960s little was known about the density structure of the universe on large scales. In retrospect we were influenced by de Vaucouleurs' claim of significant density structures on scales of hundreds of megaparsecs. In any case, we concluded that $\delta T/T \approx 0.005$.

Publication of our result (Sachs and Wolfe 1967) had a mixed reception. The major figures in the field of cosmology were very interested. During a trip to Moscow in 1971, Zel'dovich and Sunyaev told me how excited they were about our work. In the west, Peebles, Rees, and Silk turned their attention to the problem of temperature anisotropies. On the other hand, most astrophysicists showed little enthusiasm for this subject: 40 years ago astrophysical research centered on topics such as stellar evolution and the physics of radio sources and quasars rather than the large-scale structure of the universe. A revival occurred in the 1980s with the advent of dark matter cosmologies. In 1982 Peebles combined our formalism with his newly derived cold dark matter power spectrum to make a more realistic estimate of $\delta T/T$ of $\sim 10^{-5}$ on large angular scales (Peebles 1982). This was ultimately confirmed by the COBE satellite (Smoot *et al.* 1992). The result was a flood of interest in our work. While I cannot speak for Ray, I was both astonished and gratified by the amount of research our work has generated. At the time of its publication in 1967, neither of us had any idea about the impact it would have.

4.11.4 Joe Silk: A journey through time

Joe Silk is Savilian Professor of Astronomy, University of Oxford. He is an active contributor to physical cosmology and author of five books on the subject; the latest is Infinite Cosmos *(Silk 2006).*

I began my research career at a propitious time. Cosmology had been stuck in a rut for decades, but it was about to explode. I arrived at Harvard in 1964 as a beginning graduate student who was eager to become a cosmologist. This intention was nurtured by two events in my life. I had been studying for a mathematics degree at Cambridge and was not overenthused by my lectures. I was completing Part 2 of the Mathematics Tripos, so named I was told because in earlier times, the students were examined individually by their professors, while precariously perched on a three-legged stool. I accidentally stumbled into a Part 3 course given by Dennis Sciama. I heard him lecture on Mach's principle, Einstein and the origins of general relativity. I was captivated. The universe may not have been rotating that day, but my head was certainly spinning from the new vistas that were opened on a wet Cambridge morning.

Leaving Cambridge behind, I went north to Manchester to continue my studies by enrolling in a fourth year course in physics. The next event occurred when I was studying in the library and getting progressively more and more bored. Perusing at random the pages of *The Astrophysical Journal*, I was impressed by the choice of the first article of each issue, invariably on cosmology. And one of these fascinated me further. The article in question applied the virial theorem to the universe and to the growth of structure (Layzer 1963). The very notion of a cosmic virial theorem captured my imagination. The author was a cosmologist on the faculty at Harvard. Many years later, his theorem was to form the core of an important cosmological probe for weighing the amount of dark matter in the universe. This required data, which did not then exist. So it was to theory that my attentions turned, and I set about getting a fellowship from ESRO, the research-orientated predecessor of the European Space Agency. I took the fellowship to Harvard to work with my idol, Professor David Layzer.

Layzer agreed to supervise my research on the topic of how galaxies formed in the expanding universe. I soon discovered that Layzer was an arch proponent of the cold big bang. It did not take me long to explore the possibilities of galaxy formation in an initially cold universe. Indeed, I found the outcome for galaxy formation was entirely satisfactory. However the issue of data soon posed a serious challenge. The CMBR, the fossil radiation from the beginning of the universe, was discovered by Arno Penzias and Robert Wilson the year I started graduate school, 1964. The problem was that the CMBR argued strongly for a hot big bang. The timing was truly optimal for a confrontation of theory and data.

But first, there were confrontations between the rival theorists. Many refused to accept the cosmological nature of the CMBR. Local origins were strongly advocated, especially in a cold big bang. Relations between my Harvard supervisor and the leading proponent of the hot big bang, James Peebles at Princeton, were tense. News of the rivalry filtered down to the dark and dank basement office at Harvard College Observatory, where the graduate students were sheltered. I slowly migrated away from the theory of a cold big bang. My first paper struggled with Mach's principle in an unusual setting. Going back to my cosmological roots, I tackled the problem of galaxy formation in Gödel's rotating universe. But this research direction seemed to have little future; nor for that matter did the concept of a cold big bang.

To his immense credit, Layzer was remarkably open-minded and encouraged me even when I eventually became disillusioned with his increasingly baroque attempts to incorporate the newly discovered CMBR into the context of a cold universe. The CMBR did seem to be most simply interpreted as the fossil blackbody radiation from a primordial thermal fireball.

I spent part of the summer of 1965 at a summer school on the Cornell University campus in upstate New York, organized by the American Mathematical Society. Cosmology was at the transition between a branch of general relativity and one of astronomy. The theme was the rapidly emerging subject of what would now be called "physical cosmology." My fellow students included Jim Gunn, Bruce Peterson, and Arthur Wolfe, all to subsequently leave their marks in cosmology via the eponymous effects associated respectively with tracers of the ionization history of the universe in quasar spectra, and the large angular scale fluctuations in the CMBR that are associated with the observed large-scale inhomogeneity of the universe (Gunn and Peterson 1965; Sachs and Wolfe 1967).

During the following summer of 1966, I was employed as a research assistant at American Science and Engineering, an MIT spin-off company started by Bruno Rossi and Riccardo Giacconi that had recently launched an X-ray satellite to search for fluorescence X-rays from the Moon. As often happens in science, the serendipitous discoveries of the first X-ray source Scorpio X-1 and the diffuse X-ray background overshadowed the initial goal. My summer brief was to develop a theory for the X-ray background. This radiation had to be of cosmic, indeed of extragalactic, origin as a consequence of its observed isotropy on the sky. The X-ray background was relatively uniform, and so had to be produced by many distant galaxies. One could speculate freely, on the basis of one known galactic X-ray source! This was how I developed a taste for studying diffuse backgrounds, a topic that was ripe for investigation and was to play a central role in much of my future research.

At this point in time, I almost became an observer. Harvard in those days required its budding theorists to undertake an observational project. Armed with the approximate coordinates of a new X-ray source in the constellation of Cygnus, I spent many cold nights that winter at Harvard's Agassiz Observatory, Massachusetts. My mission was to use the 36-inch reflector to photograph the star field several times per night at the location of the X-ray source. I would develop the plates myself, then the following day I would bury myself in the depths of the Harvard College Observatory, huddling over a blink comparator device to search for short timescale variability. The recently discovered Scorpio X-1 counterpart, a bright blue star, varied and flickered on timescales of nights and perhaps even hours, and the logic went, so should Cygnus X-1. There was a theoretical argument, based on the scaling with the ratio of X-ray to optical luminosity of Scorpio X-1, that suggested one should be seeing a variable, blue, 12th magnitude star.

Of course, my mission failed, and I was eventually scooped by professional observers who had the advantage of clear skies, the world's largest telescopes, and, most importantly, extensive experience. Theoretical prejudice was found, not for the first or last time, to be detrimental to the observer's health. By way of consolation, I was not alone in being led astray: Allan Sandage, who had previously identified Scorpio X-1 as a flickering 13th magnitude blue star, was searching for an 18th magnitude counterpart to Cygnus X-1 (Giacconi *et al.* 1967)!

Cygnus X-1 turned out to be the brightest star (9th magnitude) at the center of my plates. It was even a previously cataloged star, HD 226868. My plates were certainly well centered, but somehow I missed the variations. In fact, the images I took were mostly in terrible seeing, trailed and out of focus. Little surprise that I could hardly compete with the experienced astronomers on the Mt. Palomar 200-inch telescope.

I was increasingly frustrated from my attempts at astronomical observations, and felt that I most likely suffered from a version of the Pauli principle: whatever could go wrong in an experiment that I undertook did go wrong, with even my proximity seemingly having a malign influence on the outcome and even the functioning of the experiment. So I resolved to become a theorist. My doctoral thesis was to be entitled *The Formation of the Galaxies*. But I still had to write it (Silk 1968a).

Meanwhile, the debate on the interpretation of the CMBR intensified. This was still at a time when the steady state universe had a vocal band of supporters. Much of the debate came to a climax at the second conference I attended, in early 1967 at the Goddard Institute for Space Studies in New York. This for historical reasons was the *Third Texas Symposium on Relativistic Astrophysics*, following earlier meetings in the series at Dallas and Austin. Those were heady days. Quasars as superstars highlighted the first Texas meeting in 1963, but their true distance and nature were still being hotly debated. What stole the show for me, however, was the question of the origin and nature of the CMBR. I even recall encountering George Gamow surrounded by a small crowd and declaiming in his curiously high-pitched voice that he had lost a penny, Penzias and Wilson had found a penny, and was it his penny?

Despite the new developments in cosmology being pioneered by Jim Peebles and that further developed the theory of the hot big bang, I found no better solution to understanding the origin of galaxies until in the summer of 1967 I found myself at Woods Hole, Massachusetts. The occasion was my enrollment as a student in the Woods Hole Oceanographic Institute Summer School. Traditionally, WHOI held an annual summer school on applications of fluid dynamics. That year, the chosen field was astronomy. The topic was astrophysical fluid dynamics and I was fortunate to hear lectures by such luminaries as Richard Michie and Ed Spiegel. But my true inspiration came from George Field, who lectured on galaxy formation. The summer project that I chose under Field's direction was to incorporate the newly discovered cosmic fossil radiation into galaxy formation theory. I was inspired, and worked day and night. I studied the coupling of the matter and radiation in the early universe, and in particular, the transition from optically thick to thin regimes at a redshift of about 1000. I used the adiabatic mode of density fluctuations, described by sound waves in the baryon–photon plasma prior to matter-radiation decoupling, to evaluate the associated radiation density fluctuations. Within a few months, I had produced my first paper on this topic, presciently entitled *Fluctuations in the Primordial Fireball* (Silk 1967). To form the galaxies, the initial density fluctuations must have had a finite amplitude that left a potentially observable trace in the CMBR via the acoustic imprint in the temperature fluctuations on subdegree angular scales.

There was one initial hiccup. I was almost scooped again, so I felt when I first saw the paper by Sachs and Wolfe (1967) that appeared later that same year. But my spirits lifted when I realized that their predictions of large angular scale fluctuations were based on an extrapolation of the observed large-scale irregularity of the universe. This was an observation with no accompanying theoretical explanation. The irregularity was seen in the observed large-scale structure of the galaxy distribution, but did not have to be there. By studying the coupling and growth of primordial density irregularities, the temperature fluctuation strength could be predicted. It was a phenomenological prediction.

My predictions, on the contrary, were focused on the theory of galaxy formation. This, after all, was the title of my doctoral thesis. But how was one to test such a theory, in the era before the advent of the very large telescopes and the space telescopes? The solution came from the prediction of small angular scale temperature fluctuations. These provided a crucial missing link in the connection between the initial conditions and the formation of the galaxies. The irregularities arose from a fundamental theoretical argument. The fluctuation strength could be predicted via the requirement that galaxies must have formed by the gravitational instability of tiny density fluctuations whose amplitude was calculated from the theory laid down in the pioneering paper of Lifshitz (1946). Fluctuations grew in strength via the effects of gravity in the expanding universe. Without such fluctuations there would be no galaxies. Not that it was particularly clear at the time, or indeed for decades later to anybody beyond a select handful of cosmologists, but I had come up with a theoretical prediction that was fundamental to our understanding of the big bang as a cosmological model of the observed universe. In fact, my results were entirely complementary to those of Sachs and Wolfe, who had concentrated on the superhorizon scales where the primordial density ripples are imprinted. I studied the interaction of matter and radiation on subhorizon scales, where the physics of acoustic waves modifies the primordial fluctuation spectrum and boosts its amplitude. I evaluated the characteristic angular scale of the fluctuations that seeded galaxies and galaxy clusters. My predictions were further refined a year later (Silk 1968b) when I evaluated the minimum scale of surviving adiabatic density fluctuations due to the coupling with the radiation field. There was a corresponding minimum angular scale above which the temperature fluctuations could survive and be detectable.

Of course, history had the last word in 1992 when the COBE satellite verified, to within a factor of 2, the Sachs–Wolfe prediction on angular scales in excess of $7°$ (Smoot *et al.* 1992). It was to take almost another decade before the fine angular scale anisotropy predictions on subdegree scales were confirmed.

The idea was straightforward. If galaxies formed by gravitational instability from primordial infinitesimal density fluctuations, the inferred radiation dominance of the early universe meant that they had to have a finite amplitude given the limited time available in the matter-dominated regime for fluctuation growth. So I predicted, initially very naïvely, that the required amplitude of temperature fluctuations on angular scales of tens of arcminutes or less in order to form large-scale structure such as clusters of galaxies had to be about 3 parts in 10,000. The fluctuations could not be any smaller, otherwise the galaxies and galaxy clusters would not have had time to form. The argument was remarkably simple. The growth factor since last scattering was 1000 in a flat, matter-dominated universe. Hence the initial density fluctuations to form clusters by today had to be of order 0.1%. The temperature fluctuations were correspondingly of order a third of this, for the adiabatic mode in which density is proportional to the cube of the temperature. One byproduct of the calculations was the damping of the fluctuations as the last scattering surface was traversed. The angular scale associated with the so-called "last scattering surface" denotes the angle subtended by the transverse projection of the finite time of recombination, converted to a comoving length scale. Below the minimum angular scale of a few arcminutes, where the damping sets in, one would not expect much in the way of primordial temperature fluctuations in the CMBR.

Over the next two decades, there was a very small and select group of theorists who pursued and refined these calculations, pioneered by Peebles, Sunyaev, Zel'dovich, and collaborators. One early critique was that the adiabatic density fluctuations would be erased by the finite thickness of the last scattering surface and that velocity modes would predominate. This turned out not to be the case once a more sophisticated treatment of fluctuations was developed. The notion of acoustic peaks in the matter was developed in a classic paper by Peebles and Yu (1970), and independently proposed that same year by Sunyaev and Zel'dovich (1970c), and in the radiation intensity by Doroshkevich, Zel'dovich, and Sunyaev (1978). The latter paper improved on the earlier discussion by Sunyaev and Zel'dovich in 1970, but was itself later substantially corrected and refined in the first rigorous treatment of the subject by Silk and Wilson (Silk and Wilson 1981; Wilson and Silk 1981; Wilson 1983). I had taken up a faculty position at Berkeley in 1970, and with my student Michael Wilson, I developed the first modern relativistic treatment of temperature fluctuations by solving the coupled Boltzmann and Einstein equations in a curved background. From now on, one could hope, at least in principle, to measure the curvature of the universe by studying a map of the sky.

Of course, the ultimate verification was to take a long time and painstaking effort. There were several generations of CMBR experiments. A prolonged period followed when the improved experimental limits were above the progressively refined theory (Figure 4.50). Each time there was a major experimental improvement, as happened with the pioneering attempts of Partridge and Wilkinson (1967), then of Uson and Wilkinson (1982), the theoretical hurdle was raised with the advent of more precise calculations. Until the mid-1980s, the select band of theorists who worked on the CMBR were lonely voices in the wilderness. I distinctly recall at one conference during this period how Geoff Burbidge labeled us the "background brigade," arguing confidently that the absence of detectable temperature fluctuations proved that the gravitational instability theory of structure formation was wrong, and thereby cast doubt on the big bang itself.

The final theoretical refinements came with the introduction of cold dark matter. The large-scale CMBR anisotropy in the CDM model was computed by Peebles (1982), and the small-scale anisotropy was computed

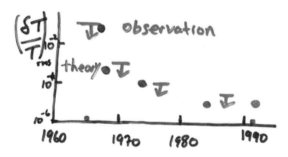

Fig. 4.50. Upper limits on the fractional CMBR temperature anisotropy $\delta T/T$, demonstrating that theory remained ahead of the observations for some three decades. From a transparency dating from 1992.

independently by myself and Nicola Vittorio at Berkeley (Vittorio and Silk 1984) and by Dick Bond and George Efstathiou at Cambridge (Bond and Efstathiou 1984). Nor was it long before the cosmological constant was probed via these predictions (Vittorio and Silk 1985). The weakly interacting cold dark matter allowed fluctuations to grow despite the tight baryon–photon coupling once the universe was matter-dominated. The prediction of temperature fluctuations arising from structure formation was now an order of magnitude or so lower than the early predictions, 3 parts in 100,000 at the first acoustic peak at an angular scale of about 30 min of arc, and substantially lower on smaller angular scales where the damping played a role. It was to take another five or six years before a ground-based experiment (TOCO) and the balloon-borne experiments (BOOMERANG, MAXIMA) provided strong confirmation of the elusive signal.

Nor even then was the solution completely definitive. Refined data were needed for the next step. This was the prediction that one could measure the curvature of the universe (Sugiyama and Silk 1994) in the sky. It turns out that in the CMBR alone, there are significant parameter degeneracies (Efstathiou and Bond 1999). Indeed, the simple addition of a Hubble constant as measured by the Hubble Space Telescope key project ($72 \, \mathrm{km \, sec^{-1} \, Mpc^{-1}}$) leads to the highly significant inference that the universe has close to zero spatial curvature. This result was greeted with joy by many theorists who regarded it as a prediction of inflationary cosmology. I personally am less convinced by the predictive power of inflation, recalling the equally vocal band of inflationary theorists in the 1990s who welcomed the low density, spatially curved, universe then favored by observational cosmologists with suitably tuned inflationary models.

However, while one can always find inflationary models to explain whatever phenomenon is represented by the flavor of the month, it is certainly true that the generic predictions, associated with the vast majority of the models of inflation on the market, have had two immense successes. One of these is the verification of the flatness of space. Another stems from an achievement of the three-year data from WMAP, which has succeeded in eliminating one of the rival hypotheses to inflation, the Harrison–Zel'dovich prediction of the scale-invariant nature of the primordial density fluctuations. This asserts that the spectral index of the scalar fluctuation power spectrum is $n_s = 1.0$, on the basis of simple but compelling scaling arguments. However, this is one situation where simplicity has to be abandoned when confronted with reality. The new result from the WMAP satellite (Spergel *et al.* 2007) is that $n_s = 0.95 \pm 0.02$. This is expected as a consequence of the finite duration of inflation with smaller and smaller fluctuations exiting the horizon later and later as inflation peters out and the fluctuation distribution gradually rolls over in power.

Nowadays, cosmology seems rather boring. All measurements converge on the standard cosmological model with hypothesized ingredients of dark matter and dark energy that are themselves poorly understood. It requires immense hubris to be confident that we have found the final solution, given our woefully inadequate mastery of the first instants of the big bang. The ultimate theory of cosmology will surely include our standard cosmological model as a component.

4.11.5 *George F. R. Ellis: The cosmic background radiation and the initial singularity*

George Ellis is Professor Emeritus of Mathematics at the University of Cape Town, where he has run a relativity and cosmology research group since 1973. He continues to write on relativity theory and cosmology, but also nowadays writes on the emergence of complexity and the way the human mind works.

The DAMTP at the University of Cambridge was lucky to get Dennis Sciama as a University Lecturer in 1961. His passionate love of physics, astrophysics, and cosmology was matched by his enthusiastic understanding that what mattered most in a research group was finding and supporting bright students, who were where the future of the subject lay. Furthermore, as was true also of John Wheeler, he believed that some of the deepest advances would come from describing physical effects through precise mathematical formulations, combining a good understanding of physical effects with a knowledge of the latest mathematical techniques. Thus as well as encouraging the kind of approximation techniques that lie at the heart of much physical understanding, he also encouraged the search for exact mathematical theorems that could express important physical results. But he insisted that theory should have relevance to the real world, so one should explore all possible observational aspects and subject them to rigorous test. Theorems on black holes or cosmology were of little use without some link to possible testing by astronomical observations. He would always push one on this point: How is it observable? How do you test it?

A key issue at that time was the possible existence of spacetime singularities: whether they occurred at the start of the universe on the one hand, and at the endpoint of gravitational collapse of astronomical objects on the other. In both cases simple general relativity models with exact symmetries indicated there was indeed a spacetime singularity, and John Wheeler

(1964) in particular emphasized that this was a major crisis for theoretical physics, because such a singularity indicated a beginning or end not just to space, time, and matter, but also to physics itself, and hence represented the limit of physical understanding. The question was whether the indication of a singularity was a result just of the simplified models used, which excluded rotation for example, and so would go away if more realistic models were used. This was precisely the kind of area where the combination of new mathematical techniques with deep physical insight could be expected to pay off. One route was investigating the occurrence of singularities in spatially homogeneous but anisotropic universe models, and (inspired by Dennis and informed particularly by Engelbert Schücking when he visited King's College, London) Stephen Hawking and I obtained useful results in this regard (Hawking and Ellis 1965), as did Larry Shepley (1965), who was working under Wheeler's guidance. However, these were rather special models and did not include the effects of inhomogeneities.

The *annis mirabilis* for the subject was 1965 – the same year the CMBR was discovered – when Roger Penrose (1965) published his extraordinarily innovative paper on the existence of singularities at the end of gravitational collapse. Working very much on his own, he combined methods from topology, geometry, and analysis to show singularities would occur under realistic astrophysical conditions, provided a generic energy condition was satisfied by all matter and fields present. The key geometric concept he introduced was a *Closed Trapped Surface*: a two-sphere in spacetime such that light emerging and expanding outward from the sphere had an area that decreased with time, instead of increasing as happens in flat spacetime. This would generally be associated with the existence of an event horizon, whose formation would show a black hole had come into being. When such a closed, trapped surface existed and some auxiliary conditions were satisfied, it was inevitable that a singularity would occur; and these conditions would be likely to occur in gravitational collapse situations, because they occur in Schwarzschild spacetimes. Because the requirements of satisfying the energy conditions and existence of a closed trapped surface are both *inequalities*, they can occur in realistic real-world situations: they are stable to perturbations of the model. Thus his theorem gives the needed kind of generalization of previous results from special geometries to generic situations: these conditions implied existence of a singularity, a spacetime boundary.

The paper, though clearly written, was obscure both because it was very brief and because it introduced a combination of new mathematical techniques into general relativity studies that were not in common use at the time. This led to a flurry of activity in which relativity research groups like that at DAMTP in Cambridge, including Sciama, Brandon Carter, Stephen Hawking, and myself, and that at King's College London, including Hermann Bondi and Felix Pirani, scrambled to get up to speed. We ran a series of joint seminars in London and Cambridge where we explained to each other what Penrose had done and the underlying mathematical ideas, with useful input from others such as Bob Geroch and Charles Misner. Penrose himself of course also gave seminars on the topic, but one needed more background than we had at that time to comprehend fully what he had done.

Stephen Hawking's first major insight was that a closed trapped surface would occur in a time-reversed sense in the early universe, and this could be used to extend Roger Penrose's theorem to the cosmological context. This followed from Fred Hoyle's discovery that there would be a minimum angular diameter for the observed size of an object of fixed size as it was moved back to earlier and earlier times in cosmology. The result was codified in Allan Sandage's (1961) magisterial paper on using the 200-inch telescope at Palomar for cosmological observations, and so was by then well known. This feature is obscured in the usual conformal diagrams used to indicate causal relations clearly; it is obvious when one uses proper distance coordinates instead, revealing the true onion-like shape of the past null cone shown in Figure 4.51 (from Hawking and Ellis 1968; see also Ellis and Rothman 1993).

The figure shows that the past light cone refocuses as one goes back into the past, hence there are indeed time-reversed, closed trapped surfaces in the standard Friedmann–Lemaître models of cosmology. But this means they will also exist in perturbed models, implying closed trapped surfaces exist also in these more realistic universes, and so imply an initial singularity in these cases too. Stephen produced a series of theorems that applied specifically to the cosmological situation: under various slightly different sets of conditions, a universe that is expanding and filled with matter and energy obeying a physically acceptable equation of state must have had a beginning sometime in the past, regardless of any lack of symmetry today. In his words, "...time has a beginning."

This is a pretty important conclusion, so one should consider it carefully: this is a convincing argument for an initial singularity, but it is not an observational proof. What would be a good observational link? Enter here the CMBR and its recently discovered importance for cosmology. I do not recall specific seminars on the topic, but obviously it would have been widely discussed. One of the key things that Dennis insisted on is that a research group should have a fixed coffee time each day when they could regularly meet together, providing a natural setting for discussing the latest

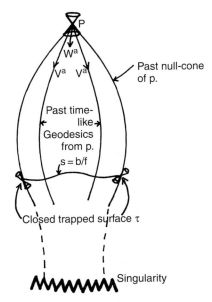

Fig. 4.51. Past-directed time-like geodesics from p starting to converge again before the surface $= b/f$ (which may be chosen so that the primeval plasma has combined to atomic hydrogen and helium, making space close to transparent to the CMBR).

papers and data with each other. Coffee tables were provided with white tops intended to be written on: at hand would be pens and table-top cleaners. Dennis would often come in at coffee time clutching some new scientific paper in his hand and ask, Did you see this? What does it mean? And the same would have happened with the CMBR. Dennis immediately understood the significance for cosmology of the CMBR and its interactions with matter, and in particular how it provided evidence for a hot early state of the universe.

But did it provide evidence of an actual singularity? The realization came from discussions *inter alia* with Dennis, Martin Rees, and Bill Saslaw that, unlike matter, the CMBR permeated *all* spacetime and so was a dominant dynamical feature not only in the early universe but in empty space in the recent universe. Hence one could see if it alone would imply existence of a singularity, a great advantage being that its nature and density were well understood, while that of the matter was much less clear: was it hot or cold, molecular or ionized for example? And just what was its density and atomic composition? What fraction was uniformly distributed and what fraction clumped? The beauty of theory based on the CMBR was that its nature was so simple: being blackbody, just one number sufficed to determine all its properties. And its high degree of isotropy showed it was very smoothly distributed.

Stephen asked me to help work this out. He had developed a singularity theorem where by imposing a slightly stricter local condition than existence of a closed trapped surface (*the reconvergence condition*: there was reconvergence in the past of all time-like geodesics through one spacetime point), he relaxed a global requirement of Penrose' original theorem: existence of a Cauchy surface (a space-like surface that intersects every time-like and null curve in the universe). This auxiliary requirement was reasonable in the context of local gravitational collapse, but not in the context of cosmology. The question was if we could show the reconvergence condition was true in a realistic universe model. We knew a lot about the geometry of time-like curves and null surfaces, and particularly the focusing power of matter – this had been made clear by Rainer Sachs, Jürgen Ehlers, and Roger Penrose. So could this radiation alone imply sufficient focusing to imply the reconvergence condition? Yes it could. Its near isotropy showed the universe was almost Robertson–Walker in the observable region,[25] so one could base one's calculations on such a geometry. Then where would reconvergence take place: before the last scattering surface, and so in the domain we can actually see, or beyond it, and so in the domain inaccessible to direct observation, where the universe might be quite unlike a Robertson–Walker geometry? Stephen's insight was to note that either (if the matter density was low) the radiation itself would cause the needed refocusing at a low enough redshift, or else (if the matter density was high) the matter that thermalized the radiation, as implied by its very accurate blackbody spectrum, would do so. In both cases reconvergence would occur within the universe domain we can see, and so where we pretty much understand the broad nature of the geometry; and this would be true even if the radiation were rescattered by reheated intergalactic gas at relatively recent times. The result, being an inequality, would remain true if the universe were not exactly Robertson–Walker, but something like it in the observable domain, and possibly quite unlike it in the hidden domains at very early times (rotation and shear might dominate there, for example). So the mere existence of the CMBR with near isotropy and a thermal spectrum would do the job of ensuring reconvergence, without requiring exact symmetries of the spacetime and without knowing details about the matter present. Thus Hawking's exact mathematical theorem applied to the real universe if the matter obeys the energy conditions, because the existence of the CMBR shows its geometrical conditions are satisfied. The CMBR alone would, therefore, show that a cosmological singularity – a start to spacetime – must exist (Hawking and Ellis 1968).

---

[25] See Section 4.2.2 of Ellis (2006) for a recent discussion and references.

Dennis was very pleased with this result, which was followed by a similar but more elegant calculation given in the summary volume *The Large Scale Structure of Space-time* published in 1973 (see Hawking and Ellis 1973, Section 10.1). It was of course recognized at the time that this result must not be taken too literally: it is based on general relativity theory, which will break down if a quantum gravity epoch is encountered in the early universe. Indeed the implication was generally taken to be that what was predicted in physical terms was that a quantum gravity epoch could not be avoided in the early universe: effects like rotation could not turn the universe around at fairly early times so that it never entered such a domain. Whether quantum gravity effects would avoid a singularity in the quantum gravity epoch was unknown (and remains so to this day, but with some evidence from loop quantum cosmology that this is possible).

Does the result still hold today? Not really. Guth's (1981) introduction of the inflationary universe idea established that an effective scalar field could be a plausible dominant form of energy in the very early universe, which would violate the energy conditions of the singularity theorems; indeed this is what makes the accelerating expansion of the inflationary epoch possible. This has become the dominant paradigm of present-day cosmology (e.g. Kolb and Turner 1990; Dodelson 2003). The implication of singularity existence, therefore, no longer follows, because one of the major conditions for the singularity theorem to hold is no longer believed to be true: it fails at times when quantum field effects dominate. Indeed it is possible not merely that there was no singularity, but even that there was no quantum gravity domain: the universe might never have reached the densities needed for such a domain to occur. Explicit examples of Eddington–Lemaître type universes where this is the case can be found in Ellis and Maartens (2004). These examples have been criticized because they start off in a rather special state (at very early times they are asymptotic to an Einstein static universe); and indeed the choice may ultimately be between a spacetime singularity or a very special initial state. Which is more undesirable is a philosophical argument; whether one can observationally discriminate between them is an open question.

However that argument works out, acceptance of the inflationary universe kind of dynamics means that the existence of the CMBR no longer necessarily implies the existence of a singular initial state at the start of the universe. Its existence does, however, still imply not only that there was a hot big bang era in the early universe, but that this era extended back till times when quantum fields became dynamically dominant. That remains an important conclusion.

4.12 Measuring the CMBR anisotropy

4.12.1 Ronald N. Bracewell and Edward K. Conklin: Early cosmic background studies at Stanford Radio Astronomy Institute

Ron Bracewell was at Stanford University from 1955 and at the time of his death in August 2007 was Professor Emeritus. Ned Conklin was a graduate student at Stanford at the time described here. After a number of years at NRAO and Arecibo, he cofounded FORTH Inc. and worked in the field of scientific computer programming until retiring recently.

When the existence of the CMBR was announced in 1965, we discussed various kinds of measurements that might be made. Ned had been looking for a thesis topic (Conklin 1969), and this was an interesting and brand new field. The most obvious measurement was the absolute amplitude at one or more new wavelengths, but that would have been an extremely difficult experiment. It is relatively easy to make a measurement of the apparent temperature relative to that of a known absorber when an antenna is pointed at empty sky, but that is only the start. Then all the other possible sources of antenna temperature such as atmospheric losses, losses in the system and the antenna, unwanted pickup in the side lobes, etc. must be accurately estimated or measured and subtracted from the observation to yield the residual (if any) due to the CMBR itself. In nearly all cases these unwanted sources of radiation are substantially higher than the few degrees of the CMBR, and so the subtraction process is prone to errors. Further, we recognized that, once a measurement had been published at one frequency and a thermal spectrum was posited, there would be the unconscious bias toward confirming it at other frequencies, leading to a situation where one might think of extraneous radiation sources contributing to a measured value, subtract their effect until the CMBR temperature was reached and then stop looking quite as diligently, so that a secondary measurement would not be truly independent.

Discarding the absolute value left measurements of the angular structure (if any) of the radiation field. In their original paper Penzias and Wilson (1965a) reported that the CMBR was isotropic, but the precision of their measurements was low. Two possibilities presented themselves (Bracewell 1966; Conklin 1966), each with its own set of experimental problems – measuring the fine scale on the order of arcminutes (inhomogeneity), and measuring the large scale on the order of degrees (anisotropy). In the end we pursued both.

The weak CMBR detected by Penzias and Wilson was reported as isotropic. So it was with cosmic rays, but they were quickly found to vary with altitude, latitude, and season, so we thought it likely that the cosmic background also might prove to be not isotropic. Any anisotropy of the remote cosmic background would be hard to address but in the summer of 1967, when these measurements were undertaken, Stanford Radio Astronomy Institute was by chance well equipped to look for inhomogeneity in the newly announced microwave radiation. A 735-ft minimum-redundancy microwave array of five interconnected 60-ft paraboloids and associated electronics was already under construction (Bracewell *et al.* 1973). With particular attention to symmetry and matching at a frequency of 10,690 MHz, a pair of identical feed horns was installed at the focus of one 60-ft dish, one horn pointing to the zenith, the other down on the level paraboloid. The paraboloid was tilted very slightly off zenith so as to cause the radio source Cygnus A to pass through the center of the fixed antenna beam and serve as a calibration source.

A waveguide tee junction, built to connect the two horns, incorporated a two-state ferrite circulator: in one state the signal received from the radiometer was that from the upward-looking horn whose beamwidth was about 80°, and in the other state the signal was from the horn that fed the paraboloid forming a resultant beamwidth of about 13′. The circulator was subjected to square-wave switching at 37 Hz. The signal delivered to the output arm of the tee, after preamplification, mixing, intermediate frequency amplification, and detection thus consisted of a noisy square wave jumping between the antenna temperatures in the 80° beam and the 13′ beam formed by the paraboloid. One can understand that in the state of electronics of the day extreme care was needed to deal quantitatively with such a small jump.

The response to the upward-pointing horn was expected to be 900 K (the noise temperature of the radiometer) plus the mean intensity of the cosmic background, now known to be ~2.7 K, plus a few more degrees from any atmospheric radiation and ground radiation in the antenna side lobes. The response of the horn looking into the reflector and focused into a pencil beam 13′ wide would be nearly the same and the difference would reveal any local departure from isotropy as the sky rotated overhead. Clearly, constancy with time of the 900-K noise temperature of the receiver is also of the essence of the experimental design, as well as constancy of the atmospheric radiation.

The experimental differential results did show both a long-term trend (over a period of hours), which was evidently caused by thermal effects in the radiometer and switched circulator, and short-term fluctuations, which did not repeat from day to day and correlated fairly well with changing atmospheric conditions. The long-term trend was removed by subtracting a half-hour running average from the data; the effect of this was to limit the inhomogeneity analysis to beamwidths less than about 7.5° (30 min of right ascension), not a serious limitation. The short-term atmospheric fluctuations, being random, could not be removed and simply added to the effective rms antenna temperature. Out of 65 days of observation, the data from 29 days were rejected entirely as having excessive short-term fluctuations, and the remaining 36 days of data were averaged and analyzed.

The results did not show any noticeable inhomogeneities in a strip of sky located at 40.6° N declination and from 10^h to 19^h of right ascension. The mean observed rms antenna temperature of the data, although somewhat higher than expected, decreased as $N^{-1/2}$ up to the limits of our data, where N was the number of days in the average, indicating that the results were purely statistical and that no floor of intrinsic CMBR inhomogeneity was limiting the observed results. At the $13'$ beamwidth of the paraboloid, the 3σ limit to any intrinsic inhomogeneities in our data was about 7.9 mK. By integration of the data, even lower limits could be set for larger angular scales Θ up to a few degrees, and limits could also be set for angular scales smaller than $13'$, recognizing that their amplitudes had been reduced by the spatial smoothing in the antenna beam. The final published results were $\sigma = 103/\Theta$ for $0' < \Theta < 13'$, and $\sigma = 28\Theta^{-1/2}$ for $13' < \Theta < 120'$, where σ is the rms intrinsic inhomogeneity in millikelvins (Conklin and Bracewell 1967). These limits were low enough to be useful constraints on various competing theories of the CMBR.

The very low noise radiometers widely available today were not common in the late 1960s and since statistical rms fluctuations unfortunately decrease only as the square root of the integration time, improving the inhomogeneity limits below the values quoted above would have required excessive observing time. Also, the borrowed equipment being used for the experiment was needed for the completion of the five-element array, so at this point we turned our attention to the anisotropy question. Here we had at least one likely positive result on theoretical grounds, albeit at a very low level. If the Earth were moving with respect to the rest frame of the CMBR, then there would be an increase in the CMBR temperature in the direction of motion, and a similar decrease in the opposite direction. (The theory is discussed below.) A speed of $300\,\mathrm{km\,s^{-1}}$ would result in a temperature change of only one part in a thousand, or about 2.7 mK, quite an experimental challenge in that era (and still not easy).

Measuring this kind of effect with any type of moving or scanning antenna was obviously out of the question. Varying ground pick-up in antenna side lobes and changes in atmospheric reradiation would far outswamp the few millikelvins from the CMBR itself. The only hope was to construct some kind of system that was fixed with respect to the ground and let the sky sweep through the beam. The dipole anisotropy would then appear in the antenna temperature as a sine wave with an amplitude proportional to the velocity and a period of one sidereal day.

The apex of the Earth's motion (if any) with respect to the CMBR was completely unknown. But unless it were closely aligned with the north celestial pole, there would be an equatorial component which could be measured most easily. As with the inhomogeneity experiment, a radiometer would need to be switching between two sources with antenna temperatures as nearly equal as we could devise, and yet be sensitive to the dipole anisotropy. Because of the extremely low amplitude, it was desirable that no integration time be spent on reference loads or reference patches of sky such as the north celestial pole.

We ended up with an antenna system consisting of two identical horns (Figure 4.52) with 14° beamwidths directed east and west respectively at a zenith angle of 30° and enclosed in a truncated conical screen to intercept radiation from the ground. To limit reception of unwanted thermal radiation from the surroundings these horns incorporated, at their rims, the short-circuited quarter-wave transmission line chokes familiar from microwave radar practice.

As the sky passed overhead any temperature difference between the two patches of sky pointed at would contribute a component to the recorded

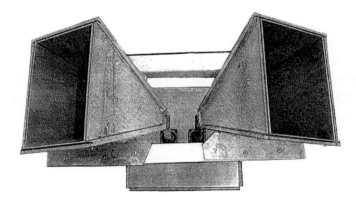

Fig. 4.52. Historic horns through which flowed the radiant energy (estimated at roughly half an erg during the integration time of 425 hours) that revealed the Sun's motion through the cosmos.

system temperature. As Earth rotates a given region of sky will pass first through the eastern antenna beam and then, several hours later, through the western beam. Since the polarity of the antenna temperature is opposite for the two antennas, this configuration is measuring the finite difference of the temperature distribution in the sky. It can be shown that the original sky distribution is recoverable from the finite difference record except for the mean value, which is not of interest here. The amplitude of any dipole anisotropy is reduced somewhat by the finite differencing, but the sensitivity remains higher than for a fixed reference system.

An extremely sensitive and stable radiometer was essential for this experiment. Receivers with very low system temperatures (such as masers) were just not in the budget, and it was difficult to keep them stable. Since the CMBR is essentially an extremely wide bandwidth thermal signal, an alternative for increasing the sensitivity was to increase the bandwidth of the receiver. We were fortunate in obtaining a very wide bandwidth, fairly low noise tunnel diode amplifier on loan from NASA, and the radiometer system was built around it. The wide bandwidth made the radiometer susceptible to airborne radar, but that was sporadic, and easily detected on a chart recorder and eliminated from the data.

The noise figure of the tunnel diode amplifier was such that it was theoretically capable of reaching an rms output fluctuation of 5 mK for a 1-min average. It took months of patient experimentation to construct a complete radiometer system that would stably operate at that level, but eventually we succeeded. A key feature that was added at this time was that the entire radiometer including the antennas and front-end electronics was mounted on a turntable. At 5 minute intervals the turntable rotated 180°, interchanging the east and west antennas. This second differencing removed the last small asymmetries and drifts in the electronics and left us with a system that appeared capable of detecting millikelvin variations in the CMBR.

Just a few days of observation near sea level at Stanford was enough to show that operation there was hopeless because of fluctuating atmospheric absorption and reradiation at the 8-GHz frequency of our system. We needed an observing site with extremely low water vapor, and that meant in general high altitude and ambient temperatures below 0 °C. Again we were fortunate in finding a reasonably accessible high-altitude facility at the University of California's White Mountain Research Station. The entire radiometer and associated data-taking electronics were installed in a small trailer and towed up to the Barcroft Station at latitude 37° N in October 1968. Here, at an altitude of 12,500 ft and atmospheric pressure less than two-thirds of that at sea level, the receiver performed acceptably close to its theoretical limit (although the human observer had difficulties!).

Because of Solar radiation in the antenna side lobes, useful data could be taken only at night time. Two month-long observations were made in October 1968 and April 1969 and were combined to obtain a complete 24-h record. It was immediately evident that nonthermal galactic radiation was affecting the data; this was removed by extrapolating an all-sky map of the galactic antenna temperature (from Pauliny-Toth and Shakeshaft 1962) to our 8 GHz observing frequency and subtracting it.

Details of the reduction procedure are given by Conklin (1969), along with a preliminary result based on the first observing run. The final reported result for the first detection of the dipole asymmetry was an amplitude of 2.28 mK (formal standard error ±0.68 mK, total estimated error ±0.92 mK) at a phase corresponding to right ascension 10^{h} 58^{m} in (Conklin 1972), indicating a Solar velocity in the equatorial plane of 255 ± 76 (formal), ± 103 (total) km s^{-1}. No significant smaller-scale anisotropies were seen in the data, which covered a region of sky from about 25° to 39° N declination and the full range of right ascension. The precision achieved in the light of determinations years later with much improved electronics was quite respectable.

The theoretical impact of the discovery that the Sun possessed an absolute motion through the universe was striking, both to the scientific community and to science writers (Sullivan 1969) addressing the general public. When our letter appeared in *Nature*, a letter to Bracewell from Professor Jakob L. Salpeter in Adelaide reported on a paper written by Kurd von Mosengeil (1907) where we read:

Alle Versuche, einen einfluß der Erdgeschwindigkeit auf die elektrodynamischen Erscheinungen festzustellen, haben ein negatives Resultat ergeben. Um dies zu erklären, haben H.A. Lorentz[1]) und in noch allgemeinerer Fassung A. Einstein[2]) das „Prinzip der Relativität" eingeführt, nach welchem es prinzipiell unmöglich ist, einen derartigen Einfluß aufzufinden.

All attempts to establish an influence of Earth's velocity on electrodynamic phenomena have given a negative result. To explain this H.A. Lorentz (1904), and in greater generality A. Einstein (1905), have introduced the principle of relativity, according to which it is in principle impossible to discover such an influence.

We now know more about special relativity. In 1969 textbooks of thermodynamics did not mention moving observers; one could only wonder what appearance the microwave sky would present. The two reports (Bracewell 1968) deduced that the brightness observed in the forward direction would increase and that the spectral distribution would still be that of a blackbody, but with a temperature apparently higher than that seen by a stationary observer. Naturally the spectral components would be shifted to higher frequencies by Doppler effect, but that alone would not result in a Planck spectrum; two other effects are involved. Stellar aberration would reduce the solid angles subtended by sky elements on or near the apex of Solar motion and this would result in an increase in brightness as measured in watts per square meter per hertz per steradian. Finally, the electromagnetic field strengths of both the electric and magnetic fields would be increased a little by the relativistic Lorentz transformation. Combining these three effects we found (Bracewell and Conklin 1967) that the spectrum would preserve Planck's blackbody form. In the forward direction, if the observer was moving at one thousandth of the light velocity c, the apparent temperature would be greater by one-thousandth than the temperature T seen by an observer at rest. For an observer moving with velocity v we found that, in a direction making an angle θ with the direction of motion, the observed temperature would be $T[1 + (v/c)\cos\theta]$. If the observer's velocity was not negligible with respect to c then the observed temperature would be $T[1 + \beta\cos\theta/\sqrt{1 - \beta^2}]$, where $\beta = v/c$.

Though a literature search did not uncover this result, it was reasonable to assume that it was known; the internal reports (Glints) were for the edification of the graduate students. It was, therefore, a surprise when Condon and Harwit (1968) reported that the spectrum seen by a moving observer would not be that of a blackbody. The internal memorandum was dusted off and submitted for publication to the appropriate journal but, being rejected, was resubmitted to *Nature* (Bracewell and Conklin 1968). Shortly after that we learned from Professor Salpeter's letter, that the same conclusion had been reached by von Mosengeil in 1907. Our discovery of the Lorentz–Einstein undiscoverable naturally made a wide impression. In due course Corey and Wilkinson (1976) of Princeton University launched many balloon flights and extended Conklin's results to a range of declinations, and Smoot, Gorenstein, and Muller (1977) of the University of California at Berkeley took data from many flights of the NASA-Ames Earth Survey (U-2) Aircraft. These more detailed endeavors neatly bracketed the original Stanford discovery (Figure 4.53). The remarkable detail discernible in the COBE satellite images of later years have continued to grip the lay imagination.

The Sun's motion might reveal itself in other ways: for example it should be evidenced by a dependence of density of distant galaxies on θ; they should be less tightly packed in the general direction of Pisces than in the direction

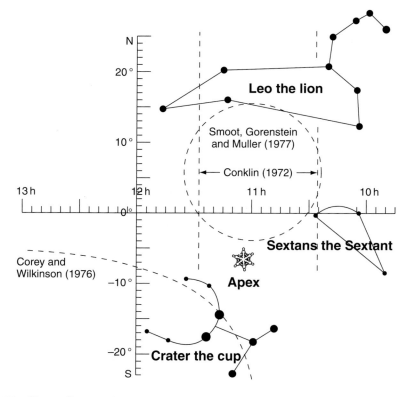

Fig. 4.53. Constellation chart showing a naked-eye observer on the bridge of Spaceship Sun that we are heading toward the point labeled "Apex." The coordinates have been refined by spacecraft observations. Also shown are the earliest reports from Stanford, Princeton, and Berkeley.

of Sextans. Counts of faint galaxies in the zones $\theta < 15°$ and $165° < \theta < 180°$ should differ by about a part in a thousand.

Meanwhile, the effort to discern a dipole component against a spatial noise background had left open the question whether spatial fluctuations existed, on a fine angular scale, in the background radiation. Careful observation did not reveal any such pattern but did allow an upper limit to be placed on the size of any such departures, as averaged over small solid angles. A technically remarkable low limit of 0.005 K for beamwidths broader than 10 arcmin was reported (Bracewell and Conklin 1967).

This was less than exciting for an aspiring PhD candidate but caught the attention of theoretical cosmologists and was quoted for several years as a constraint on assorted cosmological theories.

Thus in these early days, three significant contributions were made to the future and continuing studies of the fundamentally significant CMBR.

When the NSF withdrew support from universities in order to fund the NRAO the staff of the Stanford Radio Astronomy Institute dispersed, some to NRAO, some to radio astronomy elsewhere, and some to industry. The remaining graduate students eked out their dissertations with funding support for medical imaging, Solar thermovoltaic energy, theory of dynamic spectra, etc. The mothballed observatory was torn down on March 11, 2006.

Postscript: On Being the First to Know (EKC). In the summer of 1969 I was reducing all the data that had been taken in our two-month long observing runs looking for CMBR anisotropy. This involved hundreds of IBM cards filled with five-digit numbers, each one indicating the value of a one-minute integration from the radiometer, and a complex program that combined the observing runs, aligned and averaged the data, subtracted the estimated galactic radiation and Fourier-transformed the result. Because the amplitude of the dipole component was expected to be very small, there was no way to get a preliminary indication along the way of any statistically significant result; it was all or nothing. Late one evening I was at the Computer Center, ready at last with the run that would have the final result. I submitted the deck of cards to the mainframe computer operator, and in about half an hour, back came the stack of output paper. This would have two important numbers in it that represented about 18 months of effort: the dipole anisotropy and its standard error. Did I have something significant? Yes!

At that point I recalled an article I had read recently, I believe by Philip Morrison, that one of the joys of research like this is that for a moment, until you choose to tell someone else, there is something you know that no one else in the world knows. I went home that night elated to have had that happen to me, and although all of this occurred over 35 years ago I have never forgotten it.

4.12.2 Stephen Boughn: The early days of the CMBR – An undergraduate's perspective

Steve Boughn is Professor of Physics and Astronomy at Haverford College. His expertise includes the theory and detection of gravitational waves, extragalactic astronomy, and cosmology.

I'm not sure why Peebles and Partridge invited me to participate in this collection of remembrances since I was only an undergraduate in the 1960s and most of what was swirling around above my head was just that, over my head. Of course, I've learned a great deal about big bang cosmology and the

CMBR since then, and I have been involved in CMBR research off and on for the last 40 years. However, I'll endeavor to relate as accurately as possible what was going on in my mind back then in the hope that the recollections of a novice might be of some interest or at least provide some amusement.

I enrolled as an undergraduate at Princeton University in 1965, the same year as the announcement of the discovery of the CMBR. I had been enticed by special relativity early in high school and so arrived at Princeton committed to becoming a physics major. Astronomy had also been a source of fascination for me since I was very young. It was an exciting time in astronomy; quasars had just been discovered and pulsars were soon to follow. However, my entire knowledge of cosmology consisted of Hubble's discovery of the expansion of the universe and the notion that Einstein's theory was capable of "explaining" what was going on. Still that was what excited me most. Even so, for my first three years at Princeton I busied myself studying physics (no astronomy – even though today I'm a professor of astronomy, I've never actually taken a course in astronomy!). Then in 1968, at the end of my junior year, it was time to choose a senior thesis topic. Of course, I immediately pestered the people in Dicke's Gravity Group for possible projects. I still remember the suggestions. Dicke suggested a project having to do with Solar oblateness and its relation to Brans–Dicke theory. John Wheeler suggested two projects, one having to do with the dragging of inertial frames. (At Stanford, Francis Everett and my future PhD advisor Bill Fairbank were already deeply involved in what is now known as Gravity Probe B to test this effect.) Wheeler knew I was from Wyoming and so described this to me in terms of a cowboy's lariat. The second project he suggested was experimental. He thought it would be interesting to try to measure the advanced potential implied by the Feynman–Wheeler absorber theory of radiation using one of the CMBR radiometers of Dave Wilkinson and Bruce Partridge. Partridge and Wilkinson suggested a project also involving one of their microwave radiometers, but used in a more standard way to attempt to measure the dipole and quadrupole moments of the CMBR. This is what I chose and that decision had a lasting impact on my professional career. I hope I realized at the time how fortunate I was to have the choice (as an undergraduate) to work with such wonderful scientists, but I probably didn't.

I began reading about the CMBR right away but didn't know about general relativity or about cosmological models for very qualitative descriptions. So for me, my isotropy project soon became more a matter of getting the apparatus to work and to make a reliable measurement than of thinking about the cosmological consequences. I'm sure that others will make this same point, that is, that while the motivation of an experiment is extremely important, getting the experiment to work and making the best possible measurement usually take over and determine the measure of success – at least in one's own mind. If this weren't the case, I suspect that experiments would not be as successful as they are nor would science advance as rapidly as it does. I do remember being extremely careful to track down all the possible sources of systematic error, something I'm sure I acquired from Partridge and Wilkinson, and that has held me in good stead as an experimentalist.

The instrument was an 8.6-mm radiometer used by Dave to measure the CMBR spectrum and the plan was to compare the temperature of two points in the sky separated by 90° on the celestial equator. The 90° separation was picked to maximize the sensitivity to a quadrupole signal in the CMBR – I believe that one motivation was a possible anisotropic expansion of the universe, but to an undergraduate these were just words. The real motivation was to do an isotropy experiment at a shorter wavelength than other experiments to see if the previously measured CMBR isotropy was independent of wavelength. At the time, extragalactic sources for the CMBR had not been completely ruled out and one might expect that these sources would exhibit anisotropy at higher frequencies. I'm sure some big bang enthusiasts might say we were wasting our time; however, such null tests are all part of the important "network of measurements" that validate any scientific model. It now seems hard to believe that the data were recorded on many, many rolls of Esterline-Angus, pen and ink, strip chart paper that were painstakingly analyzed by hand. Of course, we found no anisotropy at our level of sensitivity, about 0.4%, a respectable limit but certainly not the best at the time. The design of the size and shape of the reflector was left entirely up to me. Since the beam was required to switch by 90° on the sky, the reflector should be switched by ±22.5°, or so I thought. That would only be true if the beams were reflected in a direction perpendicular to the axis of the reflector. They were not. As a result the beam throw on the sky was 64°, a value that prompted several inquiries after our paper was published (Boughn, Fram and Partridge 1971). It was both an embarrassing lesson for me and a testament to the involvement that Partridge and Wilkinson expected from their undergraduate students.

I assembled the radiometer in the Gravity Group haunt, a large area in the northwest corner of the basement of Palmer Laboratory, the home of the Physics Department. The place was an absolute maze with endless piles of equipment punctuated by several desks supporting stacks of papers that often spilled onto the floor. One evening I brought my two-month-old daughter with me while I worked on the radiometer. I tucked her in her carrier in a safe place among the jumble of apparatus as I worked on the radiometer. Some time later Dicke wandered in to search for something and uttered, "Well, what have we here!" as he happened, with delightful surprise, upon my daughter. During the day the place was a beehive of activity with three other undergraduates (Mike Smith, David Payne, and Bill Baron) working on CMBR-related projects, one of which was the first attempt to measure the polarization of the CMBR; two senior graduate students (Paul Henry and Karl Davis), who were building more radiometers, one of which was to be the first balloon-borne instrument; three more junior graduate students (Ed Groth, Jim Cambell, and Dave Fram), who had just arrived at Princeton; three postdocs (Jer Yu, Paul Boynton, and Neil Rasband), who were also working on the CMBR; and our leaders, Dicke, Wilkinson, Peebles, and Partridge. The excitement was palpable and the community spirit of the quest ever present as everyone helped with each other's projects. It was no wonder that I had a somewhat inflated notion of the importance to physics of what was going on there.

By the time I left Princeton in the spring of 1969, I was finally beginning to learn a little general relativity and began to think a little more deeply about the CMBR. After reading some of the fundamental papers on gravitational lensing I began to wonder if perhaps gravitational lensing of the CMBR by massive galaxies might not result in some anisotropy. I don't know if I mentioned this to Wilkinson or Partridge, but if I did I suspect they would have told me (with a smile) to go away and think about it some more. I did and after a laborious calculation was surprised that everything canceled out and lensing could not, in fact, generate any anisotropy whatsoever. It seems there is something called the "brightness theorem" that is valid even in the presence of gravity. Even though I was beginning to think more deeply I was, alas, still a novice. It turns out that collapsing or expanding concentrations of mass can generate anisotropies in the CMBR via the gravitational redshift, as was predicted by Sachs and Wolfe (1967), but it wasn't until three years ago that I and others detected this "integrated Sachs–Wolfe effect" and found its amplitude to be consistent with the existence of a cosmological constant, another thread in the "network of measurements."

I was involved in two more rounds of anisotropy experiments: one in the late 1970s and the other in the late 1980s. These took place well after the decade that is the subject of this book; however, I think there is an important point regarding these and other null CMBR anisotropy measurements of that 25-year period. It's understandable that the measure of success of a given measurement was, in part, determined by the upper limit it set on the level of the fluctuations in the CMBR. However, it seemed that some people (usually theorists) in the field took these limits very seriously, distinguishing between experiments that yielded upper limits that differed by as little as 20 or 30%. Most experimentalists realize that differentiating observations on this basis is nonsense. I once conferred with a statistician about the best and most robust statistic that I should use to set an upper limit to CMBR fluctuations. He seemed mystified by my use of statistics and responded that what I should properly do is to report the sensitivity of my instrument and then say whether or not I detected anything. Setting upper limits, he maintained, is not a proper use of statistics. Hmmm.... I once asked Wilkinson about what I described as over-interpreting the statistics of null results. He said not to worry. This situation was just the result of anxious cosmologists biding their time until CMBR fluctuations were actually observed. Sure enough, this came to pass. However, I do see a hint of the same problem returning, with some people judging cosmological observations by their usefulness in reducing the errors on the various parameters of the currently favored cosmological model and paying scant attention to the diversity of those observations, a diversity that will be absolutely necessary in ushering in any new understanding of our universe.

I know now that cosmology certainly wasn't considered to be one of the important areas of physics research in the 1960s and I was well aware then that there were other great discoveries being made. The professor of my very first physics course at Princeton was Val Fitch, who had the year before discovered CP violation in particle physics, a discovery for which he and Jim Cronin would later receive the Nobel Prize. The second semester of the course was taught by M. L. Goldberger, another notable figure in particle physics and future President of Caltech. Yet, from my limited point of view (in the biased environment of Princeton's Gravity Group) I perceived cosmology as one of the most important and fascinating areas of all fundamental science. I still do.

4.12.3 Karl C. Davis: Going the "easy" direction – and finding a lot of the wrong thing

Karl Davis, after a short stint with Princeton's Astrophysical Sciences Department, joined the staff of what is now the Pacific Northwest National Laboratory, where for more than 20 years he worked primarily in more "down to earth" remote sensing, from stratospheric to subterranean. He is now semiretired; a self-employed "tinkerer."

I'm reminded lately by a popular television show of events four decades and as many million meters distant (to within what I might then have called "cosmological accuracy"[26]) of a time and place when, while listening to learned debate on helium abundance, I would occasionally be very confidently told it had become too scarce at a nearby location, at least in liquid form.

The referenced location was a former pigeon coop atop a turret on the roof of Guyot Hall, shown in Figures 4.22 and 4.23. As a newly arrived graduate student in 1965, familiar with both microwaves and cryogenics, who'd "gravitated" to R. H. Dicke's research group, it became my task to haul liquid helium and nitrogen from the basement of the (old) physics building to the radiometer on Guyot. That is not far as a pigeon might have flown, but the last stages required manhandling the 20- or 30-l dewars up a long flight of steps to the roof, with care to minimize loss of expensive helium, then raising them 10 m or so by block and tackle. It comes back to me vividly when I see aerial views of the buildings that appear as the purported setting of Fox TV's series "House."

To this day when people ask: "what's physics?" I picture the sisal rope and wooden blocks we used as I reply: "ropes and pulleys, magnets, ..." As a small-boat sailor I was well aware of better alternatives, but graduate students suffer in silence, and this was about the origin of the universe, not just the structure of another molecule. (My previous work with microwaves and cryogenics had been in coupled electron–nuclear spin resonance.)

The person requesting the delivery and often assisting was Dave Wilkinson, who would later become my thesis advisor.

I think I was at least dimly aware even then that I was drawn to observational cosmology, and other activities of the Gravity Group, as being as close as I could get to philosophy while working with real, tangible things: hardware. "Fundamental Questions" and playing with elegant instrumentation both fascinate me. I say "playing" without apology, believing with Martin Buber: "Play is the exaltation of the possible."

With growing confidence that the CMBR had the predicted blackbody spectrum, the most logical approach for further measurements was to use higher frequencies, both to confirm the spectrum and to work with a stronger signal. Also there was already interest in mapping its spatial distribution, so the angular resolution of antennas began to be a consideration.

There were two very significant practical considerations against higher frequencies, however. The state of microwave technology was then such that receiver noise increased rapidly with frequency so it wasn't always clear an

---

[26] I at that time understood that to mean something like: "the order of magnitude is probably right, maybe plus or minus a few." It's a testament to the enormous progress in the field that this no longer seems even an amusing caricature.

improvement in signal-to-noise ratio would be achieved. More fundamentally, atmospheric absorption and consequent emission, particularly from water vapor, is also stronger at higher frequencies, and water distribution is notoriously variable.

It was not obvious higher frequencies were the only or even best way to go, particularly for spatial mapping. Professor Wilkinson and I decided my thesis research would be a measurement of the angular variation at 5 GHz (6-cm wavelength) – a factor of 2 down from the initial Princeton measurements. Not only were low-noise receivers readily available, but water vapor absorption is substantially less. We were both well aware that radiation from the galaxy, at least looking through the plane of the disk, is comparable to 3 K at this frequency, and higher at lower frequencies. 5 GHz seemed a sensible minimum.

I was particularly pleased, for esthetic as well as practical reasons, that the frequency allowed use of the same "Dicke switch" microwave configuration used in the 10-GHz radiometer, an arrangement not then possible at higher frequencies since a critical component wasn't available for them. This configuration allows perfect geometrical symmetry of the inputs of the differential receiver – certainly the most elegant and arguably productive arrangement for measuring small differences between relatively large signals. (I was amused when the manufacturer of the key component insisted it wouldn't work as I wanted it delivered. We agreed on a configuration they could test "their way" which I could reassemble and use as I intended.)

A further consideration was that, with the hoped-for relative immunity from atmospheric absorption, the 5-GHz mapping might well be done from the ground, even conveniently near Princeton, despite atmospheric water there. (Thought was given to relocating the instrument to a high-altitude location if the approach proved useful. It was designed to withstand and even operate exposed to high winds.)

We also planned to run the measurement for a continuous 12-month period so solar, lunar, and local terrestrial radiation would move uniformly across the celestial sphere. There was also a hope, perhaps naive, that effects of local variations in water vapor might cancel to a helpful degree over that time. It was also a reasonable amount of time for "integrating down" measurement noise to achieve what was thought a useful sensitivity.

The scanning plan seemed pleasingly simple and direct. The radiometric difference between the north celestial pole and the surrounding areas would be measured by fixing one antenna on the pole while scanning the second at the same elevation, thus through the same length of atmosphere. At Princeton's latitude the resulting coverage is a little more than a hemisphere.

(The antennas were designed for use as far south as Hawaii, with thoughts of a possible volcano summit site, above much of the troposphere.)

Scanning was accomplished by rotating an off-axis parabolic reflector, the single moving component. The reflector was deep ("fast") enough that the opening of the feed antenna was almost immersed in the combination of the reflector and the ground screens affixed to its sides. The fixed, north-pointing antenna was identical. Circular polarization was used so, ideally, beam pattern, including side lobes, would be invariant with rotation. In fact it soon became clear there was a small, variable, azimuth-dependent offset which ultimately prevented measurement of any polar component of anisotropy. (Figure 4.54 shows the complete instrument, which is ∼3 m tall at the center.)

With the intent of operating for a full year automation was an obvious concern. For its time I thought the final system was impressive. Data were recorded directly on digital magnetic tape, which Princeton's main-frame computer could read. I could thus view my data within hours of taking it to the computer center, as I did most Monday mornings. Even the chain printer-produced density map seemed delightfully convenient at the time. Scanning was programmed to compensate efficiently for the variation in area of circumpolar regions with celestial latitude, albeit by plug-board.

In contrast to the adventurous observers who took themselves and fragile instruments off to exotic and frequently uncomfortable locations, or worried over the vagaries and frustrations of balloon flights, I only had to stop by

Fig. 4.54. View from north and slightly east of the 5-GHz "Isotropometer." Photo: Robert Matthews.

Princeton's Forrestal campus every morning and check my equipment. Other than the azimuth drive occasionally shearing a pin, the only interruptions were due to my being the sole user of the building housing the electronics, so power, heat, or air conditioning could and did fail unnoticed at times for as long as a weekend, or on the few occasions when the parabolic reflectors collected snow.

A few months into the planned yearlong run it became obvious I was getting a nice, low-resolution map of our Galaxy. The final result was, most clearly, an even more precise galaxy map. I did, through various schemes of attempting to subtract known galactic radiation and/or excising or de-weighting affected areas, convince myself and perhaps even Dave Wilkinson that I had measured a real, statistically significant dipole term in the CMBR, consistent with known motions and other microwave measurements then beginning to appear. However, in the face of the obviously strong galactic contamination I couldn't provide what either of us felt was a clearly persuasive argument for a positive result.

In the early stages of investigating a new territory, or research area, it's appropriate to explore every reasonable-seeming path, and it's inevitable that some will prove less useful. I think I showed clearly that, whatever the challenges of technology and atmospheric interference, higher frequencies are preferable for studying the CMBR ... at least until we can get well away from the Milky Way.

4.12.4 Paul S. Henry: Driven to drink – Pursuit of the cosmic microwave background radiation

Paul Henry is a member of the Technical Staff of AT&T Laboratories. His thesis experience at Princeton sparked a lifelong career in telecommunications research, which continues to this day.

The last thing I remember from that day is Dusty Rhoads and Gene DeFreece depositing me at my motel room. Dave Wilkinson, my thesis advisor, was waiting for me and was none too happy. "Where in hell have you been?" he demanded. "I've been looking all over for you. We have an interview with that reporter from the Hobbs paper in half an hour."

"Not now, Dave, please. You can cover for me." And at that point I passed out.

So that's what it had come to. After four years of grad school – most of them spent laboring in Bob Dicke's Gravity Research Group (the "GRG"), preparing for what was supposed to be a definitive experiment in observational cosmology – I had ended up in a Hobbs, New Mexico motel, tanked to the gills with 190-proof Everclear grain alcohol.

The trouble had begun about two years earlier, when Dave Wilkinson and Paul Boynton (a postdoc in Dicke's group) recognized that ground-based measurements of the isotropy of the CMBR would forever be plagued by atmospheric effects, especially emissions due to water vapor. High altitude was the key to success. After rejecting satellite- and aircraft-based platforms, they concluded that a balloon-borne experiment, flying above 99% of the water vapor in Earth's atmosphere, could be a cost-effective approach. A low-noise, wideband radiometer, slowly rotating to scan the sky, could collect enough data in a 10-h flight to yield a good estimate of the 24-h (dipole) anisotropy of the CMBR. (Ned Conklin and Ron Bracewell at Stanford University had already done a lovely experiment to measure the equatorial component of the anisotropy, reported in Conklin 1969, but the polar component, and therefore the total magnitude, was still unknown.) No one in the GRG had ever worked with research balloons before, but Dave had discovered that a federal agency, the NCAR, could provide the equipment and services needed to conduct high-altitude research experiments. He asked me to dig a bit deeper.

I had come to Princeton solely because a college professor had mentioned Dicke's work in cosmology as proof that you didn't need to be in the then-fashionable mainstream of big-time, high-energy physics in order to do exciting research. Believing then, as I do now, that less is more, this comment sounded like high praise to me. I had zero knowledge of cosmology (who's Hubble?); it was the possibility of doing small-scale experimental physics, not cosmology *per se*, that excited me. The experiment proposed by Dave and Paul was just what I had been looking for.

The next thing I knew, I was up to my eyeballs in the details of microwave radiometers, thermal design, telemetry, and failure-mode analysis. Dave and Paul apparently trusted me to manage the project on my own, or suspected that it was going to be a huge sink of time they would rather not invest. Either way, they gave me more than enough rope to hang myself.

As I labored through the endless details of design, construction, and testing of my equipment, I began to sense that my little project, though extremely modest by the usual standards of experimental physics, was part of a much bigger picture, a brewing revolution in cosmology. Time and again, I would overhear the big shots of the GRG (to me, anyone with a PhD qualified as a big shot) discussing things like scalar–tensor gravitation, helium formation, and radiation decoupling. Something momentous was clearly in the works. They would invite me to join their conversation, but I always demurred. With the myopia typical of so many grad students, I was happy to let the experts debate deep cosmological matters – all I wanted to do was get on with my experiment so I could satisfy Dave and get my degree. Truth be told, I was in heaven with my thesis project and wanted nothing more. Every circuit design, every test run was a labor of love. Long hours in the lab, deep into the night, gave me a sense of accomplishment that I had never known before. Leave the big picture to the experts; in my own small way, by performing the GRG's first balloon experiment, I was going to be a pioneer too. That was enough for me.

As with any experiment, almost nothing went smoothly. In most cases I managed to deal with problems as they arose, but occasionally I would get stuck. That's when Dave would step in to save me. My gondola design, for example, was as light as I could make it, but even so it was still far too heavy for the balloon we planned to use. I was stumped. Dave made some calls and found an expert in lightweight structures, who showed me how to use the elements of airframe design to cut the gondola weight in half. One problem solved; alas, countless others, from antenna design to radiometer calibration to battery management, still remained.

One by one, the problems yielded. The most memorable part of my graduate experience was about to begin. In late fall of 1969 I packed up my experimental apparatus along with a bunch of test gear and took it to the NCAR balloon base in Palestine, Texas, where I was greeted by Dusty Rhoads, the facility supervisor. He was friendly, but obviously very busy with other research groups preparing their own balloon experiments. He showed me to my assigned workspace, assured me that if I needed help, I could ask any of the staff, and then left to tend to one crisis or another. Despite Dusty's assurances, and despite the dozens of people working around me in that same building, I felt alone and, for all intents and purposes, totally lost. Luckily for me, Paul Boynton had arranged to travel to Palestine to help me get started. He met me the next morning, and together we unpacked my equipment and ran some initial tests that confirmed that the radiometer was in good shape. He could only stay a couple of days – for the rest of my time in Palestine I would be on my own – but before he left he introduced me to Rai Weiss and Dirk Muehlner, who had the workspace next to mine, and arranged to have them "babysit" me, to be available to help should an emergency arise. They were preparing their own CMBR experiment, an effort far more sophisticated and ambitious than mine (Muehlner and Weiss 1973a). On the one hand, I was delighted to be hooked up with guys who, unlike me, actually knew what they were doing. But on the other, I was utterly intimidated by their clear command of just about everything I wished I knew, but didn't. In any case, I needn't have fretted. Not only were Rai and Dirk always eager to offer advice and assistance whenever I needed it, it turned out that they were delightful dinner companions as well. We were all staying at the Sadler Motel, and as there weren't many eateries in Palestine, we usually had dinner together at the motel restaurant. The menu was limited – basically your choice of fried chicken, chicken-fried steak or chicken-fried pork chop – but the meals were a treat all the same. Rai was a gifted raconteur. Drawing from a bottomless well of stories about goings-on in the MIT physics department – tales too scandalous (and delicious!) to repeat here – he kept us enthralled for hours.

As I continued preparations at the balloon base, I was delighted to discover that, despite my hurried introduction to the facility by Dusty Rhoads, both he and his entire staff were, in fact, absolutely committed to "customer service" for their visiting researchers. One day, for example, Earl Smith from the electronics shop stopped by to ask how things were going. I was generally in pretty good shape, I told him, but I had discovered that a telemetry interface box I had built back at Princeton wasn't compatible with NCAR's equipment. I would have to redesign and rebuild much of it on-site. He asked a few more questions, wished me well, and went back to his shop. The next morning he presented me with a new telemetry package that he had specially modified to be compatible with my experiment. I didn't have to do any redesign at all – just plug and play. Earl was typical of the entire NCAR staff. They would recognize when help was needed, step in without overstepping, and do whatever it took to move a project forward.

After a week of fixing one problem only to find yet another, flight day arrived, though not by my choice. Dusty told me the evening weather forecast was for clear skies and calm surface winds – conditions too good to pass up; I was to be ready for launch at sunset. (I needed a night flight to avoid microwave radiation from the Sun.) He introduced me to Gene DeFreece, the launch director, who ran through a final check-list. He nodded with approval when he saw that I had followed Rai's advice and secured a six-pack of beer in the telemetry section of the gondola. Over the course of the night at 75,000 ft, the beer would chill down to a temperature just right for drinking by the recovery crew, who would have followed the path of the balloon in a pickup truck, in order to retrieve the gondola after it parachuted to Earth. It was a little gesture to repay – or at least acknowledge – the courtesy that the NCAR staff had shown me over the past week.

Shortly before sunset Gene and his crew mounted my gondola on the launch truck and secured it to the huge, uninflated balloon stretched out along the ground. From the truck, he ordered his team to start filling the "bubble" at the top of the balloon with helium. I stood with Dusty at the edge of the launch pad as he directed the operation via walkie-talkie, alternating between Gene on the truck and the telemetry staff in the control room, who were synchronizing their equipment with the signals from my radiometer. When Dusty got word from Gene that the balloon was ready, he turned to me:

"Are you ready, Paul?"

"I don't know."

"We can't launch till you say you're ready."

"How do I know when I'm ready?"

... A long, tense silence ...

"You're ready now," he decided for me. "Let it go," he radioed to Gene.

Seconds later the bubble slowly started to lift the huge polyethylene bag connecting it to my gondola on the truck. As it climbed above the truck, the driver followed underneath until Gene sensed that the bubble was pulling up hard enough to support the payload, at which point he cut the gondola free. Liftoff!

After the stress of the launch, the rest of the flight was an anticlimax. I worked in the telemetry room, monitoring the radiometer, performing periodic calibration runs, and following the track of the balloon as it drifted east. The experiment functioned beautifully throughout the night. As expected, shortly after dawn Solar radiation appeared in the radiometer output; no more useful data could be taken. Dusty ordered cutdown. The pickup crew found the gondola immediately after it landed, reported that the beer was just right, and delivered the payload to Palestine that evening.

Dave had built enough money into our project budget to pay for a second flight if it should be needed, so I left my equipment in Palestine – just in case – and returned to Princeton to analyze the data. It didn't take me long to discover that all was not well. I traced an unexpected periodic component in the data to radiation from the Moon. The screens I had designed to shield the radiometer from this effect had apparently been inadequate. The effect was too small to have been seen in sea-level preflight tests, but in the near-perfect conditions at altitude it was clearly visible and strong enough to render much of my isotropy data useless. I asked Dave for permission for another flight. He agreed, and also volunteered to come along to help. (Or maybe to ensure that I didn't screw up again!) This time, we scheduled our flight for a moonless night.

I designed a new screen system and did some quick tests. Meanwhile, the NCAR balloon operation had moved to its winter site in Hobbs, New Mexico, a few hundred miles west of Palestine, in order to accommodate stronger high-altitude winds. They had brought my apparatus from Palestine, so when Dave and I arrived in Hobbs, everything was ready for us. I constructed a screen according to the design that I had developed in Princeton, but as soon as we started testing, Dave spotted problems with it. Moonless night or not, the screen had to be right. I couldn't explain why it wasn't working, but Dave suspected diffraction from the screen edge – a problem he had seen in some of his earlier CMBR experiments. He repositioned the edges and the trouble disappeared! A problem that could have held up launch for days was dispatched in an hour.

Launch day for flight #2 arrived. I was as stressed for this one as I had been for flight #1, but this time I at least had the courage to say "Go." And go we did. The flight was uneventful – minor problems with thermal control and telemetry, but the backup systems took over and got us through the night. The next morning, after cutdown, we all felt pretty well satisfied. The balloon crew wanted to celebrate, as they often did after a successful flight, but Dave and I were exhausted and just wanted to sleep. Dave managed to excuse himself and headed back to our motel, leaving me at the mercy of Dusty & Co. Half asleep, I vaguely recall riding with them for several miles in their NCAR pickup truck until we came to a roadhouse that appeared to be one of their favorite spots. Inside, at what I guess was their usual table, we were welcomed by the waitress, who was clearly glad to see us and greeted the crew members by name. Pointing to me, Gene told her, "He'll have an orange blossom." I had no idea what that was, but when it arrived it looked like a glass of orange juice, so I downed it in a gulp and asked for another. A few minutes later the table exploded in laughter when I reported that the right side of my face was numb. I don't know how long I continued to provide amusement for the group, but apparently Dusty and Gene finally decided that they had done enough to help me celebrate, and took me back to my motel, where Dave was seething. As he greeted me with a few choice words of disapproval, he no doubt realized that my plea to skip the interview with the *Hobbs Daily News* was in everybody's best interest, so he went without me.

One hangover and several airline connections later I was back in Princeton with stacks of telemetry tape, followed shortly by my radiometer, which arrived via truck. For the next nine months I processed data, analyzed instrumental quirks that I should have spotted long before, and tracked down anomalies. I had to throw out some suspect data, but fortunately there was enough left over for my thesis. And what did I have? I had the first measurement of the polar component of the anisotropy of the CMBR, as well as its total magnitude. As Conklin and Bracewell had concluded before me, the radiation is not isotropic. There is a hotspot in the sky. But it is no hotter than could be accounted for by the likely rotation of the local supercluster of galaxies, of which our Galaxy is a member. Our confidence that the CMBR is a cosmological phenomenon, not just a local effect, notched up a bit.

My work was good enough for a thesis and a brief piece in the journal *Nature* (Henry 1971). I felt relieved, and lucky to be finished. I knew my experiment had been far from perfect – there were many things I could have done better. I was thankful to Dave for showing mercy in my time of need. It wasn't until a decade later that I sensed that maybe my work had not been so bad after all. I was visiting the Electrical Engineering department at Stanford, where I passed a display case featuring Ron Bracewell's long career. There was a large collection of papers, propped open to key pages, showing a few of his many contributions. One of them showed the isotropy measurements that he had made with Ned Conklin. And there, right beside it, opened to Figure 1, was my *Nature* paper. I was in good company.

5

Cosmology and the CMBR since the 1960s

This chapter was written in collaboration with J. Richard Bond

In 1970 we knew that space is filled with a near-uniform sea of microwave radiation, the CMBR. That was interesting, an addition to the list of what is known about the universe. But, as we have been discussing, there was reason to think that this was a particularly important find, a fossil left nearly undisturbed from a time when our universe was very different from now – dense, hot and rapidly expanding. The main piece of evidence was the spectrum – the variation of the intensity or energy in this radiation with wavelength – which was known to be close to the thermal form one would expect for radiation left undisturbed from hot early stages of an expanding universe. It was encouraging also for this hot big bang picture that it offered an explanation of another observation, the large abundance of the element helium, which could be another fossil remnant from the early universe. But in 1970 this big bang interpretation was a large conclusion about the nature of the universe to draw from the exceedingly limited set of evidence we could bring to bear. Perhaps the CMBR originated in some other way, possibly by processes operating in the universe as it is now, as in the steady state cosmology. Or perhaps the radiation came from a time when conditions were different from now but different also from standard ideas about the big bang.

The early exploration of alternatives to the idea of a hot big bang is reviewed beginning on page 34. Debates such as this are a normal and healthy part of science, and there is a standard procedure for resolving them: sift through the evidence, set about gathering more of it, and explore how it all might fit together. The process is a learning curve: by trial and error seek ways to develop improved trials that may on occasion show how to do even better, and then still better. This process in the study of the

CMBR commenced in the 1960s and continued for the next four decades. It was motivated in part by the fascination of making difficult measurements that require invention of new ways of doing things. A few of the important CMBR measurements were done using existing general-purpose telescopes and detectors, but for the most part the measurements required instruments designed for specific purposes. Many of the studies of the background radiation were motivated also by the prospect of aiding resolution of a specific issue: is the CMBR really a fossil from the early stages of expansion of the universe? We believe another important, though seldom articulated motivation, was the more vague feeling that better measurements of this radiation might teach us something unexpected about the world around us. And indeed the continued improvements in the methods of measurement of the CMBR and in the methods of interpreting the measurements have led to deep advances in our understanding of the large-scale nature of the physical universe.

In 1970 there were reasonably convincing – though largely theoretical – reasons to believe that if the universe were expanding from a hot big bang, and the evolution were described by the physics in textbooks, then the remnant sea of radiation would be quite smoothly distributed and have an energy spectrum that is quite close to the thermal blackbody form shown in textbooks of thermal physics. The argument is based on the fact that in the early universe the energy density in this radiation would have been very much larger than what was in the matter. The consequence is that the behavior of the matter, however complicated, was not likely to have had much effect on the smooth distribution and thermal equilibrium of the much larger energy in radiation. But we have noted on page 29 that as the universe expands and cools the energy in the matter decreases less rapidly than the energy in the radiation: at the present epoch the energy in matter – and its equivalent in mass – dominates. Matter is capable of absorbing and emitting radiation, and these processes must have had some effect on the distribution and energy spectrum of the CMBR. Also, the distribution of matter around us is lumpy, and the gravitational pull of these mass concentrations also had to have affected the spatial distribution of the radiation. The theme for this chapter is the progress in the study of these two phenomena, the disturbances to the energy spectrum of the CMBR and to its spatial distribution as measured by the variation of the radiation received across the sky.

The studies of these disturbances required new methods of measurement and the interpretation of what is measured. The latter rests on fundamental theory that has proved to be strikingly durable. It includes Einstein's general relativity theory, which he completed in 1915, with the addition of the cosmological constant two years later. In the early 1930s one could have explained to Richard Tolman at Caltech central elements of the hot big bang picture, including the behavior of thermal radiation in an expanding universe and the gravitational growth of clustering of the distribution of mass. The physics is in his book, *Relativity, Thermodynamics and Cosmology* (Tolman 1934), though not emphasized in the way we do today.[1] The nuclear physics of light element production in the early universe would have been confusing in the early 1930s, but later in the decade Bethe and Bacher were discussing properties of the isotopes of hydrogen and helium that enter these reactions (Bethe and Bacher 1936).

The measurement techniques have gone well beyond what would have been familiar in the 1930s. Perhaps even less clearly foreshadowed by what was happening in the 1930s are the methods of numerical computation that came to play such an important role in handling data and interpreting the results.

In the 1970s the measurements heavily depended on the methods of visualizing, analyzing, and controlling electromagnetic radiation that were developed as part of weapons research during World War II. In the USA this war research was concentrated at the Radiation Laboratory – the Rad Lab – at MIT. Some of the greatest American physicists of that generation – Hans Bethe, Bob Dicke, Robert Pound, Ed Purcell, Julian Schwinger to name a few – laid the foundation for devices of practical importance, such as radar, using remarkable inventions such as the "Bethe hole coupler" and the "Magic Tee." When these methods were brought over to the study of the CMBR two of the principal actors were Dicke and Wilkinson, and some of the most worn books on their shelves were in the *MIT Radiation Laboratory Series*, particularly volume 8, *Principles of Microwave Circuits* (Montgomery, Dicke and Purcell 1948), in which so many devices and techniques are carefully and beautifully explained.

The Rad Lab series taught the CMBR community how to think about the properties of incoming radiation at wavelengths near 1 cm.[2] One needed to shepherd the electric and magnetic fields of the extraterrestrial radiation into a device capable of measuring the strength of the field oscillations in a selected range of frequencies. Using techniques developed by the 1950s one could accurately compute main beam patterns of antennas, side-lobe levels,

---

[1] The literature on the gravitational growth of mass clustering in an expanding universe was particularly confused, but in retrospect we can see the effect, computed in spherical symmetry, in Tolman's book.

[2] We follow the historical trend in specifying the wavelength in early parts of this story and the frequency in later parts. The relation is $f_{\mathrm{GHz}} = 30/\lambda_{\mathrm{cm}}$.

the radiation bandwidth, and the absorption and emission of microwave radiation in its interaction with materials.

The key to many of the earlier measurements is that the instruments dealt with a single spatial mode of oscillation of the radiation. But the exploration of properties of the CMBR demands observations at higher frequencies, wavelengths near and shorter than 2 mm. Here the methods of coherent detection described in the Rad Lab series, where the radiation is treated as an oscillating electromagnetic field, were not practical. At the time, one instead had to use cryogenic bolometers, solid-state devices that respond to energy received in radiation or any other form. There was much less practical experience with these devices; new techniques had to be developed. In the previous chapter Harwit and Weiss (commencing on pages 329 and 342) recall the challenges they encountered in learning how to shepherd the short wavelength radiation in the CMBR onto a bolometer while keeping out the much more intense radiation and other types of energy originating in local sources in and around the detector and Earth. Progress along this learning curve continued for the next several decades.

Essential to this work – and to the entire scientific enterprise – is the mutually reinforcing interaction of theory and observation. The interaction could be discerned in the study of the CMBR in the 1970s, but at a modest level (in research that was then at a very modest level of activity). As we have remarked, theory had led to the conviction that the clumpy distribution of matter that is heated by stars and by gravitational collapse must have disturbed the CMBR. The observers were very willing to search for such departures from an exactly homogeneous and thermal sea of radiation. But the theorists knew too little about how the matter grew warm and clumpy to offer specific suggestions about the effects on the CMBR that the observers might want to try to detect. The observers in turn were not in a good position to tell the theorists what effects they might best be able to observe: too little was known about the technology of detection of disturbances to the CMBR and about the challenge of distinguishing them from other sources of radiation ranging from the Milky Way and other galaxies down to the instrument itself.

In our recollection the community in the 1970s expressed no great sense of despair about the exceedingly uncertain state of understanding of what clues to the past might be present and measurable in the CMBR. It is difficult also to document any prescient and coherently expressed sense of elation about the wonderful things that would be learned if it were possible to sort out the puzzles. Rather, people saw that there are possibly interesting measurements to make and computations to be done, and they set about doing them.

5.1 The CMBR energy spectrum

Figure 5.1 (from Peebles 1971) shows what was known about the CMBR energy spectrum – the distribution of energy with wavelength or frequency – by the end of the 1960s. The solid curves in both plots (labeled "a" and almost hidden in the right-hand one) are a thermal blackbody spectrum fitted to the longer wavelength measurements. The vertical scale is logarithmic in the plot to the left and linear in the one to the right. That makes the three upper limits based on observations of interstellar molecules (in the technique reviewed in Chapter 3 commencing on page 42) appear to be much closer to the thermal spectrum in the left-hand plot.

At wavelengths longer than about 1 cm the measurements in 1970 were a good fit to the power-law form characteristic of the long wavelength Rayleigh–Jeans part of a thermal radiation spectrum. The ground-based measurements at wavelengths from 1 cm to 3 mm, and the interstellar CN measurements at 2.64 mm (shown in the left-hand figure as the small symbol on the curve and closest to the peak, and in the right-hand figure at 113 GHz as the small dot near the peak) showed the expected thermal bend down from a power law.

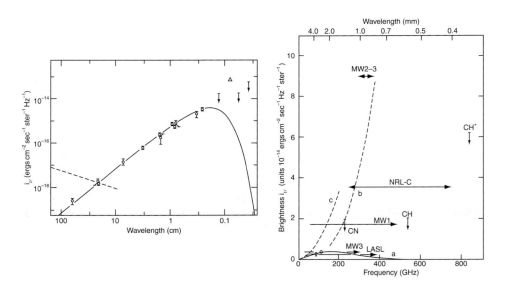

Fig. 5.1. Measurements of the CMBR spectrum in 1970. The left-hand figure shows that the data then were well fit by a thermal spectrum (the solid line) at wavelengths longer than 2 mm. The right-hand figure shows an expanded view of the short wavelength measurements. The various components and measurements are discussed in the text. The relation between frequency and wavelength is $f_{\rm GHz} = 30/\lambda_{\rm cm}$.

A demonstration of the steep drop in intensity at higher frequency characteristic of a thermal spectrum took another two decades. This requires measurements above the atmosphere, which is a strong source of radiation at these wavelengths, from a balloon, aircraft, rocket, or satellite. This is more difficult than on the ground. And we have remarked that there was less experience working with the bolometers used at these shorter wavelengths. It does not seem surprising therefore that in 1970 some of the spectrum measurements at wavelengths near 1 mm mistakenly indicated that the energy in the CMBR is significantly larger than in a thermal spectrum.

The situation at these shorter wavelengths is indicated in the right-hand plot in Figure 5.1. It is adapted from Bortolot, Clauser, and Thaddeus (1969), in a report of the three upper limits shown as downward-pointing arrows based on observations of the absorption lines from ground and excited energy levels of the interstellar molecules named in the labels. The measurement from the relative populations in the ground level and an excited level of CN at $\lambda = 2.64$ mm is shown as the open circle (at shorter wavelength and close to the thermal radiation curve). The dashed curve labeled "c" is the Rayleigh–Jeans power-law form extrapolated from the longer wavelength measurements; it did not fit the short wavelength data. In the steady state cosmology radiation produced in a narrow frequency range is spread by the expansion of the universe into a lower frequency tail with the power-law form $i_\nu \propto \nu^3$. Curve "b" is this form with amplitude adjusted to fit just under the CN limit. It did not look very promising even then.

More direct radiometer measurements of the CMBR energy spectrum are indicated by the broader horizontal lines in the right-hand plot. The lowest, labeled LASL, is from a group at the Los Alamos Scientific Laboratory (Blair *et al.* 1971). It is consistent with the energy in a blackbody spectrum. The measurement by the US NRL and Cornell collaboration, labeled NRL-C (Pipher *et al.* 1971), and the very preliminary result from Muehlner and Weiss (1970), in the three wavelength bands labeled MW, indicated larger energy densities. These measurements need not all have been inconsistent: one had to consider the possibility that radiation is present in narrow wavelength bands to which only some of the experiments were sensitive. But the apparent excess of energy over a thermal spectrum proved to be a result of systematic error in an exceedingly difficult measurement. Straightening this out occupied a large fraction of the experimental CMBR community up to 1990, when the COBE/FIRAS and the University of British Columbia (UBC) COBRA experiments (Mather *et al.* 1990; Gush, Halpern and Wishnow 1990) at last showed that the CMBR energy spectrum is very close to thermal, as we will discuss.

The bolometers used in the energy spectrum measurements at short wavelengths require a method of wavelength selection. Some of the early CMBR measurements employed filters designed to admit radiation in a band of wavelengths while adequately rejecting radiation at all other wavelengths. Others used a Fourier transform spectrometer (FTS) method. Moving mirrors cause radiation at different wavelengths to interfere at different rates. A bolometer detects this interference pattern. The Fourier transform of the output from the bolometer separates wavelengths, yielding the spectrum of the incident radiation.

Harwit, Weiss and their colleagues used bandpass filters in the work they recall in Chapter 4 on pioneering measurements of the CMBR energy from rockets and balloons. Both emphasize a great challenge: learn how to deal with the contamination of the measurement by radiation originating from Earth and from the many parts of the instrument. Muehlner and Weiss (1973a) showed that the CMBR energy spectrum does indeed turn over at short wavelengths, as expected for radiation with a thermal spectrum. Weiss, commencing on page 342, describes how they got there from the preliminary measurement shown in Figure 5.1.

The following year a group at Queen Mary College (Robson *et al.* 1974) reported an FTS spectrum measurement from a balloon launched from Palestine, Texas. Though there were difficulties with the calibration, and the atmospheric contribution was not well understood, their measurement indicated that the CMBR roughly follows a thermal spectrum over the peak to perhaps 50% accuracy. Later results would show their measurement was systematically high. The technique was promising, but the group did not repeat the experiment. Instead groups at UBC, Berkeley, and NASA/GSFC would improve the method. Mather's PhD thesis with Richards at Berkeley led to the design of the FIRAS FTS experiment on the COBE satellite mission.

We present in a series of figures a panoramic overview of the evolution of the art of CMBR energy spectrum measurements. The measurements are plotted as the effective thermodynamic temperature; the horizontal line is the temperature $T_0 = 2.725\,\mathrm{K}$ established by FIRAS. The expanding scale of the vertical axis, the temperature, is an indication of the progress.

Figure 5.2 shows the measurements at the end of 1975. Following a little more than a year after Muehlner and Weiss (1973a), the Queen Mary College and Berkeley groups were producing detailed measurements near 2-mm wavelength. The measurements did not agree with each other, or with the average of the lower frequency measurements. This situation set the stage for the next 15 years. Though it had become clear that the spectrum is approximately thermal near the peak, the deviations from thermal, if real, would require serious rethinking about the nature of the early universe.

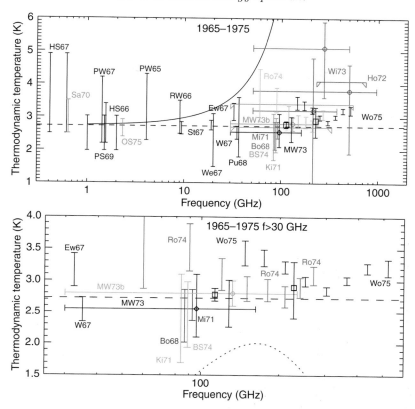

Fig. 5.2. CMBR absolute temperature measurements published between 1965 and 1975. The dashed horizontal line is the temperature established by FIRAS. The rising curve in the upper panel shows the inferred thermodynamic temperature of a Rayleigh–Jeans power law. The open square symbols are CN measurements. Bandwidths of bolometric measurements are indicated as horizontal bars. Upper limits are indicated by downward-pointing triangles at the ends of the horizontal bar. Measurements at the same frequency have been displaced slightly for clarity. Where measurements are crowded, the identification (in Tables A.1 and A.2) is placed near the bottom of the error bar. The lower panel shows an expanded view of the $f > 30$-GHz region. The dotted line is a scaled version of a 2.725-K thermal spectrum that indicates the energy peak.

Between 1976 and 1989 advances in technique produced the measurements across the CMBR spectrum shown in Figure 5.3. At frequencies between 1 and 100 GHz, a coordinated set of measurements was carried out by a Berkeley, Haverford, and Italian collaboration at the Barcroft Laboratory on White Mountain in California. Smoot et al. (1987) reported that the weighted average of these data is $T_0 = 2.7 \pm 0.05$ K. When the precise 25-GHz Johnson and Wilkinson (1987) measurement is added to the mix, one obtains 2.766 ± 0.022 K. This is high but close to what was established later, 2.725 K.

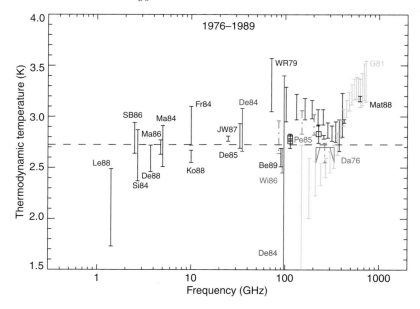

Fig. 5.3. Absolute temperature measurements published between 1976 and 1989. Note the accuracy of the 1989 CN measurements indicated by square boxes.

At higher frequencies, the second-generation Berkeley balloon-borne FTS measurement again showed an approximately thermal spectrum but at a temperature higher than that found at lower frequencies. The excess may have been from a high-emissivity section of the optics (Bernstein *et al.* 1990). In this time period first Peterson, Richards and Timusk (1985) and then Matsumoto *et al.* (1988) brought back passband-defined photometry with more sensitive detectors. Though precise these two measurements proved inaccurate.

The interstellar CN measurements are shown as the square symbols in Figures 5.2 and 5.3 (Thaddeus 1972; Hegyi, Traub and Carleton 1974; Crane *et al.* 1989; Meyer, Roth and Hawkins 1989; and others listed in the Appendix, p. 479). Steady refinement of the technique of this indirect method gave some of the then most precise estimates of the CMBR temperature. A weighted average of the data between 1976 and 1989 gives $T_0 = 2.766 \pm 0.019$ K at 113 GHz and $T_0 = 2.82 \pm 0.08$ K at 227 GHz.

In the same year as the Queen Mary College balloon flight, Gush (1974), at the University of British Columbia, reported attempted FTS measurements in rocket flights launched from Churchill, Manitoba, in 1971 and late 1972. Gush felt that it was necessary to get above the atmosphere to make a clean detection. The spectrometer was cooled to 4 K. The germanium bolometer detectors were cooled to 0.37 K with ^{3}He, a technique that would be adopted by other CMBR groups. In the first flight the instrument did not work properly. In the second the instrument was found not to be adequately shielded from radiation from Earth. Gush (1981) reported another attempt, launched from Churchill in 1978. This time the shielding of Earth radiation was successful. Unhappily, after release of the instrument package the rocket motor "slowly overtook the payload because of a residual after-burnout thrust." With the rocket motor in the field of view of the instrument, the correction for radiation from this source beclouded interpretation of the measurement. The corrected results are shown in Figure 5.3.

After the problematic 1978 flight, Gush worked with Mark Halpern, Vittorio De Cosmo, and Ed Wishnow to better understand how incident radiation couples to the instrument, how to make reference blackbody loads that are truly black (less than 0.1% reflective), how to make the reference loads change temperature and relax back to thermal equilibrium on times short compared to the 9-minutes rocket flight, how to control residual reflections, how to use bolometers at 0.3 K on a rocket in free-fall, and how to null an interferometer over a broad frequency band. Their instrument, COBRA, broke new ground on multiple fronts.

The Gush group recalls that

During the early era Gush launched experiments on Black Brant sounding rockets from the Canadian rocket range in Churchill Manitoba. The rocket was taken by rail, but people and equipment, including liquid helium, flew on a DC3 airplane from Winnipeg to Churchill. Now there is a somewhat rare phenomenon, which nearly every low-temperature physicist encounters at some time, where the neck tube of a helium dewar freezes with ice. This can be dangerous. In Gush's case, at altitude, the pressure within the storage dewar came to equilibrium with the atmosphere. As the plane descended, the relief valves on the dewar leaked, atmospheric gases were sucked into the dewar and water ice clogged the dewar neck tube. The clog is below any relief valves and there is a great danger of pressure building up from the boil-off of liquid helium. Since the plane wasn't very big and didn't have a lower level for baggage, people sat next to equipment. Gush noticed that the helium was no longer venting in the normal way, and managed to open the dewar valve assembly and jam a copper rod through the ice blockage. He thereby avoided an explosion that might have brought down the airplane.

Another story he told us about this expedition illustrates the importance of hockey in Canada at that time. There was a big hockey game on television, Churchill did not receive this transmission, and flights were booked by people flying to Winnipeg where they rented hotel rooms to watch the game. So Gush and his crew (his wife) had to stay a couple of extra days in a vacant Churchill.

In the summer of 1989, in preparation for the successful flight, we took COBRA to Bristol Aerospace in Winnipeg for vibration tests. A team from SED systems in Saskatoon, who built the command and telemetry part of the payload, was also there. The main thing to learn from the vibration test was whether our ^{3}He refrigerator would remain cold through the launch, so the test took place with the main instrument at a temperature of 2 K and the detectors at 0.26 K. A low-level sweep of frequencies showed that the resonance formed by the rigidity of the SED module and the mass of COBRA lay within the range of plausible rocket motor vibrations. The decision was to sweep in frequency at the maximum amplitude the table could provide. Our ^{3}He refrigerator survived the test, which was good news. However, most of the mechanical support for the spectrometer did not survive. Had this been the launch we would have collected useful data, but it was not the launch and we had to go home and rebuild the cryostat.

The final testing site was the David Florida Laboratory near Ottawa. This facility had a table capable of random noise vibrations. We believed this would better test the effect of the mechanical resonance between spectrometer and SED module. The test level was set at the acceleration seen with smaller payloads, not the force the motor would provide, but this time we were ready. During this test the phase of our scanning motor slipped a bit, but everything else was fine, and we launched the payload on 20 January 1990 without even opening the cryostat again.

After the successful flight, which was much gentler than either test, we discovered the reason our motor had shifted phase at DFL: the drive mechanism had completely torn, and was only coupling the motor to the mirrors by accidental friction. But the friction held through the launch, flight and landing. As John Mather said when he saw a photo of our shredded coupling, "good luck is the best kind."

In May 1990, Gush, Halpern, and Wishnow announced the results of their flight and published the result later that year. This measurement yielded the beautiful fit to a thermal blackbody spectrum indicated by the solid line in the spectrum in Figure 2.2.

As it happened, the NASA COBE satellite mission, which carried the FIRAS experiment designed to make the same energy spectrum measurement, also using a Fourier transform spectrometer, was launched on November 18, 1989, two months before the UBC launch. It too had been in preparation since 1974. One week before the launch of the UBC rocket, John Mather, speaking for the COBE team, announced the successful demonstration by the FIRAS experiment that the CMBR energy spectrum is close to thermal. Again, their measurement[3] follows the solid curve in Figure 2.2.

The COBE satellite project, including the FIRAS energy spectrum measurement, grew out of a meeting at the Goddard Institute for Space Studies near Columbia University. Michael Hauser recalls that

The meeting, held on September 27, 1974, was triggered by the NASA Announcements of Opportunity to propose Explorer mission science definition teams (AO-6,

---

[3] The UBC and COBE experiments compared the CMBR spectrum to the radiation from a black reference load held at temperatures close to that of the CMBR. That is, they tested the difference between the CMBR energy spectrum and a blackbody spectrum, not the energy distribution in a blackbody spectrum directly.

Delta or Scout-class) or Scout Explorer-class mission implementation (AO-7). Scientists attending the meeting included John Mather and Patrick Thaddeus from the NASA Goddard Institute for Space Studies; Rainer Weiss and Dirk Muehlner from MIT; David Wilkinson from Princeton; Michael Hauser from the NASA Goddard Space Flight Center (GSFC); and Joe Binsack, Deputy Director of the MIT Center for Space Research. The state of CMBR spectrum and anisotropy measurements and the desirability of searching for the expected cosmic infrared background (CIB) were reviewed, along with what could be accomplished with instruments in space. This group, except for Joe Binsack, and along with Robert Silverberg from GSFC, submitted a mission definition study proposal for a "Cosmological Background Radiation Satellite" in October 1974. The proposal was a Delta-class mission to measure the CMBR spectrum from 0.1 mm to 3 mm wavelength, and to measure the CMBR anisotropy using microwave radiometers at wavelengths of 3, 5, 9, and 16 mm and a broad-band 0.5-mm to 3-mm radiometer. The proposal also included an infrared instrument to search for the CIB. At the same time, a group from the Jet Propulsion Laboratory led by Samuel Gulkis submitted a proposal for a "Cosmic Microwave Experiment," a Scout-class mission to measure CMBR anisotropy with radiometers at 5.5, 9.6, and 16 mm. A group from the University of California Berkeley Space Sciences Laboratory, led by Louis Alvarez and including George Smoot, also submitted a proposal, "Observational Cosmology: The Isotropy of the Primordial Black Body Radiation," for a Scout-class mission to measure CMBR anisotropy with radiometers at 9 mm and another in the 2.5 to 3.8 mm range.

NASA's initial response to these proposals was to add Mather and Weiss to the already-formed mission definition team for a different science project, the Infrared Astronomical Satellite (IRAS), to investigate whether the CMBR spectrum could be measured on the same mission. Mather and Weiss reported in 1976 that the spectrum measurement could not be accomplished on the IRAS mission. NASA then appointed a Cosmic Background Explorer (COBE) Mission Definition Team consisting of members from all three "cosmology" proposals – Mather, Weiss, Wilkinson, Hauser, Gulkis and Smoot – to define a Delta-class mission concept that combined the CMBR and CIB investigations. By that time, the broad-band anisotropy instrument had been dropped because of the expected strong interference by radiation from interstellar dust. The initial report of the Mission Definition Team included the CMBR spectrum measurement, the CMBR anisotropy measurements at 3.3-, 5.7-, 9.6-, and 12.8-mm wavelengths, and the CIB search. The 12.8-mm radiometer was dropped before the flight development phase was started in 1984 to free up resources to improve the sensitivity of the 3.3- and 5.7-mm radiometers by operating them at 140 K. This change turned out to be essential for the discovery of the primordial anisotropy. In place of the 12.8-mm radiometer on COBE, NASA funded a balloon-borne anisotropy survey at 15.6-mm wavelength (Cottingham 1987).

In Mather and Boslough (1996), Mather recalls that he had to travel to Cambridge and spend two days talking with Rai Weiss before Rai would agree to participate. Rai felt the project would be too problematic, too expensive, and he already had too many irons in the fire. (I recently learned that he was already working on early LIGO gravitational wave detector concepts at the time.) David Wilkinson was pretty reluctant to get engaged with NASA. Since I had been hired by NASA only weeks before, Mather had no such problem with me.

Hauser mentions the three experiments in the COBE satellite mission. FIRAS, with Mather as Principal Investigator, showed that the energy spectrum of the CMBR is very close to thermal. DMR, with Smoot as Principal Investigator, was designed to look for variation of the CMBR temperature across the sky. As will be discussed, this yielded an important hint to the theory of the gravitational disturbance to the spatial distribution of the radiation by the growing clustering of the matter. Mather and Smoot shared the 2006 Nobel Prize in Physics for these accomplishments. This is the second Nobel Prize recognizing the importance of CMBR studies. The third COBE experiment, DIRBE, with Hauser as Principal Investigator, was designed to measure the energy of extraterrestrial radiation at wavelengths ranging from $0.24\,\mathrm{mm} = 240\,\mathrm{\mu m}$ to $1.25\,\mathrm{\mu m}$, and to attempt to discern the extragalactic component of that radiation. The DIRBE team successfully isolated for the first time the extragalactic background at 140 and 240 μm, and provided useful upper limits at the other wavelengths (Hauser *et al.* 1998). This has proved to be a valuable probe of the amount of emission of energy by stars, much of which is absorbed by interstellar dust local to the sources and reradiated at the long wavelengths to which DIRBE was sensitive. The FIRAS data independently revealed the submillimeter spectrum of the extragalactic background due to galaxies at wavelengths longer than the DIRBE range (Puget *et al.* 1996; Fixsen *et al.* 1998). This was a very successful satellite science mission.

Work on the UBC COBRA and COBE FIRAS experiments commenced in the early 1970s and succeeded in 1990. The differences in dates of launch and of announcements of results are negligibly short compared to the development time, more than 15 years. The UBC measurement lasted about five minutes. The FIRAS experiment took data for roughly ten months, permitting multiple systematic checks of the results. It detected the CMBR dipole and even the intrinsic anisotropy discussed in the next section (Fixsen *et al.* 1997). But first results from both experiments made the central point: the CMBR energy spectrum is very close to thermal over the peak. Both projects are to be honored for this historic advance. The considerable imbalance in the assignment of prizes and awards illustrates the capricious nature of that operation.

Figure 5.4 shows the CMBR energy spectrum measurements through 2006. The figure is a beautiful example of the maturation of experimental

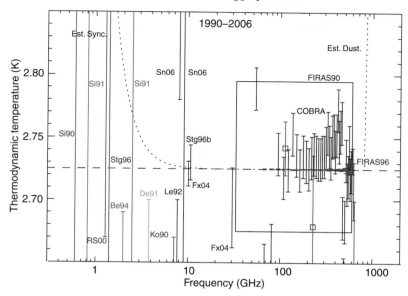

Fig. 5.4. Absolute temperature measurements published between 1990 and 2006. The box FIRAS90 is from the preliminary analysis of the COBE data (Mather *et al.* 1990), which indicated $T_0 = 2.735 \pm 0.06$ K. The tight band of points terminating at FIRAS96 is from the 1996 analysis by Fixsen *et al.* (1996). In the final analysis (Mather *et al.* 1999), FIRAS gives $T_0 = 2.725 \pm 0.001$ K. The COBRA points, and the unlabeled points near the FIRAS90 box, are from Gush, Halpern, and Wishnow (1990); they indicate $T_0 = 2.735 \pm 0.017$ K. Staggs *et al.* (1996b) and Fixsen *et al.* (2004) respectively report 2.721 ± 0.010 K and 2.730 ± 0.014 K near 10 GHz. The dotted line that curves up at 10 GHz is an estimate of the galactic synchrotron emission near the galactic poles (Fixsen *et al.* 2004). The dotted line near $f = 800$ GHz is an estimate of the dust emission (Fixsen *et al.* 1998). The square boxes show the more recent CN measurements (Roth and Meyer 1995, and others listed in Table A.1). Table A.2 lists references keyed to the symbols.

technique. Whereas for years the millimeter-wave region presented the greatest experimental challenge, it is now by far the best measured part of the spectrum. And at wavelengths where once the CN measurements gave the most accurate results the direct measurements have come to dominate.

The dotted lines in Figure 5.4 indicate estimates of synchrotron on the left and dust emission on the right from our Galaxy near the galactic poles. They can be compared to the early estimate by Doroshkevich and Novikov (1964) shown in Figure 4.5 on page 102. The error bars at $f < 10$ GHz are relatively large in part because of the problem of correcting for radiation from the galaxy. The fortunate situation for cosmology is that the bulk of the CMBR energy is between these two foreground sources. That is what allowed these precision spectrum measurements and the anisotropy measurements described below.

Prior to the COBE and UBC measurements the last experiment that indicated a possible anomaly in the CMBR energy spectrum, by Matsumoto *et al.* (1988), found that the energy in the CMBR at wavelengths less than 1 mm is 8% larger than would be expected from a thermal spectrum fitted to the longer wavelength measurements. That is not a large fractional deviation, but the amount of excess energy is large compared to what might be thought could be readily transferred from matter to the microwave background. What theorists made of this is illustrated in the proceedings of the conference, *The Cosmic Microwave Background: 25 Years Later*, held in L'Aquila, Italy, in June 1989. (As we have discussed in Chapter 3 the date for this nominal 25th anniversary of the discovery of the CMBR had to be somewhat arbitrary because the discovery of this fossil was a process, not an event.)

The review at this conference by Danese *et al.* (1990) shows that the evidence for excess energy was taken seriously, and it gives a good picture of what people who were attempting to understand the excess were thinking. The two most popular ideas were that the excess was produced by radiation from dust or else by scattering of CMBR photons by fast-moving electrons in hot plasma. In the latter process a photon–electron scattering event tends to add energy to the photon, increasing its frequency and decreasing its wavelength. This takes photons from the long wavelength part of the CMBR spectrum and adds them to the short wavelength part.[4]

What would be the source of the energy that heats the dust or accelerates the electrons? Starlight production in the universe as it is now is not great enough to produce the apparent energy excess in the CMBR. But it is not difficult to imagine that young galaxies were considerably stronger sources of starlight that warm dust. The origin of enough hot plasma to produce the apparent distortion of the CMBR spectrum also would require a new hypothesis. A popular thought was that the dark matter particles that would account for the dark mass in the outer parts of galaxies may tend to decay or annihilate, producing energetic particles in our visible sector that end up as fast-moving electrons. The decay would have to be slow enough that most of the dark matter is still present, but rapid enough that there are enough electrons to significantly disturb the CMBR. In their review paper Danese *et al.* (1990) show that the fast electron picture offers a good fit to the CMBR spectrum that seemed to be reasonably well determined by

---

[4] The analysis of this process was published by Kompaneets (1956). Longair (2006, p. 434) describes its origin in the Soviet nuclear weapons program. Weymann (p. 194) recalls early exploration of the effect on the CMBR. Zel'dovich and Sunyaev (1969) worked out the effect on the CMBR in detail; it is appropriately termed the Sunyaev–Zel'dovich or SZ effect.

Matsumoto *et al.* (1988). But by the time of publication of the proceedings of this conference the COBE and COBRA measurements had demonstrated that there is no excess to be explained: the relatively large amounts of fast electrons or warm dust are not wanted.

This is a beautiful example of the healthy – and seldom easy – progress of science. Experimentalists learned by trial and error and consultation with each other how to make exceedingly difficult measurements. Theorists played their part by exploring ideas that by the nature of the situation had to have been speculative. The notion of decaying or annihilating dark matter is exceedingly speculative, but it is not silly. The nature of the dark matter is a long-running mystery, and there is a long record of discussions of the idea that the dark matter may be unstable, decaying. The change in 1990 was the demonstration that the decay cannot have been fast enough to have produced a substantial perturbation to the CMBR. We can be sure that the effect of fast-moving electrons on the energy distribution in the CMBR is real. At the conference mentioned above Birkinshaw (1990) gave observational evidence that CMBR photons that pass through a rich cluster of galaxies are boosted in energy by scattering by the hot electrons in the intra-cluster plasma (as computed by Zel'dovich and Sunyaev 1969). This effect has become a valuable diagnostic of the plasma in clusters. We can also be sure that radiation from interstellar dust heated by starlight adds energy to the short wavelength part of the CMBR spectrum: the effect is observed in the Milky Way. That is, the processes Danese *et al.* (1990) were considering are real, though they operate at a lower level than was considered possible at that time.[5]

Mather recalls (in Mather and Boslough 1996) presenting the COBE measurement at the January 1990 meeting of the American Astronomical Society. His concentration on presenting a clear explanation was broken by the standing applause of the audience, an unusual event. But the measurement was an exceedingly important demonstration that the CMBR has the distinctive blackbody spectrum characteristic of relaxation to thermal equilibrium. This relaxation could not have happened in the universe as it is now: we know space is close to transparent at CMBR wavelengths because sources are observed at these wavelengths at the Hubble distance. The universe had to have expanded from a hot dense state in which it could have relaxed to equilibrium.

---

[5] For example, the energy transferred to the CMBR by its interaction with electrons in warmer plasma, in the SZ effect noted in footnote 4, is no more than about 1% of the present mean energy density in starlight. That is, the energy to make an observable perturbation to the CMBR spectrum is available, but it is in forms that do not significantly interact with the CMBR.

This is a deeply important result. But beyond that the spectrum is informative only in the sense that processes that could have disturbed the spectrum have proved not to be all that energetic.

5.2 The aether drift

The second part of our theme is the variation of the CMBR temperature across the sky that reflects the effects of departures from an exactly uniform universe. The variations are of two kinds: one produced by our motion and the other intrinsic to the CMBR. We consider the former in this section.

An observer moving through a uniform – homogeneous and isotropic – sea of thermal radiation sees that the radiation received from the direction toward which the observer is moving is warmer than the radiation received from the opposite direction. This is in effect a wind, like the wind in the face of a runner in still air. It is sometimes termed the Compton–Getting (1935) effect, after an analysis of the effect of motion through a gas of cosmic ray particles, or the aether drift (Peebles 1971) or, now most commonly, the dipole anisotropy. The last refers to the description of the variation of the CMBR temperature across the sky expressed as a sum over spherical harmonic functions.[6] At velocities of interest, motion through the CMBR produces only the first term in this sum, the dipole. The older name, aether drift, recalls the fact that the CMBR defines a preferred rest frame in which the dipole anisotropy is observed to vanish. It recalls the pre-relativity concept of an aether that defines an absolute rest frame that is independent of the contents of the universe. The measurement of our motion relative to the CMBR is a very different situation, of course: the relativity principle allows definition of motion relative to something, in this case the CMBR.

The CMBR spectrum is quite close to thermal. This means an observer moving relative to the preferred rest frame defined by this radiation sees that the radiation incident from any direction has a thermal spectrum with radiation temperature that varies with the direction of observation as[7]

$$T(\theta) = \frac{T_0 \left(1 - \frac{v^2}{c^2}\right)^{1/2}}{\left[1 - \left(\frac{v}{c}\right)\cos\theta\right]} \simeq T_0 \left[1 + \left(\frac{v}{c}\right)\cos\theta\right]. \tag{5.1}$$

Here v is the speed of the observer relative to the preferred frame defined by the CMBR, c is the speed of light, θ is the angle from the direction of

---

[6] Spherical harmonics are the analogs on the sphere of the plane waves used in a Fourier expansion in flat space; the sum is written out in equation (5.10) on page 443.

[7] This was demonstrated in Peebles and Wilkinson (1968) and Bracewell and Conklin (1968).

observation to the direction of motion relative to the preferred frame, and T_0 is the CMBR temperature observed in the rest frame defined by the radiation. The second expression is a good approximation if v is much less than c, as observed. In this limit the pattern of variation with direction is the dipole term $\ell = 1$ in the sum in equation (5.10).

It was not known in the 1960s but later found that the dipole anisotropy from our peculiar motion is much larger than the intrinsic anisotropy of the CMBR. There were in the 1960s reasonably good estimates of the velocity v in equation (5.1) that suggested the precision to aim for in an experiment to find the dipole anisotropy due to our motion. The Sun is moving around the center of the Milky Way Galaxy in a near circular orbit at a speed of about $200\,\mathrm{km\,s^{-1}}$, the center of the Milky Way is moving toward the nearest large galaxy, the Andromeda Nebula, at about $100\,\mathrm{km\,s^{-1}}$, and we and the Andromeda Nebula are falling toward the nearest large concentration of galaxies, the Virgo Cluster, at a few hundred kilometers per second relative to the general expansion of the universe. Thus it was not unreasonable to suppose that the motion of the Sun relative to the CMBR would be in the neighborhood of $v \simeq 300\,\mathrm{km\,s^{-1}}$. At this velocity the maximum departure of the temperature from the mean in the dipole would be

$$\delta T = Tv/c \simeq 0.003\,\mathrm{K} = 3\,\mathrm{mK}, \qquad (5.2)$$

or about a tenth of a percent. This amplitude was a reasonable – though not widely discussed – goal for the early searches for departures from an exactly isotropic sea of radiation.

Progress to detection of the dipole anisotropy pattern and then to its precision measurement is shown in Figure 5.5, from Lineweaver (1997). The top graph shows the convergence to a dipole amplitude that is close to equation (5.2): Earth is moving through the sea of microwave radiation at $370\,\mathrm{km\,s^{-1}}$. In the lower graph, showing convergence to the direction of our motion, the numbers increase with increasing date of the measurement. This can be compared to the plot on page 392 in which Bracewell and Conklin show early ideas about the direction of our motion. Lineweaver lists references keyed to the numbers in the figure.

Table 5.1 tabulates the progress in the measurement of the dipole. It depended on advances in detectors and in the methodology of their use. That was an important step toward the still more demanding exploration of the intrinsic anisotropy to be measured on smaller angular scales (and to be discussed in the next section).

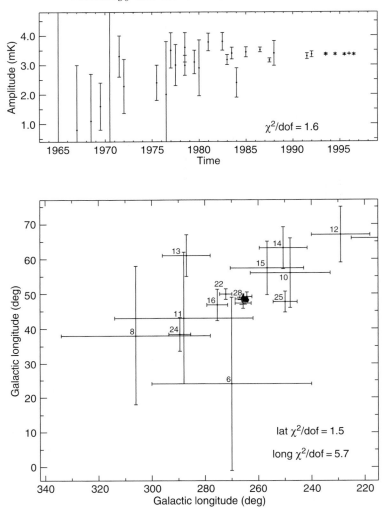

Fig. 5.5. Thirty years of dipole measurements (Lineweaver 1997). The numbers in the lower figure indicate chronological order. For example, the U-2 measurement is the tenth entry in Table 5.1. Early measurements could not pin down the direction.

The first measurement that constrained the dipole anisotropy, by Wilson and Penzias (1967), indicated $\delta T/T < 0.03$. They obtained it by comparing measured antenna temperatures in different directions in the sky. The precision of this method is limited by the difficulty of monitoring the inevitable drifts of sensitivity of the telescope and drifts of the noise originating in the instrument and from the ground. Comparing temperatures in different parts of the sky by rapidly moving the telescope so as to reduce the time between measurements would reduce the effect of these instrumental drifts, but it would exacerbate variations in instrumental noise caused by changes

Table 5.1. *Measurements of the CMBR dipole anisotropy*

Measurement	Frequency GHz	δT mK	α hours	δ degrees
Wilson & Penzias (1967)	4	<100	–	–
Partridge & Wilkinson (1967)	9	3 ± 6	–	–
Conklin (1969)	8	2.3 ± 0.7	10.3	–
Henry (1971)	10	3.2 ± 0.8	10.5 ± 4	-30 ± 25
Boughn et al. (1971)	35	7.5 ± 11.6	–	–
Davis (1971)	5	2.5 ± 1.5	10 ± 2	–
Conklin (1972)	8	2.3 ± 0.9	11	–
Corey & Wilkinson (1976)	19	2.5 ± 0.6	13 ± 2	-25 ± 20
Muehlner (1977)	60–300	~2.0	$\simeq18$	~0
Smoot et al. (1977)	33	3.5 ± 0.6	11.0 ± 0.6	6 ± 10
Smoot & Lubin (1979)	33	3.1 ± 0.4	11.4 ± 0.4	9.6 ± 6
Cheng et al. (1979)	19–31	2.99 ± 0.34	12.3 ± 0.4	-1 ± 6
COBE/DMR	30–90	3.353 ± 0.024	11.20 ± 0.02	-7.06 ± 0.13
WMAP	22–90	3.358 ± 0.017	11.19 ± 0.003	-6.9 ± 0.1

in mechanical stresses on the instrument. This is not a problem for observations in space, where part of the established strategy is to scan the sky by stably spinning the spacecraft.

Partridge (commencing on page 221) describes the strategy he and Wilkinson introduced: fix the instrument and scan the sky by reflecting the antenna beam using a moving mirror – a reflector consisting of a large metal sheet. Their setup is shown in Figure 4.24. The reflector swings between two positions. One directs the beam toward the north celestial pole, close to Polaris. The other swings the beam about 90° to the highest point in the sky along a circle 8° below the celestial equator. As Earth rotates the instrument scans the sky along this circle. The mirror switches the beam between pole and equator in a 30-minutes cycle. Also, following the scheme we discussed on page 46, a Dicke switch rapidly compares the antenna temperature to that of a smaller horn pointed to the zenith. That is, the large vertical beam from the smaller antenna serves as a reference load that suppresses the effect of instrumental drifts. Because the two directions of the tighter beam from the larger antenna are at the same distance from the zenith, the difference of antenna temperatures on the celestial circle and the reference point at the celestial pole aids subtraction of radiation from the atmosphere.

Switching strategies in this spirit appear in subsequent CMBR anisotropy measurement. Some of the details of the Partridge–Wilkinson experiment, along with similar comments on the later experiments, are entered in the overview of experimental advances in the Appendix commencing on page 487.

The Partridge and Wilkinson (1967) bound on the dipole anisotropy is $\delta T/T < 0.003$. The precision was limited by radiation from the atmosphere, which inevitably varies with time and position in the sky and so cannot be entirely cancelled by the switching strategy. The most troublesome source at wavelengths of interest is water vapor, which tends to be particularly clumpy, and tends to be abundant in the atmosphere where the data were taken, at Princeton University. The search for more suitable sites for anisotropy measurements is another recurring theme of anisotropy measurements. Partridge recalls (commencing on page 225) the move to Yuma, Arizona, which is dry at ground level. Unfortunately for this experiment, Yuma proved to have considerable atmospheric water vapor. Bracewell and Conklin (p. 385) moved their experiment from Stanford to the Barcroft Station at the White Mountain Research Station, where the thinner, cooler air contains much less water vapor. On page 401 Henry recalls his experience in moving the instrument still higher, in a balloon.

Conklin and Bracewell pioneered another important strategy: use Dicke switching to compare antenna temperatures of horns that are identical (as far as can be arranged) and, in their experiment, directed 75° apart at the same zenith distance. That allows rapid comparison of the CMBR temperatures at two well-separated points in the sky with minimal disturbance to the instrument. This method has been used in many subsequent experiments. As shown in Figure 5.6, their antenna arrangement, pictured in Figure 4.52, resembles the DMR instrument on the COBE satellite that detected the large-scale ripples in the CMBR. It also resembles the WMAP instrument that yielded the accurate anisotropy measurements that will lead us to the close of this story.

The Stanford White Mountain measurement, which was Conklin's doctoral dissertation (Conklin 1969),[8] sampled in the sky in a circle at 32° above the celestial equator. This measurement was designed to detect the velocity component perpendicular to Earth's axis. The amplitude in Table 5.1 is corrected to what would be measured around the celestial equator. There is a reasonable case that this was the first detection of the dipole. It agrees with the satellite measurements entered at the bottom of the table, which show that the component of the amplitude along the celestial equator is $\delta T = 3.34\,\mathrm{mK}$.

---

[8] We enter the numbers from his thesis; slightly different numbers are in his 1969 *Nature* article. The reanalysis in Conklin (1972) is entered lower in the table.

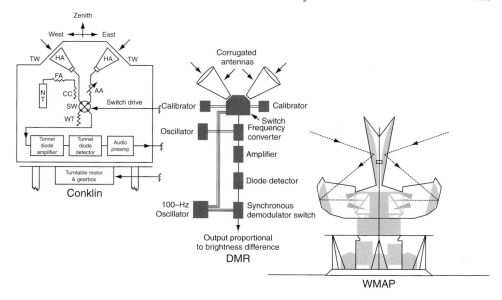

Fig. 5.6. The upper left figure shows the ground-based apparatus used by Conklin in his thesis. The central figure shows one of the DMR instruments aboard the COBE satellite. The lower right figure shows WMAP. All used the differential technique. The separation between horns (HA) in Conklin's apparatus and in DMR is 40 cm, and in WMAP the separation between the two outermost reflectors is 250 cm. Figure courtesy of Angela Glenn.

The Stanford dipole measurement required correction for radiation originating in the Milky Way Galaxy. The major contribution is synchrotron radiation (from relativistic electrons accelerated by the interstellar magnetic field; it is sketched as the dotted curve on the left side of Figure 5.4). The dashed curve in Figure 5.7 is Conklin's estimate of this contribution to the anisotropy signal from the 0.404-GHz Pauliny-Toth and Shakeshaft (1962) map. It is an extrapolation from what is measured at longer wavelengths, and it agrees reasonably well with the more modern estimate based on the Haslam et al. (1981) map. The top solid curve in the figure is the measurement. The CMBR dipole appears as the difference between the measurement and the galactic component and is shown in the bottom of the plot. The correction for the galaxy is large, and the reliability of the extrapolation was difficult to assess then. In 2008 the dominant sources of emission at 10 GHz are still not well quantified. We know there must be a contribution from thermal (free-free) plasma radiation and there might be radiation from spinning dust. One could not have assessed these effects in the early 1970s. Nevertheless the agreement with the WMAP dipole is strikingly close. Here we see yet another of the themes of the anisotropy measurements: devise strategies to reduce the effect of the galaxy.

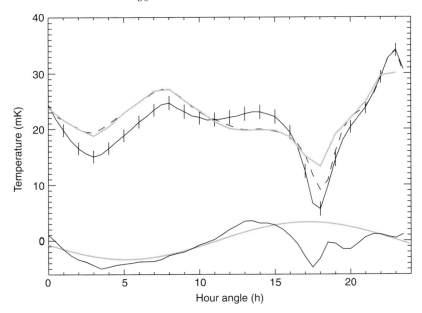

Fig. 5.7. The Stanford dipole measurement. The solid line near 20 mK shows the data from Conklin's thesis sampled at each half hour in right ascension along the circle sampled in this experiment. The dashed line is Conklin's model of the galactic contribution. It agrees reasonably with the extrapolated 0.408-GHz Haslam map, as shown by the thin gray line, except in the region of high emission. The lower solid black curve is Conklin's determination of the dipole anisotropy from his thesis. The lower gray line is the component of the WMAP dipole along the circle of Conklin's observations.

While Conklin and Bracewell went to White Mountain to minimize the effect of the atmosphere, Henry (1971), in his PhD thesis research with Wilkinson, flew his apparatus on a balloon. This experiment used the Conklin–Bracewell method of Dicke-switching between horns, here pointed at 45° to the zenith in opposite directions, as shown in Figure 5.8. The important new departure in this experiment is the rotator shown in the sketch: the instrument in its gondola rotated at one revolution per minute around the vertical axis so as to scan a circle in the sky. The Earth rotation during the 12-h flight at night allowed mapping roughly half the northern hemisphere, permitting a view of the dipole orientation. This strategy of rotation reappears in future measurements from balloons and in space. For example, COBE spun at 0.8 revolutions per minute and WMAP spins at one revolution per two minutes.

One might have thought that an instrument hanging from a balloon would not be significantly disturbed by rotation. But the Dicke switch that compared the antenna temperatures in this experiment was a magnetic device,

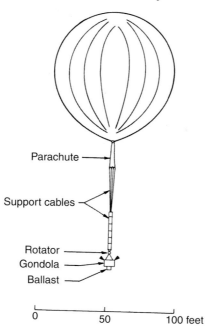

Fig. 5.8. The Henry (1971) anisotropy experiment. The arrangement of the two horns sketched as small black triangles at the top of the gondola can be compared to the two horns in the Conklin–Bracewell experiment shown in Figures 4.52 and 5.6.

and Earth's magnetic field caused an offset of sensitivity to the two antennas. As the instrument rotated this offset changed, producing an apparent anisotropy. This artificial 10–15 mK antenna temperature difference was synchronized to the angular position of the gondola relative to Earth's magnetic field. There may have been a similar sensitivity to Earth's magnetic field in the Stanford experiment, but that would not have mattered because the switch did not move relative to Earth. In the Henry experiment the magnetic field affected the north–south difference in sky temperature but allowed a measurement of the east–west celestial signal. In modern terms we would call this "projecting a mode out of the data," which is done frequently. But in Henry's case the bad mode was close to the one he was after, separated only in part by sky rotation and in part by knowledge of the source of the problem.

Like Conklin, Henry assessed the level of contamination by extrapolating the Pauliny-Toth and Shakeshaft map. Henry observed at 10 GHz, a slightly higher frequency than did Conklin. That reduced the galactic contamination by 40%. Also, a smaller fraction of his coverage was contaminated by regions of strong emission from the Milky Way. Although there were lingering doubts about the Henry measurement, it seems clear now that he also observed the real dipole amplitude and orientation. And the community learned to beware of systematics even in the apparently benign environment of the upper atmosphere.

Boughn, Fram and Partridge (1971) moved to a still higher frequency, 38 GHz, where the radiation from the galaxy is weaker relative to the CMBR than in earlier experiments. This ground-based experiment suffered from large atmospheric emission. Their result, a dipole amplitude less than about 0.5%, accordingly is less tight than the two earlier experiments. But it was an important early check on whether the dipole might vary with wavelength. Boughn, commencing on page 393, recalls the experience.

In a PhD experiment concurrent with Henry, Davis (1971) explored measurement strategies in a ground-based detector. The approach, which Davis recalls on page 397, had elements in common with Partridge and Wilkinson (1967). Here, though, antenna beams were formed by two off-axis steerable parabolic reflectors. One reflector was aimed at Polaris, which served as a reference. The other observed the sky through the same air mass (zenith distance) but at a different azimuth and thus swept out an arc as Earth rotated. The dipole would be evident in the difference between the two. The site, Princeton, is far from ideal, but the longer integration time enabled the experiment to approach Henry's sensitivity.

Building on Henry's balloon experience, Brian Corey, yet another of Wilkinson's graduate students, employed better magnetic shielding in experiments in four balloon flights (Corey 1978). The measurements at three frequencies (19, 25, and 31 GHz) allowed better separation of a CMBR signal from radiation from the galaxy and residual atmosphere above the balloon. It also allowed a test of the prediction that the dipole amplitude δT is independent of frequency if the CMBR spectrum is thermal. Following the first of the flights, in May 1975, they reported their measurement of the dipole at the June 1976 meeting of the American Astronomical Society in Haverford, Pennsylvania (Corey and Wilkinson 1976). This is a reasonably convincing detection.

Muehlner (1977) reported results the MIT group obtained from a series of balloon flights. As in Henry's experiment, the instrument detected differences of antenna temperatures in opposite directions at the same zenith distance in a rotating balloon gondola. The advance was the short wavelength that offers a check of the possible effect of radiation from interstellar dust, a concern at the time. Our Galaxy proves to be clear enough to allow precision CMBR anisotropy measurements. It took later instrumental developments, including narrowband radiometers, to show this.

Smoot, Gorenstein and Muller (1977), who were based at the University of California, Berkeley, also used a short wavelength, 9 mm (33.3 GHz), so they had the advantage of small interference from the galaxy, and they moved the instrument to a high-flying U-2 aircraft, so it was above most of the radiation from the atmosphere. Like others, they used Dicke-switching between near-identical horn antennas, in this case pointed 60° apart. Smoot and Lubin (1979) completed the sky coverage with flights from the southern hemisphere. The signal-to-noise ratio of the U-2 measurements was a significant advance that gave a clear demonstration of the dipole. The result agreed with the earlier measurements.

Table 5.1 illustrates the rate of progress in the art of measurement of the CMBR anisotropy. By the end of the 1970s it was clear that the anisotropy is dominated by the dipole pattern in equation (5.1), and that the dipole is not very sensitive to the wavelength at which it is measured. In an interesting check of consistency Halpern *et al.* (1988) noted that if the CMBR did not have a thermal spectrum then the amplitude of the dipole would depend on the frequency at which it is measured. They used their dipole anisotropy measurements at short wavelengths to limit the possible departure from a thermal spectrum.

A related test came from the COBE/DMR experiment entered in the second line from the bottom of Table 5.1. If the CMBR spectrum is thermal the annual change δv of Earth's velocity changes the dipole amplitude in equation (5.1) by the amount $\delta T = T_0 \delta v/c$ (since $v \ll c$). DMR detected this effect (Kogut *et al.* 1996). Since Earth's annual motion is well known, the DMR measurement of the dipole annual variation (with the preflight DMR calibration) predicts the value of T_0 if the spectrum is thermal. Kogut *et al.* showed that the result agrees with the FIRAS absolute measurement to about 0.03 K. This annual change is now so well established that it is used to calibrate the WMAP satellite measurement.

To summarize, the evidence is that the intrinsic CMBR anisotropy is small compared to the anisotropy produced by our motion through this sea of thermal radiation.

The dipole measurement did produce a surprise: the motion of our Galaxy through the radiation is larger than expected from the gravitational attraction of the two nearest masses, the galaxy M31 and the Virgo Cluster of galaxies. Wilkinson, on page 207, recalls his reaction. But our large motion is not so surprising when one considers that the galaxy distribution is clumpy on scales larger than the distance to the Virgo Cluster, and that these large-scale clumps can produce a considerable gravitational attraction. But the analysis of that effect of the large-scale clustering of mass is another story.

5.3 The CMBR intrinsic anisotropy spectrum

We come now to the last aspect of the theme for this chapter: the discovery and interpretation of the disturbance to the spatial distribution of the CMBR associated with the growth of mass concentrations in the structures we observe as galaxies and concentrations of galaxies. The disturbance is measured by the variation – the anisotropy – of the CMBR across the sky. One must of course remove the dipole from our motion through the radiation. What remains is the intrinsic part.

The intrinsic anisotropy amounts to a CMBR temperature variation across the sky of about three parts in one hundred thousand. Measuring this tiny effect was a challenge, and learning how to interpret what was measured was a challenge too. The invention of the theoretical basis for the now well-tested interpretation was informed by what the improving anisotropy measurements were indicating, and the advances in theoretical ideas in turn influenced developments of the measurement strategies when theory and observation started converging.

We shall review a considerable sequence of experiments. They commenced in early attempts in the 1960s, continued through the accumulation of experience and the development of technical resources in the 1970s and 1980s, and reached maturity in the "Decade of Discovery" of the intrinsic anisotropy in the 1990s. We offer two panoramic surveys of what was done. The first is the sequence of figures commencing on page 448 that show the growing precision of the measurements. The second is Table A.3 beginning on page 487 that presents essential details of the sequence of anisotropy projects in work that spanned more than three decades. This was the work of research groups of increasing size that required increasing institutional investment. This effort was driven by the growing realization that successful measurements would teach us something of interest about the world around us. And there is of course the ever-present fascination of measurements that are exceedingly difficult but not manifestly impossible.

Though the interplay of advances in theory and practice is an important part of the story of the discovery and interpretation of the intrinsic CMBR anisotropy, we find it more efficient to offer a separate account of the development of theoretical considerations before presenting in Section 5.3.2 the progress toward the anisotropy measurement and its interpretation.

5.3.1 Theoretical concepts

The story of how the community arrived at the now well-tested ΛCDM model (named after key postulates, Einstein's cosmological constant and cold dark matter) for the intrinsic anisotropy of the CMBR merits a book of its own. This account is abbreviated but we hope it makes two points clear. First, the theory was not wholly a product of logical thinking. That certainly was a factor, but there were also inspired guesses and the occasional denial of inconvenient evidence. Second, the theory was chosen to agree with early indications from the observations: the early fit of theory and measurement is hardly surprising. The theory became a believable approximation to reality because it passed the subsequent development of a tight network of tests.

The first thing to consider on the theory side is that, in general relativity, the distribution of mass in the expanding universe is gravitationally unstable.[9] A region where the mass density is slightly in excess of the mean tends to gravitationally attract matter from nearby places where the density is slightly lower: overdense regions grow more dense relative to the mean at the expense of underdense regions that grow increasingly underdense. Thus one might imagine that slight departures from an exactly uniform mass distribution in the very early universe grew by this instability into the clumpy distribution of matter we see around us, in the process gravitationally disturbing the distribution of the CMBR.[10] In the 1960s there was no empirical evidence that the theoretical basis for this picture, general relativity, is a valid approximation on the enormous scales of cosmology. But the picture clearly was worth considering, and it has borne fruit.

A second consideration is the role of the sea of thermal radiation that fills space. Until the universe has expanded and cooled to temperature $T \simeq 3000\,\mathrm{K}$ (at redshift $z \simeq 1000$) the radiation is hot enough to ionize the baryons, and the scattering of photons by the free electrons causes the radiation to exert a strong drag force on the plasma. That makes matter and radiation move together, behaving like a single fluid.[11] This matter-radiation fluid has pressure, that of the radiation. Where the density of this fluid is larger than average, the pressure is above average.[12] The pressure pushes the

[9] Lifshitz (1946) first analyzed this instability. Early doubts about its reality trace to the fact that the perturbation to spacetime curvature associated with the departure from homogeneity is constant in the early universe (Novikov 1964; Peebles 1967). That is, the early universe is close to permanently wrinkled by a mass distribution that is close to smooth in the early universe and grows increasingly clumpy as the universe expands.

[10] The meaning of a departure from homogeneity requires a definition. We will use synchronous (time-orthogonal) coordinates, $\vec{x}$ and t, in which the time–time component of the metric tensor is unity and the time–space components vanish. This can be defined by a family of freely moving observers, each with a label $\vec{x}$ and a clock synchronized to its immediate neighbors. An event in spacetime is specified by the label $\vec{x}$ of the observer who passes through the event and the observer's clock reading t at the event. If the observers are moving with the matter in the early universe then a comparison of their records of the variation of the density with time along each world line shows the growing departures from homogeneity.

[11] The coupling of plasma and radiation was pointed out in Peebles (1965).

[12] There is a complication of historical interest. The gravitational instability of the expanding universe means the total mass density had to have started out very close to uniform in order

fluid away from regions of higher pressure and density and toward regions of lower pressure and density. That is, pressure acts in the opposite direction to gravity. Which dominates, pressure or gravity, depends on the length scale – the wavelength – of the departure from a uniform mass distribution. At long wavelengths gravity is the dominant force, and it causes departures from a uniform distribution of the baryon-radiation fluid to grow rather than oscillate. At short wavelengths pressure wins and causes a departure from a uniform mass distribution to oscillate. The Jeans length marks the boundary between these cases.[13] In the established model the transition between the two cases for the plasma-radiation fluid is at a length that subtends an angle of a few degrees. That causes the peak in Figure 5.18.

The oscillation at shorter wavelengths is the same effect as a sound or acoustic wave in air. The acoustic oscillation of the baryon-radiation fluid in the early universe abruptly terminates when the radiation temperature in the expanding universe drops to $T \simeq 3000\,\mathrm{K}$, cool enough to allow the plasma to combine into neutral atoms, largely hydrogen and helium.[14] These atoms are not appreciably dragged by the radiation. That means the baryons, now bound in atoms, are set free to gravitationally gather into

to have ended up with the approximate homogeneity we observe around us. But that allows a clumpy early distribution of matter that is balanced by opposite fluctuations in the distribution of the thermal radiation, making the total mass density uniform. This situation is unstable because radiation pressure moves the plasma, producing an inhomogeneous mass distribution that gravity can cause to grow more clumpy. This initial condition is termed "isocurvature" because the total mass density is uniform, so spacetime curvature is not perturbed. The initial condition that proves to fit the tests, and which is assumed in the consideration presented in this paragraph, is said to be "adiabatic," a term taken from thermal physics. It means that conditions in a region of above-average mass density in the early universe could be obtained from conditions in a locally low-density region by slowly – adiabatically – compressing both matter and radiation. Zel'dovich (1966) gave an early discussion of the distinction.

13 Gamow (1948a) discussed the Jeans length in an expanding universe; Bonnor (1957) gave a useful derivation. An intuitive approach allows us to see the main elements. In the early universe the pressure is so large that the speed of an acoustic oscillation is close to the speed c of light. When the age of the expanding universe reckoned from some exceedingly early time is t, an acoustic wave will have traveled a distance of about ct. If the wavelength is much smaller than ct then the wave has had time to oscillate and pressure wins. If the wavelength is much larger than ct then the wave has not oscillated: gravity wins and the wave amplitude grows. In effect, if the wavelength is much larger than ct the wave cannot "feel" that it is a wave rather than a part of a uniform universe.

14 Analysis of the conversion of plasma to atoms requires several considerations. The ground level of atomic hydrogen is labeled $1s$, the lowest excited levels, $2s$ and $2p$ (where, for historical reasons, s means zero angular momentum and p, one quantum of angular momentum). Combination of an electron and a proton to the $1s$ level produces a photon that is energetic enough to promptly ionize another atom, so nothing is gained. Combination of an electron and a proton to the $2p$ level releases a lower energy photon, but the decay from $2p$ to $1s$ releases a Lyman-α photon that soon finds an atom in the $1s$ level and promotes it to $2p$, where there is a reasonable chance the thermal radiation can ionize it. But the cosmological redshift reduces the frequencies – and energies – of the Lyman-α photons so they slowly become harmless. Also, the combination to $2s$ produces an atom with no angular momentum, and since a photon has one quantum of angular momentum the atom decays from $2s$ to $1s$ slowly, by the emission of two harmless photons. These considerations were independently worked out in Zel'dovich, Kurt and Sunyaev (1968) and Peebles (1968).

concentrations that may eventually become galaxies. The disappearance of the plasma accordingly is termed the epoch of "decoupling." The radiation moves nearly freely from the "last scattering surface" at the time of decoupling.

It is of great interest to this story that the abrupt termination of the acoustic oscillations leaves distinctive patterns in the distributions of radiation and baryons. The situation can be compared to the sound produced by blowing air across a set of pipes of different lengths, each with one end closed and the other open. The closed end may be compared to the condition in the very early universe that the initial departures from an exactly uniform mass distribution are small and growing. The open ends can be compared to decoupling. Each pipe produces sound at a resonant frequency determined by the quarter-wavelength that fits the pipe. The early mass distribution may be described as a sum of acoustic – sound – waves. Each wave has an oscillation frequency. The frequency may be such that the amplitude of a wave is approaching zero at the time of decoupling; this wave is suppressed. At another wavelength the frequency may be such that the amplitude is growing at decoupling; this wave has a head start for growth after that. The resonance in a pipe is determined by the fit of the wavelength of sound to the length of the pipe. In the hot big bang model the resonance is the fit of the period of oscillation of a wave in the matter-radiation fluid up until the time to decoupling. The result in either case is a distinctive pattern of favored wave amplitudes or frequencies. The effect on the matter distribution at decoupling is observed in the present-day large-scale galaxy distribution; it is termed the "baryon acoustic oscillation." The pattern of favored and unfavored waves also impresses a distinctive pattern on the intrinsic CMBR anisotropy, including the prominent anisotropy peak at an angular scale of about $1°$ ($\ell = 220$) shown in Figure 5.18 on page 462. The observational pursuit of this effect is the topic for the next section.

The oscillation of a sound wave in air is damped by diffusion of the air molecules. The acoustic oscillation of a small-scale irregularity in the distribution of the baryon-radiation fluid in the early universe is also damped, in this case by the diffusion of the radiation through the plasma. This damping effect is observed in the intrinsic CMBR anisotropy discussed in Section 5.3.2.[15] It causes the decrease in the anisotropy spectrum at angular scales $\lesssim 0.3°$ ($\ell \gtrsim 500$) shown in Figure 5.18.

The hiatus in the growth of the small-scale clustering of the baryons prior to decoupling, when the baryon-radiation fluid was oscillating rather than

---

[15] The suppression of small-scale density fluctuations by radiation diffusion is termed "Silk damping," after the derivation in Silk (1967) and recalled by Silk on page 376.

growing more clumpy,[16] contributed to early concerns about how large the CMBR anisotropy might have to be to account for the existence of galaxies. That in turn contributed to the early acceptance of the successful dark matter hypothesis, as follows.

It will be recalled that prior to decoupling the baryons and radiation moved together, as a fluid. That means the distributions of baryons and radiation would have been about the same at decoupling. The density fluctuations in the matter had to have been large enough then – about a tenth of a percent – to have grown into galaxies and clusters of galaxies by the present epoch. But if the fluctuations in the distribution of the radiation were similarly large, the intrinsic CMBR anisotropy would have to be about $\delta T/T \sim 0.001$, comparable to the dipole anisotropy. As the improving experiments showed that the intrinsic anisotropy could not be this large, theorists did the right thing: they cast about for more promising ideas, such as the condition mentioned in footnote 12.

Silk's sketch of the situation, on page 377, gives a useful impression: experimentalists who were attempting to find the CMBR anisotropy were being asked to aim for a target that moved as the theorists tried and abandoned ideas. But on the theoretical side the evolving experiments offered a not dissimilar situation: theorists on occasion found themselves shaping their theories to fit evolving measurements.[17]

Cowsik and McClelland (1973) and Szalay and Marx (1974, 1976) pointed to the now well-tested solution to the puzzle of the small CMBR intrinsic anisotropy: postulate that the dark matter on the outskirts of galaxies and in clusters of galaxies is not baryonic, but rather some other kind of matter that does not interact with electromagnetic radiation. The lack of interaction means radiation is not scattered by the dark matter. That means there is no radiation drag to prevent the early growth of its clustering. This would allow the formation of the clumpy present distribution of matter with considerably

[16] There is another contribution to this hiatus, which Gamow (1948a) seems to have been the first to note in the literature. When the mass density in radiation is larger than in matter the gravitational instability does not operate on scales smaller than the Hubble length because matter is not an important factor in the dynamics and pressure keeps the thermal radiation close to smooth. We know of no early publication of the computation of this effect, but Alpher (in Alpher and Herman 2001) recalls working on problems of this sort. A related effect Gamow and Teller (1939) noted is that if the mass density is well below the Einstein–de Sitter value then the gravitational instability is suppressed.

[17] An example in the early 1980s is the evidence of detection of intrinsic anisotropy $\delta T/T \simeq 10^{-4}$ (Fabbri *et al.* 1980b; Boughn, Cheng and Wilkinson 1981; see also page 491). That invited the model proposed in Peebles (1981a). The Boughn *et al.* result was withdrawn; the source of the Fabbri *et al.* signal was not understood. When it became clear that the anisotropy had been overestimated, the model that explained it was replaced by the CDM model in Peebles (1982). As will be described, this has proved to be a good approximation.

less disturbance to the thermal radiation than if all the matter were baryonic and tightly coupled to the radiation before decoupling.

Broad interest in this concept of nonbaryonic dark matter was triggered by a different development: the apparent experimental evidence that the electron neutrino has a mass of about 30 eV, or about 5×10^{-32} g (Lubimov *et al.* 1980). It was exciting because at this mass, and with the number density of neutrinos that would have been produced in the early universe at thermal equilibrium with the electromagnetic radiation, there would be enough mass in neutrinos to account for the missing mass (or as one now says, dark matter) in galaxies and clusters of galaxies. Doroshkevich *et al.* (1981) described these considerations in a lead paper in one of a series of influential conferences, the *Tenth Texas Symposium on Relativistic Astrophysics*, in Baltimore, December 1980. The laboratory evidence for massive neutrinos was withdrawn – as indicated in Table 2.1 the mass density in neutrinos is now known to be small – but the interest in nonbaryonic dark matter remained.

This ~30 eV neutrino candidate for the dark matter is termed "hot dark matter" (Bond *et al.* 1984) because at high redshift the neutrinos move about rapidly, as in a hot gas. That would have had the effect of smoothing out small wavelength irregularities in the mass distribution, to the extent that the first bound objects to form by the gravitational instability process would have been larger than a single galaxy. That led to interest in a "top-down" picture for structure formation, in which mass concentrations the size of clusters form and then fragment into galaxies (Zel'dovich, Einasto and Shandarin 1982). But this is not promising: the Milky Way Galaxy is old and it is part of a modest concentration – containing another large spiral galaxy, the Andromeda Nebula M31, and some dozens of dwarf satellite galaxies – that is coming together now. This looks more like a "bottom-up" process in which galaxies form and then collect into groups and clusters of galaxies. But that presented no conceptual problem. At the *Eleventh Texas Symposium on Relativistic Astrophysics*, in Austin, December 1982, Pagels (1984) reviewed how ideas in particle physics offer a variety of other maybe more promising dark matter candidates.

Bond, Szalay and Turner (1982) and Blumenthal, Pagels and Primack (1982) took a logical step: postulate that the dark matter particle mass is about ten times what had been under discussion for neutrinos. The larger mass of these "warm dark matter" particles would lower their thermal velocities, reduce the smoothing effect, and so lower the mass of the first generation of gravitationally bound mass concentrations. The particle mass could be chosen so the first generation would have masses characteristic of galaxies.

But the community soon adopted the simplifying postulate that smoothing by streaming is quite unimportant: the dark matter began cold. This is the cold dark matter in what became the CDM cosmology and later, with the addition of Einstein's cosmological constant, the ΛCDM model.

The CDM postulate was not forced by a systematic evaluation of viable alternatives; it was adopted as a working hypothesis, in part because it is a conceptually simple way to reconcile the smooth distribution of the CMBR with the lumpy distribution of matter, and in part because it is relatively easy to analyze the gravitational growth of clustering of CDM. The postulate was validated by passing searching tests, but that happened much later.

Let us move on to another consideration. We have remarked that the expanding universe is gravitationally unstable: slight departures from a perfectly uniform mass distribution grow larger and can eventually become gravitationally bound concentrations. The flow of oil in a pipeline also is unstable: any deviation from exactly smooth flow grows into turbulence. There is a serious difference, however. Turbulent flow loses all "memory" of its initial conditions. The gravitational growth of a clumpy mass distribution in an expanding universe preserves signatures of its initial conditions. That means we have to specify the character of the mass distribution in the early universe: how big was the typical departure from uniformity on each length scale?

The mean square value of the departure from a uniform mass density as a function of length scale, or wavelength, is measured by the mass fluctuation power spectrum.[18] Gravitational instability causes this fluctuation spectrum to grow. On large scales, where the effect of the thermal radiation pressure may be ignored, the present spectrum as a function of wavelength is easily computed from the primeval form, and it can be compared to the observed

---

[18] The mass density $\rho(\vec{x})$ as a function of position $\vec{x}$ and time t in the coordinates defined in footnote 10 is

$$\rho(\vec{x}, t) = \bar{\rho}(t)(1 + \delta(\vec{x}, t)). \tag{5.3}$$

The mean mass density is $\bar{\rho}(t)$ and $\delta(\vec{x}, t)$ is the fractional departure from the mean. The correlation function is the mean value,

$$\xi(x) = \langle \delta(\vec{x} + \vec{y}) \delta(\vec{y}) \rangle, \tag{5.4}$$

averaged over all positions $\vec{y}$. The mass distribution is assumed to be a spatially stationary – statistically independent of position – isotropic random process, meaning ξ is a function only of the distance x between the two points in the average. The mass fluctuation power spectrum $\mathcal{P}(k)$ is the Fourier transform of the mass correlation function. In a standard normalization,

$$\xi(x) = \int d^3k \, \mathcal{P}(k) e^{i\vec{k}\cdot\vec{x}}, \qquad \langle \delta^2 \rangle = \xi(0) = \int d^3k \, \mathcal{P}(k) = \int 4\pi k^3 \mathcal{P}(k) \, d\ln k. \tag{5.5}$$

The mean square value – the variance – of the density contrast is $\langle \delta^2 \rangle$. One sees that the variance per logarithmic interval of the wavenumber k, or wavelength $\lambda = 2\pi/k$, is $4\pi k^3 \mathcal{P}(k)$.

large-scale distribution of the galaxies. On somewhat smaller scales, the primeval mass fluctuations grow into gravitationally bound concentrations that can be compared to the observed number density of clusters of galaxies. This comparison was an important early test of the CDM picture (as discussed on page 457). The test on still smaller scales that compares the abundance of galaxies to what might be expected from a given primeval mass fluctuation spectrum is more difficult because galaxy formation is a strongly nonlinear complex process that at the time of writing is not completely understood.

The point of these comments is that the theory of the intrinsic CMBR anisotropy has to specify the nature of the early departure from exact uniformity. Part of the confusion we mentioned about the expected amount of the intrinsic anisotropy reflected explorations of alternative initial conditions, including the consideration mentioned in footnote 12, that might have helped reconcile the present clumpy matter and smooth CMBR. But, as was the case for the idea of nonbaryonic dark matter, the successful initial condition was under discussion before it was seen to be indicated by the CMBR anisotropy measurements, as follows.

In general relativity theory a disturbance to a uniform mass distribution disturbs the uniform curvature of spacetime. When the disturbance to the curvature is large, it tends to evolve into black holes. The evidence is that there are black holes – in the centers of galaxies, and in remnants of massive dead stars – but the fraction of all mass in black holes is small. We do not want to postulate initial conditions that promote wholesale black hole formation. This is avoided by the elegant – in the eyes of some – postulate that the spacetime curvature fluctuations are statistically the same, and small, on all length scales. The virtue of this "scale-invariant" initial condition was recognized in the early 1970s. It became popular a decade later, with the realization that the inflation scenario (discussed on page 520) for what was happening in the exceedingly early universe, before the relativistic big bang model could have applied, naturally (but not inevitably) produces close to scale-invariant departures from exact homogeneity.[19] It is not surprising

---

[19] The scale-invariant initial condition is

$$\mathcal{P} \propto k^{n_s} \quad \text{with} \quad n_s = 1. \tag{5.6}$$

Early discussions of this initial condition are in Harrison (1970), Peebles and Yu (1970), and Zel'dovich (1972). Near scale-invariance in simple versions of the inflation scenario was recognized by Mukhanov and Chibisov (1981), Hawking (1982), Starobinsky (1982), Guth and Pi (1982), and Bardeen, Steinhardt and Turner (1983). Bardeen *et al.* calculated the tilt from pure scale-invariance, showing that n_s tends to be slightly less than unity. The self-consistent quantum analysis of linear fluctuations in gravity and the field that is perturbing it is in Fischler, Ratra and Susskind (1985).

that the tilt, or departure, from pure scale-invariance tends to be small because the density fluctuations we see are a narrow slice of what is produced during inflation. The cosmological tests discussed in Section 5.4 fit a near scale-invariant initial condition, tilted in the direction expected in simple implementations of inflation.

The model for the CMBR anisotropy also requires the choice of a cosmological model. General relativity theory applied to a homogeneous and isotropic universe gives the relation between the expansion rate, the mean mass density $\bar{\rho}$, the cosmological constant Λ, and the space curvature in equation (G.1). This relation at the present epoch is

$$H_0^2 = \frac{8}{3}\pi G\bar{\rho} - (a_0 R)^{-2} + \frac{\Lambda}{3}. \tag{5.7}$$

Recall that the Hubble parameter is the ratio $H_0 = v/r$ of the galaxy recession velocity, computed from the measured redshift, to its distance r. The measure $(a_0 R)^{-2}$ of space curvature can be positive, negative, or zero. (This notation and the numerical factor in the last term are historical remnants.) If $(a_0 R)^{-2} = 0$, space sections of constant time are on average flat, with Euclid's spatial geometry (but, although space sections are flat, the mass does curve spacetime). The density parameters are defined by

$$\Omega_m H_0^2 = \frac{8}{3}\pi G\bar{\rho}, \quad \Omega_k H_0^2 = -(a_0 R)^{-2}, \quad \Omega_\Lambda H_0^2 = \frac{\Lambda}{3}. \tag{5.8}$$

That is, Ω_i is the fractional contribution to H_0^2 by the ith term on the right-hand side of equation (5.7). Because these three parameters are much discussed, it is well to display a simple identity,

$$\Omega_m + \Omega_k + \Omega_\Lambda = 1. \tag{5.9}$$

This model of spacetime has one parameter, H_o, that fixes the length scale and two free dimensionless density parameters.

Einstein and de Sitter (1932) pointed out that of the three terms in equation (5.7) the only one we can be reasonably sure is present is the mass density, so it is logical to consider first the case where space curvature and Λ may be neglected. This came to be termed the Einstein–de Sitter case. A more recent argument from the inflation scenario reinforces the idea that space curvature may be neglected: it is proposed that rapid early expansion[20] irons out space curvature.[21] A nonzero cosmological constant is an arguably

---

[20] In this scenario an exceedingly large early mass density, held close to constant by a large and negative effective pressure, drives very rapid expansion that stretches out and so smooths curvature variations. This would account for the observed strikingly uniform large-scale structure of the universe.

[21] There are alternatives: a variation on the inflation theme allows a low-density universe with significantly negative space curvature (e.g., Gott 1982; Lyth and Stewart 1990; Ratra and

natural outcome of quantum physics, but a natural value is absurdly large compared to what is allowed in relativistic cosmology.[22] A hopeful remedy for this conundrum was that some symmetry forces the vacuum energy density to the only natural and acceptable value, zero.[23]

In short, it was not unreasonable in the 1980s to suppose space curvature and Einstein's Λ both are negligibly small in equation (5.7). Now we have evidence that space curvature is small and Ω_Λ is significant, as we will describe.

The next thing to consider is the choice of language by which the anisotropy measurements are reduced to statistics that may be compared to what the model predicts. Early choices tended to follow such experimental setups as beam switching that measures the mean square temperature difference as a function of the angular separation of the beams. An example is shown on page 233. The practice that became the established convention in the mid-1990s is to subtract the mean and dipole from the CMBR temperature as a function of position in the sky and express the remaining fluctuating part as the sum[24]

$$\delta T(\theta, \phi) = \sum_{\ell \geq 2, \, -\ell \leq m \leq \ell} a_\ell^m Y_\ell^m(\theta, \phi). \tag{5.10}$$

The two angles, θ and ϕ (polar and azimuthal), specify the position in the sky. The spherical harmonic $Y_\ell^m(\theta, \phi)$ fluctuates between positive and negative values in a regular pattern across the sky. The angular scale of the fluctuation is defined by noting that, except near the poles where Y_ℓ^m is small (if $m \neq 0$), the lines in the sky along which Y_ℓ^m vanishes approximate rectangular boxes with narrower width

Peebles 1994; Linde 1995). This approach, which was motivated largely by evidence that Ω_m is well below unity, does not pass later cosmological tests.

[22] The cosmological constant may be written in the Einstein field equation as the term $\rho_\Lambda g_{\mu\nu}$, where the first factor is a constant and the second is the metric tensor. Considered as part of the energy–momentum tensor, this term represents mass density ρ_Λ and pressure $p_\Lambda = -\rho_\Lambda$, a situation similar to what is envisioned in the inflation scenario in the previous footnote. Since a velocity transformation does not change $g_{\mu\nu}$ this term has no preferred velocity, a symmetry one might expect of the quantum vacuum energy density. That is, Einstein's cosmological constant arguably is a natural description of the quantum vacuum. Its energy has contributions from the zero-point energies of all fields, an exceedingly large sum. In the 1930s the physicist Wolfgang Pauli is reported to have remarked that the sum is so large that the resulting curvature scale of spacetime would not even reach to the Moon.

[23] At the time of writing a hopeful remedy introduced in Peebles and Ratra (1988) and Wetterich (1988) is that Λ is evolving to its "natural" value, zero, but slowly. This usually is characterized as an effect of an "equation of state" parameter $w = p_\Lambda/\rho_\Lambda$ that is slightly different from -1, where the dark energy pressure is defined in the previous footnote.

[24] Early applications in cosmology are in Yu and Peebles (1969) and Peebles (1973), as a measure of the distribution of matter, and Zel'dovich, Rakhmatulina and Sunyaev (1972), as a measure of the CMBR anisotropy. The notation as it appears here was introduced and the Sachs–Wolfe (1967) quadrupole term computed in Peebles (1982). Bond and Efstathiou (1987) used the notation in their CMBR anisotropy analyses, but general adoption took another decade (Wright *et al.* 1994b).

$$\varphi \simeq \pi/\ell \text{ radians.} \tag{5.11}$$

The standard measure of the CMBR anisotropy on the angular scale belonging to ℓ (through equation 5.11) is conventionally written as

$$(\delta T_\ell)^2 = \frac{\ell(\ell+1)}{2\pi}\langle|a_\ell^m|^2\rangle, \tag{5.12}$$

where the angular brackets indicate the average over m.

Equation 5.12 is called the "anisotropy spectrum" (named after the spatial power spectrum discussed in footnote 18). This spectrum is plotted in Figures 5.9–5.13 and 5.15–5.18 that chronicle the progress in the anisotropy measurements.[25]

Now we are in a position to list the assumptions of the CDM cosmological model. They include Einstein's general relativity theory applied to a close to uniform mass distribution, expansion from a hot big bang, matter density dominated by cold dark matter, scale-invariant adiabatic initial departures from homogeneity (as discussed in footnote 12 on page 435, and on page 441), and the Einstein–de Sitter cosmological model (with negligibly small values of space curvature and the cosmological constant). The model was assembled in Peebles (1982) and applied to the relation Sachs and Wolfe (1967) had found between departures from an exactly uniform mass distribution in the early universe and the attendant gravitational disturbance to the present distribution of the thermal radiation. Sachs and Wolfe recall (commencing on pages 364 and 368) how they hit on this important result. They ignored the scattering of radiation by free electrons in the plasma at $z > 1000$, but that is a good approximation on large angular scales.

The CDM model was meant to demonstrate that the upper limits we had in the early 1980s on the anisotropy of the CMBR were not necessarily in conflict with the gravitational instability picture for galaxy formation: this model predicts large-scale anisotropy well below the measurements we had then. Only the quadrupole term $\ell = 2$ in equation (5.12) was presented because the analysis was motivated by the improving bounds on how large

---

[25] The square of $\delta T(\theta, \phi)$ in equation (5.10) averaged over the sphere is

$$\langle\delta T(\theta, \phi)^2\rangle = \sum_{\ell \geq 2} \frac{2\ell+1}{4\pi}\langle|a_\ell^m|^2\rangle. \tag{5.13}$$

This follows because the Y_ℓ^m are orthogonal, $|Y_\ell^m|^2$ is normalized to unit integral over the sphere, and the $|a_\ell^m|^2$ are statistically independent of m. One sees that at large ℓ, where the sum approximates an integral, the mean square fluctuation, or variance, of the temperature per logarithmic interval in ℓ is $\ell(2\ell+1)\langle|a_\ell^m|^2\rangle/4\pi$. This is the analog of the variance per logarithmic interval of wavelength in equation (5.5). It is an historical accident that $2\ell+1$ is replaced by $2(\ell+1)$ in equation (5.12).

this term might be.[26] At the time it was not even clear that this model was a step in the right direction, or if it were whether the measurements could be good enough for an interesting test of the calculation. The situation was changing, however, and Bond and Efstathiou (1987) were led to write down the Sachs–Wolfe δT_ℓ for larger ℓ. The Sachs–Wolfe effect was convincingly detected in 1992 by DMR on COBE. As one might expect, that focused community attention.

In the CDM model the anisotropy spectrum δT_ℓ at larger ℓ (smaller angular scales) shows the effect of the acoustic oscillations mentioned on page 437. We discussed this in terms of a fluid model for the behavior of the thermal radiation and the plasma prior to decoupling. The picture is helpful, but we need to do better by analyzing the diffusion of radiation through the plasma and the effect that has on the evolving distributions of the radiation, baryons, and dark matter. Apart from the effect of the dark matter, this too was worked out well before it was needed, in Peebles and Yu (1970). The application of this analysis to the CDM model in Peebles (1982), presented in the language of equation (5.12), appears for the first time in Bond and Efstathiou (1987).

We should pause here to consider why it became of pressing interest to measure in all possible detail the anisotropy spectrum δT_ℓ. Imagine a set of cosmological models, each of which has a different value of the ratio of mass densities in baryons and dark matter, while the total matter density, baryons plus dark matter, is the same in every model. All other parameters also are the same. It will be recalled that the baryon plasma and the thermal radiation act as a fluid before decoupling when the plasma combines into mostly neutral atoms. The dark matter is only weakly affected by the acoustic oscillation of this plasma-radiation fluid, through the gravitational interaction. The different weight of the baryons in different models means the velocity of sound is different. That affects the relative sizes of the oscillations of the anisotropy spectrum as a function of ℓ. (An early example of the effect is shown in Kamionkowski, Spergel and Sugiyama 1994.) A measurement of this effect thus could fix the baryon mass density. This offers a fascinating prospect: compare the baryon density derived from the CMBR intrinsic anisotropy spectrum to the baryon density derived from the theory in Chapter 3 for the origin of the light elements. This comparison of what is

---

[26] The computed value of the quadrupole anisotropy term is close to the measurement we have now. That is because the Λ term does not much matter in this calculation. One could have produced an example with even smaller quadrupole anisotropy by postulating the isocurvature initial condition mentioned in footnote 12 on page 435, because the initially uniform mass distribution suppresses spacetime curvature fluctuations and hence the Sachs–Wolfe effect. Efstathiou and Bond (1987) discussed this case.

derived from very different phenomena would be an exceedingly demanding test. How the cosmology passed this test, and others, is to be discussed.

To understand another of the tests, imagine another set of models. All have the same present values of the mean mass densities in baryons and in dark matter, the same CMBR temperature, and the same initial conditions for the CDM model for density fluctuations. But each model has different values of the cosmological constant and space curvature. The relation between the Hubble parameter H_0, the mean mass density, the cosmological constant Λ, and the space curvature measure R^{-2} is shown in equation (5.7). This relation tells us that the models in this set have different values of H_0. But in the early universe these models behave in a very similar way. That is because the mass density decreases as the universe expands faster than the curvature term (as one sees in equation G.1). Since the mass density term makes an appreciable contribution to H_0^2 at the present epoch we know that at decoupling, at redshift $z_{\text{dec}} \sim 1000$, the curvature and Λ terms were quite small compared to the mass density. This set of models thus behave in very nearly the same way in the early universe and up to decoupling. This means the fluctuations imprinted on the distributions of matter and radiation are very nearly the same in all these models. The CMBR fluctuations appear on the sky at different angular scales because Λ and the present value of space curvature fix the distance from decoupling to the present epoch. A measurement of the CMBR anisotropy gives us a measure of this distance, and hence of space curvature and Λ. This can be compared to other measurements of these parameters, giving us another demanding test.[27]

Before considering how the network of tests grew[28] we should take note of one more point: up to the end of the 20th century people were debating

---

[27] There is the complication that the distance depends on two parameters, Λ and space curvature. This "geometrical degeneracy" means that CMBR anisotropy measurements give one tight constraint on two parameters. The WMAP measurement to be discussed (Spergel *et al.* 2007) shows the constraint is well approximated by the linear expression,

$$\Omega_m + 1.41\Omega_\Lambda = 1.30 \pm 0.03. \tag{5.14}$$

Lewis and Bridle (2002) showed this relation in what became its standard form. It appears earlier, expressed in different ways, in Bond, Efstathiou and Tegmark (1997), Zaldarriaga, Spergel and Seljak (1997), and Efstathiou and Bond (1999). We have seen that the CMBR anisotropy constrains the present matter density and hence the combination $\Omega_m h^2$ (equation 5.8). If h is known that gives Ω_m, equation (5.14) gives Ω_Λ, and that gives a measure of space curvature.

[28] Early explorations of how CMBR anisotropy measurements might test cosmology were beclouded by the combination of uncertainties in the parameters for the cosmological model and in the model for structure formation. Discussions that addressed both issues include Bond and Efstathiou (1984), Vittorio and Silk (1984), Sugiyama and Gouda (1992), Bond *et al.* (1994), Kamionkowski, Spergel and Sugiyama (1994), Hu and White (1996), Jungman *et al.* (1996), and the references in footnote 27. Important to these studies was the demonstration by Seljak and Zaldarriaga (1996) of efficient methods of computation of the spectrum, and the public release of their computation package CMBFAST. This made it relatively easy to explore fits of the improving anisotropy measurements to model and parameter choices.

other ideas, alternatives to the list of postulates in the CDM model. We mentioned an example, an isocurvature CDM model. Another more influential idea came from the concept in condensed matter and particle physics that energy in a field can find itself tied up in a defect which manifests itself as a linear structure that behaves like a string under tension. An example is the vortex tubes in superfluid helium. The effect naturally appears as part of ideas about how particles acquired masses, in a situation that could be interesting for cosmology. If the mass per unit length in this kind of "cosmic string" had a value that seemed reasonable from the point of view of ideas in particle physics, then as the cosmic strings move about they would disturb an initially homogeneous mass distribution and trigger the formation of concentrations that might look like galaxies. Kibble (1976) proposed this idea; Zel'dovich (1980) and Vilenkin (1985) gave influential discussions. It allows fascinating elaborations, as in the exploding magnetized superconducting string scenario (Ostriker, Thompson and Witten 1986). Kibble's concept led to the consideration of other patterns of field structure known as "textures" (Turok 1989; Crittenden and Turok 1995) that at the time also seemed promising as seeds for structure formation.

As it happened, these alternative ideas about how the galaxies formed have proved to be unable to account for the improved CMBR anisotropy measurements. That does not vitiate the concept: at the time of writing it is not known whether cosmic strings or textures or some other exotic phenomenon might be present and produce observationally interesting adjustments of the ΛCDM model. It is only prudent to bear in mind that the universe is large, and has room for many things. But let us turn now to the story of how measurements sorted out ideas and established the particularly simple ΛCDM model as a convincing approximation to the real world.

5.3.2 *Advances in the anisotropy measurements and analysis*

In retrospect we can identify three parts to the lines of evidence that established the ΛCDM cosmological model. The first was the demonstration that there is no known viable alternative. That was important, if only to community psychology. More convincing was the demonstration that the model can be adjusted to fit the anisotropy spectrum that became measured with considerable precision over a large range of angular scales. More convincing still was the demonstration that the model that fits the anisotropy measurements passes a tight network of tests based on observations of a suite of quite different phenomena. The state of these cosmological tests at the turn of the century is summarized in Table 4.3.

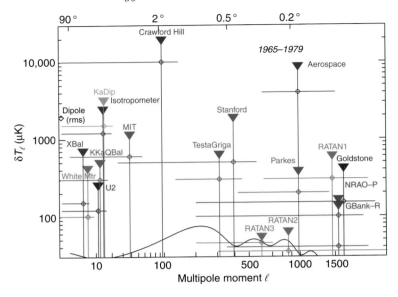

Fig. 5.9. Early probes of the CMBR anisotropy spectrum through the decade most noted for the identification of the dipole. The curve at the bottom of the figure shows where the measurements ended up, at the ΛCDM model. The measurements are described and their labels identified in Table A.3.

Progress in the experiments along the learning curve to the precision anisotropy measurements is shown in the series of figures commencing with Figure 5.9. We supplement that with the information on the experiments presented in Table A.3 in the Appendix. In the figures and this table the measurements are ordered by the date of publication. That can be confusing because the measurements typically were known outside the group well before publication. In the 1980s and earlier the usual medium was mailed preprints; a decade later the community had adopted the internet medium familiarly known as "astro-ph" (and now termed "arXiv").

Beginning in 1965, and continuing through 1979, the searches for anisotropy on angular scales smaller than the dipole yielded only upper limits on how rough the microwave sky might be. Figure 5.9 shows the situation. The upper limits, and in later figures the measurements, of the anisotropy spectrum δT_ℓ (defined in equation 5.12) are plotted on the vertical axis. The upper horizontal axis shows the angular scale along the spectrum, and the lower horizontal axis shows the multipole index ℓ.[29] The dipole anisotropy

---

[29] The relation between ℓ and angular scale is shown in equation (5.11). In each figure the horizontal axis is proportional to the 0.3 power of ℓ. This is a convenient way to scale the spectrum to spread out the data. We need a logarithmic scale for δT_ℓ in Figure 5.9 because the errors in these earliest data are so large. Linear scales on the vertical axis accommodate the later measurements.

would appear on the left side of the figure, at $\ell = 1$, but this term, which is dominated by the aether drift, is not relevant for the search for the intrinsic roughness of the CMBR.

The name near each measurement here and in the following figures refers to Table A.3. The triangles mark upper limits at one standard deviation. This means, roughly, that each measurement established that there is a 66% chance that the CMBR anisotropy actually is below the top of the downward-pointing triangle representing the experimental result.[30] The crossbar for each measurement shows the width of the window in ℓ within which the experiment probed the anisotropy spectrum. In Figure 5.9 and the next figure the vertical positions of the crossbars have no significance because all the measurements are upper limits. The solid curve is the ΛCDM model fitted to the WMAP data. It is placed here to show where the measurements were heading.

Figure 5.10 shows the anisotropy measurements in the next decade. To give a picture of the progress in the search for the CMBR anisotropy, we show in the light gray boxes here and in the following figures the weighted averages

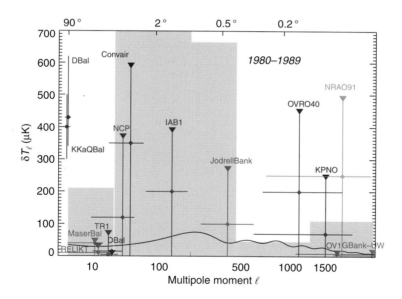

Fig. 5.10. Anisotropy results for all measurements published in the 1980s. Here and in what follows the light gray boxes correspond to weighted averages over values in a given range of ℓ from the previous measurements. Because of the rapidity of the advances, most of the weight comes from data in the plot just prior to the one being viewed.

---

[30] There is an additional uncertainty from calibration error. This is not represented here or in the subsequent plots; it is typically 10–15%.

of the spectrum bounds from the compilations in the previous figures.[31] It is noteworthy that the experiments had sufficient sensitivity that had they chosen to observe degree angular scales (ℓ near 200) they could have detected the anisotropy near its peak. At the time observers were not thinking in terms of the anisotropy spectrum, and they were not yet focused on the CDM model.

The relative heights of these gray boxes and the data points in Figure 5.10 illustrate the progress in the measurements in the 1980s at small and large angular scales. This progress is reflected also in the considerable difference of vertical scales in Figures 5.9 and 5.10. (The change in vertical scale from logarithmic to linear changes the shape of the solid curve; the theory is the same.)

The point labeled TR1 was at first considered a possible anisotropy detection but later reported by the team to be contaminated by foreground emission. The left-hand side of this figure shows two measurements early in the decade (entered in the table of experiments in the Appendix on page 491) that appeared to have detected the quadrupole term δT_2 (that is, $\ell = 2$). As we remarked in the footnote 17 on page 438, these results were not confirmed by subsequent observations. The improved limits on the anisotropy motivated the introduction of the CDM model.

Apart from the fact that the CDM model predicts CMBR anisotropy below what could be measured in the 1980s, the model largely attracted attention in that decade as a picture for structure formation: the origin of galaxies and clusters of galaxies. An influential study by Davis *et al.* (1985) demonstrated that the CDM model shows promise. The CDM prediction of the CMBR anisotropy spectrum was studied (Bond and Efstathiou 1984; Vittorio and Silk 1984), but the prediction did not necessarily inspire confidence as a guide to how experimentalists might best tune measurement strategies. As we mentioned, there were other ideas about how structure formed. And within the CDM model there could have been interesting complications. An example is the long-discussed notion that the dark matter may be decaying, producing ionizing radiation. If that, or ionizing radiation from early generations of stars, produced significant numbers of free electrons after the nominal epoch of decoupling, then the large-scale Sachs–Wolfe effect would have been about the same as in the standard model but the small-scale anisotropy would have been smoothed by scattering by free

---

[31] The box widths in ℓ are selected from the WMAP bins in the last figure in this series. The weighted averages – the box heights – are computed from δT_ℓ and the error in this quantity. Though it is more rigorous to average in δT_ℓ^2 and its error, the resulting plots are not significantly different. We find that the δT_ℓ averaging tends to slightly downweight outlying points, resulting in a more accurate visual impression of the progress.

electrons (e.g., Bond and Efstathiou 1984). An experimentalist who accepted this picture would have concentrated on measuring the anisotropy on large scales and avoided degree-scale measurements where it was later found that the spectrum peaks.

The modern field of CMBR anisotropy studies had roots in the workshop "Delta T over Tea," the first to focus on the search for the intrinsic CMBR anisotropy. It was hosted by CITA (the Canadian Institute of Theoretical Astrophysics) in Toronto in May 1987. It included theorists and representatives from nearly all the experimental groups. There were models under discussion for interpretation of the intrinsic anisotropy, if discovered. Instruments with sensitivities capable of testing models were coming on line. The measurement techniques – interferometry, direct mapping with spinning balloon-borne payloads, ground-based beam switching – were as varied as the research groups. This also was the time when the community started to agree on how to quote experimental results and uncertainties, though the language on page 443 did not become standard for another decade. The widely accepted feeling that meaningful anisotropy measurements may be just around the corner was tempered, however, by the uncertainty in how troublesome emission from sources in the foreground might be.

The tremendous activity by many groups during the 1990s, the "Decade of Discovery," leads us to show progress in this decade in three figures. Figure 5.11 shows the anisotropy spectrum measurements published during

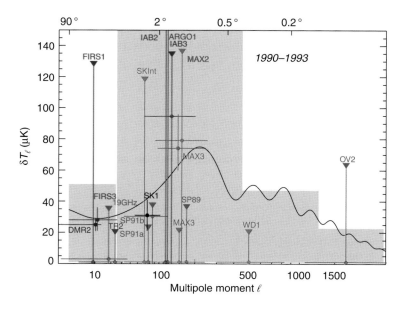

Fig. 5.11. Anisotropy measurements published between 1990 and 1993.

1990 to 1993. The two reported detections of the quadrupole term are not included in the average over prior data shown as the gray boxes in this figure.

The very influential advance in 1990 was the first detection of the intrinsic anisotropy of the CMBR (that is, after subtraction of the dipole term produced by our peculiar motion). It is labeled DMR2 (the second report of DMR data), at the lower left-hand part of the figure. Here and in the subsequent figures detections are indicated by solid circles, as opposed to downward-pointing triangles for upper limits.

The DMR measurement on the COBE satellite (Smoot *et al.* 1992) showed that the spectrum at $\ell \lesssim 10$ (in the language later adopted[32]) agrees with the gravitational perturbation to the CMBR produced by the scale-invariant adiabatic initial condition in the CDM model. This was demonstrated by Wright *et al.* (1992). It very effectively focused interest in what proved to be the right direction.

The DMR measurement was convincing because it was done from the near-ideal environment of space and it was designed with many internal consistency checks. In the drive to cancel out systematic instrumental errors each separate receiver was made as symmetric as possible and used the inherent stability of Dicke-switching (Section 3.5) to compare the outputs from the two horns. This strategy traces back to the Conklin and Bracewell (1967) system illustrated in Figure 5.6. The Berkeley U-2 experiment (Smoot, Gorenstein and Muller 1977) used the strategy in its clear demonstration of the dipole anisotropy. That was influential in the choice of the DMR design. As indicated in Figure 5.6, the WMAP space mission also followed this strategy. The WMAP design was not directly taken from DMR or other experiments. But when the design was complete and heads were lifted from the paper, the similarity was striking. It is a testament to the robustness of a good idea.

---

[32] The DMR discovery papers presented the CMBR temperature correlation function $C(\alpha) = \langle \delta T(1) \delta T(2) \rangle$, where "1" and "2" refer to different spatial directions. This is the average of products of the departure of the temperature from the mean over all pairs of directions in the sky with angular separation, or lag, α. The correlation function $C(\alpha)$ is the Legendre transform of the $|a_\ell^m|^2$ in equations (5.12) and (5.13). The analog for a random process in flat space is the relation between the correlation function and the power spectrum in equation (5.5). Blackman and Tukey (1958) explained why, although the correlation function and the power spectrum are formally mathematically equivalent, their use in data analysis may not be equally effective. They argued that the power spectrum may be the more suitable, affording reduction of sensitivity to correlation among terms and improved sensitivity to signatures of physical processes. Tukey (1966) put it that "the tragic accident that killed H. R. Seiwell and his family destroyed the only man who could usefully look at a plot of autocorrelation against lag." The advantage of the power spectrum over its transform, the correlation function, certainly applies to CMBR anisotropy analyses, and it is true also of the analysis of the large-scale galaxy distribution. There are exceptions: the correlation function is more useful in the analysis of the nonlinear clustering of galaxies on smaller scales.

As was the case for the FIRAS and UBC COBRA measurements of the energy spectrum, the competition to space anisotropy missions from lower orbit measurements was not far behind. In 1985 Cheng, Meyer, and Page began building a balloon payload with four frequency bands based on 0.3-K bolometers developed for this purpose by Pat Downey and Rai Weiss at MIT. This project came to be called FIRS. During the time between the COBE design[33] and the COBE/DMR anisotropy detection there were considerable advances in detector technology, especially with bolometers. In one night of measurement the FIRS sensitivity was comparable to that of one year of the COBE/DMR measurement. The first flight, in October 1988, ended after eight minutes when the balloon burst. The instrument was repaired and reflown in October 1989, two months before the COBE launch. It mapped a third of the sky. Anisotropy was detected and published before the DMR announcement (Meyer, Cheng and Page 1991), but it could not be uniquely attributed to the CMBR. At the time the authors could not unambiguously rule out contamination from radiation from the galaxy or atmosphere, or systematic error in the instrument. But, after the DMR anisotropy announcement, the comparison of the DMR and FIRS sky maps showed clear correlation.[34] The consistency of these two independent experiments was a reassuring check that the intrinsic CMBR anisotropy had been detected and had similar amplitudes at 50 GHz (DMR) and 170 GHz (FIRS).

One sees in Figure 5.12, and in the average over this and earlier data that is shown as gray boxes in Figure 5.13, that by the mid-1990s the anisotropy spectrum seemed to peak at an angular scale somewhere between 2° and 0.2°. Important to this case were the upper limits on the anisotropy on still smaller angular scales in the experiments indicated in the lower right-hand part of Figure 5.10 (Appendix, pp. 493 and 496). Though the quality of these early experiments was variable, and calibration errors were typically 15%, the case that the anisotropy spectrum peaks in the range of ℓ shown here was not unpromising. That was interesting because the CDM model predicts a peak about where the measurements were indicating the anisotropy is largest.

By the late 1990s the anisotropy spectrum shown in Figure 5.13 (and in gray boxes in Figure 5.15) was seen to have a reasonably well-defined peak. The Saskatoon experiment (SK) had demonstrated the rise of the spectrum

---

[33] The COBE design was proposed in 1974, as Hauser recalls on page 418. In 1986 COBE was redesigned to accommodate the change from a shuttle to Delta Launch in response to the Challenger disaster. Chuck Bennett recalls that it was a sprint from then to the 1989 launch.

[34] Ganga *et al.* (1993), with important assistance from Steve Boughn of Haverford and J. R. Bond at CITA.

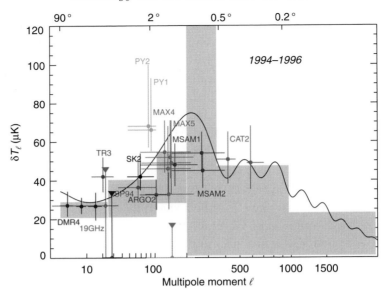

Fig. 5.12. Anisotropy measurements published in the mid-1990s.

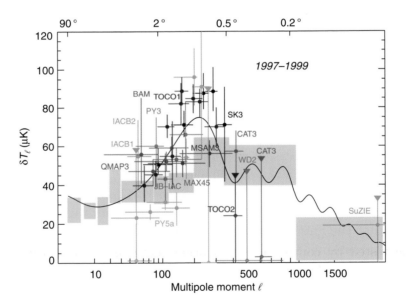

Fig. 5.13. Anisotropy measurements published between 1997 and 1999.

with increasing ℓ to this prominent first peak by measuring the anisotropy spectrum in multiple ℓ-bands. This experiment benefited from the low water vapor in the still arctic air that in the winter tends to settle onto the plains near Saskatoon, Saskatchewan. Some details are noted on page 501 in the

Appendix (where one may also see origins of other experimental group names that are useful to the experts but confusing to the rest of us). The technical advance that made SK and many other experiments possible was a low-noise and broadband amplifier designed by Marian Pospieszalski at NRAO (Pospieszalski 1992; Pospieszalski *et al.* 1994). They were first fielded by Phil Lubin's group at UC Santa Barbara (Gaier *et al.* 1992). The amplifiers were based on high electron mobility transistors, or HEMTs. They were sensitive, required cooling to just $\sim 20\,\mathrm{K}$, and were stable enough to permit new types of observing strategies. SK was important for another reason. The experience gained with the instrumentation was the foundation for the design of the WMAP satellite radiometers we discuss below.

Later in the decade, TOCO (Appendix, p. 505), using experimental techniques similar to SK, showed both the rise to the first peak and the fall toward smaller angles. This experiment avoided most of the noise from the atmosphere by observing from an altitude of $5.2\,\mathrm{km}$ (17,000 ft) in northern Chile. At about the same time, the North American BOOMERANG experiment (detailed on page 506) reduced interference by the atmosphere by raising the experiment in a balloon to $38\,\mathrm{km}$. It also showed the anisotropy peak. These measurements with the others shown in Figure 5.13 localized the anisotropy peak to $\ell_{\mathrm{peak}} \approx 210 \pm 15$. The argument we outlined in the last section indicates that in the CDM model this is where the peak would be if the curvature of sections of constant time were small.[35] That attracted attention because, as we noted on page 442, the inflation scenario argues for small space curvature. In the language of equation (5.9) it means the density parameters satisfy (Dodelson and Knox 2000; Knox and Page 2000; Lange *et al.* 2001, Table 1, lines 4b and 5b; see also the retrospective accounting in Table 4 in Bond, Contaldi and Pogosyan 2003)

$$\Omega_k = 0.06 \pm 0.13. \tag{5.15}$$

This is an interesting constraint on the cosmological model, and one that can be tested by consistency with other methods of measuring the curvature parameter Ω_k (e.g., Bahcall *et al.* 1999). The state of the tests of consistency of this and other measures at the time of writing is summarized in Section 5.4.

At this point we may ask – and perhaps people then should have been asking more often than they did – whether we should accept this constraint.

---

[35] This assumed the distance scale, h, is not very far from astronomers' estimates, a consideration required by the degeneracy mentioned in footnote 27 on page 446. In this case space curvature is the primary factor determining ℓ_{peak}. If Λ may be neglected then the arguments outlined on page 445 show that $\ell_{\mathrm{peak}} \simeq 200\,\Omega_m^{-1/2}$ (Kamionkowski, Spergel and Sugiyama 1994).

It assumes the CDM model. Can we be sure this is an adequate approximation? By the end of the 1990s the improved anisotropy measurements and methods of computation had excluded all proposed alternatives (e.g., Albrecht, Battye and Robinson 1997; Pen, Seljak and Turok 1997). That had an important effect on community thinking, though its weight as evidence is more difficult to judge: how hard had people looked for alternatives? The expectation from inflation that space sections are flat certainly sped acceptance of the evidence that this is so. There was the variant mentioned in footnote 21 on page 442, which allows a significant negative value of Ω_k, but the community feeling generally was that conventional inflation is more elegant, and the accumulating evidence from the CMBR anisotropy peak that Ω_k is small reinforced that.

Opinions on the role of Einstein's cosmological constant Λ were mixed. We noted on page 442 reasons for thinking that Λ ought to be negligibly small: any other plausible value is absurdly large. If Λ vanished equation (5.15) would leave us with the Einstein–de Sitter model. This was generally considered to be the elegant case. And it was consistent with the measurements of the CMBR anisotropy spectrum. There was a modest problem, that this requires that the galaxy distribution is smoother than the mass distribution, and a serious problem, that equation (5.15) with $\Lambda = 0$ requires $\Omega_m = 1$, an uncomfortably large mass density according to other ways to estimate it.

The established solution is to learn to live with Λ; the network of evidence is outlined in Table 5.3. The situation was less clear in the mid-1990s, however. An example is the analysis by Bartlett *et al.* (1995) of the case for the Einstein–de Sitter model.

The high mass density in the Einstein–de Sitter model causes rapid slowing of the rate of expansion of the universe, reducing the time since the big bang. The time also depends on the present rate of expansion, as measured by the Hubble parameter H_0. Bartlett *et al.* remarked that the rapid deceleration could be reconciled with astronomers' measures of ages of the oldest stars if the present expansion rate were slow enough, with H_0 about half the astronomers' estimates. Another constraint comes from the large-scale distribution of the galaxies. An argument similar to the one for the peak of the anisotropy spectrum shows that, in the CDM model, a similarly defined feature in the power spectrum of the galaxy distribution (defined in footnote 18 on page 440) fixes the value of the product $\Omega_m H_0$. The measurement (Efstathiou, Sutherland and Maddox 1990) could accommodate $\Omega_m = 1$, but again only if H_0 were well below the range of astronomers' estimates. There is yet another constraint from clusters of galaxies (White and Frenk 1991; White *et al.* 1993). These systems are large enough that gravity might be expected to have gathered in them a fair sample of the ratio of baryonic to dark matter. The baryon mass M_b in a cluster could be estimated from the X-ray radiation from the plasma (with a modest correction for baryons in stars), and the total mass M_m from the gravity needed to balance the plasma pressure. One has an estimate of the mean baryon density ρ_b from the big bang model for the origin of helium and deuterium. So an estimate of the mean matter mass density is

$$\rho_m \simeq \rho_b M_m / M_b. \tag{5.16}$$

This method also depends on the distance scale, that is, H_0. White *et al.*, using a conventional value of H_0, found a low mass density. Thinking at the time is indicated in the title for their paper: *A Challenge to Cosmological Orthodoxy*. Bartlett *et al.* pointed out that this catastrophe too would be avoided if H_0 were small enough. Lowering H_0 could not accommodate other tests listed in Section 5.4, however. Bahcall and Cen (1992), and Eke, Cole, and Frenk (1996) showed that the abundance of rich clusters of galaxies is a serious problem in a CDM universe with Ω_m close to unity. The same is true of estimates of peculiar velocities of galaxies (relative to the general expansion), which seemed low if $\Omega_m = 1$ and mass is as clustered as the galaxies, independent of h. Fall (1975) first made this point. It was confirmed by the small relative velocities of the galaxies obtained from the first complete galaxy redshift survey (Davis and Peebles 1983; Huchra *et al.* 1983; and further analyzed in Peebles 1986). But one could hope that these measures that could not be finessed by lowering h were misleading. It also required cavalier disregard of the astronomers' measurements of H_0, of course, but the issue was worth considering.

Other papers considered a low-density model without Λ (footnote 21 on page 442) and still others took what proved to be the successful path: keep space sections flat and accept Λ. The arguments (e.g., Peebles 1984; Krauss and Turner 1995; Ostriker and Steinhardt 1995) cited much the same data as Bartlett *et al.*, but with different emphasis, including acceptance of the astronomers' H_0. We take as an indication of the reluctance to go in this direction the influence of preliminary results from a new application of a classical cosmological test, the redshift–magnitude relation (mentioned on page 57). Early results of the application of the test to a class of supernovae seemed to support the Einstein–de Sitter model (Perlmutter *et al.* 1997). Lineweaver *et al.* (1997) reviewed the evidence that pointed to density parameters $\Omega_m \simeq 0.3$ and $\Omega_\Lambda \simeq 0.7$ (numbers close to what were later established), but concluded that this is "excluded by the strict new ... limits from supernovae." When the supernovae results matured (Riess *et al.*

1998; Perlmutter *et al.* 1999), they became an important component in
the web of evidence in Table 5.3 for what now may be named the ΛCDM
model. This is the old CDM picture, but with the addition of Einstein's
cosmological constant, and the later adjustment of a slight departure from
scale-invariance.

Measurements published in the year 2000 include a new class of anisotropy
experiments that used high-sensitivity "spider-web" bolometers. These
detectors were the product of many years of development, first at UC
Berkeley and then in a Caltech–JPL collaboration. They were used in the
BOOMERANG (Appendix, p. 506) and MAXIMA (p. 506) measurements
from balloons that obtained impressively clear maps of the CMBR temperature on pieces of the sky. Examples are shown in Figure 5.14. The maps
show fluctuations around the much larger mean that has been subtracted.
The technology that gave this memorable advance is part of the lineage of
the detectors on the Planck CMBR anisotropy satellite mission.

These and other measurements obtained by the year 2000 fixed the spectrum shown in Figure 5.15 well enough to establish an important new test,
from consistency of two quite different constraints on the mass density in
baryons. First, we remarked (p. 445) that the weight of the baryons affects
the shape of the CMBR anisotropy spectrum. That means the fit of the

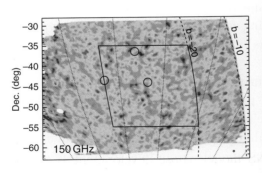
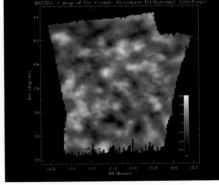

Fig. 5.14. Maps of the CMBR sky in the year 2000 from the BOOMERANG (left;
de Bernardis *et al.* 2000) and MAXIMA (right; Hanany *et al.* 2000) experiments.
The radiation temperature differs from the mean, 2.725 K, by ±0.0003 K in the
hottest and coldest spots. The instrument resolution is finer than the spots and
the instrument noise is small compared to the observed pattern. The color schemes
are different in these two images, as are the positions in the sky, but the data are
consistent. Though previously seen in the power spectrum, these are the first direct
images of the acoustic oscillations. (Reproduced by permission of the AAS and
Nature Publishing Group.)

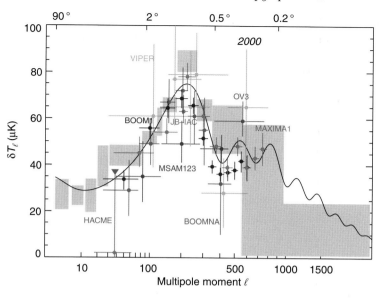

Fig. 5.15. The publications in the year 2000.

ΛCDM model to the measured anisotropy spectrum predicts the baryon mass density, assuming this cosmological model. Second, we discussed in Section 3.1 light element formation in a hot big bang. The fit to the measured abundances, corrected for estimates of what the stars had done, also predicts the present baryon density, assuming this cosmology.[36] The density is measured by the parameter combination $\Omega_b h^2$ (equation 5.8, where the dimensionless Hubble parameter h is defined by $H_0 = 100h\,\mathrm{km\,s^{-1}\,Mpc^{-1}}$). Analyses of data obtained by the year 2000 (Bond $et\ al.$ 2000; Jaffe $et\ al.$ 2001; Lange $et\ al.$ 2001) typically indicated

$$\Omega_b h^2 = 0.030 \pm 0.005,\ \mathrm{CMBR};\quad \Omega_b h^2 = 0.019 \pm 0.002,\ \mathrm{BBNS},\qquad (5.17)$$

from the fits of the big bang model to anisotropy measurements and light element measurements (in the BBNS model whose development is traced beginning on page 23). This was an exciting demonstration that we might actually be able to test the hot big bang model by comparing what is deduced from phenomena that probe very different aspects of the structure and evolution of the physical universe as it is now and was in the remote past.

The fit of the model to the anisotropy measurements in Figure 5.15, with the assumption that the Hubble parameter is close to the astronomers' value within a range that was thought to be conservative (and proves to be so),

[36] There were the complications we noted in Section 3.2 – a convincing test requires a network of checks – but this was an excellent start.

also required that the space curvature parameter is (Bond *et al.* 2000; Jaffe *et al.* 2001; Lange *et al.* 2001)

$$-0.05 \lesssim \Omega_k \lesssim 0.15. \tag{5.18}$$

This tightened the earlier indications for small space curvature.

The years 2001 and 2002 saw refined analyses of the BOOMERANG and MAXIMA data set and a report of the new DASI (Appendix, p. 508) results from a year of data taken with a dedicated 30-GHz interferometer at the South Pole. Here attention was shifting to the existence of the predicted secondary anisotropy peaks at larger ℓ (smaller angular scale). One sees in Figure 5.16 that the combined measurements gave good evidence for a second peak.

One should bear in mind that in the ΛCDM model the CMBR temperature anisotropy – the measured variation of the thermodynamic radiation temperature across the sky – is independent of wavelength. A ground-based 143-GHz bolometric experiment (Romeo *et al.* 2001) observing from Tenerife, combined with earlier HEMT-based measurements at lower frequencies, showed consistency with this condition at $\ell \lesssim 150$.

Figure 5.17 shows important new results in 2003 at both large (Archeops) and small (VSA, CBI, and ACBAR) angular scales. Also shown are results from further refinement of the analysis of the original BOOMERANG

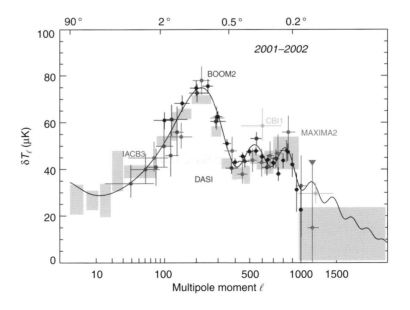

Fig. 5.16. The measurements published in 2001 and 2002.

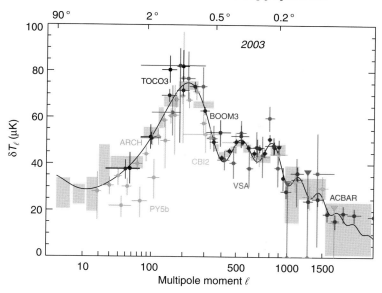

Fig. 5.17. The situation in 2003. As in the previous figures in this sequence, the measurements are discussed and their labels identified in Table A.3 in the Appendix.

experiment. The second acoustic peak is clear, both in the data published that year and in the average over previous years shown as the gray boxes. One also starts to see the third peak and the general decrease in the spectrum at larger ℓ. The latter is in line with what is expected from Silk damping (footnote 15 on page 437).

The fit of the ΛCDM model to the greatly improved measurements in Figure 5.17 (Bond, Contaldi and Pogosyan 2003) required baryon density

$$\Omega_b h^2 = 0.0217 \pm 0.002, \text{ CMBR}, \quad \Omega_b h^2 = 0.0214 \pm 0.002, \text{ BBNS}, \quad (5.19)$$

and space curvature parameter

$$\Omega_k = 0.03 \pm 0.04. \quad (5.20)$$

These are tighter than in equations (5.17) and (5.18), and a clear and demanding check of theory and practice.

The next big advance came from the WMAP satellite; the data from the first year are shown in Figure 5.18. It was clear by 1990 that the best precision in the range of angular scales shown in these figures required the best conditions, well above the atmosphere. Satellite missions were discussed at a NASA workshop on March 27, 1991, by Phil Lubin, UC Santa Barbara (for intrinsic anisotropy and the SZ effect), George Smoot, Andrew Lange, and

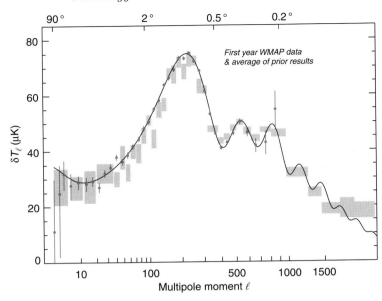

Fig. 5.18. The WMAP measurements compared to an average of all previous results.

Paul Richards, UC Berkeley (for the SZ effect), and David Wilkinson, Princeton University (for intrinsic anisotropy). This was before the COBE/DMR detection shown in Figure 5.11;[37] it was only known then that the intrinsic anisotropy is well below the dipole. There was no guarantee that precision anisotropy measurements could even be interpreted in terms of some stable theory. But our lumpy universe had to have perturbed the CMBR in some fashion, and we are conditioned to believe that good measurements sort out theories.

All three papers at the 1991 workshop mentioned the new technology of HEMTs (High Electron Mobility Transistors, from NRAO) that have considerably better sensitivity than the receivers on COBE/DMR. Wilkinson, arguably the most optimistic, proposed that HEMTs might be stable enough that a single beam in a spinning satellite could scan a circle in the sky fast enough to suppress instrument drifts. Lyman Page (a recent addition to the Princeton Gravity Group) and Wilkinson submitted a proposal to NASA in August 1991 for studies of the CMBR anisotropy with HEMTs, with particular attention to system stability. This proposal returned to a two-beam design. Elements of this proposal became part of the MAP design. In 1992, Norm Jarosik, Page, Wilkinson, and Ed Wollack (then a graduate

[37] Wright sent his first confidential email to the DMR science working group reporting his analysis indicating that the intrinsic anisotropy had been detected on August 17, 1991.

student at Princeton) went to Martin Marietta[38] to talk with Sandy Weinreb and colleagues about building a satellite mission. Later in 1992, the Princeton group met with Mike Jensen, Charles Lawrence, and the JPL group to discuss other possible designs, and then met with Chuck Bennett and his colleagues from the NASA GSFC group. The latter meeting grew into the Microwave Anisotropy Probe (MAP) collaboration, with NASA's Bennett as the Principal Investigator.

In the early 1990s, Bennett had been arguing for a MIDEX-class program within NASA, and also for GSFC support for a CMBR mission proposal. Other groups were also laying foundations for measurements in space. NASA's 1995 call for proposals yielded three candidate CMBR anisotropy satellites. In addition to MAP, a JPL team with Charles Lawrence as Principal Investigator submitted a proposal for the Primordial Structure Investigation (PSI) mission, which also would use NRAO HEMTs, and another Caltech/JPL team with Andrew Lange as Principal Investigator submitted FIRE, which would be based on cryogenic bolometers. In February 1996 a European team proposed to the European Space Agency the COBRAS/SAMBA satellite, which used both HEMTs and bolometers. They were later joined by the PSI and FIRE teams. This became the Planck satellite mission.

NASA selected MAP in April 1996; it was built over the next five years. The photograph in Figure 5.19 was taken on June 30, 2001, the day of the MAP launch. That was just over a decade after Dave's white paper on an anisotropy satellite mission, a relatively short time as such things go. Data taking began in August 2001. In September 2002, Dave died of the cancer he had been battling for 17 years. But he had seen the preliminary maps and knew what MAP could do. When the data were released in February 2003, the satellite was renamed the Wilkinson Microwave Anisotropy Probe in his honor.

The WMAP results in Figure 5.18 may be compared to the weighted but essentially unedited compilation of all previous results shown as the gray boxes in this figure. WMAP substantially reduced the error bars at $\ell \lesssim 500$. The consistency with earlier measurements is excellent – and perhaps even surprising given that no calibration or other systematic errors are included in the compilation of earlier measurements. This is a shining example of

---

[38] This is a company with strong aerospace roots. Names in the alphabet soup in this and the next paragraph include NRAO, the National Radio Astronomy Observatory; JPL, the Jet Propulsion Laboratory; GSFC, the Goddard Space Flight Center; and MIDEX, a mid-size NASA satellite mission, a mid-size NASA satellite mission. At Martin Marietta, Weinreb was already advocating millimeter-wave integrated circuits (MMICs) that later came to the forefront of millimeter-wave coherent detection.

Fig. 5.19. Dave Wilkinson, Dick Bond, and Rashid Sunyaev standing in front of a Saturn V rocket on the day of the WMAP launch.

collective work by the experimental community. The measurement of the anisotropy spectrum at large angular scales, to the left of the first peak in Figure 5.18, requires observations of a large part of the sky. Since Earth interferes with these measurements from the ground and balloons, it is not surprising that the average of prior results most notably departs from WMAP at these scales. The WMAP measurements at small scales, large ℓ, are limited by the angular resolution of its horn antennas. The gray boxes at the right-hand side of the figure show the importance of other methods of observation.

There was no one single thing on WMAP that set the mission apart from previous experiments. Rather it was a combination of attributes designed to work in concert to take full advantage of an orbit beyond the Moon with the spacecraft positioned to allow shielding of the Sun and Earth at the same time (near the Sun–Earth Lagrange point L2). As the instrument scans the sky the Sun is at a fixed angle to the sunscreen. That keeps temperatures of key components close to constant, varying by less than $200\,\mu K$. That meant each radiometer could be calibrated over three years' observation to better than 0.5% accuracy with just three parameters per radiometer to be derived from the measurements. One sees in Figure 5.6 symmetries in optical design that reduce sensitivity to instrumental drifts. The same philosophy informed the strategy of the pattern of scanning the sky. The mission design

-200   T (µK)   +200

Fig. 5.20. The 61-GHz ("V band") full-sky map from WMAP. Emission from matter in the plane of the galaxy produces the red band across the middle. The effective temperature of this foreground component changes with wavelength. Just off the plane this foreground emission is a minor contamination to the fossil radiation. The thermal temperature of the CMBR increases by 400 µK from blue to red. The CMBR anisotropy is seen as the variations of color across the sky. The measurement noise is not significant in this figure. Figure 5.18 shows a statistical measure of this temperature variation across the sky. (Image courtesy of NASA and the WMAP Science Team.)

and the relatively benign conditions in outer space allowed for unprecedented control of systematic errors. The consistency checks in the first year included comparisons of 20 full maps of the sky spanning five wavelengths from ten radiometers. An example of what the instrument obtained is the map of the microwave sky in Figure 5.20.

The elegant image of the microwave sky and the beautiful fit of theory and measurement of the CMBR anisotropy spectrum shown in Figure 5.18 offer a good place to end our story of the CMBR. But an assessment of what these measurements teach us must include an account of the results from other traditions of research in cosmology.

5.4 The cosmological tests

Other lines of research have been examining other phenomena that have yielded their own measures of the universe. These histories merit their own reviews, but that cannot be done here; our goal instead is to draw attention to a striking result. These independently obtained measures of quite different phenomena point to a picture of the large-scale nature and evolution of the universe that agrees with the model that gives the close fit to the CMBR anisotropy spectrum measurements in Figure 5.18. All of these different measurements are difficult and their interpretations hazardous: we are attempting to draw large conclusions from exceedingly limited data.[39] But the consistency of the considerable network of theory and observation we will review in this section argues that we are not likely to have been misled by an inadequate theory or by systematic errors in difficult measurements: we have reason to be confident that the cosmology that fits the evidence, the ΛCDM model, is a useful approximation to reality. That is not to say that the model is reality – we make progress by successive approximations – but that when it is replaced by some deeper theory the successor will predict a universe that behaves much like the early 2008 ΛCDM model.

Our strategy[40] differs from what is done to obtain precision measurements of cosmological parameters under the assumption of the ΛCDM model. That approach tests the model but not the full picture, because the emphasis is on the nominally tightest constraints. We want to present the full variety of phenomena whose measurements probe different aspects of the model, and could have falsified it. We simplify this operation, at the expense of loss of information, by tabulating reductions of the measurements to a relatively few independently measured quantities that can be compared to what is expected from a convenient reference cosmological model. Each of these comparisons is a separate and independent challenge to the model. A difference of measured and expected values greater than what is allowed by the measurement uncertainty would mean either that there is a systematic error in the measurement or that there is something wrong with the reference model. The latter might be remedied by a modest parameter adjustment, or the problem could go deeper.

---

[39] The hazard of basing a cosmology on the CMBR temperature anisotropy alone is illustrated by Turok's (1996) example in which a good approximation to the observed anisotropy spectrum is produced by postulated stresses that causally rearrange the mass, reducing the horizon problem to the origin of large-scale homogeneity. There is no known physical source of the stresses, but this is a cautionary example. Spergel and Zaldarriaga (1997) pointed out that the example may be tested: in the standard model the temperature anisotropy on scales of degrees is present at decoupling, and scattering of the radiation by free electrons during decoupling polarizes the radiation, leaving a characteristic anticorrelation between the CMBR temperature fluctuation and polarization at separations of a few degrees. The effect is observed (Kogut et al. 2003; Nolta et al. 2008), consistent with the standard model. But we are not aware of a search for a refinement of the cautionary example that could fit this measurement.

[40] For other ways to assess the cosmological tests and guides to the literature of the measurements and their interpretation, see Bennett et al. (2003), Spergel et al. (2003, 2007), Page et al. (2003b), Reichardt et al. (2008), and Ratra and Vogeley (2008).

Table 5.2. *Reference cosmological model*

Parameter	Reference value
Distance scale	$h = 0.72$
Expansion time	$t_0 = 13.9$ Gyr
Density parameters	
Dark energy	$\Omega_\Lambda = 0.74$
Dark matter	$\Omega_{CDM} = 0.21$
Baryons	$\Omega_b = 0.044$
Neutrinos	$\Omega_\nu < 0.02$
Radiation	$\Omega_r = 10^{-4.33}$
Space curvature	$\Omega_k = 0$
Primeval mass fluctuations	
Amplitude	$\sigma_8 = 0.80$
Spectral index	$n_s = 0.96$
Opacity after decoupling	$\tau = 0.09$

Our reference model in Table 5.2 is from the fit of the ΛCDM model to the WMAP CMBR anisotropy measurements (in the tabulation released in March 2008 by Dunkley *et al.*) under the condition of zero space curvature and constant Λ.[41] The distance scale refers to the relation between redshift and distance, $cz = H_0 r$, normalized to $H_0 = 100\,h\,\mathrm{km\,s^{-1}\,Mpc^{-1}}$. The density parameters Ω_i for the important dynamical actors are defined in equation (5.8) (and equation G.1); they translate to mass densities $\rho_i = 1.88 \times 10^{-29}\,\Omega_i h^2\,\mathrm{g\,cm^{-3}}$. The Ω_i sum to unity. Several tests are sensitive to the measure Ω_m of all low-pressure matter, the sum of CDM, baryons, and massive neutrinos. In the reference model, $\Omega_m = 0.26$.

The constants σ_8 and n_s fix initial conditions for the evolution of the mass distribution out of initially adiabatic Gaussian random mass density/curvature fluctuations. The measure of the departure from a uniform mass distribution is the present value σ_8 of the root-mean-square fractional fluctuation of the mass within randomly placed spheres with radius $8h^{-1} \simeq 11$ Mpc. The power spectrum (defined in footnote 18 on page 440) of the primeval mass distribution has the power-law form $\mathcal{P}(k) \propto k^{n_s}$. The anisotropy measurements at the time of writing seem to prefer a slight tilt from scale-invariance, $n_s = 0.96$. This would be a falsification of an argument

[41] We emphasize that this is not meant to be the best fit to all constraints at the time of writing. Rather, the reference model is meant to be independent of the tests discussed in this section, to the extent that is possible, so that each serves as a separate and independent challenge to the cosmology. These challenges are not really independent, because they influenced thinking on flat space sections, but the convention is a not unreasonable approximation.

from elegance, but it finds a natural interpretation in the inflation scenario
(footnote 19 on page 441).[42]

The last entry is the optical depth τ for scattering of the CMBR by free
electrons in intergalactic plasma present now and back to redshift $z \sim 15$.
Scattering reduces the CMBR anisotropy, and it also contributes to the
polarization of the CMBR. The large-scale polarization has been measured
(Page *et al.* 2007), fixing the value of τ. The result is not inconsistent with
what is known about the sources of the radiation that reionized the baryons,
but at the time of writing this is a loose constraint.

The cosmological test results are presented – in highly condensed forms –
in Figure 5.21, representing the CMBR anisotropy measurements, and in the
fifth column of Table 5.3, representing the suite of other cosmological tests.
These plots show the difference $(M - R)/\sigma$ between a measured quantity,
M, and its predicted value, R, in the reference model, divided by a measure,
σ (the standard deviation), of the uncertainty in the measurement. In the
ideal case where the world being measured is described by a reference model
that is adequate given the precision of the measurements, and the measurement errors are reliably known (and have the simple Gaussian normal error
distribution), two-thirds of the $(M - R)/\sigma$ are between -1 and $+1$, and
95% are between -2 and $+2$. The error flags in the plots in Figure 5.21 and
Table 5.3 have unit length. This means that if the test results are satisfactory then two-thirds of the error flags cut the line at zero difference between
theory and practice.

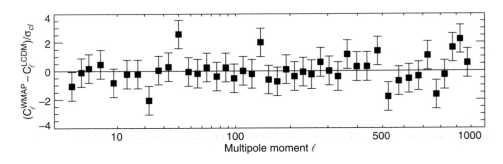

Fig. 5.21. Cosmological tests. This condenses the CMBR anisotropy test to a plot of
the difference between measurement and reference model prediction in units of the
standard deviation in the measurement, averaged over ranges of ℓ. The horizontal
axis is the same as in Figure 5.18. The fifth column of Table 5.3 shows a similar
condensation of the other cosmological tests.

---

[42] Steinhardt and Turner (1984) showed examples of simple implementations of inflation in which
$0.7 \lesssim n_s \lesssim 0.9$, and they showed how to construct implementations with n_s closer to unity.
That is, the evidence of tilt is encouraging for the inflation concept.

Table 5.3. *Cosmological tests*

	Parameter	Fiducial	Measured	$(M-R)/\sigma$
Baryon density				
BBNS	$\Omega_b h^2$	0.0227	0.0219 ± 0.0015	
Baryon budget	Ω_b	0.042	>0.005	
Stellar evolution ages	$t_*,\ \mathrm{Gyr}$	13.7	12.3 ± 1.0	
Distance scale				
Distance Ladder	h	0.72	0.69 ± 0.08	
Gravitational lensing	h	0.72	0.75 ± 0.07	
SNeIa distance modulus	$\delta\mu(z=1)$	1.00	0.99 ± 0.08	
Large-scale structure				
Matter power spectrum	$\Omega_m h$	0.187	0.213 ± 0.023	
Baryon acoustic oscillation	Ω_m/h^2	0.50	0.53 ± 0.06	
Dynamical mass estimates				
Galaxy velocities	Ω_m	0.26	$0.30^{+0.17}_{-0.07}$	
Lensing around clusters	Ω_m	0.26	0.20 ± 0.03	
Lensing autocorrelation	$\sigma_8\Omega_m^{0.53}$	0.39	0.40 ± 0.04	
Galaxy count fluctuation	$\sigma_8(g)$	0.80	0.89 ± 0.02	
Rich clusters of galaxies				
Present mass function	$\sigma_8\Omega_m^{0.37}$	0.49	0.43 ± 0.03	
Mass function evolution	σ_8	0.80	0.98 ± 0.10	
	Ω_m	0.26	0.17 ± 0.05	
Cluster baryon fraction	$\Omega_b h^{3/2}/\Omega m$	0.103	0.097 ± 0.004	
Baryon evolution	$\Omega_\Lambda + 1.1\Omega_m$	1.03	1.2 ± 0.2	
Lyα forest	n_s	0.96	0.965 ± 0.012	
Neutrino density	$\Omega_\nu h^2$	<0.02	0.001	
ISW	detected, at about the fiducial prediction			

The situation for the CMBR anisotropy measurements is satisfactory.[43] Considering that these measurements are represented by many $(M-R)/\sigma$ (that are lumped in averages over ranges of ℓ in Figure 5.21), the control and understanding of errors required to achieve this fit is an impressive accomplishment.

The situation for the other tests is not as close to ideal.[44] That is not surprising because it is difficult to understand the errors in many of these tests, in part because some are works in progress, and in part because some depend on complex processes we never will be able to analyze to a high degree of

---

[43] A standard deviation in an anisotropy spectrum measurement is the sum in quadrature of the standard deviation in the measurement and the sampling error resulting from the limited size of the observable universe. The sum of the squares of the 43 ratios plotted in Figure 5.21 is $\chi^2 = \sum(\mathrm{M}-\mathrm{R})^2/\sigma^2 = 35$. Taking account of the six free choices of parameters, this is close to what it is expected to be. For a closer statistical measure of the goodness of fit, see Nolta *et al.* (2008).

[44] The sum of the squares of the 16 entries plotted in the last column in Table 5.3 is $\chi^2 = \sum(\mathrm{M}-\mathrm{F})^2/\sigma^2 = 26$. This is 1.6 times what one would expect if the error flags were reliable.

accuracy. Whether the model fit to these tests is acceptable under these conditions is a more subjective judgment call. But that is part of science.

The first two columns in Table 5.3 list cosmological quantities, in some cases expressed in convenient approximations. The predicted values in the reference model are in the third column. The fourth column shows measured values derived from the observations by the application of standard physics. In some cases that includes the full machinery of the ΛCDM model (with constant Λ). The uncertainties assigned to the measured quantities have mixed provenance. For the purpose of this discussion, it is best to consider them to be about one standard deviation, σ. That is assumed in the fifth column in the comparison of the difference of measurement and model relative to the measurement uncertainty.

Our explanations of the entries are brief, and we offer only a few references to early work and to the situation at the time of writing. It must be emphasized that this is a brutal condensation of what was done.

The tests in Table 5.3 are grouped in categories, starting with the mean mass density in baryons. The first of the measures is from the big bang model for the formation of the light elements (BBNS). We began our story with the origin of this model, at a time when the CMBR was an often overlooked byproduct. Critical advances in the art of element abundance measurements led to the situation reviewed in Steigman (2007). We use Steigman's translation from measured abundances of deuterium, which appears to be the most useful isotope for this purpose, to the inferred abundance before stars started burning deuterium, and from there to the baryon density $\Omega_b h^2$ in the hot big bang cosmology that produces this deuterium abundance.[45] The comparison is to the reference model that produced the consistent match of theory and observation of the CMBR anisotropy in Figure 5.21. We have noted, and it is worth emphasizing again, that this comparison is critically demanding because it compares what is derived from two quite different phenomena. This is a link in the network of tests.

As indicated in the second entry, the total baryon mass observed in stars and plasma is only 10% of reference (Fukugita and Peebles 2004). It is thought that the rest is in plasma at densities and/or temperatures low enough not to be a significant X-ray source. This hypothesis is not tested at the time of writing. But the entry is relevant because the cosmology could have been falsified by evidence that there are more baryons around

---

[45] Fits of the baryon density to estimates of the primeval abundances of the isotopes produced in significant abundances in the early universe, ^{4}He, ^{3}He, ^{7}Li, and deuterium, the stable heavy isotope of hydrogen, are not demonstrably consistent within estimated uncertainties. This may be a simple consequence of the complexities of the astronomy of abundance estimates, or it is possible that new physics causes minor adjustments to the theoretical abundances.

us than is allowed by the element abundances or by the CMBR anisotropy. The baryons in rich clusters of galaxies are hot enough to be observable X-ray sources; that yields constraints from clusters of galaxies discussed below.

The comparison of evolution ages, t_*, of the oldest stars and the predicted cosmic expansion time, t_{cosmic}, became meaningful when Sandage removed the early errors in galaxy distance measurements. Sandage (1961) demonstrated that t_* and t_{cosmic} are close. That is another critical comparison of what is inferred from quite different observations, here of distant galaxies and nearby stars, and another link in the network.

In Table 5.3, we adopt the mean of t_* from Krauss and Chaboyer (2003) and Gratton *et al.* (2003). We do not correct for the time interval from the big bang to the formation of the stars they analyzed, so it is not surprising that the reference time from the big bang is the longer. The astrophysics of star formation is not well enough understood to check consistency of the time difference. If we were attempting formal statistical tests of significance we would have to do something about that, but the near consistency of t_* and t_{cosmic} makes the point for our purpose.

There is a long tradition of astronomical calibration of luminosities of Cepheid variable stars and other objects for the purpose of establishing the scale of distances to the galaxies. The first distance scale entry in the table is the mean from three recent studies (Freedman *et al.* 2001; Riess *et al.* 2005; Sandage *et al.* 2006).

The second distance scale entry (Koopmans *et al.* 2003) uses the physics of gravitational deflection of light in a particularly well-observed situation, where the mass in and around a pair of galaxies produces four observed images of one background source galaxy. A separate entry is indicated because this uses gravitational lensing on scales far larger than the precision tests of general relativity. The steps in the application are worth noting: observations of the three time delays along the four paths of radiation received from the variable radio source, along with analysis of the mass distribution in the lensing galaxies, are fitted to the reference cosmology (or close to it) by adjusting the length scale. There are two critical comparisons: to the length scale derived by the quite different observations in the first entry and to the reference model.

Another test is based on the frequency of occurrence of such multiple images of a distant quasar or radio source that happens to be close enough to the line of sight of a massive foreground galaxy. The condition for multiple images depends on the spacetime geometry (through the angular size distance defined in the Glossary). It is also sensitive to the abundance of the massive galaxies that produce most of these strong lensing multiple images. Mitchell *et al.* (2005) show that the measured rate of strong lensing favors a significant positive value of Ω_Λ, but at the time of writing this is a preliminary result so it is not entered in Table 5.3.

Spacetime geometry also affects the observed brightnesses of distant objects; the application in the table is to the redshift–magnitude relation mentioned on page 57. Tolman (1934) discussed this relation. It requires comparison of the brightness in the sky of objects at different redshifts that have close to the same intrinsic luminosity. Sandage (1961) measured this relation for the most luminous galaxies, which are close to this condition, in an attempt to distinguish among cosmological models. Riess *et al.* (1998) and Perlmutter *et al.* (1999) realized Sandage's goal using supernovae (SNeIa) instead of galaxies. In the table, the measurements of SNeIa are represented as the difference $\delta\mu(z)$ between the measured distance modulus $m - M$ and the function $m - M = 5\log_{10} z + C$ with the constant C normalized to the low-redshift supernovae. The entry is the value of $\delta\mu$ at $z = 1$ based on the measurements presented in Wood-Vasey *et al.* (2007).

The next two entries are measures of the large-scale mass distribution. In the CDM model the mass fluctuation power spectrum bends from the primeval form $\mathcal{P} \propto k^{n_s}$ at large scales to approach $\mathcal{P} \propto k^{n_s-4}$ with increasing wavenumber k (because the mass density in radiation suppresses early growth of shorter wavelength modes of the CDM distribution). The observed length scale at the bend gives a measure of the product $\Omega_m h$. Efstathiou, Sutherland and Maddox (1990) found that in their mapping of the large-scale galaxy distribution this measure of $\Omega_m h$ is too small for an Einstein–de Sitter universe (with $\Omega_m = 1$) with a conventional estimate of the distance scale h. As we noted, some at the time felt that this may mean h has been overestimated, others that the primeval mass fluctuation power spectrum has a bend that happens to mimic the effect of a small $\Omega_m h$, and others, with Efstathiou *et al.*, that we may have to learn to live with Λ. The last of these ideas became part of the reference model. The entry for $\Omega_m h$ in Table 5.3 is from Tegmark *et al.* (2004).

The second entry in this category is a measure of the effect of the acoustic oscillation of the matter-radiation fluid in the early universe. We discussed this on page 437. The oscillation produces the ripples in the CMBR anisotropy spectrum shown in Figure 5.18, and it has a similar though less pronounced effect on the present matter power spectrum, $\mathcal{P}_0(k)$. The ripples in $\mathcal{P}_0(k)$ produce a bump in its Fourier transform, which is the mass autocorrelation function. A straightforward illustration of the theory, absent the CDM, is in Peebles (1981b). Eisenstein *et al.* (2005) detected the effect.

Percival (in a private communication) points out that the length scale defined by the bump is largely sensitive to Ω_m/h^2. Percival derives the value in the table from the analysis in Percival *et al.* (2007), under the assumption that $\Omega_b h^2$ is close to the value in the first entry in this table. That correlates the tests, but the close consistency with Ω_m/h^2 in Table 5.3 argues for the security of this assumption.

The departure from a uniform mass distribution produces a gravitational field that causes galaxies to move relative to the general expansion of the universe. If galaxies adequately trace mass then observations of these galaxy peculiar velocities give measures of Ω_m independent of h. A large literature on galaxy dynamics (beginning with the studies summarized in footnote 13 on page 31, and continuing through the studies in the 1980s noted on page 457) favors a low value of Ω_m, in the neighborhood of the first of the dynamical mass estimates in the table (Feldman *et al.* 2003). Galaxies cannot precisely trace mass – different types of galaxies have different distributions – but since in the standard model gravity gathers together mass and galaxies alike it would be surprising if the distributions of galaxies and mass were very different. This means that although studies of galaxy dynamics cannot yield precise measures of Ω_m they do add a link to the network of evidence of the accuracy of the model as a description of the real world.

The second entry refers to the study of the gravitational deflections of light paths that distort – lens – images of background galaxies (Kaiser and Squires 1993). Sheldon *et al.* (2007) used observations of the gravitational distortion of images of background galaxies to estimate the ratio of mean concentrations of mass and light around clusters of galaxies. That multiplied by the cosmic mean luminosity density is the estimate of Ω_m in the second entry in this category. The third entry (Fu *et al.* 2008) compares the autocorrelation of the lensing distortion across the sky to the mass autocorrelation function, which is assumed to be the ΛCDM model except that σ_8 and Ω_m are adjusted to fit this set of measurements. The best constrained combination is shown.

The rms fluctuation $\sigma_8(g)$ of counts of normal galaxies (such as the Milky Way) found in a sphere of radius $8h^{-1}$ Mpc can be compared to the rms mass fluctuation σ_8 in the reference model. The long history of development of the art of measurement of the statistics of the galaxy distribution has led to measurements so tight (Tegmark *et al.* 2004) that the difference between σ_8 and $\sigma_8(g)$ is formally very significant: it is off scale in the fifth column in Table 5.3. As we have noted this is not a problem for the model, because galaxies are not precision tracers of mass, but the similarity of values of σ_8

and $\sigma_8(g)$ adds to the case that the reference model is broadly consistent with reality.

Rich clusters of galaxies are the largest gravitationally bound mass concentrations. Their structures are thought to be simple enough to allow derivation of measures of the conditions out of which they grew. The cluster mass function – the number density as a function of mass and redshift – gave important early indications that Ω_m is well below unity (Bahcall and Cen 1992; Eke, Cole and Frenk 1996). The first entry in this group uses masses of low-redshift clusters derived from the gravitational field needed to counter the pressure of the intracluster plasma (Reiprich and Böhringer 2002) and masses from gravitational lensing (Dahle 2007). We show the more tightly measured parameter combination. The other two entries (Bahcall and Bode 2003) add the evolution of the cluster mass function.

White and Frenk (1991) and White *et al.* (1993) made the important point that a rich cluster of galaxies is large enough that gravity might be expected to have drawn together a close to fair sample of the cosmic ratio of mass densities in baryons and dark matter. The baryon mass density $\rho_b(r)$ as a function of radius in a cluster may be estimated from the X-ray radiation of the intracluster plasma, with a modest correction for the baryons in stars. The total mass density $\rho_m(r)$ is estimated from the gravity needed to balance the plasma pressure. The ratio $f = \rho_b(r)/\rho_m(r)$ may be compared to the cosmic ratio Ω_b/Ω_m (with due correction for estimates from numerical simulations of the plasma mass lost by the cluster). That constrains the parameter combination in the next entry. The last entry in this category comes from the condition that f does not vary with time, which fixes the indicated parameter combination. The measured values for these last two entries are from Allen *et al.* (2004).

Quasar spectra show a forest of Lyman-α resonance absorption lines produced by the clumpy distribution of atomic hydrogen along the line of sight. This is what would be expected from the growth of clustering of matter in the ΛCDM model (Cen *et al.* 1994; Croft *et al.* 1998). The qualitatively new probe offered by these observations is the measure of the mass distribution at redshift $z \sim 3$ on relatively small scales, ~ 3 Mpc (in coordinates fixed to the general expansion). That compared to the mass distribution measured on larger scales at lower redshift affords an independent probe of the power spectrum index n_s entered in Table 5.3 (Seljak, Slosar and McDonald 2006).

The standard model contains a sea of neutrinos left from thermal equilibrium with the radiation at high redshift. The present number density in each family is close to $100\,\mathrm{cm}^{-3}$. That multiplied by the neutrino rest mass and summed over families is the mass density in this form. The reference value in Table 5.2 is an upper bound from the fit to anisotropy measurements (Ichikawa, Fukugita and Kawasaki 2005). The measured quantity in Table 5.3 is from lower bounds from measured neutrino mass differences. There is space between the two, but the test is to be recorded: it could have falsified the model.

The gravitational field produced by the expanding and increasingly clumpy mass distribution tends to vary with time, and that perturbs the energy of the CMBR. On large scales, where the fluctuations are small (so linear perturbation theory is good approximation), the effect on the CMBR vanishes in the Einstein–de Sitter model. If the mass density is lower than Einstein–de Sitter the perturbation produces a positive correlation between the CMBR temperature and the mass distribution along the same line of sight traced by galaxy counts and by X-ray and radio tracers of the galaxies. If the mass density is larger it produces an anticorrelation. This integral Sachs–Wolfe (1967; ISW) effect is observed as a positive correlation at about the level expected in the reference model (Boughn and Crittenden 2005; Ho *et al.* 2008). The measurement is not precise but it is a significant addition to the network of tests.

Yet another test is based on the CMBR polarization produced by free electron scattering near the surface of last scattering. We have mentioned the relation between the temperature and the polarization. The polarization power spectrum, first seen by Kovac *et al.* (2002), has been shown to agree with the reference model (Nolta *et al.* 2008).

The decision on when the weight of experimental evidence forces the transition of a useful working hypothesis into a convincingly established standard model is in part a personal one and in part a herd effect. The consistency – within reasonable uncertainties of difficult measurements – of these probes of many different aspects of the large-scale nature of the universe convinces us that the spatially flat ΛCDM hot big bang cosmological model is a good approximation to reality. Measurements of properties of the thermal sea of microwave radiation, the CMBR, played an important role in developing this remarkable advance in science. One sees from Table 5.3 that the same is true of other traditions of research. Cosmology has become big science.

5.5 Lessons

What does this story teach us? We have an example of the durability of concepts in physics. Einstein completed general relativity theory almost a century before the very successful applications shown in Figure 5.18, in the fit of theory and observation of the CMBR anisotropy, and in Table 5.3, in the fit to other cosmological tests. Einstein's test of the theory was his demonstration that it explains the departure of the orbit of the planet Mercury from the prediction of Newton's theory of gravity. The radius of the orbit of Mercury is about 10^{13} cm. Cosmology applies the theory at the Hubble length, 10^{28} cm. This is an extrapolation of some 15 orders of magnitude from Einstein's test. The success of the extrapolation is a striking demonstration of durability.

We have an illustration also of the wonderfully modest variety of fundamental concepts in physics. We mentioned that all of the elements of the relativistic ΛCDM theory for the anisotropy spectrum were discussed before ΛCDM was invented. For example, the theory of radiative transfer is important in analyses of laboratory experiments, Earth's atmosphere, the structures of stars, and the process of decoupling of matter and radiation that determined the shape of the curve in Figure 5.18. We admonish students in physical science to learn the fundamentals of physics because they tend to keep reappearing.

Good science also demands recognition that this durability of physics does not offer an entirely reliable guide to the future. Though many elements of physics were developed before they were seen to be needed for the interpretation of the CMBR measurements, many other elements that seemed to be equally interesting were in the end found not to be relevant, at the level of accuracy of the tests we discussed in the last section. An example is cosmic strings and textures. The motivation for considering them is excellent, but the CMBR data show no evidence of them, at the time of writing. An example of another sort, and the most widely celebrated surprise in cosmology in the Decade of Discovery, is the evidence of detection of the cosmological constant Λ. The concept certainly is durable – Einstein introduced it in 1917 – but its popularity has varied. Einstein rejected it as logically unnecessary. In the 1970s many particle physicists were glad to agree, and to explain why the notion of this term is quite absurd. But the evidence later showed that we must learn to live with Λ or something that acts very much like it.

Hogg (p. 70) shows us another lesson: in the symbiosis of pure science and practical engineering the benefits flow both ways. The low-noise amplifiers used in the Bell Laboratories communications experiments he describes were the product of curiosity-driven research on the quantum interaction of matter and radiation. The experiments were part of spectacular advances in the technology of communications. They also revealed a phenomenon, fossil radiation, that hugely benefited the study of the large-scale structure of the universe, a subject we suppose is as purely curiosity-driven as any.

Yet another lesson is the tenacity of scientists, which can approach the durability of concepts in science. There is nothing special about cosmology in this regard except that the subject has been so small that it is practical to look into this tenacity in some detail. Consider that in the early 1930s Hubble and Tolman were leading the drive to construct the great 200-inch telescope in Southern California, for the purpose of testing the idea of an expanding universe. Tolman's book describes what might be learned about spacetime if it were possible to observe sufficiently distant – and faint – galaxies, and compare them to nearby ones (Tolman 1934). Sandage continued this program through the 1960s, using the 200-inch telescope and other instruments. Some four decades after that the visions of Hubble, Tolman, Sandage, and others were realized: the observations – of supernovae rather than galaxies – have been deeply important to testing the theory of the expanding universe. In this book we have traced the development of a parallel set of ideas, on a hot big bang, from the first explorations of the physics by Gamow and his colleagues in the 1940s, to the burst of activity in research related to the CMBR in the 1960s, and to the enormous effort in the following four decades that brought the subject to Figure 5.18 and Table 5.3. This is the fruit of some four generations of scientists since the discovery of the idea of an expanding universe. All the work of all those generations produced great advances. The cosmology community is not satisfied; however, because one can see how the theory and measurements might be improved. But it is not likely that the communities of science will ever be satisfied.

Appendix

We want to convey an impression of the broad effort that went into the measurements of the CMBR without obscuring the story with details. To that end we present separately, in this Appendix, tabulations of the experiments that figure in the panoramic illustrations of the progress of the measurements in the figures in Chapter 5. We present first a tabulation of observations of the interstellar CN absorption lines that gave important early measures of the CMBR energy spectrum. The next, longer, tabulation is of the experiments that explored the energy spectrum of the CMBR. The progress on this front is illustrated in Figures 5.2–5.4. The third and still longer tabulation shows the experiments in the enormous effort to map out the intrinsic anisotropy of the CMBR, in the developments illustrated in Figures 5.9–5.18.

Table A.1 lists the progress in measurements of the CMBR energy spectrum at the two energy level separations in the CN molecule that happen to be conveniently located for this purpose. The general method of observing the excitation of interstellar CN and interpreting the excitation in terms of the CMBR spectrum is discussed in Chapter 3 (commencing on page 42). The essays in Chapter 4 (starting on page 74) recall what happened when this interpretation was first generally realized. The observations yield measures of the CMBR spectrum at the two frequencies indicated in the table and plotted as open boxes in Figures 5.2 to 5.4. It was very important to early studies of the CMBR that the CN observation at 113 GHz combined with the measurements at lower frequencies, which could be made from the ground, showed that the CMBR has a Rayleigh–Jeans power-law spectrum with the downward departure from the power law at 113 GHz to be expected for a thermal spectrum as it heads to its peak. The table shows the advances in tightening this measure and adding another at 227 GHz.

Table A.1. *The CMBR from interstellar CN, 1965 to 1993*

Reference	T_0 (K) 113 GHz	T_0 (K) 227 GHz
Field & Hitchcock (1966)	3.05 ± 0.35	–
Peimbert (1968)	3.7 ± 0.7	–
Hegyi et al. (1972)	2.7 ± 0.13	$2.9^{+0.5}_{-0.7}$
Thaddeus (1972); Clauser (1970)	2.78 ± 0.1	–
Hegyi et al. (1974)	–	$2.9^{+0.4}_{-0.5}$
Blades (1978)	2.80 ± 0.4	–
Crutcher (1985)	2.82 ± 0.08	–
Meyer & Jura (1985)	2.70 ± 0.04	2.76 ± 0.2
Crane et al. (1986)	2.74 ± 0.05	2.75 ± 0.24
Meyer et al. (1989)	2.76 ± 0.07	2.83 ± 0.09
Crane et al. (1989)	$2.796^{+0.014}_{-0.039}$	–
Kaiser & Wright (1990)	2.75 ± 0.04	–
Palazzi et al. (1990)	2.834 ± 0.085	2.831 ± 0.056
Roth et al. (1993)	$2.729^{+0.023}_{-0.032}$	2.656 ± 0.057
Roth & Meyer (1995)	$2.742^{+0.021}_{-0.022}$	2.679 ± 0.057

Table A.2 shows the progress in direct measurements of the CMBR energy spectrum. Section numbers are entered where we have recollections of the measurements from Chapter 4. In some of the other cases we have added a few words to indicate special aspects of the measurement.

We come next to the long sequence of experiments that yielded the CMBR anisotropy spectrum that has proved to be so very informative. Table A.3 is a list of what we believe to be all the anisotropy experiments performed between 1967 and 2003 that probed what we now term the primary anisotropy. This corresponds to spherical harmonic multipoles $\ell \lesssim 2600$, or angular scales greater than about five minutes of arc (in the notation of equation 5.10). The measurements are plotted in the sequence of figures beginning on page 448.

The experiments are listed in order of the publication date of their first CMBR science paper. A strong preference has been given to papers published in refereed journals. In addition, preference is given to those data analyses done by the core team members. In many places we note a "first" for the purpose of helping to chart the path of new ideas and trends in the field. The basis of the claim is the publication date.

The convention of plotting anisotropy results as a function of the spherical harmonic index ℓ did not catch on until the mid-1990s. To place a point on the power spectrum, one must know its sensitivity to a "window"

Table A.2. *Absolute temperature measurements between 1965 and 2006*

Reference[a]	ID	Frequency (GHz)	T_0 (K)
Penzias & Wilson (1965a) §4.5.1, 4.5.2	PW65	4.08	3.3 ± 1^b
Roll & Wilkinson (1966) §4.7.2, 4.7.3	RW66	9.0	3.0 ± 0.5
Howell & Shakeshaft (1966) §4.10.2	HS66	1.55	2.8 ± 0.6
Howell & Shakeshaft (1967b)	HS67	0.4 & 0.6	3.7 ± 1.2
Penzias & Wilson (1967)	PW67	1.42	3.2 ± 1
Welch *et al.* (1967) §4.10.3	We67	20.0	2.3 ± 0.8^c
Wilkinson (1967)	W67	35.1	$2.56^{+0.17}_{-0.22}$
Stokes *et al.* (1967) §4.7.4, 4.10.6	St67	9.37	$2.69^{+0.16}_{-0.21}$
		19.0	$2.78^{+0.12}_{-0.17}$
Ewing *et al.* (1967) Switched superheterodyne receiver with LHe cold load on White Mountain	Ew67	32.5	3.16 ± 0.26
Puzanov *et al.* (1967) Superheterodyne receiver with LN_2 cold load	Pu68	36.6	3.7 ± 1^d
Boynton *et al.* (1968) §4.10.5	Bo68	90.0	$2.46^{+0.40}_{-0.44}$
Pelyushenko & Stankevich (1969) §4.10.4 Coherent receiver with LN_2 cold load	PS69	1–2	2.5 ± 0.5^e
Stankevich *et al.* (1970) Coherent receiver using the moon as a moving calibrated screen	Sa70	0.65	3.0 ± 0.5
Kislyakov *et al.* (1971) Switched heterodyne receiver with LN_2 cold load at the 3 km altitude Shternberg Institute	Ki71	83.8	2.4 ± 0.7
Millea *et al.* (1971) Superheterodyne receiver with LHe cold load	Mi71	90.4	2.61 ± 0.25
Houck *et al.* (1972) §4.10.7 Photodetectors on a rocket-borne LHe-cooled telescope	Ho72	231	< 4.1
Williamson *et al.* (1973)	Wi73	50–380	$3.4^{+1.4}_{-3.4}$
bolometers at 2 K in		50–500	$5.1^{+0.8}_{-1.5}$
a rocket-borne photometer		50–1000	$3.8^{+0.8}_{-1.9}$

[a] We found the reviews by Danese and De Zotti (1978), Weiss (1980), and Partridge (1995) particularly useful. These authors made corrections to some of the published values that we have adopted where noted.

[b] The original result, $T_0 = 3.5 \pm 1$ K, was corrected to $T_0 = 3.3 \pm 1$ K in Penzias (1968).

[c] The original result was $T = 2.0 \pm 0.8$ K and was later corrected as discussed in §4.10.3. Weiss and Partridge convert to thermodynamic temperature and find $T_0 = 2.45 \pm 1$ K. We do not include this correction.

[d] The published $T = 2.9 \pm 0.7$ K was converted by Weiss to thermodynamic temperature $T_0 = 3.7 \pm 1$ K. Partridge additionally corrects the Puzanov *et al.* atmospheric calculation to give $T = 2.9 \pm 0.9$ K. We use only the thermodynamic correction.

[e] We have increased the errors by $\sqrt{3}$ to account for the galactic subtraction. Additionally, Partridge (1995) raises concerns about the subtraction of the atmospheric contribution.

Table A.2. (*cont.*)

Reference	ID	Frequency (GHz)	T_0 (K)
Muehlner & Weiss (1973a) §4.10.10 Balloon-borne 1.8 K InSb detector in LHe-cooled telescope with filter wheel	MW73	30–160	$2.55^{+0.45}_{-0.55}$
Muehlner & Weiss (1973b)	MW73b	30–230	2.8 ± 0.2
		30–330	< 2.7
		30–550	< 3.4
Boynton & Stokes (1974)	BS74	90.0	$2.48^{+0.50}_{-0.54}$
Robson *et al.* (1974)	Ro74	60–270	2.73 ± 0.08^f
Woody *et al.* (1975) Balloon-borne 1.65-K FTS calibrated on the atmosphere and an external load	Wo75	120–520	$2.99^{+0.07}_{-0.14}$
Otoshi & Stelzried (1975)	OS75	2.3	2.66 ± 0.26^g
Dall'oglio *et al.* (1976) Ground-based 2-K bolometric receiver with external LHe load at 3500 m Testa Grigia site	Da76	210–300	< 2.7
Woody & Richards (1979) Balloon-borne 4-K FTS with a 0.3-K bolometer	WR89	50–500	$2.96^{+0.04}_{-0.06}{}^h$
Gush (1981) Rocket-borne 4-K FTS with a 0.3 K bolometer	G81	25–720	2.7–2.9^h
de Amici *et al.* (1984)	De84	33	2.87 ± 0.21^i
		90.9	2.4 ± 1.0
Friedman *et al.* (1984)	Fr84	10.0	2.91 ± 0.19^i
Mandolesi *et al.* (1984)	Ma84	4.76	2.71 ± 0.2^i
Sironi *et al.* (1984)	Si84	2.5	2.62 ± 0.25^i

[f] This pioneering measurement was the first to report a result from an FTS, an instrument similar to that used by the UBC, Berkeley, and COBE groups. The quoted temperature comes from Robson and Clegg (1977). The atmosphere was not accounted for in the published result and there were apparent inconsistencies in the calibration.

[g] Used a maser plus superheterodyne receiver calibrated with waveguide LHe and other loads. The data were taken in 1967 and presented in a JPL progress report, and submitted for publication in 1975 "since there is still considerable current interest in the [CMBR]." We convert the quoted 3σ error of 0.77 K to 0.26 K. Partridge suggests an atmospheric correction and finds $T_0 = 2.76 \pm 0.3$.

[h] The Gush data, and to a lesser extent the Woody–Richards data, were not good fits to a blackbody curve. Additionally, Gush had to model and subtract the emission from the rocket nose cone that drifted into his field of view. See page 417.

[i] A coordinated set of measurements was performed at the Barcroft Laboratory on White Mountain, California. They used Dicke-switched superheterodyne receivers, total power receivers, and large LHe loads. Summaries are given in Smoot *et al.* (1983, 1985, and 1987).

Table A.2. (*cont.*)

Reference	ID	Frequency (GHz)	T_0 (K)
Peterson *et al.* (1985)	Pe85	85.5	2.8 ± 0.16^j
Balloon-borne 0.3-K bolometric		151	$2.95^{+0.11}_{-0.12}$
photometer with a filter wheel		203	2.92 ± 0.1
		264	$2.65^{+0.09}_{-0.10}$
de Amici *et al.* (1985)	De85	33.0	2.81 ± 0.12^i
Witebsky *et al.* (1986)	Wi86	90.9	2.57 ± 0.12^i
Mandolesi *et al.* (1986b)	Ma86	4.75	2.70 ± 0.07^i
Sironi & Bonelli (1986)	SB86	2.5	2.79 ± 0.15^i
Johnson & Wilkinson (1987)	JW87	24.8	2.783 ± 0.025
Balloon-borne coherent receiver			
with LHe-cooled internal and			
external loads			
Kogut *et al.* (1988)	Ko88	10	2.61 ± 0.06^i
de Amici *et al.* (1988)	De88	3.7	2.59 ± 0.13^i
Levin *et al.* (1988)	Le88	1.41	2.11 ± 0.38^i
Matsumoto *et al.* (1988)	Mat88	258	2.799 ± 0.018
Rocket-borne six-channel		423	2.955 ± 0.017
photometer cooled to 1.2 K		624	3.175 ± 0.027
Bersanelli *et al.* (1989)	Be89	90	2.60 ± 0.09^i
Mather *et al.* (1990)	FIRAS90	30–600	$2.735^{+0.06}_{-0.06}$
FTS with 1.5-K bolometers			
on a satellite			
Kogut *et al.* (1990)	Ko90	7.5	2.60 ± 0.07^i
Sironi *et al.* (1990)	Si90	0.6	3.0 ± 1.2^k
Gush *et al.* (1990)	COBRA	55–585	2.736 ± 0.017
FTS with 0.3-K bolometers			
on a rocket			
Sironi *et al.* (1991)	Si91	0.82	2.7 ± 1.6^k
		2.5	2.71 ± 0.21^k
de Amici *et al.* (1991)	De91	3.8	2.64 ± 0.06^k
Levin *et al.* (1992)	Le92	7.5	2.64 ± 0.06^k
Bensadoun *et al.* (1993)	Bn93	1.47	2.26 ± 0.19^k
Mather *et al.* (1994)		30–600	2.726 ± 0.005
Bersanelli *et al.* (1994)	Be94	2.0	2.55 ± 0.14^k

[j] Based on an analysis of systematic contributions to the incident flux from the instrument, Bernstein *et al.* (1990) concluded that these results were better interpreted as upper limits.
[k] A continuation of the program in footnote (i) but done from the South Pole. For Si91 the LHe cold load was in coax.

Table A.2. (*cont.*)

Reference	ID	Frequency (GHz)	T_0 (K)
Staggs *et al.* (1996a) Correlation receiver observing from NRAO at Greenbank, WVa	Stg96a	1.4	$2.65^{+0.33}_{-0.30}$
Staggs *et al.* (1996b) Balloon-borne 5-K coherent receiver with internal load	Stg96b	10.7	2.730 ± 0.014
Fixsen *et al.* (1996) FTS with 1.5-K bolometers on a satellite	FIRAS96	30–600	2.728 ± 0.002
Mather *et al.* (1999) Final FIRAS value.		30–600	2.725 ± 0.001
Raghunathan & Subrahmanyan (2000) Coherent receiver with cooled internal loads	RS00	1.28	3.45 ± 0.780
Fixsen *et al.* (2004) Balloon-borne 3-K coherent receiver with 3-K internal and external loads	Fx04	10.1 30.3	2.721 ± 0.010 2.694 ± 0.032
Singal *et al.* (2006) Balloon-borne 2–8-K coherent receiver with internal load	Sn06	8.0 8.3	2.90 ± 0.12 2.77 ± 0.16

in ℓ-space. In computing the window functions, W_ℓ, of the pre-1993 experiments a number of approximations were made as noted throughout the table.

For many years, it was customary to report 95% upper limits. To make the comparison between measurements easier, we have converted all 95% and 90% upper limits and confidence intervals to 1σ limits. In all cases we have just divided the error by two and denote this as $1\sigma_{\text{con}}$ (for converted). This practice is problematic when the likelihood is skewed, but it gives a reasonable way to visualize how the results compare. Calibration error is generally not included in the error bar and is usually near 10% in temperature. That is, if a 10–15% shift in a result slides it into accord with the model then it is a good fit. We note calibration errors if they are larger than 20% or particularly important.

Contamination by foreground emission is a continual worry for researchers. A foreground is most simply quantified with an amplitude and a spectral index so that the emission follows $T(\nu) \propto \nu^\beta$ where, in the Rayleigh–Jeans limit, β is ≈ -2.9 for synchrotron, -2.1 for free-free, 0 for the CMBR, and ≈ 2 for dust. Before the mid-1990s the level of galactic free-free emission was poorly known. Many of the early worries were laid to rest with the 1996 and subsequent measurements of the hydrogen recombination line H_α (from which the microwave emission was deduced) by the Gaustad, McCullough and van Buren (1996) and Simonetti, Dennison and Topasna (1996) groups.

At the bottom left of each entry, we give the lead institute or institutes. While this is clear early on, it became less straightforward as multi-institute collaborations became the norm. For example, WMAP was a collaboration between Brown University, the University of Chicago, NASA/GSFC, Princeton University, UCLA, and UBC at launch but was a "partnership" between NASA/GSFC and Princeton.

The history of CMBR measurements is as much one of raw sensitivity as of experimental technique and observing strategy. In the late 1960s Penzias and Wilson, Conklin, and Pariĭskiĭ had amplifiers that achieved between 1 and $3\,\text{mK}\,\text{s}^{1/2}$. That is, in one second a signal-to-noise ratio of unity could be achieved on 1- to 3-mK signals. Others had detectors that were 10 to 20 times worse and made up for the difference by observing longer or with a strategy more immune to systematic error. Even today, 40 years later, great strides can be made with detectors that achieve $0.3\,\text{mK}\,\text{s}^{1/2}$, just an order of magnitude better; albeit, modern radiometers generally use multiple detectors at once.

In the conceptually simplest measurement, one would observe N independent patches of sky with a Gaussian-shaped beam of full-width at half maximum $\theta_{1/2}$ to obtain t_i data points with error bars σ_t. The window function for such an observation is

$$W_\ell = \exp[-(\ell+1/2)^2/2(\ell_s+1/2)^2], \quad \ell_s = 2\sin(\theta_{1/2}/\sqrt{32\ln 2}) - 1/2. \quad \text{(A.1)}$$

This is the Legendre transform of the beam. In a signal-dominated measurement, one would then compute the rms (root mean square) deviation from the mean, $T_{\text{rms}}^2 = \sum_{i=1}^{N}(t_i - \bar{t})^2/(N-1)$, and find the band power (equation 5.12) through

$$\delta T_\ell = T_{\text{rms}}/\sqrt{\sum_\ell W_\ell/\ell}. \quad \text{(A.2)}$$

The uncertainty on T_{rms} would be derived from σ_t. The factor of ℓ below W_ℓ makes the band power a measure of the rms temperature fluctuation per logarithmic interval of angle or ℓ (footnote 25 on page 444).

This simple example has two major problems. The first is that the quantity one desires is the uncertainty on the intrinsic rms fluctuation of the sky temperature as opposed to the rms fluctuations of the measurements. The ability to extract the celestial rms depends critically on the signal-to-noise per observation. The analysis is properly done using the likelihood function,

$$L(t_1 \ldots t_N, \delta T_\ell) = \frac{\exp(-\sum_{ij} t_i t_j M_{ij}^{-1}/2)}{\sqrt{(2\pi)^N \det M}}. \tag{A.3}$$

Here $\mathbf{M} = \delta T_\ell^2 \mathbf{S} + \mathbf{N}$ is the covariance matrix that encodes the observing pattern in the signal component, $\mathbf{S}$, and the full instrument noise matrix in the noise component, $\mathbf{N}$. In practice, one computes this likelihood as a function of δT_ℓ and plots the maximum and the effective $\pm 1\,\sigma$ points.

The second problem with the observing strategy in our example is that it puts a heavy burden on the long-term stability of the instrument. That is, small variations in the instrument output, or "offset," can mimic a sky signal. This problem is overcome by observing differences in temperature (which are encoded in $\mathbf{S}$) more rapidly than the instrument or sky can appreciably change. This "differencing" or "chopping" is done in a wide variety of ways as discussed throughout the table. Figure A.1 shows some of the more common methods and defines some terms.

The upper left panel of Figure A.1 shows a "single difference" measurement where the signal received from the right-looking observation is

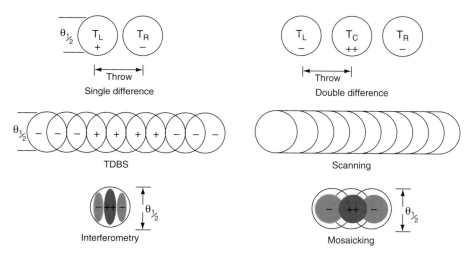

Fig. A.1. Scan strategies for CMBR observations. The outline of the beam full-width at half maximum is shown as $\theta_{1/2}$. The upper four panels show methods of "differencing" or "chopping" done with various arrangements of receivers. The bottom two panels are for interferometric observations. Note that an interferometer generally forms synthesized beams smaller than the main beam whereas other techniques form effective beams larger than the main beam. Figure courtesy of Angela Glenn.

subtracted from the left-looking observation at angular distance θ_{throw}. The result is sensitive to the sky temperature fluctuations on the scale $\ell \sim 1/\theta_{\text{throw}}$ in the spherical harmonic expansion in equation (5.10). Differencing narrows the window function, the range of angular scales to which the measurement is sensitive. The "double difference" technique, upper right, is the difference between two single-difference measurements. It is sensitive to a similar range in ℓ though with a sharper cutoff on the low ℓ side. For both of these patterns, observers often used lock-in amplifiers to averaged differences at a given spot on the sky for an extended period of time.

The middle left panel of Figure A.1 shows "time-domain beam synthesis" (TDBS). In this case the beam is scanned and sampled rapidly and then the beam pattern is synthesized in software. The synthesis is akin to the filtering in the first examples and can yield a set of well-characterized effective window functions. In the example shown, a "double difference" beam pattern is synthesized with the weighting "$- - - + + + + - - -$." The middle right shows a direct scan as is often done for mapping. With this method, the data are generally high passed filtered and the solution is more map-based than window-function based.

The bottom two panels in Figure A.1 illustrate interferometric observations. The lower left shows a "double difference" synthesized beam for a main beam of width $\theta_{1/2}$. Such a pattern can be formed with a short-baseline interferometer. Note that an interferometer forms synthesized beams smaller than the main beam whereas TBDS forms beams larger than the main beam. The lower right pattern shows a mosaicked pattern for interferometric observations. Such patterns are used to extend sensitivity to larger angular scales and to higher resolution.

There have been a number of compilations of CMBR anisotropy data over the years from which the present study benefited. Bruce Partridge's (1995) book lays out most of the basic techniques and includes many insights into the pre-1990s experiments. For those early years, Rai Weiss's (1980) review and the compilation by Yurii Pariĭskiĭ and Dmitrii Korol'kov (1986) were also particularly useful. Bharat Ratra and colleagues have worked with a number of experimental groups and reanalyzed much of the data from the 1990s. The compilation in Podariu *et al.* (2001) was especially helpful. Similarly, George Smoot and Douglas Scott (1998) compiled a list for the particle physics data book, and Bersanelli, Maino and Mennella (2002) reviewed many of the measurements in *Il Nuovo Cimento*. Our reference points for much of the analysis were Dick Bond's lecture notes from the 1993 Les Houches (Bond 1995a) and Capri (Bond 1995b) meetings and the later Bond, Jaffe, Knox (2000) analysis. However, for all entries in the table we went back to the original papers.

In this table, the first line in the header names the experiment, the detector technology, and the effective range of ℓ values – angular scales – the experiment sampled. The description is followed by the name of the institution at which the experiment is based, wholly or in large part, and references to the literature for the experiment.

Table A.3. *Anisotropy experiments*

Isotropometer   Coherent receiver at 9.4 GHz $1 \leq \ell \lesssim 20$

The first instrument designed specifically to measure the anisotropy was Partridge and Wilkinson's "isotropometer." It used the original absolute temperature radiometer modified to Dicke-switched between a zenith-pointed reference horn and a $\theta_{1/2} = 5°\!.5$ beam pointed at a $42°$ elevation above the southern horizon ($\delta = -8°$ from Princeton). With a 50% duty cycle, a large flat switched the beam from the equatorial target region to a reference region near Polaris (through an equal air mass) to permit the subtraction of long-term drifts. Data were taken intermittently over a year and a number of systematic checks were performed. Based on Penzias and Wilson's longer wavelength measurements, they argued that a galactic contribution is negligible. They searched for the equatorial component of the dipole placing an upper limit of 3.2 mK (1σ) on its amplitude. They also report a 2.5σ measurement of variation of a quadrupolar nature which is attributed to some unidentified noninstrumental source. We take $T_{\mathrm{rms}} = 5$ mK as the 1σ limit on the anisotropy corresponding to $\delta T_\ell < 2700\,\mu$K.

Princeton
Wilkinson and Partridge (1967); Partridge and Wilkinson (1967)

Stanford   Coherent receiver at 10.69 GHz $25 \lesssim \ell \lesssim 560$

The radiometer Dicke-switched between a $\theta_{1/2} = 80°$ zenith reference horn and a $\theta_{1/2} = 0°\!.2$ zenith-pointed beam from a 18.2-m parabola. The radiometer output as a function of hour angle was analyzed for anisotropy. Residual atmospheric fluctuations limited the measurement. Two analyses were reported in 1967, the first based on 11 days of observations. We give results from the more sensitive second observation where a running average over $7°\!.5$ was removed from the data, limiting sensitivity to $\ell \gtrsim 25$. The paper also introduces a scaling method to separate instrument and sky noise that was used through the 1980s and introduced the Gaussian autocorrelation function analysis that persisted in the literature until the early 1990s. With the 3.6-mK (1σ) limit, we find $\delta T_\ell < 2100\,\mu$K.

Stanford
Conklin and Bracewell (1967)

Crawford Hill   Maser at 4.08 GHz $2 \lesssim \ell \lesssim 140$

The same radiometer that discovered the CMBR was scanned across the sky over 18 h to place limits on the anisotropy. The receiver was $\approx 10x$ as sensitive as other CMBR radiometers. The reference termination warmed at a rate of 33 mK/h. From 28 measurements scattered over the sky, they conclude "there is no large-scale deviation from isotropy of more than 0.1 K." Taking this as a 2σ upper limit made with a beam of $\theta_{1/2} = 0.°8$, one finds $\delta T_\ell < 21,500\,\mu K$ ($1\sigma_{con}$).

Bell Labs
Wilson and Penzias (1967)

Aerospace   Coherent receiver at 88 GHz $600 \lesssim \ell \lesssim 1500$

Two receivers illuminated a 4.6-m dish producing two $\theta_{1/2} = 3'$ beams separated by $12.'5$. The receiver outputs were differenced, the telescope nodded in azimuth by $12.'5$ and a second difference performed. This is the first example of a "double difference" method for CMBR studies. In blank fields of sky, a limit of 15 mK (interpreted as 1σ) is reported, corresponding to $\delta T_\ell < 9600\,\mu K$ at $\ell = 900$.

Aerospace Corp.
Epstein (1967)

White Mountain   Coherent receiver at 8 GHz $1 \leq \ell \lesssim 9$

The radiometer Dicke-switched between two $\theta_{1/2} = 12°$ feeds pointed $\pm 30°$ from the zenith along an E-W baseline to search over 23 days for the CMBR dipole induced by Earth's motion. In addition, the instrument was rotated by $180°$ about the vertical axis every 5 minutes. The high site was less contaminated by atmospheric fluctuations than the Stanford site. After subtracting a $\sim$10-mK galactic contribution by extrapolating the 404-MHz Pauliny-Toth and Shakeshaft map to 8 GHz, Conklin reported a preliminary measurement of the dipole with amplitude 1.6 ± 0.8 mK in the direction $\alpha = 13^h$. After Conklin's 1969 PhD thesis (Conklin 1969), this became 2.3 ± 0.9 mK on a great circle in the direction $\alpha = 11^h$, as reported at an International Astronomical Union Symposium (Conklin 1972). Though the error bars include an estimate for a ± 0.1 uncertainty in the galactic radiation index (the statistical uncertainty was ≈ 0.02 mK), there were lingering doubts about the extrapolation (e.g., Webster 1974). With the 22-GHz WMAP data, one finds that an index of $\beta \approx -1.5$ is needed to match Conklin's model at 8 GHz. However, the value of the spectral index and the degree to which it varies between 8 and 22 GHz is still an active area of investigation. Another overall limitation to the experiment was that a single circle at $\delta = 32°$ was measured. Neglecting the correction for the Earth's motion, the WMAP dipole (3.358 ± 0.017 mK in direction $\alpha = 11.19^h$ at $\delta = -6.°90$ for the full sky) has amplitude 2.83 mK at $\delta = 32°$, in excellent agreement with Conklin's measurement. As a limit on the nondipole anisotropy, we take the fit error of 0.75 mK and find $\delta T_\ell < 460\,\mu K$ (1σ) at $\ell = 6$.

Stanford
Conklin (1969, 1972); Webster (1974)

RATAN   Parametric amplifiers at 13.2 GHz $620 \lesssim \ell \lesssim 2100$
                                                       and $300 \lesssim \ell \lesssim 980$

RATAN produced an azimuthally thin but vertically wide beam that was coupled to very low noise amplifiers. A series of measurements with a $\theta_{1/2}^2 = 1.'4 \times 20'$ beam were taken with the $3\,m \times 130\,m$ Large Pulkovo Radio Telescope (a prototype for RATAN-600). In RATAN1, two pointings separated by $5.'6$ were compared over an unspecified region of sky. Modeling the window function as a single-difference, the 0.7-mK limit becomes $\delta T_\ell < 670\,\mu K$ (1σ) at $\ell = 1450$. In RATAN2 the signal beam, $\theta_{1/2}^2 = 1.'3 \times 40'$, was drift-scanned in azimuth (in the narrow direction) and compared to a reference beam near the NCP (north celestial pole). Filtering of the data put an effective cut at $\ell \approx 300$. Converting their sky limit of $70\,\mu K$ ($1\,\sigma$), one obtains $\delta T_\ell < 70\,\mu K$ at $\ell \approx 830$. RATAN3 shared much with RATAN2 though the integration time set an effective beamwidth of $5'$ and the filtering time set a cutoff at $\ell \approx 200$. Using the 1σ limit at $10'$ length scales and a limit of $60\,\mu K$, we find $\delta T_\ell < 60\,\mu K$ at $\ell \approx 600$. In RATAN4, multiple frequencies were used to understand galactic emission. New limits were given at $3.9\,GHz$ with a $\theta_{1/2} = 0.''9 \times 10'$ beam. From the plot, source confusion enters at $\approx 50\,\mu K$ at $\ell \approx 1400$ though we did not find enough information to make a band power estimate. A review of the program and analysis is given in Pariĭskiĭ and Korol'kov (1986). There have been a number of reanalyses of and questions about these limits over the years due to the method of analyzing data from the highly asymmetric beams and due to the potential point-source contamination. The corrections could be as much as a factor of 2 upward. To estimate the maximum amplitude of the source contamination, we extrapolated the WMAP-fitted, though unidentified, source spectrum with an index of $\beta = -2.1$ to $13.2\,GHz$. At $\ell = 830$, we find $\delta T_\ell \approx 350\,\mu K$. However, judicious observing with high resolution beams can push well below this limit.

Pulkovo, USSR
RATAN1: Pariĭskiĭ and Pyatunina (1971); RATAN2: Pariĭskiĭ (1973a); RATAN3: Pariĭskiĭ *et al.* (1977); RATAN4: Berlin *et al.* (1983); Pariĭskiĭ and Korol'kov (1986)

KaDip   Coherent receiver at 35 GHz $1 \lesssim \ell \lesssim 21$

A radiometer built by Wilkinson and modified to search for the dipole and large-scale anisotropy at $35\,GHz$. The reference feed pointed toward the zenith and the $\theta_{1/2} = 4°$ signal feed was directed with a large chopping flat alternately between $\pm 32°$ of South on the celestial equator with a 20-m period. Data were averaged for $20\,m$ so the effective beam in azimuth was $\theta_{1/2} = 7°$. No signal was seen at 12 h or 24 h periods to the limits of 6.6 mK and 11.6 mK respectively. We use the former as the limit on $T_{\rm rms}$ to find $\delta T_\ell < 3530\,\mu K$ at $\ell = 15$.

Princeton
Boughn *et al.* (1971)

XBal   Coherent receiver at 10 GHz $1 \leq \ell \lesssim 7$

The balloon-borne X-band radiometer used two $\theta_{1/2} = 15°$ feeds pointed $\pm 45°$ from the zenith and Dicke-switched at $2\,kHz$ to search for the CMBR dipole anisotropy. It was the first CMBR anisotropy balloon flight. The platform rotated at 1 revolution per minute (rpm) and data were coadded. Roughly a quarter of the sky was measured. Earth's magnetic field coupled to the switch producing a 15–20 mK signal in the "sine" phase of the rotation, but $< 0.4\,mK$ in the "cosine" phase. After correcting the cosine phase for the galactic contribution by extrapolating the Pauliny-Toth and Shakeshaft map from $404\,MHz$ to $10\,GHz$, Henry reported a measurement of the dipole with amplitude $3.2 \pm 0.8\,mK$ in the direction $\delta = -30° \pm 25°$ and $\alpha = 10.5^{\rm h} \pm 4^{\rm h}$. This result, stable for galactic emission indices between $\beta = -2.65$ and -2.95, agrees with later measurements. Though the statistical significance of the fit, $\chi^2/\nu = 2.3$, is modest, it would be surprising if this and Conklin's measurement were not the dipole. As a limit on the intrinsic anisotropy, we take the fit error of $0.8\,mK$ multiplied by $\sqrt{2.3}$ to find $\delta T_\ell < 775\,\mu K$ (1σ).

Princeton
Henry (1971)

NRAO-P   Parametric amplifier at 10.7 GHz $200 \lesssim \ell \lesssim 2160$

The output of two $\theta_{1/2} = 2\rlap{.}'8$ feeds separated by an angle of 0.1° (and later 1°) were compared as the 42-m radio telescope scanned a small region of sky. No signal was detected with either throw at a level of $T_{\mathrm{rms}} = 0.08$ mK, though the details of the analysis are unclear. This value was revised to 0.48 mK in an erratum after discussions with Zel'dovich. With the more restrictive larger throw we find $\delta T_\ell < 180\,\mu$K at $\ell = 1550$.

Pulkovo, USSR
Pariĭskiĭ (1973b, 1974)

Goldstone   Maser at 8.4 GHz $200 \lesssim \ell \lesssim 2250$

A $\theta_{1/2} = 2\rlap{.}'3$ beam from the 64-m Goldstone radio telescope was repeatedly drift-scanned over a 1° stretch. Averaging smeared the effective beam to $\theta_{1/2} \approx 3'$ and the scan length puts in a cutoff at $\ell \approx 200$. The 1-σ limit was $T_{\mathrm{rms}} \approx 0.7$ mK corresponding to $\delta T_\ell < 460\,\mu$K at $\ell = 1640$.

Cal State and JPL
Carpenter *et al.* (1973)

Parkes   Correlation receiver at 2.7 GHz $420 \lesssim \ell \lesssim 1400$

A correlation receiver measured the differential output of two zenith-pointed $\theta_{1/2} = 0.133°$ feeds separated by an angle of 0.316°. The measured limit is $T_{\mathrm{rms}} = 0.42$ mK corresponding to $\delta T_\ell < 420\,\mu$K (1σ) at $\ell = 980$.

Gor'kii, USSR
Stankevich (1974)

U-2   Dicke-switched radiometers $1 \le \ell \lesssim 16$
                     at 33 GHz

The radiometer Dicke-switched at 100 Hz between two $\theta_{1/2} = 7°$ corrugated feeds pointing 60° apart. This was probably the first use of corrugated feeds for an anisotropy measurement. The geometry had much in common with Conklin's White Mountain radiometer. A specially modified U-2 airplane flew the radiometers at 20 km to reduce atmospheric emission and allow easy change of the antenna orientation. Roughly 3/4 of the sky was measured in 18 flights. The dipole was unambiguously measured after the first eight flights with amplitude 3.5 ± 0.6 mK in the direction of $\alpha = 11.0^{\mathrm{h}} \pm 0.6^{\mathrm{h}}$ and $\delta = 6° \pm 10°$, and the residual anisotropy was given as $T_{\mathrm{rms}} = 0.5$ mK ($1\sigma_{\mathrm{con}}$). By the end of the northern campaign, the amplitude was 3.6 ± 0.5 mK in the direction of $\alpha = 11.2^{\mathrm{h}} \pm 0.5^{\mathrm{h}}$ and $\delta = 19° \pm 8°$ with the same anisotropy limit. The southern hemisphere flight suffered from a larger correction for galactic emission but was shown to be consistent with the northern flights. The radiometers were the forerunners of those used on COBE/DMR. The anisotropy limit corresponds to $\delta T_\ell < 280\,\mu$K at $\ell = 11$.

Berkeley
Smoot *et al.* (1977); Gorenstein and Smoot (1981); Smoot and Lubin (1979)

TestaGriga   Bolometer at 250 GHz $130 \lesssim \ell \lesssim 450$

A custom bolometric receiver on a 1.5-m on-axis Cassegrain antenna with $\theta_{1/2} = 0\rlap{.}°42$ and a throw of $0\rlap{.}°42$ produced by a small wobbling flat mirror. Observations were carried out at 3500 m at Testa Grigia. This was the first bolometric system to report measurements on the anisotropy, though Muehlner and Weiss had been making bolometric balloon-borne absolute measurements since 1970. Group members were ultimately involved with BOOMERANG. The measured limit $T_{\mathrm{rms}} < 0.16$ mK ($1\sigma_{\mathrm{con}}$) corresponds to $\delta T_\ell < 160\,\mu$K at $\ell = 310$. In his book, Partridge points out that the result is likely given in antenna temperature and should be multiplied by 4.4 to convert it to thermodynamic temperature, yielding $\delta T_\ell < 700\,\mu$K.

Florence, IT

Caderni *et al.* (1977)

GBank-R   Parametric amplifier at 15 GHz $350 \lesssim \ell \lesssim 2100$

Observations were made with the NRAO 42-m Greenbank telescope to search for the Sunyaev–Zel'dovich effect in clusters and limit the CMBR anisotropy. The flux from two $\theta_{1/2} = 3'$ beams separated by $17\rlap{.}'4$ was switched at 10 Hz into a dual-input receiver. The reported limit is $T_{\mathrm{rms}} < 310\,\mu$K ($1\sigma_{\mathrm{con}}$) corresponding to $\delta T_\ell < 150\,\mu$K at $\delta T_\ell \approx 1600$.

Princeton and JPL

Rudnick (1978)

MIT   Bolometer at 210 GHz $2 \lesssim \ell \lesssim 56$
                and 450 GHz

A custom differential bolometric receiver with $\theta_{1/2} = 1\rlap{.}^\circ6$ beams separated by $126°$–$135°$ was flown multiple times from a balloon. A 300-Hz mechanical chopper alternately directed radiation from one beam and then the other into the receiver as the gondola rotated at 2 rpm. For one flight only one beam was scanned. The 1.5 K cryogenic techniques trace back to the Muehlner and Weiss (1970) balloon-borne absolute measurements. The refereed paper focuses on understanding and mapping dust emission for future anisotropy studies, though Muehlner and Weiss give a dipole analysis in a conference proceeding and report that in the submillimeter to millimeter range the anisotropy is "no more than a part per thousand over large scales." This statement roughly corresponds to $\delta T_\ell < 1280\,\mu$K at $\ell = 38$.

MIT

Muehlner (1977); Owens *et al.* (1979)

KKaQBal   Dicke-switched radiometers $1 \lesssim \ell \lesssim 17$
                    at 19, 24.8, and 31.4 GHz

Four balloon flights used a configuration similar to that of Henry's (1971). Radiometers at 19 GHz (K-band, $\theta_{1/2} = 10°$), 24.8 GHz ($\theta_{1/2} = 8°$), and 31.4 GHz (Ka band, $\theta_{1/2} = 6°$) all measured the dipole, confirmed its thermal spectrum, and, because of the high altitude and high frequency, distinguished it from atmospheric and galactic contamination. They report: "early measurements ... were hampered by atmospheric noise, large corrections for galactic radiation, and limited sky coverage ... it seems likely that the Solar motion was seen 10 years ago (Conklin 1969; Henry 1971)." Numerous systematics checks were described. They found dipole amplitude 2.99 ± 0.34 mK in the direction $\alpha = 12.3^{\mathrm{h}} \pm 0.4^{\mathrm{h}}$ and $\delta = -1° \pm 6°$. The 1σ limit on the residual anisotropy is 1 mK ($1\sigma_{\mathrm{con}}$) which we give as $\delta T_\ell < 550\,\mu$K at $\ell = 12$ assuming an effective $\theta_{1/2} = 6\rlap{.}^\circ5$ beam. Though the results were published after the U-2 measurement, the first discussion of them, based on a May 1975 flight, appeared in the *Bulletin of the American Astronomical Society*, where it was reported that at 19 GHz the dipole was detected with amplitude 2.9 ± 0.7 mK in the direction $\alpha = 12.3^{\mathrm{h}} \pm 1.4^{\mathrm{h}}$ and $\delta = -21° \pm 21°$. After a subsequent flight in January 1980, team members reported that the "Quadrupole anisotropy is detected at the 4σ level ... we believe [it is] intrinsic to the 2.7° radiation." This was not confirmed. An additional flight and a new measurement with a new instrument (Fixsen *et al.*) showed that the result was due to a combination of ground radiation and radiometer noise at the rotation frequency.

Princeton
Corey and Wilkinson (1976); Cheng *et al.* (1979); Boughn *et al.* (1981); Fixsen *et al.* (1983)

KPRO   NRAO radiometer at 31 GHz $600 \lesssim \ell \lesssim 2000$

Two $\theta_{1/2} = 3\rlap{.}'6$ beams from the 11-m Kitt Peak radio telescope were switched in azimuth at 2.5 Hz to produce a single-difference with 9′ separation. After a number of controlled corrections, the largest of which was 0.4 mK for telescope efficiency, and different analyses, a limit $T_{\mathrm{rms}} < 0.27$ mK ($1\sigma_{\mathrm{con}}$) is reported. This corresponds to $\delta T_\ell < 256\,\mu\mathrm{K}$ at $\ell = 1500$.

Haverford
Partridge (1980)

Convair   Bolometers at 240 GHz $25 \lesssim \ell \lesssim 85$

A Michelson interferometer operated as a simple receiver with $\theta_{1/2} \approx 2\rlap{.}°5$ was flown aboard a Convair 990 airplane as part of NASA's ASSESS II mission. A wobbling flat scanned the beam 4° at 20 Hz through the airplane window. The paper reports difficulty with electronic and microphonic pickup but limits the anisotropy to $T_{\mathrm{rms}} < 0.8$ mK (1σ), corresponding to $\delta T_\ell < 1000\,\mu\mathrm{K}$ at $\ell = 55$.

Florence, IT
Fabbri *et al.* (1980a)

DBal   Bolometers at 170 GHz $1 \leq \ell \lesssim 2$

The $\theta_{1/2} = 5\rlap{.}°2$ beam of a balloon-borne Michelson interferometer (operated as a simple receiver) sinusoidally oscillated at 30 Hz and 6° throw by a large flat. The gondola rotated at $\approx 1/3$ rpm. Only data with galactic latitude $|b| > 20$ were analyzed for anisotropy to avoid the galactic contamination seen by the team and reported in Muehlner (1977). A $2.9^{+1.3}_{-0.6}$ mK dipole in the direction $\alpha = 11.4^{\mathrm{h}} \pm 0.7^{\mathrm{h}}$ and $\delta = 3° \pm 2°$ is reported, significantly enhancing knowledge of the dipole spectrum. The measurement is especially impressive given the small throw. The fit, $\chi^2/\nu = 5.3$, was poor due to the presence of a quadrupolar signal. They report "our $\cos 2\theta$ term should not be immediately identified with the quadrupole moment, but is suggestive of a quadrupolar anisotropy in the [CMBR] with amplitude $\sim 1/3$ the dipole." Among other things, the limited sky coverage limited confidence in a quadrupolar detection. Before the paper was submitted to *Physical Review Letters*, Fabbri *et al.* (1982) reported a significant detection of anisotropy at 6° angular scales at the 2nd Marcel Grossmann conference. This detection, in addition to the dipole and quadrupole, was hinted at being cosmological though the possibility of galactic sources was not ruled out. In Melchiorri *et al.*, after analyzing and rejecting various possible systematic sources, this became "suggest the existence of cosmic anisotropy" at $T_{\rm rms} = 0.3 \pm 0.05$ mK. Or, in the case of an incorrect assumption of the dust-to-gas ratio, it can be interpreted as an upper limit 20 μK ($1\sigma_{\rm con}$), corresponding to $\delta T_\ell < 18$ μK at $\ell = 23$. We determine the quadrupole value by scaling the dipole value. Ceccarelli *et al.* refined the analysis and explored various explanations of the signal. In retrospect, the source of the 6° anisotropy is not clear, but a cosmological interpretation is not supported.

Florence, IT
Fabbri *et al.* (1980b); Fabbri *et al.* (1982); Melchiorri *et al.* (1981); Ceccarelli *et al.* (1982)

NRAO91   Coherent receiver at 4.8 GHz $715 \lesssim \ell \lesssim 2400$

A dual channel receiver was fed by two $\theta_{1/2} = 3.'2$ beams from the 91-m NRAO telescope. The beams were separated by 7.'34 and switched. After averaging over a number of different pointings in right ascension, it was found that $T_{\rm rms} \lesssim 0.8$ mK ($1\sigma_{\rm con}$). This corresponds to $\delta T_\ell < 500$ μK at $\ell = 1800$.

Virginia Poly and NRAO
Ledden *et al.* (1980)

OVRO40   Parametric receiver at 10.6 GHz $680 \lesssim \ell \lesssim 1800$

A parametric amplifier-based dual channel receiver was fed by two beams from the OVRO 40-m telescope. One beam had $\theta_{1/2} = 3.'03$ and the other was a lower-grade reference beam with $\theta_{1/2} \approx 4'$. Several differential measurements were combined to produce a double difference pattern with a beam separation of 11'. Fifteen such observations were made. A limit was given as $T_{\rm rms} < 0.69$ mK (1σ). Assuming a double-difference pattern, this corresponds to $\delta T_\ell < 460$ μK at $\ell = 1100$.

CalTech and NRAO
Seielstad *et al.* (1981)

GBank-UW   NRAO Maser at 19.5 GHz $1640 \lesssim \ell \lesssim 4000$

A $\theta_{1/2} = 1.'5$ beam was switched by 4.'5 at 3.33 Hz using the nutating subreflector on the 140-ft Greenbank Telescope. This pattern was differenced with one shifted by 4.'5 to produce a "double difference." The straight $T_{\rm rms}$ over 24 fields was 56 ± 73 μK. Their analysis gives a 1σ limit of 160 μK corresponding to $\delta T_\ell < 115$ μK at $\ell = 2450$. More observations were taken in 1982, 1983, and 1984 resulting in $T_{\rm rms} < 33$ μK ($1\sigma_{\rm con}$) or $\delta T_\ell < 24$ μK. Observations ceased when the systematic error limit was reached.

Princeton
Uson and Wilkinson (1982, 1984a,b)

MaserBal   JPL Maser at 24.5 GHz $2 \lesssim \ell \lesssim 16$

Three balloon flights covered $\approx 85\%$ of the sky with $\theta_{1/2} \approx 7°$ using the same layout as with Cheng *et al.* (1979). Earth's motion was detected in the dipole, the quadrupole was not seen at $T_{\ell=2} < 85$ μK ($1\sigma_{\rm con}$), and the anisotropy was limited to 100 μK rms between 10° and 180°. This corresponds to $\delta T_\ell < 56$ μK at $\ell = 11$.

This was the first presentation of a map of the CMBR sky, and it was the first analysis of the temperature correlation function.

Princeton
Fixsen *et al.* (1983)

WBal   NASA/UVa Schottkey mixer $2 \lesssim \ell \lesssim 16$
                        at 90.9 GHz

The balloon-borne Dicke radiometer used a rotating flat to alternate the $\theta_{1/2} \approx 7°$ beam between $\pm 45°$ of the zenith as the gondola rotated at 1 rpm. Foreground emission was known to be minimal near 90 GHz. The first two balloon flights shared the Wilkinson and Weiss gondolas respectively. Subsequent flights were made with a new dedicated gondola spinning at $1/2$ rpm. The last two of three flights covered over half the sky. The results were published in the same *Physical Review Letter* as Fixsen *et al.* (1983) and placed the limit $T_{rms} \approx 100 \, \mu K$ (1σ) on the quadrupole. The payload was flown again in the southern hemisphere. From the combined data, a 90-GHz map was made of 85% of the sky using matrix methods similar to those later used for COBE. Again, no quadrupole was detected at roughly the same limit. Based on the noise in the experiment, we estimate that a limit of $\delta T_\ell < 20 \, \mu K$ at $\ell = 11$ (not plotted) could have been obtained in the absence of foreground emission and instrumental effects.

Berkeley
Lubin *et al.* (1983, 1985)

JodrellBank   Dicke-switched radiometers $250 \lesssim \ell \lesssim 700$
                        at 5 GHz

Two feeds illuminated the MkII telescope to produce two $\theta_{1/2}^2 = 0.°13 \times 0.°17$ beams separated by $0.°5$ in azimuth. The detector outputs were differenced at 143 Hz. The telescope was wagged by $0.°5$ to produce a "double difference" pattern. The analysis included a careful point-source subtraction (especially required due to the low frequency) and a search for systematic effects in the baseline subtraction. It also brought the statistical analysis of CMBR data to new levels, introducing unbiased estimators. They report $T_{rms} < 410 \, \mu K$ ($1\sigma_{con}$) at $10'$ scales, corresponding to $\delta T_\ell < 280 \, \mu K$ at $\ell = 400$.

Jodrell Bank, UK
Lasenby and Davies (1983)

RELIKT   Parametric amplifiers at 37.5 GHz $1 \leq \ell \lesssim 19$

RELIKT was the first CMBR satellite experiment. A differential, parametric amplifier-based radiometer compared radiation from a reference antisolar direction (with $\theta_{1/2} = 10°$) to that from a perpendicular ring with a $\theta_{1/2} = 5.°8$ beam that was scanned by satellite rotation. The reference position was stepped by $7°$ in the ecliptic after averaging over rotations for a week. Six months of data were taken covering nearly the full sky. The initial analysis of one averaged ring put $T_{rms} < 600 \, \mu K$ (1σ). A limit on $90°$ scales of $0.2 \, mK$ was also presented but this was not converted to a quadrupole limit because of the small sky area covered. Subsequent analyses covering $\approx 70\%$ of the sky accounted for contamination by Sun, Earth, and Moon emission into the antenna side- and back-lobes. After rejecting contaminated data, and cutting data with galactic latitude $|b| < 15°$, they found the quadrupole amplitude $< 63\,\mu K$ ($T_{rms} < 40\,\mu K$), the octopole $<170\,\mu K$ ($T_{rms} < 100\,\mu K$), and through a correlation analysis, $\delta T_\ell < 40\,\mu K$ at $\ell = 13$ ($1\sigma_{con}$). The dipole was measured to be $3.16 \pm 0.12\,mK$ in the direction $\alpha = 11.3^h \pm 0.2^h$ and $\delta = -7.5° \pm 2.5°$. In 1992, an analysis identified a feature (the "blamb") at $\approx 3\sigma$ significance that was ascribed to the CMBR. This feature was not confirmed in the COBE data (Bennett *et al.* 1993).

Moscow, USSR
Strukov and Skulachev (1984); Strukov *et al.* (1987); Klypin *et al.* (1987); Strukov *et al.* (1992); Bennett *et al.* (1993)

| **NCP** | Dicke-switched radiometer at 10 GHz | $9 \lesssim \ell \lesssim 50$ |

Two $\theta_{1/2} = 2°\!.5$ beams separated by $10°$ horizontally drift-scanned a ring at $\delta = 80°$. The radiometer Dicke-switched between one beam and a larger reference beam pointed at the NCP. The input of the radiometer was alternated between the east and west beams at $1/6$ to 4 minutes intervals and then the east- and west-looking data were differenced. Data from Norwegian and Italian campaigns (the first in the arctic) with somewhat different observing parameters were combined. They find $T_{rms} < 750\,\mu K$ ($1\sigma_{con}$) corresponding to $\delta T_\ell < 380\,\mu K$ at $\ell = 34$.

Bologna, IT, and Haverford
Mandolesi *et al.* (1986a)

| **Tenerife** | Dicke-switched radiometers at 10.45, 14.9, and 33 GHz | $14 \lesssim \ell \lesssim 29$ |

The observations were carried out over a number of years at the Observatorio del Teide, Tenerife. In TR1, two 10.45-GHz, $\theta_{1/2} = 8°\!.5$ feeds were pointed at a large chopping flat and differenced at 63 Hz. The beam separation was $8°\!.2$. Also, a flat wobbled the beams by $\theta = 8°\!.2$ resulting in a double-difference pattern. The team reported detecting anisotropy. Arguments based on external data sets at 1.4 GHz to track galactic synchrotron, and Hα to track free-free, were given to suggest that the detection was the CMBR. They state: "we believe a substantial part of the detected signal may be due to CMB anisotropy, given reasonable assumptions about the spectrum of the galactic background," and give a detection at $T_{rms} = 100\,\mu K$. They also warn us that the level should "strictly be interpreted as an upper limit." In TR2, a second 14.9-GHz system with $\theta_{1/2} = 5°\!.6$ feeds, but similar beam separation, was added. They found that all the features at 10.45 GHz and 14.9 GHz could be explained as foreground emission and reported an upper limit $T_{rms} = 26\,\mu K$ ($1\,\sigma_{con}$). In Readhead and Lawrence (1992), a private communication from Lasenby shows that the TR1 signal can be explained as foreground emission. In TR3, a 33-GHz $\theta_{1/2} = 5°\!.5$ system was added and the noise per observation significantly reduced. The 10.5-GHz data are used to show that there is minimal contamination of the 15- and 33-GHz data and that the signal derived from those bands, $\delta T_\ell = 42^{+10}_{-8}\,\mu K$, is predominantly that of the CMBR. It is commented that extrapolations from lower frequency observations (as done for TR1) have serious problems that "essentially render such predictions useless." In 1995, Lineweaver *et al.* showed that the TR3 data correlated with the COBE/DMR data with amplitude $\delta T_\ell = 34^{+13}_{-15}\,\mu K$.

Manchester, UK, Cambridge, UK and IAC
TR1: Davies *et al.* (1987); Readhead and Lawrence (1992); TR2: Watson *et al.* (1992); TR3: Hancock *et al.* (1994); Lineweaver *et al.* (1995)

IAB-I   Italian Antarctic Base; ^{3}He-cooled $60 \lesssim \ell \lesssim 130$
                 bolometers at 150 GHz

Site testing for CMBR observations was done at the Italian Antarctic Base (IAB) in the 1986–1987 austral summer. In IAB1, observations were made by scanning the $\theta_{1/2} = 1°$ beam (1°.3 field of view, FOV) with an under-illuminated, 1-m diameter oscillating offset parabolic primary, and then synchronously detecting the bolometers. The beam was thrown with p-p (peak-to-peak) amplitude 1°.2 at 8 Hz. (We approximate the 1°.3 Winston cone-defined FOV as a 1° Gaussian and the throw as a square wave.) They report $T_{\rm rms} < 340\,\mu\rm K$ ($1\,\sigma_{\rm con}$) corresponding to $\delta T_\ell < 400\,\mu\rm K$ at $\ell = 135$. Large-scale Galactic emission correlated with IRAS satellite images was detected and "good atmospheric transmission and stability" reported. This was the first published CMBR anisotropy result from the Antarctic. The next campaign, IAB2, added a second frequency channel, decreased the FOV to 1°, increased the throw to 5° p-p, and tested different strategies. They report a similar limit, $T_{\rm rms} < 400\,\mu\rm K$ ($1\,\sigma_{\rm con}$), but now corresponding to $\delta T_\ell \approx 230\,\mu\rm K$ at $\ell = 105$ because of the smaller beam and larger throw.

"La Sapienza," Rome, IT
IAB1: Dall'Oglio and de Bernardis (1988); IAB2: Andreani *et al.* (1991)

MITBal2   Bolometers between 200–2000 GHz $45 \lesssim \ell \lesssim 90$

Between 1974 and 1981, the MIT group flew six flights with differential $\theta_{1/2} = 17°$ beams, and a variety of bolometers, gondola rotation periods, and frequency channel configurations. Though the spectrum of the dipole was measured up to ≈ 380 GHz, thereby complementing absolute temperature measurements, the team was "unable to set new limits on a possible quadrupole moment anisotropy of the CMBR [or others] because we have not been able to remove the effects of diffuse Galactic emission adequately." It eventually took better spatial and spectral resolution in the bolometric measurements to separate the CMBR from dust with bolometers. This was the first multi-channel bolometric balloon-borne anisotropy experiment.

MIT
Halpern *et al.* (1988)

OVRO   Maser at 20 GHz $1150 \lesssim \ell \lesssim 2430$
              NRAO HEMTs at 15 and 32 GHz $361 \lesssim \ell \lesssim 756$

A series of coordinated measurements involving a large number of checks for systematic errors and foreground contamination were done at the Owens Valley Radio Observatory. The first, OV1, was done in response to Uson and Wilkinson and aimed directly at constraining the anisotropy at cluster scales. The receiver employed a 20-GHz JPL maser Dicke-switched between two $\theta_{1/2} = 1'.8$ beams separated on the sky by $7'.15$. A three-beam or "double difference" pattern was obtained by observing again with the telescope offset by $7'.15$. OV1 ("NCP") gave an upper limit $\delta T_\ell = 9 \pm 13\,\mu\rm K$ at $\ell = 1786$ after one datum, clearly contaminated by a point source, was excised. Building on that, OV2 ("RING40m") observed 40 interlocked fields in a ring around the NCP. A signal above noise was clearly detected but attributed to uncleaned radio sources. Interpreting the detection $T_{\rm rms} \approx 100\,\mu\rm K$

as a $1\,\sigma$ limit we find $\delta T_\ell \leq 65\,\mu\text{K}$. We show the first limits in the plots but note that the choice depends on one's prior knowledge. In OV3 ("RING5m"), Leitch *et al.* used NRAO HEMTs at 15 and 32 GHz and the VLA to separate diffuse and point-source contamination from the CMBR in the 32-GHz channel. The observation again involved double differencing but this time with $\theta_{1/2} = 7\rlap{.}'5$ and a $22'$ amplitude throw. After identifying that 88% of the variance in the 32-GHz band was due to the CMBR, they reported $\delta T = 59^{+8.6}_{-6.5}\,\mu\text{K}$ (including calibration error) at $\ell = 589$.

Caltech
OV1: Readhead *et al.* (1989); OV2: Myers *et al.* (1993); OV3: Leitch *et al.* (2000)

SKInt	NRAO mixer-based interferometer at 43 GHz	$45 \lesssim \ell \lesssim 90$

A two-element interferometer produced a double-difference pattern that was wobbled in azimuth by $\pm2\rlap{.}°5$ on the sky. The window function is complicated because of the rectangular primary beams and the east–west differencing. Our approximation comes from the transform of the double-difference pattern given in the paper. We convert $T_\text{rms} < 150\,\mu\text{K}$ ($1\,\sigma_\text{con}$) at a correlation angle of $1\rlap{.}°1$ to $\delta T_\ell \lesssim 120\,\mu\text{K}$ at $\ell = 67$. SKInt was the first interferometer designed specifically to measure the CMBR; it foreshadowed the close-packed arrays of the 2000s.

Princeton
Timbie and Wilkinson (1990)

FIRS	Balloon-borne 0.3-K bolometers at 167, 273, 476, and 680 GHz	$1 \leq \ell \lesssim 25$

Radiation from a single multi-moded $\theta_{1/2} = 3\rlap{.}°8$ feed was compared to a stabilized cryogenic thermal load with a cryogenic switch at 4.5 Hz. The first flight covered just a ring but placed a limit $\delta T_\ell \approx 130\,\mu\text{K}$ ($1\sigma_\text{con}$) on the anisotropy. In the first fully successful flight, FIRS2, roughly 50% of the sky was mapped. In a back-to-back letter with DMR1, a limit $T_\text{rms} = 44\,\mu\text{K}$ (95% cl) was reported. However it was noted that: "There is a clear detection of some sort of anisotropy. There are several possible sources [galactic emission, atmosphere, systematic errors, or the CMB]." Analyses of the higher frequency channels in 1991 and 1992 singled out the CMBR as a likely source. Additionally, a bug was found in the code that had skewed the confidence intervals. (We report only the quoted value, without conversion.) After the DMR discovery, Ganga *et al.* (FIRS3) showed a better than $4\,\sigma$ positive correlation between DMR and FIRS at a level of $\delta T_\ell = 28^{+8}_{-7}\,\mu\text{K}$. John Mather of the COBE team later commented "we had confirmation [for DMR]" (Mather and Boslough 1996). In FIRS4 and Bond (1995c) the data were analyzed on their own in addition to place limits on n_s. The reanalysis agreed with the 1993 analysis.

MIT, Chicago, Princeton, NASA/GSFC
FIRS1: Page *et al.* (1990); FIRS2: Meyer *et al.* (1991): FIRS3: Ganga *et al.* (1993); FIRS4: Ganga *et al.* (1994); Bond (1995c)

ARGO	Balloon-borne 0.3-K bolometers at 150, 250, and 375 GHz, and 1.5-K bolometer at 600 GHz	$60 \lesssim \ell \lesssim 170$

ARGO used a 1.2-m on-axis Cassegrain telescope with a chopping secondary that sinusoidally switched the $\theta_{1/2} = 52'$ beam 1.8 deg p-p on the sky. A reimaging dichroic system split the incident beam into the four frequency bands. The detector output was synchronously demodulated with the 14-Hz chopping secondary, producing a series of differential measurements at discrete points across the sky. In the first flight, ARGO1, a dust-correlated signal was observed and the 150-GHz channel was hampered by RFI (radio frequency interference). After correcting for dust emission, there was a residual signal but it was not attributed to the CMBR. The team reported an upper limit in the 330-GHz channel ($\theta_{1/2} = 25'$, $60 \leq \ell \leq 200$) corresponding to $\delta T_\ell \approx 180\,\mu K$ at $\ell \approx 120$. In ARGO2, the team measured an excess variance above the detector noise at approximately the $6\,\sigma$ and $9\,\sigma$ level in the 250- and 150-GHz channels respectively. The spectrum of the signal was consistent with that of the CMBR. Though the correlation coefficient between the channels was just 0.5, the cross correlation was significant at $4.2\,\sigma$. This indicates a common CMBR signal with a low level of contamination. Converting the 150-GHz result to a band power and adopting a conservative $4.2\,\sigma$ detection gives $\delta T = 32.6 \pm 7.8\,\mu K$ at $\ell = 107$. An IRAS dust-correlated component was seen in the 600-GHz channel and thus, assuming a power-law extrapolation, dust emission could be excluded as the source of the signal at 150 and 220 GHz. Free-free emission could not be ruled out but was not believed to be the source of the excess variance.

La Sapienza, Rome
ARGO1: de Bernardis *et al.* (1990); ARGO2: de Bernardis *et al.* (1994)

SP/ACME   Ground-based, South Pole $36 \lesssim \ell \lesssim 106$
                     NRAO HEMTs at 30 and 40 GHz

The group's series of measurements began with a balloon-borne SIS mixer-based radiometer at 91 GHz that was subsequently placed at the South Pole in the 1988–1989 season. SP89 gave a limit $T_{\rm rms} = 45\,\mu K$ ($1\,\sigma_{\rm con}$) corresponding to $\delta T_\ell < 38\,\mu K$ at $\ell = 167$. SP91a used 30-GHz NRAO HEMT amplifiers with a resulting $\theta_{1/2} = 1°\!.5$ beam. The beam was chopped with a $3°$ p-p amplitude and the signal synchronously demodulated. The central optical axis was then stepped across the sky in $2°\!.1$ increments. The full HEMT bandwidth was broken into four sub-bands. The SP91a analysis measured a signal with a spectral index characteristic of synchrotron emission, and only consistent with the CMBR with 2% probability. The team chose to set an upper limit on the anisotropy, $\delta T_\ell = 25\,\mu K$ ($1\,\sigma_{\rm con}$) by selecting the least contaminated channel. A later analysis by Gundersen *et al.* of all channels gave $\delta T_\ell = 30.2^{+8.9}_{-5.5}\mu K$ though still with galactic spectral index $\beta = -5.1^{+1.2}_{-5.1}$. Notably, SP91a was the first experiment to use NRAO HEMTs for CMBR anisotropy measurements. SP91b was similar to SP91a. Again a signal was detected, this time with a spectrum 2σ away from a CMBR spectrum, again closer to a free-free or synchrotron spectrum. Though various cuts through the data are considered, it is concluded that "identification of the source of the signal is not possible on the basis of the measured spectrum." Additionally, correlations between bands were not considered in the original analysis. Because of the nonCMBR spectral index, we treat SP91a and SP91b as upper limits on the CMBR anisotropy. Based on the WMAP maps, it does not appear that foreground emission was the source of the observed signal suggesting the culprit was an instrumental systematic effect. By SP94, the team was building their own amplifiers. SP94 measured a signal with the spectral index of the CMBR and accounted for the correlated noise, reporting a measurement $\delta T_\ell = 36.3^{+13.6}_{-6}\,\mu K$ at $\ell = 68$.

UCSB
SP89: Meinhold and Lubin (1991); SP91a: Gaier *et al.* (1992); SP91b: Schuster
et al. (1993); SP94: Gundersen *et al.* (1995)

COBE/DMR   Dicke-switched radiometers $1 \leq \ell \lesssim 13$
at 31.5, 53, and 90 GHz

DMR used three, dual, $\theta_{1/2} = 7°$, differential radiometers aboard the COBE satellite. The opening angle between feeds was 60°. The satellite spun at 0.8 rpm as
it mapped the sky from its polar orbit. In DMR1, a new measurement of the
dipole was given and new limits were placed on the anisotropy, $T_{\mathrm{rms}} = 55\,\mu\mathrm{K}$
($1\,\sigma_{\mathrm{con}}$) or $\delta T_\ell = 32$. The paper on the discovery of the anisotropy, DMR2, gave
$T_{\mathrm{rms}} = 30 \pm 5\,\mu\mathrm{K}$ for a $\theta_{1/2} = 10°$ smoothing of the map. This corresponds to
$\delta T_\ell = 24.7 \pm 4.1\,\mu\mathrm{K}$ at $\ell = 10$ according to the Wright *et al.* (1994a) window function. The second-year release, DMR3, improved this to $\delta T_\ell = 25.1 \pm 2.2\,\mu\mathrm{K}$ and
the four-year release, DMR4, gave $\delta T_\ell = 23.9 \pm 0.8\,\mu\mathrm{K}$. Different analyses (Gorski
et al.) with different assumptions showed the basic result is robust but depends on
selection and frequency channel by up to 15%. The power spectrum is presented in
Hinshaw *et al.* and Wright *et al.* (1996) A second measurement of the anisotropy
came from an analysis of FIRAS. Fixsen *et al.* found correlation between spatial
fluctuations in the FIRAS absolute temperature map and the DMR data.

NASA/GSFC
DMR1: Smoot *et al.* (1991); DMR2: Smoot *et al.* (1992); Wright *et al.* (1994);
DMR3: Bennett *et al.* (1994); DMR4: Bennett *et al.* (1996); Gorski *et al.* (1996);
Hinshaw *et al.* (1996); Wright *et al.* (1996); Fixsen *et al.* (1997)

19GHz   Balloon-borne JPL Maser $1 \leq \ell \lesssim 36$
at 19.2 GHz

The radiometer consisted of a 19.2-GHz JPL ruby maser that was Dicke-switched
between a thermally controlled cryogenic waveguide reference load and a single
$\theta_{1/2} = 3°$ feed oriented 45° from the zenith. A calibrated noise pulse was injected
during the switching cycle. The radiometer was flown on a balloon-borne gondola
that rotated at 1 rpm. With two flights from Palestine, Texas, and two from Alice
Springs, Australia, 95% of the sky was mapped. This frequency band was originally
planned for COBE/DMR but was descoped (see page 419). A sophisticated analysis
introduced the full map likelihood (the computation of which was intractable at
the time), the "Boughn–Cottingham" statistic, and other methods. Though the
data were acquired before COBE was launched, the first scientific analysis was not
published until just before the COBE/DMR discovery. An analysis of the northern
hemisphere flights gave $\delta T_\ell < 37\,\mu\mathrm{K}$ at $\ell = 25$ ($1\,\sigma_{\mathrm{con}}$). The analysis of the full
map was redone by Ken Ganga after a coding bug was uncovered (see FIRS entry
above). Ganga found $\delta T_\ell = 27^{+19}_{-27}\,\mu\mathrm{K}$ at $\ell = 25$ ($1\,\sigma_{\mathrm{con}}$), fully consistent with
COBE. It is intriguing that the reported upper limit is roughly a factor of 2 higher
than the instrument noise level. The full map was first used in Jahoda and Boughn
and published in de Oliveria-Costa *et al.* It is still used for foreground emission
studies.

Haverford and Princeton
19GHz: Boughn *et al.* (1992); Boughn and Jahoda (1993); Ganga (1994); de
Oliveira-Costa *et al.* (1998)

MAX   Balloon-borne 0.3-K bolometers $90 \lesssim \ell \lesssim 250$
                            at 180, 270, and 360 GHz

There were five MAX flights with successive improvements. Most measurements were made with a $\theta_{1/2} = 0°\!.5$ beam and a sinusoidal $1°\!.3$ p-p throw on the sky and thus measured in a single broad window. The center of this chop pattern was discretely stepped over $\approx 6°$ in azimuth. MAX2 reported detecting "signals in excess of random ... but concerns about possible systematic errors prevent the interpretation of the signal as being uniquely cosmological in origin." There were difficulties with RFI and variable baseline drifts. MAX3 reported detecting dust, and set a limit $\delta T_\ell < 23\,\mu K$ $(1\,\sigma_{con})$ on the CMBR at $\ell \approx 150$ and a detection, $\delta T_\ell = 74^{+22}_{-14}\,\mu K$, from the same flight with observations in a less dusty (MAX2) region. Possible synchrotron or free-free contamination was deemed unlikely based on the extrapolation of lower frequency maps. MAX4 reported refined measurements with improved control of systematics and better rejection of dust. Devlin *et al.* conclude that "it has become increasingly difficult to construct alternative hypotheses for the observed [CMB-like] signals." MAX5 extended the program and Tanaka *et al.* reanalyzed (increasing by $\approx 20\%$) the MAX4 measurements to those we give here. Because of the uniqueness of the scan pattern, only the same areas (not the same places) could be remeasured on subsequent flights. Ganga *et al.* present the result from the combined analysis of MAX4 and MAX5 as $\delta T_\ell = 51^{+8}_{-7}\,\mu K$.

Berkeley/UCSB
MAX2: Alsop *et al.* (1992); MAX3: Meinhold *et al.* (1993); MAX3: Gundersen *et al.* (1993); MAX4: Devlin *et al.* (1994); MAX4: Clapp *et al.* (1994); MAX5: Tanaka *et al.* (1996); MAX5: Lim *et al.* (1996); Ganga *et al.* (1998)

IAB-II   Italian Antarctic Base $60 \lesssim \ell \lesssim 200$
                            0.35-K bolometers at 150 GHz

The observations were made by scanning the $\theta_{1/2} = 50'$ beam with a 45-cm diameter oscillating primary mirror and then synchronously detecting the bolometers. The p-p beam chop was $1°\!.4$. The observing pattern consisted of 12 points around the SCP, following a strategy used by Readhead *et al.* (1989) about the NCP. The measured signal gave $\delta T_\ell = 95 \pm 42\,\mu K$. Though it is possible that the experiment measured the CMBR anisotropy, a strong case could not be made. The authors could "not exclude contamination by low spatial frequency atmospheric noise" and note "If residual systematic effects of cold galactic dust emission are responsible for the signal detected, our result should be considered as an upper limit to CBR anisotropy." Nevertheless, IAB was a pioneering experiment. It was one of the early bolometric anisotropy experiments in the Antarctic to report a result.

ESA and Rome
IAB3: Piccirillo and Calisse (1993)

WD   100-mK bolometers at 90 GHz $300 \lesssim \ell \lesssim 830$
                            at the South Pole

White Dish (WD) used a single-moded 90-GHz bolometer cooled below 100 mK at the focus of a specially built on-axis Cassegrain telescope at the South Pole. The resulting $\theta_{1/2} = 0°\!.2$ beam was scanned in the sky in a circle of diameter $0°\!.4$. WD was the first experiment to use single-moded bolometers and the first to use adiabatic demagnetization to cool the bolometers to $<100\,mK$. Both these innovations were used by later experiments. The original results gave $\delta T_\ell = 22\,\mu\text{K}$ ($1\sigma_{\text{con}}$). The analysis was later redone to include marginalization over the 30% calibration uncertainty and the offset and gradient subtraction. Ratra *et al.* give the now accepted value, $\delta T_\ell = 48\,\mu\text{K}$ ($1\,\sigma_{\text{con}}$), or $\delta T_\ell = 75\,\mu\text{K}$ including 30% calibration error at $\ell = 477$.

Princeton
WD1: Tucker *et al.* (1993); WD2: Ratra *et al.* (1998)

SASK Ground-based NRAO HEMTs $100 \lesssim \ell \lesssim 350$
at 30 and 40 GHz

SASK1 used 30-GHz HEMTs to illuminate a parabolic primary and in turn reflect off an oscillating flat plate. The $\theta_{1/2} = 1.°44$ beam was swept $4.°9$ p-p throw on the sky in a sinusoidal pattern and sampled rapidly. A three-beam chop was synthesized by weighting the data in software. Two orthogonal polarizations were measured to search for possible systematic effects. By breaking up the band into three frequency bins, both the spectrum and amplitude of the signal were found. The frequency spectrum was that of the CMBR, but free-free could only be excluded at the $2\,\sigma$ level. Numerous tests for foreground contamination and instrumental systematics were presented. The issue of correlated noise was introduced in the analysis. SK1 concludes "the most parsimonious description of the data is that fluctuations in the CMBR dominate our sky signal." The measurements were confirmed at 40 GHz in SK2, increasing the rejection of free-free contamination to $3\,\sigma$. It was the first time subsequent ground-based experiments observed the same celestial signal. SK2 also introduced the method of shaping the beam by rapid sampling followed by weighting the sample in software (called "time-domain beam synthesis," TDBS). In SK3, the TDBS method was extended to measure the CMBR power spectrum in multiple bands and showed the rise to just past the top of the first acoustic peak. It was the first measurement of nontrivial structure in the angular power spectrum. In addition, the TDBS method permitted the confirmation of the MSAM experiment. This was the first time two different medium-scale anisotropy measurements were shown to be consistent.

Princeton
SK1: Wollack *et al.* (1993); SK2: Netterfield *et al.* (1995); SK3: Netterfield *et al.* (1997)

MSAM Balloon-borne 0.3-K bolometers at $40 \lesssim \ell \lesssim 450$
168, 270, 495, and 675 GHz (modified FIRS receiver)

MSAM used a $30'$-beam with a three-position chop of $120'$. The switching was decomposed into a "single difference" result and a "double difference" result permitting measurements in two broad windows. In the analysis, all frequency bands were analyzed together and the CMBR component was solved for, as opposed to the more common frequency-by-frequency comparison. In the first flight, two "sources" with a spectrum of the CMBR were noted. Though the anisotropy was measured, the "sources" clouded the interpretation and free-free contamination could not be ruled out with the MSAM data alone. The second flight observations overlapped the first flight and the double-difference result was confirmed (a first for degree-scale balloon-borne measurements). Measurements of H_α emission (see Introduction) showed that free-free emission did not contaminate the result. More detailed statistical analysis showed that the source-like appearance of the measurement was consistent with Gaussian statistics as reported in Kowitt *et al.*, and the SASK experiment confirmed that the fluctuations had the same amplitude at 40 GHz. A third flight observed a different region of the sky. The conclusion based on all three flights was that the data "suggest that there is a rise in the power spectrum from $\ell \sim 160$ to $\ell \sim 270$, but the statistical significance is modest." The combined results were reanalyzed and extended in Wilson *et al.*, with similar conclusions.

NASA/GSFC and Chicago
MSAM1: Cheng *et al.* (1994); MSAM2: Cheng *et al.* (1996); Kowitt *et al.* (1997); MSAM3: Cheng *et al.* (1997); MSAM123: Wilson *et al.* (2000)

PYTHON	Ground-based 50-mK bolometers at 90 GHz; NRAO HEMTs at 40 GHz, at the South Pole	Bol: $50 \lesssim \ell \lesssim 100$ HEMT: $40 \lesssim \ell \lesssim 260$

Python began with a 2×2 spatial array of single-moded bolometers at 90 GHz cooled to 50 mK and placed at the focus of a parabolic primary. Not only was it the first spatial array of detectors, but it was the first time bolometers clearly detected the anisotropy from the ground. The $0°.75$ beam was formed by an offset 0.75-m parabola and switched in discrete $\pm 2°.75$ steps with a large flat. Two triple-beam patterns were then overlapped to make a four-beam pattern. The PY1+PY2 observations gave $\delta T = 52.1^{+19.5}_{-13.1}\mu$K at $\ell = 92$, with an error bar largely limited by sample variance. PY3 confirmed and extended PY1 and PY2. Based on external data sets, the foreground emission was argued to be small, which was later confirmed. Early results had a 20% calibration error, which was steadily improved throughout the observations. For PY4, a two-feed Q-Band NRAO HEMT system replaced the bolometers and was used to show that the PY1–PY3 measurements had a CMBR-like spectrum. The TDBS scheme was adopted permitting a measurement of the power spectrum. Early results covering $\approx$600 deg^2 were given in PY5a, where the team reported seeing a "sharply increasing slope starting at $\ell = 150$." However, the data points were significantly correlated. An updated and expanded analysis in PY5b confirmed the PY3 result and showed the more gentle rise we now observe.

Princeton/Chicago
PY1: Dragovan *et al.* (1994); PY2: Ruhl *et al.* (1995); PY3: Platt *et al.* (1997); PY5a: Coble *et al.* (1999); PY5b: Coble *et al.* (2003)

CAT	Interferometer at 13.5–16.5 GHz	$330 \lesssim \ell \lesssim 680$

The Cambridge Anisotropy Telescope consisted of three dual-polarization elements operating at sea level. The synthesized beam was $\approx 25' \times 22'$, though dependent on frequency. It was the first interferometer to detect the anisotropy. The first measurements, CAT1, were made at 13.5 GHz. After removing radio sources identified by the Ryle telescope, a residual signal was seen and, through a variety of tests, shown to be on the sky. Since synchrotron emission can be large at these low frequencies, the team was cautious in their interpretation. CAT2 observed the same region at 15.5 and 16.5 GHz. A signal consistent with the CAT1 results was observed. The spectral index was close to that of the CMBR. In the analysis, the data were marginalized over a possible synchrotron component. Additionally, they report the combination of CAT2 and other prior experiments was "beginning to support the existence of a 'Doppler Peak' [between $\ell \approx 100–200$]." When CAT2 was combined with the later SK3, there was "further evidence for a downturn ... for $\ell > 300$." CAT3 added a second $2° \times 2°$ patch of sky and followed the same analysis pipeline, strengthening the case presented in CAT2.

Cambridge, UK
CAT1: O'Sullivan *et al.* (1995); CAT2: Scott *et al.* (1996); CAT3: Baker *et al.* (1999)

BAM Balloon-borne 0.3-K bolometers $30 \lesssim \ell \lesssim 90$
in a FTS at 110–255 GHz

BAM was unique. It used the differential FTS from the Gush *et al.* (1990) absolute measurement to measure the anisotropy. The two beams had $\theta_{1/2} = 0.°7$ and were separated by $3.°6$. The raw data were a set of interferograms from the difference in field strength entering the two inputs. A series of technical malfunctions curtailed the goal of directly measuring the anisotropy frequency spectrum. Nevertheless, BAM measured significant fluctuations in the signal channel and only low levels in the quadrature channel, as expected for a celestial (as opposed to instrumental) source. "However, the spectrum has not been measured with sufficient sensitivity to attribute these fluctuations definitively to cosmic origin." When interpreted as a CMBR detection, one finds $\delta T_\ell = 56\,\mu\text{K}$ at $\ell = 58$.

UBC
Tucker *et al.* (1997)

SuZIE Array of 0.3-K bolometers $1330 \lesssim \ell \lesssim 3670$
at 142 GHz

A receiver containing a 2×3 array of ^{3}He-cooled bolometers was mounted on the Caltech Submillimeter Observatory to produce six $\theta_{1/2} = 1.'7$ beams. The three elements in a row were oriented along the azimuthal scan direction and separated by $2.'3$. Each row member was electronically differenced to produce two $2.'3$ single-differences and one $4.'6$ single-difference. Two single differences were then differenced to produce a double-difference and a single-difference for each row. Observations were made by drift-scanning. It was the first time electronic differencing was used for anisotropy measurements. The technique is akin to chopping between the bolometers at the frequency equal to the bandwidth of the system. An area free of known sources was observed to place a limit $\delta T_\ell = 19^{+15}_{-19}\,\mu\text{K}$ at $\ell \approx 2400$. Because of the broad window function, the interpretation of the result is sensitive to the assumed cosmological model. The high frequency observations nicely complemented the lower frequency OV1 and OV2 results at similar ℓ.

CalTech and Stanford
Ganga *et al.* (1997); Church *et al.* (1997)

IAC-BAR Bolometers at 97, 143, $40 \lesssim \ell \lesssim 80$
231, and 273 GHz

A 4-channel ^{3}He-cooled bolometric radiometer was coupled to a 45-cm telescope to produce a $\theta_{1/2} = 2.°4$ beam at the most sensitive (143 GHz) channel. Observations were made from the Observatorio del Teide at Tenerife, Spain. Data taken with a sinusoidally wobbling primary at 4 Hz and $5.°2$ p-p throw were weighted in software to produce a double-difference pattern. Data were dominated by atmospheric rather than detector noise. In IACB1, the galaxy was clearly seen at the expected level. Though a signal was detected at 143 GHz it was not clearly demonstrated to be celestial and so the authors present an upper limit $\delta T_\ell = 59\,\mu$K ($1\,\sigma_{\rm con}$) at $\ell = 53$. After taking more data, the analysis was considerably extended in IACB2. Based on the multiple measured frequencies, the dust and atmosphere were rejected as possible contaminants. For IACB3, the beam was reduced to $\theta_{1/2} = 1.35°$, the throw was increased to $5.7°$ p-p and the TDBS technique was adopted. The results show the low-ℓ rise to the peak and are notable as the lowest-ℓ ground-based bolometric (143 GHz) measurements.

IAC and Bartol Research Institute
IACB1: Piccirillo *et al.* (1997); IACB2: Femenia *et al.* (1998); IACB3: Romeo *et al.* (2001)

QMAP   Balloon-borne NRAO HEMTs $50 \lesssim \ell \lesssim 150$
at 30 and 40 GHz

QMAP used a spatial array of one dual-polarization 30-GHz feed ($\theta_{1/2} = 0.°9$) and two dual-polarization 42-GHz feeds ($\theta_{1/2} = 0.°6$). The experiment produced a map over a limited region of sky (530 deg^2) that was directly analyzed for CMBR anisotropy. In particular, there was no recourse to "demodulation" or TDBS techniques that effectively filter unwanted ill-behaved modes. Cross-linking of the map was achieved by observing on both sides of the NCP. It was the first measurement to make such a map and the first balloon-borne focal plane array. First flight results were reported in QMAP1, second flight results in QMAP2, and the combined analysis in QMAP3. The combined map detected the CMBR at $>35\sigma$. A more detailed analysis of free-free contamination (QMAP4) led to a 2.3% and 1.3% reduction at 30 and 40 GHz respectively in originally reported results (we do not plot it a second time). Subsequently, QMAP was combined with SASK to produce the $\approx 650\,{\rm deg}^2$ QMASK map (confirming both measurements), which remained one of the largest degree-scale resolution maps until Archeops. It is of historical interest that QMAP, HACME, and BOOMERANG were proposed and initiated before the WMAP and PSI/FIRE satellite proposals.

Princeton
QMAP1: Devlin *et al.* (1998); QMAP2: Herbig *et al.* (1998); QMAP3: de Oliveira-Costa *et al.* (1998); QMAP4: de Oliveira-Costa *et al.* (2000); QMASK: Xu *et al.* (2001)

HACME   Balloon-borne NRAO HEMTs $20 \lesssim \ell \lesssim 60$
at 40 GHz

HACME used one single-polarization 40-GHz feed at the focus of the ACME Gregorian telescope. A 1.2-m rotating canted flat mirror moved the $\theta_{1/2} = 0.°8$ beam in an ellipse with 10° major axis. The gondola scan pattern linked together all the elliptical scans. The mapmaking analysis had much in common with QMAP, and slightly preceded it; both analyses were led by Tegmark and de Oliveira-Costa. A 630 deg^2 map was made. However, the removal of time-dependent spin-synchronous offset (1/f noise) in each sub-band of the HEMT amplifier markedly reduced the effective sensitivity of the map, resulting in an anisotropy limit $\delta T_\ell < 38\,\mu\text{K}$ $(1\,\sigma_{\text{con}})$ at $\ell = 38$.

UCSB
Staren *et al.* (2000); Tegmark *et al.* (2000)

MAT/TOCO	Ground-based NRAO HEMTs at 30 and 40 GHz and NRAO SIS mixers at 150 GHz	$50 \lesssim \ell \lesssim 400$

The Mobile Anisotropy Telescope on Cerro Toco used the QMAP gondola and focal plane but added two 150-GHz mixer systems with $\theta_{1/2} = 0\overset{\circ}{.}2$ and $\theta_{1/2} = 0\overset{\circ}{.}25$. There were two seasons of operations. The mixers did not work until the second season and provided the sensitivity and resolution to observe on the high-ℓ side of the first peak. The data were analyzed using TDBS. In TOCO2, the data showed "(1) a rise in the angular spectrum to a maximum with $\delta T_\ell \approx 85\,\mu\text{K}$ at $\ell \approx 200$ and a fall at $\ell > 300$, thereby localizing the peak near $\ell \approx 200$, and (2) that the anisotropy at $\ell \approx 200$ has the [frequency] spectrum of the CMB." Subsequent more accurate fits (Knox and Page, and then with the WMAP recalibration of Jupiter, Page *et al.*) gave $\delta T_{\text{peak}} = 83^{+9}_{-8}\,\mu\text{K}$ at $\ell = 200^{+15}_{-12}$. This compares favorably with WMAP's $\delta T_{\text{peak}} = 75 \pm 0.16\,\mu\text{K}$ at $\ell = 220.7 \pm 0.7$. Corrections for foreground emission are given in Miller *et al.*, and then, in combination with more data, in TOCO3. TOCO is the only experiment to detect the anisotropy with SIS mixers, and it was the first experiment to localize the first peak. However, the location was apparent from the combination of SASK plus any other $\ell > 300$ measurement prior to TOCO.

Pennsylvania and Princeton
TOCO1: Torbet *et al.* (1999); TOCO2: Miller *et al.* (1999); Knox and Page (2000); Miller *et al.* (2002); Page *et al.* (2003b); TOCO3: Nolta *et al.* (2003)

JB-IAC	HEMTs at 33-GHz interferometer	$90 \lesssim \ell \lesssim 130$ $190 \lesssim \ell \lesssim 230$

The Jodrell Bank–Instituto de Astrofísica de Canarias collaboration used a two-element 33-GHz interferometer to measure the anisotropy near $\ell = 109$. The main beam was $\theta_{1/2} = 2\overset{\circ}{.}4$ by $\theta_{1/2} = 5\overset{\circ}{.}3$ with the 15.2-cm (16.5λ) baseline set along the east–west direction resulting is a fringe period in right ascension of $3\overset{\circ}{.}5$. It was the first HEMT-based CMBR-dedicated interferometer, and only the third CMBR interferometer. An interferometer acts like a single-difference measurement switched at the observation frequency. To observe, the interferometer orientation was fixed parallel to $\delta = 41°$ and the sky drifted through the resulting interference pattern. The Moon in conjunction with an internal source set the calibration. The sine and cosine phases were coadded for 100 days, slightly filtered to remove the baseline, and edited to select regions of low foreground emission. The team reports $\delta T_\ell = 45^{+13}_{-12}\,\mu\text{K}$ with an error bar dominated by sample variance. Point sources and diffuse foreground emission were ascertained from other measurements and argued to be strongly subdominant. In a subsequent observation, the baseline was extended to 32.9λ resulting in the measurement $\delta T_\ell = 63^{+7}_{-5}\,\mu\text{K}$ at $\ell = 208$.

Jodrell Bank, UK
Dicker *et al.* (1999); Harrison *et al.* (2000)

Viper   NRAO HEMTs at 40 GHz $100 \lesssim \ell \lesssim 600$

Viper used a specially built 2.35-m CMBR anisotropy telescope at the South Pole. The two feeds produced a $\theta_{1/2} = 0°\!.26$ beam that was scanned $3°\!.6$ across the sky with a chopping flat. The novel position of the flat, at the image of the primary, kept the beam relatively stationary on the optics. The data were analyzed in six windows using TDBS techniques. To constrain potential diffuse foreground contamination, the experiment relied on external data sets (e.g., Python, IRAS). Though the analysis showed that point source contamination was rather unlikely, such a possibility could not be completely ruled out. Nevertheless, the data were interpreted as the CMBR anisotropy: "The increase above COBE anisotropy levels ... is evident in the data, as is a lower anisotropy level near the expected position of the first null at $\ell \sim 400$." Subsequent to this measurement, the ACBAR receiver was put on the telescope.

Carnegie-Mellon
Peterson *et al.* (2000)

BOOM   Balloon-borne bolometers at $100 \lesssim \ell \lesssim 1000$
                 90, 150, 240, and 400 GHz

The BOOMERANG team developed a new bolometer format (the "spiderweb") that was read out using a novel alternating current biasing scheme. For the most significant channels at 150 GHz, $\theta_{1/2} = 10'$. The first test flight (BOOMNA) lasted one night and measured the first peak. The second long-duration balloon flight (BOOM1) around Antarctica covered ≈ 700 deg^2 of the sky in a compact area with a cross-linked pattern. The combination of multiple sensitive pixels and long integration times gave a high signal-to-noise CMBR sky map with enough resolution to allow one to visualize the acoustic scale. A multi-frequency analysis demonstrated foreground contamination was not significant. The initial beam determination and pointing solution somewhat limited the interpretation of the first release. Subsequent more refined analyses (BOOM2 and BOOM3) began to delineate the higher-order acoustic peaks. For BOOM2, de Bernardis *et al.* give values for the peaks in excellent agreement with those from WMAP but also note that a flat line fit for the $401 \leq \ell \leq 1000$, $401 \leq \ell \leq 750$, and $726 \leq \ell \leq 1025$ regions is excluded at just the $2\,\sigma$ level. In the BOOM3 analysis, the second peak had been isolated from the first and second dips by better than 3σ and thus was the first single experiment to clearly show higher-order structure in the power spectrum. Another important aspect of the BOOMERANG and MAXIMA efforts was that it forced researchers to greatly improve analysis methods to deal with large volumes of high-sensitivity data. Appropriate for the beginning of the new millennium, BOOMERANG and MAXIMA ushered in the era of "high precision" anisotropy measurements.

CalTech and "La Sapienza," Rome
BOOMNA: Mauskopf *et al.* (2000); BOOM1: de Bernardis *et al.* (2000); BOOM2: Netterfield *et al.* (2002), de Bernardis *et al.* (2002); BOOM3: Ruhl *et al.* (2003)

MAXIMA   Array of balloon-borne $40 \lesssim \ell \lesssim 1200$
                   100-mK bolometers

MAXIMA observed with 16 spiderweb bolometers (8 at 150 GHz, 4 at 240 GHz, and 4 at 410 GHz) with $\theta_{1/2} = 10'$. The bolometers were cooled to 100 mK using an adiabatic demagnetization refrigerator and thereby achieved a remarkable noise figure, $75\,\mu K\,s^{1/2}$ in the most sensitive channels. This is still a record. Beam scanning was accomplished by nutating the 1.3-m primary at 0.45 Hz with an amplitude of $\pm 2°$ while also scanning the gondola in azimuth at about 0.02 Hz. Using the rotation of the sky about the NCP, MAXIMA mapped a square 124 deg² region with cross-linked scans in a single balloon flight (MAXIMA1). Though the flight was in August 1998, results were not published until May 2000 because the large amount of data required the development of new software. This was a common theme of new-millennium observations. At the time, MAXIMA1 gave the highest resolution maps of the anisotropy. By comparing frequency bands, potential foreground contamination could be constrained to negligible levels. As can be seen in the plots, the initial power spectrum is in excellent agreement with that from WMAP. In MAXIMA2, a subset of the MAXIMA1 data were rebinned with smaller pixels, reanalyzed, and the spectrum extended to $\ell \approx 1200$. Maps from a second flight (Abroe *et al.*) showed good agreement with those from the first flight and also with WMAP.

Berkeley
MAXIMA1: Hanany *et al.* (2000); Lee *et al.* (2001); Abroe *et al.* (2004)

CBI	HEMTs at 35 GHz in an interferometer	$300 \lesssim \ell \lesssim 3500$

CBI is a 13-element compact interferometer that measured the anisotropy in ten 1-GHz wide frequency bands between 26 and 36 GHz. This resulted in 780 visibility measurements. The primary beam is $\theta_{1/2} = 44'$, limiting the low-ℓ coverage to about 400 before any mosaicking. The high ℓ limit is set by the maximum dish separation. By CBI2, 40 deg² had been mapped and analyzed in a couple of ways. Considerable effort went into accounting for possible point-source contamination, the largest foreground at these scales and wavelengths. The CBI observations opened up precision measurements of the $\ell > 1000$ region of the spectrum. For example, they were the first to show the Silk damping tail in a single measurement. They were also particularly important for pinning down the correct cosmology for the first WMAP release. When combined with WMAP at $\ell < 900$ and ACBAR at $\ell > 1000$, the CBI data broke a number of parameter degeneracies. Analyses are ongoing at the time of writing and now include the detection of polarization and "excess power" near $\ell \approx 2500$ that may be due to the Sunyaev–Zel'dovich effect. Six years after the results were reported, the community was still trying to assess the degree to which the Sunyaev–Zel'dovich effect in clusters of galaxies contribute to the power spectrum at high ℓ.

Caltech
CBI1: Padin *et al.* (2001); CBI2: Pearson *et al.* (2003); CBI2: Mason *et al.* (2003)

DASI	HEMTs at 35 GHz in an interferometer	$100 \lesssim \ell \lesssim 900$

DASI was a sister experiment to CBI. It was also a 13-element compact interferometer that measured in ten 1-GHz frequency bands between 26 and 36 GHz. Its amplifiers followed the NRAO design. With its primary beam of $\theta_{1/2} = 3°\!.4$ and smaller spacing, it measured larger angular scales than CBI. Unlike CBI, DASI observed from the South Pole. Both the CBI and DASI experiments marked a large jump in sensitivity for coherent systems. Not only did they have more amplifiers than previous systems, but they used the intrinsic stability of their interferometers at good sites to observe for many months. The DASI bandwidth was large enough to limit the spectral index of the fluctuations to $\beta = -0.1 \pm 0.2$, consistent with the CMBR. Of the pre-WMAP measurements, DASI was the most accurate with its 4% calibration uncertainty, relative immunity to atmospheric fluctuations, and measured long-term stability. The DASI results were announced concurrently with those from BOOM2. As with BOOM2 there was evidence for a second peak but a flat line was also a good fit to the $\ell \gtrsim 400$ spectrum. The case was strengthened when both experiments were combined. DASI went on to make the first measurement of the polarization in the CMBR (Kovac *et al.* 2002).

Chicago
DASI: Halverson *et al.* (2002)

Archeops	Balloon-borne 100-mK bolometers at 143, 217, 353, and 545 GHz	$20 \lesssim \ell \lesssim 350$

Archeops was a balloon-borne test bed for the Planck HFI experiment. It was based on a similar receiver design (with spiderweb bolometers) and optics. The balloon payload spun at 2 rpm and mapped out 30% of the sky with $\theta_{1/2} \approx 11'$. The best channel achieved $90\,\mu\mathrm{K}\,\mathrm{s}^{1/2}$ by cooling with a dilution refrigerator, a first for CMBR balloon experiments. The 12.6% of the data with $b > 30°$ in the 143- and 217-GHz channels were analyzed for anisotropy. Whereas most of the new-millennium experiments measured the anisotropy with precision at $\ell > 200$, Archeops excelled on the low-ℓ side of the peak. Its ability to constrain the first peak was comparable to that of BOOM3. Prior to WMAP, Archeops produced the largest high-sensitivity subdegree resolution map.

France
ARCH: Benoît *et al.* (2003)

VSA	Interferometer at 34 GHz	$160 \lesssim \ell \lesssim 1500$

The very small array (VSA) was the culmination of many years of pioneering work on CMBR interferometry in the UK. It used 14 antennas with 2° primary beams. The system bandwidth was 1.5 GHz. The interferometer was used in a couple of configurations. Unlike DASI and CBI, individual antenna elements moved so that fringes not associated with the sky were washed out. Numerous tests were done to show that foregrounds did not significantly contaminate the results. As with BOOM2 and DASI, the initial data showed evidence for individual peaks though not quite at the $3\,\sigma$ level. As pointed out by the VSA team, it is very reassuring that multiple different types of measurements find the same power spectrum out to $\ell \approx 1500$.

UK
VSA: Grainge *et al.* (2003); Scott *et al.* (2003)

WMAP	NRAO HEMTs at 22, 33, 40, 62, and 94 GHz	$1 \leq \ell \lesssim 750$

WMAP was the third CMBR satellite and NASA's second. What set it apart from all other measurements is that it had a very well-defined systematic error budget with measurements to back up the budget, it covered the full sky, and it measured foreground emission with the same sensitivity as the CMBR. In particular, the errors associated with beam uncertainty, calibration, knowledge of the noise, variations in gain, and electronic transfer function were all known roughly an order of magnitude better than in previous experiments. Additionally, there were multiple levels of redundancy that permitted cross-checks on nearly every space and timescale of importance. It was the measurement that combined "high precision" with "high accuracy" enabling detailed cosmological tests.

NASA/GSFC and Princeton
Bennett *et al.* (2003)

ACBAR   Array of bolometers at 150 GHz $200 \lesssim \ell \lesssim 3000$
at the South Pole

ACBAR used a 16-element spatial array of spiderweb bolometers coupled to corrugated feeds. The highest signal-to-noise CMBR results come from the eight 150-GHz channels. ACBAR scanned the $5'$ beams with a $3°$ p-p triangle wave chop on the 2.1-m VIPER telescope. The scans were done in azimuth without vertical cross-linking. Data from the whole array were "double differenced" to remove the chopper synchronous offset. In addition, to remove atmospheric fluctuations, polynomials up to 10th order were removed from each scan. The differencing and filtering were accounted for in the analysis with the noise covariance matrix. In the first release, ACBAR mapped 24 deg^2 of the sky. This led to a clear measurement of the Silk damping tail that nicely complemented the lower frequency measurement by CBI. The ACBAR and CBI data were combined with the first-release WMAP data to break a number of parameter degeneracies. The ACBAR observations and analysis continued after the initial release and as of March 2008 provided the highest S/N measure of the Silk damping tail.

Berkeley and CWRU
ACBAR: Kuo *et al.* (2004)

Glossary

absolute magnitude Measure of the luminosity – the rate of emission of energy – of an astronomical object. The measure is logarithmic, and scaled such that a difference of five magnitudes is a factor of 100 difference in luminosity. The sign is set so that the fainter the object the larger the absolute magnitude. At 10 parsecs distance the absolute magnitude of an object is equal to its apparent magnitude.

acceleration parameter Early measure of the departure from a constant rate of expansion of the universe, $q_0 = \ddot{a}a/\dot{a}^2$, where $a(t)$ is the expansion parameter.

acoustic peaks Acoustic (sound wave) oscillation of the plasma-radiation fluid (discussed on page 436) in the early universe left ripples in the present large-scale matter distribution and the more prominent peaks in the CMBR anisotropy spectrum in Figure 5.18. Analyses of these effects in the 1960s are recalled by Sunyaev (p. 128), Peebles (p. 194), and Silk (p. 375).

active galactic nucleus Source of radiation and energetic particles in the center of a galaxy, likely powered by flow of matter around a massive black hole. *See* quasar, source counts.

adiabatic initial conditions Density fluctuations in the very early universe that would be produced by gently compressing or decompressing all components of an exactly homogeneous universe. The entropy per conserved particle thus is homogeneous.

aether drift Motion of Earth relative to the rest frame defined by the CMBR and galaxies, observed in the variations across the sky of galaxy counts (Blake and Wall 2002) and the CMBR temperature (Section 5.2).

AGN *See* active galactic nucleus.

angular size distance Coordinate distance r in the Robertson–Walker coordinates in equation (G.4). An observer sees that an object with physical diameter d at angular size distance r subtends angle $\phi = d/(a(t)r)$, where the object is seen as it was at world time t.

anisotropic homogeneous cosmological models Solutions to Einstein's general relativity theory that are spatially homogeneous but have nonzero rotation and/or shear.

anisotropy spectrum *See* CMBR anisotropy.

antenna pattern Response of an antenna to incident radiation as a function of the direction of incidence of the radiation.

antenna temperature Rate of flow of radiation – power – detected in a microwave or millimeter-wave receiver, measured in terms of the equivalent Rayleigh–Jeans temperature in the observed frequency band.

apparent magnitude Logarithmic measure of the brightness in the sky (the rate of arrival of energy per unit area) of an astronomical object. Neglecting relativistic corrections and obscuration, the apparent magnitude of an object with absolute magnitude M at distance D measured in megaparsecs is $m = M + 5 \log_{10} D + 25$. At $D = 10^{-5}$ Mpc, $m = M$.

Arzamas-16 At this Soviet facility near the town of Sarov, Zel'dovich worked on nuclear explosions and on the hot big bang cosmology.

atmospheric absorption Absorption of electromagnetic radiation in the atmosphere accompanies atmospheric emission of radiation. Water vapor and oxygen are the strongest absorbers at CMBR frequencies.

atmospheric emission In this book, electromagnetic radiation originating in the atmosphere, measured by tipping experiments.

autocorrelation function *See* correlation function.

back and side lobes In radio astronomy, smaller peaks in the diffraction pattern of an antenna that lie in the opposite hemisphere from the main beam, or to the side of the main beam. Unwanted response to radiation entering the detector by the back and side-lobes is a recurring issue in CMBR measurements. *See* antenna pattern.

baryon acoustic oscillation Pattern in the present large-scale galaxy distribution left by the acoustic oscillation of the plasma and radiation in the early universe.

baryon mass density The present value of the cosmic mean density of baryonic matter is $\rho_b = 4 \times 10^{-31}\,\mathrm{g\,cm^{-3}}$. This is one of the entries in the contents of Table 2.1, and it figures in the cosmological tests in Table 5.3

baryonic matter Neutrons and protons present in the nuclei of hydrogen and the heavier chemical elements. Associated with the baryons are enough negatively charged electrons (leptons) to balance the positive electric charges of the nuclei. Other unstable forms of baryons need not concern us.

BBNS Big Bang NucleoSynthesis. *See* helium.

beam-switching *See* Dicke microwave radiometer, phase-sensitive detection.

Bell Laboratories Bell Laboratories' communications experiments at the Crawford Hill Laboratory, Holmdel, New Jersey, played a key role in the identification of the CMBR.

Bell Laboratories 20-ft horn-reflector at Crawford Hill *See* horn antenna.

big bang cosmology Near homogeneous and isotropic expanding universe described by the Friedmann–Lemaître solution in general relativity theory with standard local physics. The name (whose origin is noted on page 14) is unfortunate: a "bang" suggests a localized event such as an explosion, while the cosmology actually describes evolution of the observable universe. Other names are "primeval atom" (p. 14) and "big squeeze" (p. 52). Big crunch and big freeze (p. 52) refer to the possible end of the universe as we know it, respectively, in gravitational collapse to exceedingly large density or in expansion continuing into an indefinitely remote and empty future. The hot big bang ΛCDM cosmology has the material contents listed in Table 2.1. Its tests are discussed in Section 5.4.

black hole General relativity theory prediction of the singular state approached by the collapse or merging of mass concentrations such as those present in dying stars and in the centers of galaxies. *See* Schwarzschild solution.

blackbody radiation *See* thermal radiation.

bolometer Device that responds to energy received in radiation, essentially independent of frequency, and to energy incident in any other form. Since thermal noise competes with the incoming signal, bolometric receivers and their surroundings are usually cooled to cryogenic temperatures. Because of their large bandwidth and low noise, bolometers are favored for CMBR observations at frequencies above 100 GHz.

Boltzmann constant The physical constant, $k = 1.38 \times 10^{-23} \, \mathrm{J \, K^{-1}} = 1.38 \times 10^{-16} \, \mathrm{erg \, deg^{-1}}$, that relates temperature T to its characteristic energy kT.

Bose–Einstein spectrum In this book, the radiation energy spectrum at statistical equilibrium when the photon number is conserved, obtained by replacing $h\nu/kT$ with $h\nu/kT + \mu$ in equation (G.5). Here μ is the chemical potential (up to the factor kT). If $\mu < 0$ the excess photons are pushed to long wavelengths where they are absorbed by the plasma. At $\mu > 0$ the effect on the spectrum is shown in Figure 4.7. The COBE/DMR bound based on the CMBR spectrum is $\mu < 9 \times 10^{-5}$ (Fixsen *et al.* 1996).

Brans–Dicke theory *See* scalar–tensor gravity theory.

brightness theorem In standard physics, the flow of energy in a beam of free radiation per unit area, time, solid angle, and frequency interval varies as $i_\nu \propto (1 + z)^{-3}$, where the frequency of a packet of the radiation varies as $\nu \propto (1+z)^{-1}$. Manifestations include the CMBR dipole anisotropy and the Sachs–Wolfe effect.

bulk flow In this book, the streaming motion of the Local Group of galaxies and its neighbors at a speed of $600 \, \mathrm{km \, s^{-1}}$ toward right ascension in the standard form $\alpha = 10^\mathrm{h}$ and declination $\delta = -25°$.

burning In this book, reactions of atomic nuclei in stars or the early universe that change the abundances of the chemical elements.

calibration In this book, conversion from instrument reading to a physical quantity such as temperature. *See* load.

Cameron, Alastair G. W. Pioneered, with Burbidge *et al.* (1957), the theory of nucleosynthesis in stars.

Cassiopeia A, Cas A galactic supernova remnant: the strongest celestial radio source outside the Solar System.

C-band Microwave radiation in the frequency range 6–8 GHz.

CDM cosmological model A relativistic big bang cosmology. The baryonic mass is less than the mass in nonbaryonic near-collisionless cold dark matter. The primeval departure from homogeneity is adiabatic, Gaussian, and near scale-invariant. With the addition of Einstein's cosmological constant this is the standard ΛCDM model with the contents in Table 2.1.

celestial equator and pole Projection of Earth's equator and polar axis onto the celestial sphere.

Climax Site of a high attitude observatory in Colorado.

closed universe Spacetime geometry with $R^{-2} > 0$ in equation (G.4).

clusters of galaxies Largest gravitationally bound concentrations of galaxies. In Abell's (1958) original definition, the mean number density of galaxies within a sphere of radius 2 Mpc centered on a rich cluster is greater than about 100 times the global mean density of galaxies.

CMBR Cosmic Microwave Background Radiation, also termed the "CBR" and "CMB:" the sea of thermal microwave radiation at temperature $T_0 = 2.725\,\mathrm{K}$ that nearly uniformly fills space. Outside opaque bodies the universe now contains 410 CMBR photons per cubic centimeter. Half of the energy density in this radiation is at wavelengths $< 1.5\,\mathrm{mm}$.

CMBR anisotropy Variation $\delta T/T \sim 10^{-5}$ of the CMBR temperature across the sky, an effect of the departure from an exactly homogeneous and isotropic observable universe. The intrinsic anisotropy is a measure of the spatial distribution of the CMBR observed as the variation of the CMBR temperature across the sky after removal of the dipole term largely due to our peculiar motion.

CMBR anisotropy dipole Motion of Earth relative to the general expansion of the universe causes the CMBR received from the direction we are moving to be warmer than the mean by about one tenth of one percent, while radiation from the opposite direction is cooler by the same fractional amount. The three components of our motion are represented by the three $\ell = 1$ dipole terms in the expansion of the temperature as a function of position in the sky in equation (5.10). Theory and measurements of this effect are reviewed on pages 424–433.

CMBR anisotropy multipole expansion Representation of the CMBR temperature variation across the sky in the spherical harmonic expansion in equation (5.10). Footnote 24 on page 443 notes the early history of its use.

CMBR anisotropy quadrupole The five $\ell = 2$ terms, following the dipole, in the expansion in equation (5.10).

CMBR anisotropy spectrum Measure in equation (5.12) of the departure from an exactly homogeneous and isotropic sea of radiation in terms of the squares, $|a_\ell^m|^2$, of the coefficients, a_ℓ^m, of the spherical harmonic expansion in equation (5.10). Panoramic views of the progress in measuring the anisotropy spectrum with its distinctive acoustic peaks are presented in Figures 5.9–5.18 and in Table A.3.

CMBR energy spectrum CMBR radiation energy density as a function of frequency or wavelength. The measured spectrum shown on page 16 is close to thermal. To see why the expansion of the universe cools the radiation while preserving its thermal spectrum, imagine a universe periodic in some volume so large that it can have no observational effect, and decompose the electromagnetic field into the discrete modes of oscillation that fit this volume. At thermal equilibrium at temperature T the occupation number $\mathcal{N}$ (the mean number of photons) in a mode with frequency ν (wavelength $\lambda = c/\nu$) is the function of the quantity $h\nu/kT$ shown in equation (G.5). As the mode expands with the general expansion of the universe, the mode wavelength is stretched as $\lambda \propto a(t)$, where a is the expansion factor (equation 2.3). When we can ignore the interaction of the radiation with matter – an excellent approximation – $\mathcal{N}$ does not change: the photons are stuck in the mode. So as λ increases the mode temperature decreases as $T \propto 1/a(t)$ (as in equation 2.7). Since all modes cool in this same way the radiation remains thermal.

CMBR energy spectrum distortion From the late 1960s to 1990, evidence later dismissed that the CMBR spectrum is significantly different from blackbody. *See* Bose–Einstein spectrum, Sunyaev–Zel'dovich effect.

CMBR local source models Proposal discussed in the 1960s that galaxies or other extragalactic objects are the sources of the sea of microwave radiation. The cosmological tests summarized in Section 5.4 convincingly rule out the idea.

CMBR measurements before identification The review on pages 47–49 and 63–65 leads us to conclude that, prior to the Bell Labs communications studies, no experimental result suggested detection of an isotropic sea of microwave radiation. Some likely could have detected the radiation if there had been the motivation to improve treatment of radiation from the ground and atmosphere. Wall (p. 280) recalls a late chapter in this story. Hogg (pp. 70–73), Wilson (pp. 167–169) and Burke (p. 179) comment on whether the earlier Bell communications experiments could have been taken to suggest the presence of a sea of radiation. This is reviewed on pages 49–51. Penzias (commencing on page 144) and Wilson (p. 157) describe their work in establishing detection. Burke (p. 180), Turner (p. 184), Peebles (p. 191), Wilkinson (p. 204), and Roll (p. 214) add recollections of completion of the identification. The effect of the radiation was observed in absorption lines of interstellar CN. That story is reviewed in Section 3.4 and recalled on pages 74–85.

CMBR polarization Scattering by free electrons produces a small linear polarization of the CMBR, with slightly different electric field strengths in perpendicular directions.

CN *See* interstellar cyanogen.

COBE An account of the origin of the COsmic Background Explorer satellite mission and its experiments commences on page 418.

COBRA Rocket-borne University of British Columbia COsmic Background RAdiation experiment that demonstrated the thermal CMBR energy spectrum essentially coincidentally with COBE. Its history is described beginning on page 416.

cold dark matter Nonbaryonic dark matter that is initially cold in the sense that the relative velocity dispersion is negligibly small. In the standard cosmology CDM dominates the mass in clusters of galaxies and the mass in the outer parts of galaxies that are not in clusters. *See* CDM cosmological model, dark matter.

cold load *See* load.

comoving coordinates Spacetime coordinate labeling where an observer at fixed coordinate position sees close to isotropic CMBR and isotropic mean recession of the galaxies, in the construction in footnote 10 on page 435.

Compton scattering Electron–photon scattering at energies large enough to produce a significant shift of the photon energy (and frequency) in the electron initial rest frame.

Comptonization parameter *See* Sunyaev–Zel'dovich effect.

continual creation Postulate of continual spontaneous creation of matter and maybe radiation at a rate that is on average independent of time and position. *See* steady state cosmology.

correlation function Defined in equation (5.4) on page 440. For Tukey's opinion of this statistic, see footnote 32 on page 452.

corrugated feed *See* feed.

cosmic background radiation Nearly uniform sea of radiation between the galaxies, including the CMBR and radiation from sources at lower redshifts.

cosmic strings Linear concentrations of energy, not to be confused with super-string theory.

cosmological constant The constant Λ Einstein (1917) added to his general relativity theory to produce a static universe. It appears in equation (2.5), and it has the effect illustrated in Figure 2.1: when a positive Λ term dominates, the rate of expansion of the universe increases. *See* dark energy.

cosmological density parameters *See* density parameters.

cosmological principle Einstein's (1917) postulate that the universe is close to homogeneous and isotropic in the large-scale average. The measurements summarized in Section 5.4 indicate that the mass distribution averaged over the Hubble length departs from homogeneity by about one part in 10^5.

cosmological redshift The measure z in equation (2.2) of the wavelength shift of light from a distant object. In standard physics the redshift of a packet of light as it moves from source to detector may be described as the integrated effect of the Doppler shifts seen by a sequence of observers who sample the light packet at closely spaced intervals along its path.

cosmological singularity In general relativity theory, a singular start of expansion of the universe. *See* singularity theorems in general relativity.

cosmological tests Network of astronomical and laboratory tests of models of the large-scale nature of the universe summarized on pages 465–475.

cosmology In this book, the study of the structure and evolution of the universe on scales ranging from the Hubble length, 4000 Mpc, down to about 100 kpc. In the 1960s the most widely discussed cosmologies were the steady state and varieties of the big bang model. Bondi (1952, 1960a) discusses other proposals, including the possibility that the cosmological principle fails, as in fractal cosmologies. By the mid-1960s this last proposal was seriously challenged by the nearly isotropic distributions of radio sources and of the microwave and X- to gamma-ray radiation. The cosmology established four decades later is described in Chapter 2 and its tests are summarized in Section 5.4.

count–magnitude relation *See* source counts.

cryogenics In this book, the art of cooling amplifiers, detectors, and reference loads to the temperature of liquid nitrogen, 77 K, or liquid helium, 4.2 K at atmospheric pressure, or lower.

cyanogen *See* interstellar cyanogen.

Cygnus A This galaxy is the strongest radio source in the sky and outside the Milky Way.

Cygnus X-1 A binary stellar-mass X-ray source in the plane of the Milky Way.

dark energy Einstein's cosmological constant, Λ, or a form of uniformly distributed energy that behaves much like it. The equation of state parameter is the ratio $w = p_\Lambda/\rho_\Lambda$ of the dark energy pressure and energy density, where $w = -1$ for Einstein's cosmological constant. An example of negative pressure is a stretched rubber band. The gravitational action of dark energy is illustrated in Figure 2.1 and discussed in footnote 9 on p. 19. *See* rubber band.

dark matter The new name for missing mass that apart from gravity interacts with ordinary matter and radiation weakly if at all. As indicated in Table 2.1 most of the baryons also are dark, that is, not observed in stars or in the gas and plasma found to be concentrated near galaxies and in clusters of galaxies.

declination Component of position in the sky measured as an angle from the celestial equator.

decoupling At redshift $z_{\text{dec}} = 1100$ the primeval plasma combined to neutral hydrogen and helium with trace amounts of ionized and molecular hydrogen. The disappearance of almost all free electrons greatly reduced scattering of the radiation and hence reduced radiation drag on the matter, effectively decoupling baryons and radiation.

degeneracy A space-filling sea of fermions (such as neutrinos) is degenerate if the kinetic energies are low enough and the number density large enough that the exclusion principle forces occupation of almost all allowed states up to a Fermi degeneracy energy. *See* lepton degeneracy. In a second use of the word, a degenerate constraint is a measure of a combination of interesting parameters. *See* geometric degeneracy.

density parameters Measures of the cosmic mean mass density and of the other contributions to the rate of expansion of the universe based on the first integral of equation (2.5); convenient forms are

$$
\begin{aligned}
\left(\frac{1}{a}\frac{da}{dt}\right)^2 &= \frac{8}{3}\pi G\rho - \frac{1}{a^2 R^2} + \Lambda \\
&= H_0^2 \left((1+z)^4\Omega_{\text{r}} + (1+z)^3\Omega_{\text{m}} + (1+z)^2\Omega_{\text{k}} + \Omega_\Lambda\right).
\end{aligned} \tag{G.1}
$$

The measure R of the radius of curvature of space sections at fixed world time appears in equation (G.4). Hubble's constant H_0 is a measure of the present rate of expansion of the universe. The density parameters are the time-dependent fractional contributions to the square of the present expansion rate: Ω_{m} represents the mass in nonrelativistic matter, Ω_{r} the mass in radiation, Ω_Λ the dark energy, and Ω_{k} the effect of space curvature.

Department of Terrestrial Magnetism A research department of the Carnegie Institution of Washington.

deuterium Heavier stable isotope of hydrogen; the atomic nucleus contains one proton and one neutron.

dewar Container for liquid nitrogen or helium.

Dicke, Robert H. With Hoyle and Zel'dovich, one of the three central figures in the recognition in the 1960s of the fossil radiation and light elements from the hot big bang. Biography: Happer and Peebles (2006).

Dicke microwave radiometer Device for measurement of radiation by rapid comparison to a reference source at known temperature. The phase-sensitive comparison of the detector response to antenna and reference source suppresses the effect of receiver noise and gain fluctuations.

Dicke switch *See* phase-sensitive detection.

DIRBE Diffuse InfraRed Background Experiment on COBE.

dish-reflector antenna Described in footnote 28 on page 45; an image is shown in Figure 4.9.

distance scale In this book, a measure of physical distances to extragalactic objects. Ratios of distances may be estimated from differences of apparent magnitudes of objects that appear to be similar enough that their luminosities are nearly the same. The more challenging conversion to a scale of physical distances is noted on page 471.

DMR Differential Microwave Radiometer instrument on the COBE satellite.

Doppler effect Shift of the wavelength of light caused by relative motion of source and observer, or of the wavelength of sound in air caused motion of source or observer. *See* redshift.

DTM Department of Terrestrial Magnetism.

Eddington–Lemaître cosmological model Lemaître's (1927) solution for an expanding matter-filled universe that asymptotically traces back in time to Einstein's (1917) static solution. Lemaître (1931) turned to a primeval atom (big bang) cosmology, but Eddington (1931) continued to prefer the idea of expansion from a nonsingular state.

effective temperature In this book, the measure of flux of electromagnetic radiation energy at a given frequency expressed as the Rayleigh–Jeans or blackbody radiation temperature that would produce the same energy flux at that frequency.

Einstein, Albert Biographies: *Subtle is the Lord,* Pais (1982); *Einstein in Love,* Overbye (2000).

Einstein–de Sitter model Friedmann–Lemaître model with negligible space curvature and cosmological constant ($\Omega_{\rm k} = 0$ and $\Omega_\Lambda = 0$ in equation G.1). Proposed by Einstein and de Sitter (1932) as the simplest version that fits the observations, it became the paradigm from the early 1980s to the mid-1990s.

energy spectrum Radiation energy density as a function of frequency or wavelength. *See* CMBR energy spectrum.

entropy perturbation *See* isocurvature initial condition.

Eötvös experiment Precision check that the gravitational acceleration of a small test particle is independent of its composition.

erg Unit of energy: 10^7 ergs = 1 joule = 1 watt of power applied for one second.

evolving dimensionless physical parameters Dirac's (1938) idea that the strength of the gravitational interaction may change as the universe evolves, generalized to the notion that dimensionless parameters of physics may vary with time. A measure of the strength of the gravitational interaction is $Gm_{\rm p}m_{\rm e}/\hbar c \simeq 10^{-41.5}$, where G is Newton's gravitational constant, $m_{\rm p}$ and $m_{\rm e}$ are the masses of the proton and electron, and $\hbar$ is Planck's constant of quantum physics. The strength of the electromagnetic interaction is measured by the fine structure constant, $\alpha = e^2/\hbar c \simeq 1/137$, where e is the electric charge of a proton or electron. Equation (3.12) is the ratio of these two numbers. Perhaps these dimensionless numbers evolve as the universe expands, and perhaps $m_{\rm e}/m_{\rm p}$ evolves, but at the time of writing there is no convincing detection. *See* scalar–tensor gravity theory.

excess antenna temperature Term Penzias and Wilson (1965a) used for the detection of what has come to be called the CMBR. *See* antenna temperature.

exclusion principle *See* degeneracy.

expanding universe Evolving spacetime illustrated in Figure 2.1.

expansion parameter Measure $a(t)$ of the history of expansion of the universe in equation (G.4). When galaxies are not created or destroyed the mean distance between galaxies increases in proportion to the expansion parameter. The wavelength of freely propagating light stretches as $\lambda \propto a$ (equation 2.3).

feed Radiation from the sky or from the surface of a dish or parabolic reflector may be collected by a feed horn connected to a receiver. Smooth-sided feeds used in early CMBR observations were replaced by corrugated feeds with circular grooves on their inside walls that minimize the flow of electromagnetic energy along the walls, reducing loss and side lobes. *See* horn antenna, Winston cone.

FIRAS Far InfraRed Absolute Spectrophotometer on COBE.

flat universe In cosmology, an informal name for spacetime geometry that is curved but has spatial curvature $R^{-2} = 0$ in equation (G.4).

flux density In this book, rate of arrival of energy from a discrete source per unit of collecting area and frequency interval.

fossil In this book, remnants of processes operating when the universe was younger and denser than now. Remnants include most of the isotopes of hydrogen and helium, the spectrum and distribution of the CMBR, and the galaxies.

fractal cosmology Postulate that matter appears in a hierarchy of concentrations within concentrations continuing to the largest observable scales. It is convincingly ruled out by the evidence summarized in Section 5.4.

free-free radiation Electromagnetic (bremsstrahlung) radiation produced in collisions of free electrons with the ions in a plasma, as in interstellar or intergalactic matter. *See* galactic radiation.

Friedmann–Lemaître solutions Homogeneous and isotropic solutions of Einstein's field equation, including the standard big bang cosmology. The spacetime geometry is described by equation (G.4), and the expansion history by equation (G.1).

Friis, Harald T. The autobiography, *Seventy Five Years in an Exciting World* (Friis 1971), recalls Friis' role in leading the research at Bell Laboratories that led to the detection of the CMBR.

gain pattern *See* antenna pattern.

galactic radiation Radiation originating in our Milky Way Galaxy and outside the Solar System. In this book, radiation at radio to far-infrared wavelengths is of main interest for its interference with measurements of the CMBR, as indicated in Figure 5.4. Sources at these wavelengths include free-free radiation from interstellar electrons at temperature $\sim 10^4$ K accelerated by ions, synchrotron radiation from relativistic electrons accelerated by the interstellar magnetic field, far-infrared emission by interstellar dust grains heated by absorption of starlight, the 21-cm line from atomic hydrogen, and line radiation from other atoms and interstellar molecules. Penzias and Wilson (pp. 147 and 164) describe how their search for galactic radiation at 7-cm wavelength was one factor leading them to the CMBR.

galaxy A gravitationally bound "island universe" of stars with half the observed starlight contained in a radius less than about 10 kpc.

galaxy formation *See* gravitational instability, structure formation.

Gamow, George Leading developer of the hot big bang cosmology in the 1940s. Autobiography: *My World Line*, Gamow (1970).

Gamow condition Constraint on the baryon matter density at a given temperature in the early expanding universe from the condition that the reaction $n + p \leftrightarrow d + \gamma$ produces a significant but not excessive abundance of deuterium that could then fuse to form heavier elements.

Gaussian scale-invariant initial conditions The primeval departure from an exactly homogeneous mass distribution is particularly simple: a Gaussian random process produces about the same space curvature fluctuations on a broad range of length scales.

general relativity theory Einstein's (1915) theory of gravity as an effect of spacetime curvature. It has passed demanding tests on length scales ranging from the laboratory to the Solar System, and, as discussed in Section 5.4, to the scale of the Hubble length.

geometric degeneracy In this book, combinations of parameters that are usefully constrained by a cosmological test, as in equation (5.14).

GHz Unit of frequency: $1\,\mathrm{GHz} = 10^9\,\mathrm{Hz} = 10^9$ cycles per second.

gigaparsec Unit of length: $1\,\mathrm{Gpc} = 10^3\,\mathrm{Mpc} = 10^9$ parsecs $= 3.09 \times 10^{27}\,\mathrm{cm}$.

graybody radiation A mixture of blackbody radiation spectra belonging to different temperatures. That includes the dilute blackbody radiation from a source that is not fully black. A graybody spectrum generally has the Rayleigh–Jeans form, $u_\nu \propto \nu^2$, at long wavelengths but peaks at a higher frequency than blackbody radiation at the same Rayleigh–Jeans temperature.

gravitational instability picture In this book, the gravitational growth of departures from an early near homogeneous mass distribution into the mass concentrations observed as galaxies and clusters of galaxies.

gravitational radiation In general relativity theory, free fluctuations of spacetime curvature that propagate at the speed of light.

ground noise In CMBR measurements, radiation from the relatively hot ground that finds its way to the detector through the antenna side and back lobes. Screens – conducting surfaces – are used to reflect antenna side lobes away from the ground toward the sky.

Guyot Hall Princeton University site of the Roll and Wilkinson (1966) search for the CMBR and the Partridge and Wilkinson (1967) search for possible departures from its uniformity.

Harrison–Zel'dovich initial condition *See* scale-invariant initial condition.

helium The evidence developed in this book is that helium largely is remnant from the hot big bang. Helium also is produced in stars by thermonuclear burning of hydrogen, but most of this burns to heavier elements. The light helium isotope ^{3}He also is remnant (at mass fraction about 10^{-4}) from thermonuclear reactions in the early universe, and is produced in stars, by nuclear burning of deuterium, in mineral deposits, by radioactive decay of heavy elements, and in nuclear weapons, by radioactive decay of tritium. For helium cold load, *see* cryogenics.

HEMT High Electron Mobility Transistor that can be incorporated into a device (colloquially also called an "HEMT") to directly amplify centimeter-wave radiation. They have excellent noise properties. Most versions have their roots in the NRAO designs by Pospieszalski (1992), Pospieszalski *et al.* (1994), and colleagues.

Herzberg, Gerhard Notable in this story for the commentary in Herzberg (1950) on the spin temperature of interstellar CN molecules. This spin temperature was later seen to be a good measure of the CMBR temperature. Biography: *Gerhard Herzberg: An Illustrious Life in Science*, Stoicheff (2002).

H I Neutral atomic hydrogen. The H I 21-cm line radiation is an important tracer of atomic hydrogen in and around galaxies.

H II region Interstellar plasma, largely ionized hydrogen with a mix of heavier elements.

hierarchical structure formation Growth of mass concentrations such as galaxies and clusters of galaxies by a sequence of merging of gravitationally bound systems into successively larger systems. This is predicted by the standard cosmological model and it agrees with the observed tendency of matter to appear in a hierarchy of concentrations within concentrations. *See* gravitational instability picture.

Holmdel Laboratory *See* Bell Laboratories.

horizon In general relativity theory an event horizon is the boundary of events in spacetime that can in principle be seen somewhere along the world line of a chosen observer. A particle horizon marks the set of (conserved) particles that can be seen – again in principle – from a chosen event in spacetime. These terms often are informally applied to spacetime subsequent to inflation. The horizon problem mentioned on page 265 is that, in Friedmann–Lemaître solutions with $\Lambda = 0$ and pressure that is not negative, an observer can see and compare parts of spacetime that have never been in causal contact (within the spacetime, which is incomplete, that is, singular) yet have closely similar local properties. The mixmaster picture also mentioned on page 265 was a pioneering attempt to remedy this apparent violation of causality. It is now thought that the remedy is an early inflation epoch in which pressure was large and negative.

horn antenna Illustrated in many variations in Figures 3.2, 4.1, 4.2, 4.12, 4.16, 4.17, 4.22, 4.27, 4.32, 4.38, 4.52 and 5.6, and described in footnote 28 on page 45, an antenna or feed shaped like a horn or funnel that matches incident radiation to a receiver. *See* feed.

hot dark matter An early dark matter candidate was a sea of neutrinos with rest mass $\sim$30 eV relict from thermal equilibrium in the early universe, named by Bond *et al.* (1984). The sea of relict neutrinos acts as hot dark matter, but its mass density is subdominant.

Hoyle, Fred With Dicke and Zel'dovich, one of three central figures in the recognition in the 1960s of the fossil radiation and light elements from the hot big bang. Autobiography: *Home is Where the Wind Blows: Chapters from a Cosmologist's Life* (1994); biographies: *Fred Hoyle: A Life in Science*, Mitton (2005); *Fred Hoyle's Universe*, Gregory (2005); *The Scientific Legacy of Fred Hoyle*, Gough (2005).

Hubble length At the Hubble length or distance $c/H_0 \sim 4000$ Mpc Hubble's law formally extrapolates to recession velocity equal to the velocity of light. In the standard cosmology this sets the order of magnitude of the largest observable distances, and the Hubble time $H_0^{-1} \sim 10^{10}$ years sets the timescale for evolution.

Hubble's constant The coefficient of proportionality in Hubble's law at the present epoch. Steps toward measuring its value, $H_0 \simeq 72$ km s^{-1} Mpc^{-1}, are noted on page 31 and in Section 5.4.

Hubble's law Also termed "Hubble flow," the linear relation between recession velocity and distance in equation (2.1), named after Edwin Hubble's (1929) early evidence. Motions of galaxies cause a scatter $\sim$300 km s^{-1} around the mean. The deviation from the linear relation at redshifts $z \gtrsim 1$ – the redshift–magnitude relation – depends on the cosmological model.

Hz Unit of frequency: 1 Hz = one cycle per second.

inflationary cosmology Scenario (Guth 1981, 1997; Linde 1982; Albrecht and Steinhardt 1982) for what happened in the very early universe, before the standard big bang model could have been an adequate approximation. It postulates that a very large dark energy density in the early universe caused exceedingly rapid expansion that stretched out irregularities to produce the near homogeneous region we observe, and that the rapid expansion or perhaps the decay of the dark energy filled our space with the thermal sea of radiation, dark matter and baryons. The rapid early expansion can account for the observed large-scale homogeneity of the universe (*see* horizon), the small space curvature (p. 442) that was later observed (p. 461), and the observed near scale-invariant departure from exact uniformity as a natural outcome of simple models for inflation (p. 441).

Infrared Astronomical Satellite IRAS produced a widely used catalog of sources detected in wavelength bands centered at 12, 25, 60, and 100 μm.

insertion loss Absorption of radiation, as in the components of a microwave receiver system, and the consequent production of thermal radiation.

interferometry Method of observation based on the correlation between outputs of multiple spatially separated feed horns or antennas to improve angular resolution or to suppress atmospheric noise and help control systematic errors in CMBR anisotropy measurements.

intergalactic dust Hypothetical dust particles or whiskers between the galaxies that would thermalize the microwave background at relatively low redshifts. The cosmological tests in Section 5.4 convincingly rule out the idea.

interplanetary dust *See* zodiacal light.

interstellar cyanogen The molecule CN – a carbon atom bound to a nitrogen atom – has energy levels at 113 and 227 GHz (2.6 and 1.3-mm wavelength) above the ground level that made observation of spin temperatures of interstellar CN an important early measure of the energy spectrum of the CMBR.

interstellar dust Particles of condensed heavy elements in the interstellar medium absorb starlight and reradiate the energy at longer wavelengths, peaking at about 100 μm. *See* galactic radiation.

island universe cosmology Proposition that the material universe is a concentration of matter – a galaxy or system of galaxies – outside of which spacetime is asymptotically flat.

isocurvature initial condition Postulate that the primeval mass distribution is exactly uniform while the composition varies with position.

Jeans instability Effect on a gas cloud of the competing roles of gravity, which tends to cause the cloud to collapse, and pressure, which tends to cause it to expand. A gas cloud larger than the Jeans length can collapse.

Jordan–Brans–Dicke theory *See* scalar–tensor gravity theory.

K-band Microwave radiation in the frequency range 18–26 GHz.

kiloparsec Unit of length: kpc; *See* megaparsec.

Kompaneets, Aleksandr Solomonovch The Kompaneets equation describes Compton–Thomson scattering in the Sunyaev–Zel'dovich effect.

Kompfner, Rudolph Member of Bell Laboratories, 1951 to 1973. Biography: Pierce (1979).

lambda The Greek letter λ is a standard symbol for wavelength (as in equation 2.2); in cosmology the upper case, Λ, and earlier λ, represent Einstein's cosmological constant or dark energy.

ΛCDM cosmology *See* CDM cosmological model.

last scattering surface Relatively narrow range of time at which most of the CMBR photons were last scattered by free electrons. The range was set by conversion of the primeval plasma to almost completely neutral atomic hydrogen and helium when the CMBR temperature had dropped to $T \simeq 3000\,\mathrm{K}$. *See* decoupling.

Layzer–Irvine equation Relation between the kinetic energy per unit mass associated with the motion of matter relative to the general expansion of the universe and the gravitational potential energy per unit mass associated with the departure from a homogeneous mass distribution (Irvine 1961; Layzer 1963).

Lemaître, Georges Proposed the relation between the redshifts of the galaxies and the expanding homogeneous solution to Einstein's field equation first found by Friedmann. Biography: *Un atome d'univers: La vie et l'oeuvre de Georges Lemaître*, Lambert (2000).

leptons Members of the three families of neutrinos that interact only by gravity and the weak interaction, or the electron, muon or tau particles or antiparticles that also have electric charge. Examples appear in the reactions in equation (3.8). Chapter 3 ignores considerations of the three families of neutrinos, but as Wagoner notes (p. 259) the number of families determines the contribution of leptons to the thermal energy density, which in turn affects the rate of expansion and cooling and thus light element production in the early universe. The present mass density in neutrinos is discussed on pages 20 and 474. The lepton number – the number of leptons minus number of their antiparticles – is conserved under conditions considered in this book.

lepton degeneracy Postulate that the magnitude of the lepton number density is comparable to or larger than the thermal radiation photon number density, so relaxation to equilibrium produces a degenerate sea of neutrinos. Degenerate neutrinos would suppress the number of neutrons relative to protons in the very early universe; degenerate antineutrinos would suppress the number of protons relative to neutrons. Either case would affect BBNS.

light cone Idealized three-surface of light rays – null paths – that converge (when one ignores scattering) on an event in spacetime.

light year *See* megaparsec.

LIGO Laser Interferometer Gravitational-wave Observatory – a gravitational radiation detector.

little bangs Theory of light element production in massive exploding stars under conditions that approximate a hot big bang (Hoyle and Tayler 1964). *See* helium.

load In this book, a reference source of radiation at known, generally low temperature for calibration of a microwave radiometer.

Local Supercluster *See* superclusters.

lock-in amplifier Device to detect and average signal difference, as in those produced in the Dicke microwave radiometer, using phase-sensitive detection.

look-back time In cosmology, the time taken for radiation to travel from source to observer.

loss *See* insertion loss.

Lyman-α resonance Atomic hydrogen is a strong absorber of radiation at wavelength 1215 Å. Ly-α photons at this wavelength are emitted by the one-photon transition from the first excited electron energy level to the ground energy level.

Mach's principle Influential proposal that local physics is related to the large-scale nature of the universe, and that local physics may therefore evolve as the universe evolves.

main sequence Relation between mass and luminosity of a star that has close to its original composition in the inner nuclear-burning region.

MAP *See* WMAP.

maser Microwave amplification by stimulated emission of radiation. The Bell Laboratories communications experiments that detected the CMBR, and some later CMBR experiments, used low noise maser amplifiers.

mass and galaxy correlation functions In cosmology statistical measures of the distributions of mass or galaxies relative to a spatially uniform universe. *See* Correlation function.

megaparsec Unit of length in cosmology and extragalactic astronomy: $1\,\mathrm{Mpc} = 1000\,\mathrm{kpc} = 10^6\,\mathrm{pc} = 3.09 \times 10^{22}\,\mathrm{m} = 3.09 \times 10^{24}\,\mathrm{cm}$, or about three million light years. *See* parsec.

micron $1\,\mu\mathrm{m} = 1\,\mu = 10^{-6}\,\mathrm{m} = 10^{-4}\,\mathrm{cm}$.

microwave radiation Electromagnetic radiation, in this book with wavelength in the range $3\,\mathrm{mm}$–$30\,\mathrm{cm}$.

Milky Way Galaxy Spiral galaxy named for the band of light across the sky from the stars in its disk. The Solar System is in the disk about $8\,\mathrm{kpc}$ from the center of this galaxy.

missing mass Old name for dark matter.

mixer A mixer multiplies the electromagnetic field of incoming radiation by a field from a local oscillator at frequency ν_0. The product contains $\cos(\nu + \nu_0)t$ and $\cos(\nu - \nu_0)t$ terms, with amplitudes proportional to the amplitude A of the incoming radiation. The sum frequency is suppressed; the difference frequency amplified and detected. This detection technique can preserve the phase of the incoming radiation (can be coherent, an important consideration for interferometry), and allows use of low-frequency electronics to detect microwave radiation. *See* receivers.

mixmaster cosmology *See* horizon.

mode For electromagnetic radiation, a fixed spatial pattern that oscillates at one frequency.

Mpc *See* megaparsec.

NCAR National Center for Atmospheric Research. *See* Climax and Palestine.

National Radio Astronomy Observatory Operations of several large open-access radio telescopes are headquartered in Charlottesville, Virginia.

neutrino *See* leptons, lepton degeneracy.

neutrino masses The cosmic mass in neutrinos – a form of hot dark matter – is entered in Tables 2.1 and 5.2.

noise temperature Radiation – electromagnetic noise – originating in a receiver system, or one of its components, or its surroundings, usually measured in terms of the equivalent temperature in a Rayleigh–Jeans spectrum.

noncosmological redshift In this book, the proposal that quasars are not at the distances indicated by Hubble's law. This was ruled out by observations of absorption lines in quasar spectra produced by galaxies and intergalactic hydrogen at lower redshift and near the line of sight from the quasar to us.

nucleosynthesis in stars Formation of chemical elements by nuclear reactions in stars.

Olbers' paradox A static spacetime uniformly filled with a steady supply of shining stars eventually fills with starlight. In the steady state cosmology this is avoided by the expansion of the universe, which dilutes away starlight and stellar remnants. In the big bang cosmology the energy density of starlight is limited by redshift and the limited time stars have been shining.

open universe Spacetime geometry with $R^{-2} < 0$ in equation (G.4). Interest in this case is noted in footnote 21 on page 442.

oscillating universe Postulate that the present expansion of the universe was preceded by collapse in a cyclic universe.

Owens Valley Radio Observatory Radio astronomical observatory located near the White Mountain Research Station.

Palestine Site of National Scientific Balloon Facility in Texas.

parabolic-reflector antenna *See* dish-reflector antenna.

paradigm In this book, a widely accepted working hypothesis that may be promoted to a standard model if the web of tests becomes tight enough to convince the community that it is a useful approximation to reality.

parsec $1\,pc = 3.09 \times 10^{16}\,m$, or about three light years. A star at one parsec distance appears to move in the sky in an ellipse with a semi-major axis of one second of arc as Earth circles the Sun. In our neighborhood of the Milky Way the stars are about a parsec apart. Perhaps the general use of a unit whose origin is growing obscure to a good fraction of the cosmology community is irrational, but we like the reminder of astronomy's history.

particle exchange reactions Nuclear reactions such as the ones shown in equation (3.4) that do not involve emission or absorption of radiation. In conditions for light element formation in the early universe, particle exchange reactions tend to be faster than the deuteron formation reaction in equation (3.2) that depends on the weaker electromagnetic interaction.

peculiar velocity Motion relative to the CMBR or the large-scale mean motion of the galaxies.

perfect cosmological principle Starting assumption for the steady state cosmology, extending Einstein's proposal that the universe is close to spatially uniform – the cosmological principle – to the proposal that the universe also is on average unchanging.

phase-sensitive detection Reduction of the effects of receiver noise and gain fluctuations by averaging the output of a detector synchronously with switching the input between a source and a reference signal. The technique has many applications in the extraction of signals well below the receiver noise. *See* beam-switching, Dicke microwave radiometer.

photon Quantum of electromagnetic radiation, with energy $\epsilon = h\nu$ at frequency ν, where Planck's constant is $h = 6.6 \times 10^{-34}\,J\,s = 6.6 \times 10^{-27}\,erg\,s$. The role of thermal photons in nuclear reactions in the early universe is discussed on pages 28–32.

pigeon droppings Early candidate for the source of excess noise in the Bell Laboratories' microwave radiation receivers.

Planck European Space Agency mission for precision measurement of the variation of the CMBR temperature and polarization across the sky.

Planck spectrum *See* thermal radiation.

polarization *See* CMBR polarization.

power law A common functional form. In the long wavelength (Rayleigh–Jeans) limit the energy density in thermal radiation per unit interval of frequency varies as the power-law spectrum $u_\nu \propto \nu^2$ (equation G.2). The power spectrum of the primeval mass distribution is well approximated as the power-law form $\mathcal{P} \propto k^{n_s}$ (equation 5.6).

power spectrum Measure of the mean square fluctuations of a function of time, angle or position. For a function of position in the sky the spectrum is based on the spherical harmonic expansion in equation (5.10), with expansion coefficients a_ℓ^m; for a function in flat space it is based on the Fourier transform expansion $\delta(\vec{r}) = \int d^3k\, \delta_{\vec{k}} e^{i\vec{k}\cdot\vec{r}}$, with expansion coefficients (Fourier amplitudes) $\delta_{\vec{k}}$. The a_ℓ^m measure fluctuations on angular scales $\varphi \simeq \pi/\ell$ radians; the $\delta_{\vec{k}}$ fluctuations on linear scales $r \simeq \pi/k$. The angular power spectrum $(\delta T_l)^2$ that characterizes the CMBR anisotropy (and is plotted in Figure 5.18) is proportional to the squares of the a_ℓ^m (equation 5.12). The power spectrum $\mathcal{P}(k)$ that is useful for characterizing the large-scale matter distribution is proportional to the squares of the $\delta_{\vec{k}}$, with a standard normalization in equation (5.5). *See* baryon acoustic oscillation, CMBR anisotropy spectrum.

primeval atom Lemaître's (1931) name for the early dense universe.

primeval fireball Wheeler's early name for the CMBR.

primeval galaxies With "young galaxies," early name for galaxies as they first formed, observed at high redshift because of the light travel time.

Project Echo Test of communications using microwave signals reflected by a balloon in Earth orbit.

pulsar Magnetized rapidly spinning neutron star that emits radio radiation received in pulses.

Q-band Microwave radiation in the frequency range 33–50 GHz.

quasar Compact luminous active galactic nucleus, likely powered by accretion onto a massive black hole. The quasar redshift distribution was an early challenge to the steady state cosmology.

quasi-steady state cosmology Variant of the steady state cosmology with constant long-term mean expansion and relatively short-term cycles of expansion and contraction. The CMBR would be produced by absorption and thermalization of starlight during the denser parts of each cycle. If the cycles are deep enough this can be equivalent to the standard cosmology.

radiation drag In cosmology, the CMBR drag force on plasma caused by Thomson scattering by the free electrons. *See* decoupling.

Radiation Laboratory Research laboratory during World War II at the Massachusetts Institute of Technology.

radiometer Device for measurement of electromagnetic radiation energy flux density. *See* Dicke microwave radiometer.

radio source counts *See* source counts.

radio stars Early term for objects detected at radio wavelengths and variously applied to radio sources in the Milky Way and in other galaxies.

Rayleigh–Jeans spectrum Limiting form of the spectrum of thermal radiation at low frequency, $\nu \ll kT/h$, where the mean energy of a mode of oscillation of the electromagnetic field is $\mathcal{N}h\nu = kT$ (equation G.5), the expression one would predict from classical mechanics, and the energy per unit volume and frequency interval is

$$u_\nu = \frac{8\pi kT\nu^2}{c^3}. \tag{G.2}$$

This relation expresses radiation energy density or flux in terms of the Rayleigh–Jeans temperature.

reality Our interpretation is on page 6.

receivers In CMBR studies, devices that convert incoming electromagnetic radiation into an electronic signal. The term can refer to coherent or bolometric (incoherent) systems. The Dicke radiometer shown on page 45 employed a coherent detector; *see* mixer. Optical elements include horns, sometimes with lenses, or other antennas, to shepherd radiation into waveguides and then through amplification and frequency filtering. The minimum detectable incoming temperature change that can be measured in a time interval Δt is

$$\Delta T = \frac{C_1 T_s}{\sqrt{\Delta t \Delta \nu}}, \qquad (G.3)$$

where $\Delta \nu$ is the bandwidth of the receiver and the constant C_1 is a number of order unity that depends on the switching scheme. Roll and Wilkinson (1967) reported a system temperature of $T_s \sim 3000\,\mathrm{K}$; Penzias and Wilson (1965a) had a system temperature of $18\,\mathrm{K}$. In bolometric systems there is no mixer and most of the optics are cryogenic. *See* bolometer, feed, HEMT.

recombination epoch In this book, the transition at redshift $z = 1100$ from plasma to almost entirely neutral atomic hydrogen and helium. The term is unambiguous but perhaps irrational because in the standard model the baryons have been ionized from creation to recombination. *See* decoupling.

redshift Wavelength shift that may be caused by relative motion, the expansion of the universe or a time-variable gravitational potential. *See* cosmological redshift, noncosmological redshift, Doppler effect.

redshift–magnitude relation A cosmological test: the relation between cosmological redshifts and apparent magnitudes of extragalactic objects that have close to the same absolute magnitude.

Rees–Sciama effect Perturbation to the CMBR by the time-variable gravitational potential of a growing nonlinear mass concentration.

relict radiation Early name for the CMBR; its origin is recalled by Sunyaev (p. 112).

right ascension Component of position in the sky measured as an angle along the celestial equator.

Robertson–Walker line element Geometry of a homogeneous and isotropic world model expressed as

$$\mathrm{d}s^2 = \mathrm{d}t^2 - a(t)^2 \left[\frac{\mathrm{d}r^2}{1 - r^2 R^{-2}} + r^2 \left(\mathrm{d}\theta^2 + \sin^2\theta \mathrm{d}\phi^2 \right) \right]. \qquad (G.4)$$

An observer at fixed coordinate position keeps proper or world time t, θ and ϕ are polar coordinates measured by an observer at $r = 0$, r is a radial coordinate, and the expansion parameter $a(t)$ appears in equations (2.3) and (G.1). The physical radius of curvature of a space section at fixed t is $a(t)|R|$. If $R^{-2} > 0$, space is curved in the fashion of the balloon analogy on page 10 and is said to be closed. The circumference c of a circle of physical radius x drawn in this space section is $c < 2\pi x$. If R^{-2} is negative space is curved, so that $c > 2\pi x$, and is said to be open. The tests in Section 5.4 indicate R^{-2} is close to zero, meaning space at fixed t has close to Euclidean geometry (though spacetime is curved). Since the Robertson–Walker form follows from the assumed space symmetry it applies to the steady state cosmology, but with $R^{-2} = 0$ (equation 3.17).

rubber band General relativity predicts that the tension (negative pressure) of a stretched rubber band slightly reduces the gravitational attraction of its mass (energy). As discussed on page 19 the extreme negative pressure of dark energy produces the net gravitational repulsion illustrated in Figure 2.1.

Sachs–Wolfe effect Variation of the CMBR temperature across the sky caused by the gravitational effect of the departure from a homogeneous mass distribution in the early universe that grew into the present large-scale clustering of mass. Sachs and Wolfe (1967) predicted the effect; the COBE/DMR experiment detected it (Smoot *et al.* 1992; Wright *et al.* 1992). The integral Sachs–Wolfe (ISW) effect is the anisotropy produced by the time-variation of the gravitational potential along the line of sight.

S-band Microwave radiation in the frequency range 2–4 GHz.

scalar–tensor gravity theory Modification of general relativity in which the measured strength of the gravitational interaction is a function of time.

scale-invariant initial condition Adiabatic primeval departure from a uniform mass distribution that produces spacetime curvature variations that are independent of length scale (Harrison 1970; Peebles and Yu 1970; Zel'dovich 1972).

scaled horn antennas Two or more antennas with size proportional to the wavelength at which they operate, so the antenna patterns are similar, an advantage when measuring the CMBR energy (frequency) spectrum.

Schwarzschild solution Karl Schwarzschild's spherically symmetric solution of Einstein's field equation in empty space that describes the spacetime geometry around an isolated spherical mass concentration, or a black hole with no angular momentum.

Sciama, Dennis W. Influential teacher, at first a proponent of the steady state cosmology and then, after the early measurements showed the CMBR has a close to thermal spectrum, of the hot big bang. Biography: Rees (2001).

Scientific Balloon Facility Now Columbia Scientific Balloon Facility. *See* NCAR.

screen *See* ground noise.

shaggy dog Dicke's early external calibration source that approximates a blackbody radiating at ambient temperature, roughly 300 K.

Silk damping Dissipation of inhomogeneities in the spatial distributions of the plasma and CMBR before decoupling by diffusion of the radiation through the plasma. *See* decoupling.

singularity theorems in general relativity In observationally relevant situations – end points of evolution of massive stars, centers of galaxies, the expanding universe – and under broad but not general assumptions about the physics of matter and radiation, demonstration that the spacetimes of general relativity theory are incomplete, that is, singular.

sky noise *See* atmospheric emission.

Solar oblateness experiment Measurement of the shape of the Sun (Kuhn, Libbrecht and Dicke 1988 and earlier references therein) to check the possible effect of the departure from a spherical Solar mass distribution on the orbit of the planet Mercury. *See* Dicke.

source counts In cosmology, counts of extragalactic objects. Counts of galaxies as a function of direction and of brightness in the sky (apparent magnitude) offered important early tests of the cosmological principle. Counts as a function of apparent magnitude, flux density, or redshift figured in tests to distinguish between the steady state and big bang models. Later applications are to measures of the large-scale galaxy distribution and of the evolution of galaxy populations.

space curvature The radius of curvature of space sections at fixed world time t is $a(t)|R|$ (equation G.4). In the standard model $a|R|$ is much larger than the Hubble length, meaning spatial geometry is close to Euclidean. *See* density parameters, Robertson–Walker line element.

special relativity Relativity theory in flat spacetime (Einstein 1905).

spectral index Parameter in a power-law model for a radiation spectrum. The spectral index α characterizes the energy per interval of frequency ν as $u_\nu \propto \nu^\alpha$. The temperature spectral index β is often written as $T \propto \nu^\beta$ where $\beta = \alpha - 2$, with $\beta = 0$ for the Rayleigh–Jeans spectrum and typically $\beta \simeq -2.9$ for galactic synchrotron radiation, $\beta = -2.1$ for free-free radiation, and $\beta \sim 2$ for dust emission at CMBR wavelengths.

spin temperature In this book, the parameter T in the measure of the ratio of numbers of molecules in two distinct energy levels in equation (3.13). At equilibrium with thermal radiation, as in interstellar CN molecules, the spin temperature is equal to the radiation temperature.

standard cosmological model *See* CDM cosmological model.

standard model In physical science, a well-tested theory. The word "model" is meant to signify openness to the possibility of a better theory to be discovered.

steady state cosmology Postulates that continual creation of matter preserves a steady mean mass density in a near homogeneous universe that is expanding at a steady rate. This model does not pass the cosmological tests summarized in Section 5.4.

structure formation In this book, formation of the mass concentrations observed in galaxies and clusters of galaxies. *See* gravitational instability picture.

Sunyaev–Zel'dovich effect Scattering by electrons in hot plasma pushes the CMBR energy spectrum up at short wavelengths and down by $\delta T/T = -2y$ at long wavelengths, where the Comptonization parameter y is defined in equation (4.2). The analysis in the context of nuclear explosions was published by Kompaneets (1956); Zel'dovich and Sunyaev (1969) led exploration of the significance for cosmology. The SZ effect is observed in the CMBR spectrum along lines of sight that have passed through the relatively hot and dense plasma in a rich cluster of galaxies. Away from rich clusters the COBE/FIRAS bound is $y < 1.5 \times 10^{-5}$ (Fixsen *et al.* 1996), corresponding to energy transfer $\delta u/u = 4y < 6 \times 10^{-5}$ to the CMBR from hot plasma.

superclusters Largest distinct concentrations of galaxies. We are on the outskirts of the de Vaucouleurs local supercluster, roughly 20 Mpc from the center.

surface brightness Flux i_ν of radiation energy per unit area, time, solid angle, and frequency. *See* brightness theorem.

synchrotron radiation Radiation by relativistic electrons accelerated by a magnetic field, the dominant source in radio-luminous galaxies. *See* galactic radiation.

Telstar Project Bell Laboratories demonstration of transmission of a television signal via a satellite.

thermal radiation Blackbody radiation that fills space when relaxed to thermal equilibrium. The electromagnetic field in a cavity bounded by reflecting walls may be expressed as a sum of the normal modes of oscillation that fit the cavity. In a mode with frequency ν (wavelength $\lambda = c/\nu$), the mean number of photons – the occupation number – at thermal equilibrium at temperature T is

$$\mathcal{N} = \frac{1}{e^{h\nu/kT} - 1}, \tag{G.5}$$

where h and k are the Planck and Boltzmann constants. The energy of a photon is $h\nu$, the mean energy of the mode is $\mathcal{N}h\nu$, and the sum over modes yields the thermal energy per unit volume at frequency ν in the interval $\mathrm{d}\nu$,

$$u_\nu \mathrm{d}\nu = \frac{8\pi h\nu^3}{c^3} \frac{1}{e^{h\nu/kT} - 1} \mathrm{d}\nu. \tag{G.6}$$

An example is the thermal spectrum in Figure 2.2.

Thomson scattering Low energy (nonrelativistic) scattering of photons by free electrons. In the electron rest frame Thomson scattering does not appreciably change the photon energy.

tilt In the standard cosmological model the departure from scale-invariant initial conditions measured by the departure from unity of the parameter n_s in equation (5.6) and in Table 5.2.

tipping experiment Measurement of the radiation emitted by the atmosphere by observation of the variation of radiation received as a function of angular distance from the zenith.

tired light Idea discussed by Zwicky (1929) that light from distant objects is shifted to the red by loss of energy rather than the expansion of the universe. This would not account for the thermal CMBR spectrum in Figure 2.2.

tritium The unstable heavy isotope of hydrogen, formed in an intermediate step in light element production in the early universe.

TWM Traveling Wave Maser amplifier.

V-band Microwave radiation in the frequency range 50–75 GHz.

VLA Very Large Array interferometer in New Mexico yields high angular resolution radio images by means of the time correlations of signals from well-spaced antennas.

waveguide Conducting channels, usually with rectangular or circular cross section, for transmission of microwave radiation.

Weber, Joe Pioneer in the theory and practice of gravitational wave detection. His interview with Gamow is recalled on pages 62 and 181. Biography: Yodh and Wallis (2001).

Wheeler–Feynman absorber theory A reconciliation of time-symmetric classical electromagnetic theory with the time-asymmetry of the world.

White Mountain Research Station University of California research facility above Owens Valley, California, it includes Barcroft Station, site of several CMBR measurements.

Wien break In this book, the departure of a thermal radiation spectrum from the Rayleigh–Jeans power-law form that applies at long wavelengths.

Winston cone A smooth-walled horn with a parabolic profile used to concentrate incident radiation directly onto a bolometer. It was originally used to concentrate light onto photomultiplier tubes. *See* feed.

WMAP Wilkinson Microwave Anisotropy Probe. Named in honor of David Wilkinson, this NASA satellite mission was placed in a near-stable orbit beyond the Moon for the precision measurement of the variation of the CMBR temperature and polarization across the sky (Bennett *et al.* 2003, Hinshaw *et al.* 2007).

world time Mean time kept by observers moving so the universe is seen to be isotropic in the large-scale average. This time appears in equation (2.5).

X-band Microwave radiation in the frequency range 8–12 GHz.

ylem Alphers' name for the state of the early universe.

Zel'dovich, Yakov With Dicke and Hoyle, one of three central figures in the recognition in the 1960s of the fossil radiation and light elements from the hot big bang. Biography: *Zel'dovich reminiscences*, Sunyaev (2004); collected works: Ostriker, Barenblatt and Sunyaev (1992).

zenith distance Angle between the vertical and a direction of observation.

zodiacal light Radiation from interplanetary dust: both scattered sunlight and absorbed sunlight reradiated at longer wavelengths.

Zwicky, Fritz Astronomer noted for perceptive and iconoclastic ideas. He appears in this book in connection with the early tired light interpretation of galaxy redshifts, an important catalog of galaxies and clusters of galaxies, and the demonstration of the presence of dark matter in clusters of galaxies.

2C and 3C catalogs Second and Third Cambridge Catalogs of radio sources.

References

Abell, G. O. 1958, *Astrophysical Journal Supplement*, 3, 211 (198, 362, 512).

Abroe, M. E., Borrill, J., Ferreira, P. G., Hanany, S., Jaffe, A., Johnson, B., Lee, A. T., Rabii, B., Richards, P. L., Smoot, G., Stompor, R., Winant, C. and Wu, J. H. P. 2004, *Astrophysical Journal*, 605, 607 (507).

Adams, W. S. 1941, *Astrophysical Journal*, 93, 11 (44).

Aguirre, A. N. 1999, *Astrophysical Journal*, 521, 17 (278).

Aguirre, A. N. 2000, *Astrophysical Journal*, 533, 1 (278).

Albrecht, A., Battye, R. A. and Robinson, J. 1997, *Physical Review Letters*, 79, 4736 (456).

Albrecht, A. and Steinhardt, P. J. 1982, *Physical Review Letters*, 48, 1220 (520).

Allen, S. W., Schmidt, R. W., Ebeling, H., Fabian, A. C. and van Speybroeck, L. 2004, *Monthly Notices of the Royal Astronomical Society*, 353, 457 (474).

Alpher, R. A. 1948a, PhD Thesis, *On the Origin and Relative Abundance of the Elements*, The George Washington University (26–38).

Alpher, R. A. 1948b, *Physical Review*, 74, 1577 (27, 31, 34).

Alpher, R. A., Bethe, H. and Gamow, G. 1948, *Physical Review*, 73, 803 (26–28, 41, 192, 268, 324).

Alpher, R. A., Follin, J. W. and Herman, R. C. 1953, *Physical Review*, 92, 1347 (33, 192, 259).

Alpher, R. A. and Herman, R. C. 1948, *Nature*, 162, 774 (30–31, 35, 49, 50, 60, 192, 259, 269, 306, 334).

Alpher, R. A. and Herman, R. C. 1949, *Physical Review*, 75, 1089 (30, 32).

Alpher, R. A. and Herman, R. C. 1950, *Reviews of Modern Physics*, 22, 153 (32, 33, 60).

Alpher, R. A. and Herman, R. C. 1953, *Annual Review of Nuclear and Particle Science*, 2, 1 (33, 191).

Alpher, R. A. and Herman, R. C. 2001, *Genesis of the big bang*, New York: Oxford University Press (24, 61, 438).

Alsop, D. C., Cheng, E. S., Clapp, A. C., Cottingham, D. A., Fischer, M. L., Gundersen, J. O., Kreysa, E., Lange, A. E., Lubin, P. M., Meinhold, P. R., Richards, P. L. and Smoot, G. F. 1992, *Astrophysical Journal*, 395, 317 (500).

Andreani, P., Dall'Oglio, G., Pizzo, L., Venturino, C., Martinis, L., Piccirillo, L. and Rossi, L. 1991, *Astronomy and Astrophysics*, 249, 299 (496).

Arp, H. C. 1958, *Handbuch der Physik*, 51, 75 (251).

Augustine, N. 2007, Chair, Committee on Science, Engineering and Science Policy 2007, *Rising above the Gathering Storm*, Washington DC: The National Academies Press (73).

Bahcall, N. A. and Bode, P. 2003, *Astrophysical Journal Letters*, 588, L1 (474).

Bahcall, N. A. and Cen, R. 1992, *Astrophysical Journal Letters*, 398, L81 (457, 474).

Bahcall, N. A., Ostriker, J. P., Perlmutter, S. and Steinhardt, P. J. 1999, *Science*, 284, 1481 (455).

Baker, J. C., Grainge, K., Hobson, M. P., Jones, M. E., Kneissl, R., Lasenby, A. N., O'Sullivan, C. M. M., Pooley, G., Rocha, G., Saunders, R., Scott, P. F. and Waldram, E. M. 1999, *Monthly Notices of the Royal Astronomical Society*, 308, 1173 (503).

Bardeen, J. M., Steinhardt, P. J. and Turner, M. S. 1983, *Physical Review D*, 28, 679 (441).

Bartlett, J. G., Blanchard, A., Silk, J. and Turner, M. S. 1995, *Science*, 267, 980 (456).

Baum, W. A. 1956, *Publications of the Astronomical Society of the Pacific*, 68, 118 (65).

Beckman, J. E., Ade, P. A. R., Huizinga, J. S., Robson, E. I., Vickers, D. G. and Harries, J. E. 1972, *Nature*, 237, 154 (360).

Bennett, C. L., Banday, A. J., Gorski, K. M., Hinshaw, G., Jackson, P., Keegstra, P., Kogut, A., Smoot, G. F., Wilkinson, D. T. and Wright, E. L. 1996, *Astrophysical Journal Letters*, 464, L1 (499).

Bennett, C. L., Halpern, M., Hinshaw, G., Jarosik, N., Kogut, A., Limon, M., Meyer, S. S., Page, L., Spergel, D. N., Tucker, G. S., Wollack, E., Wright, E. L., Barnes, C., Greason, M. R., Hill, R. S., Komatsu, E., Nolta, M. R., Odegard, N., Peiris, H. V., Verde, L. and Weiland, J. L. 2003, *Astrophysical Journal Supplement*, 148, 1 (287, 466, 509, 530).

Bennett, C. L., Hinshaw, G., Banday, A., Kogut, A., Wright, E. L., Loewenstein, K. and Cheng, E. S. 1993, *Astrophysical Journal Letters*, 414, L77 (495).

Bennett, C. L., Kogut, A., Hinshaw, G., Banday, A. J., Wright, E. L., Gorski, K. M., Wilkinson, D. T., Weiss, R., Smoot, G. F., Meyer, S. S., Mather, J. C., Lubin, P., Loewenstein, K., Lineweaver, C., Keegstra, P., Kaita, E., Jackson, P. D. and Cheng, E. S. 1994, *Astrophysical Journal*, 436, 423 (499).

Benoît, A. *et al.* 2003, *Astronomy and Astrophysics*, 399, L19 (508).

Bensadoun, M., Bersanelli, M., de Amici, G., Kogut, A., Levin, S. M., Limon, M., Smoot, G. F. and Witebsky, C. 1993, *Astrophysical Journal*, 409, 1 (482).

Berlin, A. B., Bulaenko, E. V., Vitkovskii, V. V., Kononov, V. K., Pariiskii, I. N. and Petrov, Z. E. 1983, *Early Evolution of the Universe and its Present Structure*, 104, 121 (489).

Bernstein, G. M., Fischer, M. L., Richards, P. L., Peterson, J. B. and Timusk, T. 1990, *Astrophysical Journal*, 362, 107(416, 481).

Bersanelli, M., Bensadoun, M., de Amici, G., Levin, S., Limon, M., Smoot, G. F. and Vinje, W. 1994, *Astrophysical Journal*, 424, 517 (482).

Bersanelli, M., Maino, D. and Mennella, A. 2002, *Nuovo Cimento Rivista Serie*, 25, 1 (486).

Bersanelli, M., Witebsky, C., Bensadoun, M., de Amici, G., Kogut, A., Levin, S. M. and Smoot, G. F. 1989, *Astrophysical Journal*, 339, 632 (482).

Bethe, H. A. 1939, *Physical Review*, 55, 434 (99).

Bethe, H. A. and Bacher, R. F. 1936, *Reviews of Modern Physics*, 8, 82 (410).

Birkinshaw, M. 1990, *The Cosmic Microwave Background: 25 Years Later*, Dordrecht: Kluwer, 77 (423).

Blackman, R. B. and Tukey, J. W. 1958, *The Measurement of Power Spectra*, New York: Dover (198, 452).

Blades, J. C. 1978, *Monthly Notices of the Royal Astronomical Society*, 185, 451 (479).

Blair, A., Beery, J. G., Edeskuty, F., Hiebert, R. D., Shipley, J. P. and Williamson, K. D., Jr. 1971, *Physical Review Letters*, 27, 1154 (413).

Blake, C. and Wall, J. 2002, *Nature*, 416, 150 (286, 510).

Blum, G. D. and Weiss, R. 1967, *Physical Review*, 155, 1412 (342).

Blumenthal, G. R., Pagels, H. and Primack, J. R. 1982, *Nature*, 299, 37 (439).

Bolton, J. G. and Westfold, K. C. 1951, *Australian Journal of Physics*, 4, 476 (288).

Bond, J. R. 1995a, in Proceedings, Les Houches School, Session LX, August 1993, *Cosmology and Large Scale Structure*, ed. R. Schaeffer, Netherlands: Elsevier (486).

Bond, J. R. 1995b, in Proceedings of a workshop held in Anacapri (Capri), Italy, September 20–24, 1993, ed. M. Bersanelli, S. Cortiglioni, N. Mandolesi, G. F. Smoot and N. Vittorio, New York: Gordon and Breach Science Publishers (486).

Bond, J. R. 1995c, *Astrophysical Letters and Communications*, 32, 63 (497).

Bond, J. R., Centrella, J., Szalay, A. S. and Wilson, J. R. 1984, in Proceedings of the Third Moriond Conference, *Evolution of Galaxies and Large Structures in the Universe*, 87 (439, 520).

Bond, J. R., Contaldi, C. and Pogosyan, D. 2003, *Royal Society of London Philosophical Transactions*, Series A, 361, 2435 (455, 461).

Bond, J. R., Crittenden, R., Davis, R. L., Efstathiou, G. and Steinhardt, P. J. 1994, *Physical Review Letters*, 72, 13 (446).

Bond, J. R. and Efstathiou, G. 1984, *Astrophysical Journal Letters*, 285, L45 (378, 446, 450).

Bond, J. R. and Efstathiou, G. 1987, *Monthly Notices of the Royal Astronomical Society*, 226, 655 (443–445).

Bond, J. R., Efstathiou, G. and Tegmark, M. 1997, *Monthly Notices of the Royal Astronomical Society*, 291, L33 (446).

Bond, J. R., Jaffe, A. H. and Knox, L. 2000, *Astrophysical Journal*, 533, 19 (486).

Bond, J. R., Szalay, A. S. and Turner, M. S. 1982, *Physical Review Letters*, 48, 1636 (439).

Bond, J. R. *et al.* 2000, in IAU Symposium 201, CITA, Toronto (459).

Bondi, H. 1952, *Cosmology*, Cambridge: Cambridge University Press (51, 52, 58, 61, 65, 292, 515).

Bondi, H. 1960a, *Cosmology*, second edition, Cambridge: Cambridge University Press (51–56, 59, 61, 67, 186, 222, 292, 369, 515).

Bondi, H. 1960b, *Rival Theories of Cosmology*, London: Oxford University Press (53).

Bondi, H. and Gold, T. 1948, *Monthly Notices of the Royal Astronomical Society*, 108, 252 (15, 269, 275).

Bondi, H., Gold, T. and Hoyle, F. 1955, *The Observatory*, 75, 80 (269).

Bonnor, W. B. 1957, *Monthly Notices of the Royal Astronomical Society*, 117, 104 (436).

Bortolot, V. J., Jr., Clauser, J. F. and Thaddeus, P. 1969, *Physical Review Letters*, 22, 307 (84, 84, 413).

Bortolot, V. J., Jr. and Thaddeus, P. 1969, *Astrophysical Journal Letters*, 155, L17 (85).

Boughn, S. P., Cheng, E. S., Cottingham, D. A. and Fixsen, D. J. 1992, *Astrophysical Journal Letters*, 391, L49 (499).

Boughn, S. P., Cheng, E. S. and Wilkinson, D. T. 1981, *Astrophysical Journal Letters*, 243, L113 (438, 492).

Boughn, S. P. and Crittenden, R. G. 2005, *Monthly Notices of the Royal Astronomical Society*, 360, 1013 (475).

Boughn, S. P., Fram, D. M. and Partridge, R. B. 1971, *Astrophysical Journal*, 165, 439 (395, 432, 489).

Boughn, S. P. and Jahoda, K. 1993, *Astrophysical Journal Letters*, 412, L1 (499).

Boynton, P. E. 1967, PhD Thesis, *Double Charge Exchange Scattering of Positive Pions on Complex Nuclei*, Princeton University (326).

Boynton P. E. 2005, in Proceedings of the International School of Physics "Enrico Fermi" Course CLIX, *Background Microwave Radiation and Intracluster Cosmology*, ed. F. Melchiorri and Y. Rephaeli, The Netherlands: IOS Press; Bologna, Italy: Società Italiana di Fisica, 443 (307).

Boynton, P. E. and Stokes, R. A. 1974, *Nature*, 247, 528 (328, 480).

Boynton, P. E., Stokes, R. A. and Wilkinson, D. T. 1968, *Physical Review Letters*, 21, 462 (230, 312, 320, 322, 326, 480).

Bracewell, R. N. 1959, *Paris Symposium on Radio Astronomy*, ed. R. N. Bracewell, Stanford: Stanford University Press (61).

Bracewell, R. N. 1966, Glint no. 113 (Glints are SRAI internal reports, available from the authors.) (385).

Bracewell, R. N. 1968, Glint nos. 277, 279 (390).

Bracewell, R. N., Colvin, R. S., D'Addario, L. R., Grebenkemper, C. J., Price, K. M. and Thompson, A. R. 1973, *Proceedings of the Institute of Electrical and Electronic Engineers*, 61, 1249 (386).

Bracewell, R. N. and Conklin, E. K. 1967, Glint no. 199 (391, 392).

Bracewell, R. N. and Conklin, E. K. 1968, *Nature*, 219, 1343 (391, 424).

Brans, C. and Dicke, R. H. 1961, *Physical Review*, 124, 925 (39, 76).

Brillouin, L. 1964, *Tensors in Mechanics and Elasticity*, New York: Academic Press (344).

Broten, N. W. *et al.* 1967, *Science*, 156, 1592 (285).

Burbidge, E. M. and Burbidge, G. R. 1958, *Handbuch der Physik*, 51, 134 (251).

Burbidge, E. M., Burbidge, G. R., Fowler, W. A. and Hoyle, F. 1957, *Reviews of Modern Physics*, 29, 547 (58, 91, 251, 260, 268, 512).

Burbidge, G. R. 1958, *Publications of the Astronomical Society of the Pacific*, 70, 83 (58, 60, 91, 251, 269).

Burbidge, G. R. and Hoyle, F. 1998, *Astrophysical Journal Letters*, 509, L1 (270).

Burke, B. F. 2005, in *Radio Astronomy from Karl Jansky to Microjansky*, ed. L. I. Gurvits, S. Frey and S. Rawlings, Paris: EDP Sciences (183).

Caderni, N., Fabbri, R., de Cosmo, V., Melchiorri, B., Melchiorri, F. and Natale, V. 1977, *Physical Review D*, 16, 2424 (491).

Cameron, A. G. W. 1957, *Publications of the Astronomical Society of the Pacific*, 69, 201 (58, 260).

Carpenter, R. L., Gulkis, S. and Sato, T. 1973, *Astrophysical Journal Letters*, 182, L61 (490).

Ceccarelli, C., Melchiorri, F., Pietranera, L., Dall'oglio, G. and Melchiorri, B. 1982, *Astrophysical Journal*, 260, 484 (493).

Cen, R., Miralda-Escudé, J., Ostriker, J. P. and Rauch, M. 1994, *Astrophysical Journal Letters*, 437, L9 (474).

Chandrasekhar, S. and Henrich, L. R. 1942, *Astrophysical Journal*, 95, 288 (25, 28).

Charlier, C. V. L. 1922, *Arkiv för Mathematik, Astronomi och Fysik*, 16, 1 (40).

Charlier, C. V. L. 1925, *Publications of the Astronomical Society of the Pacific*, 37, 177 (40).

Cheng, E. S., Cottingham, D. A., Fixsen, D. J., Goldin, A. B., Inman, C. A., Knox, L., Kowitt, M. S., Meyer, S. S., Puchalla, J. L., Ruhl, J. E. and Silverberg, R. F. 1997, *Astrophysical Journal Letters*, 488, L59 (502).

Cheng, E. S., Cottingham, D. A., Fixsen, D. J., Inman, C. A., Kowitt, M. S., Meyer, S. S., Page, L. A., Puchalla, J. L., Ruhl, J. E. and Silverberg, R. F. 1996, *Astrophysical Journal Letters*, 456, L71 (502).

Cheng, E. S., Cottingham, D. A., Fixsen, D. J., Inman, C. A., Kowitt, M. S., Meyer, S. S., Page, L. A., Puchalla, J. L. and Silverberg, R. F. 1994, *Astrophysical Journal Letters*, 422, L37 (502).

Cheng, E. S., Saulson, P. R., Wilkinson, D. T. and Corey, B. E. 1979, *Astrophysical Journal Letters*, 232, L139 (427, 492, 493).

Cheung, A. C., Rank, D. M., Townes, C. H., Thornton, D. D. and Welch, W. J. 1968, *Physical Review Letters*, 21, 1701 (295).

Cheung, A. C., Rank, D. M., Townes, C. H., Thornton, D. D. and Welch, W. J. 1969, *Nature*, 221, 626 (295).

Church, S. E., Ganga, K. M., Ade, P. A. R., Holzapfel, W. L., Mauskopf, P. D., Wilbanks, T. M. and Lange, A. E. 1997, *Astrophysical Journal*, 484, 523 (503).

Clapp, A. C., Devlin, M. J., Gundersen, J. O., Hagmann, C. A., Hristov, V. V., Lange, A. E., Lim, M., Lubin, P. M., Mauskopf, P. D., Meinhold, P. R., Richards, P. L., Smoot, G. F., Tanaka, S. T., Timbie, P. T. and Wuensche, C. A. 1994, *Astrophysical Journal Letters*, 433, L57 (500).

Clauser, J. F. 1970, PhD Thesis, *Measurement of the Cosmic Microwave Background by Optical Observations of Interstellar Molecules*, Columbia University (479).

Coble, K., Dodelson, S., Dragovan, M., Ganga, K., Knox, L., Kovac, J., Ratra, B. and Souradeep, T. 2003, *Astrophysical Journal*, 584, 585 (502).

Coble, K., Dragovan, M., Kovac, J., Halverson, N. W., Holzapfel, W. L., Knox, L., Dodelson, S., Ganga, K., Alvarez, D., Peterson, J. B., Griffin, G., Newcomb, M., Miller, K., Platt, S. R. and Novak, G. 1999, *Astrophysical Journal Letters*, 519, L5 (502).

Compton, A. H. and Getting, I. A. 1935, *Physical Review*, 47, 817 (424).

Condon, J. J., Cotton, W. D., Greisen, E. W., Yin, Q. F., Perley, R. A., Taylor, G. B. and Broderick, J. J. 1998, *Astronomical Journal*, 115, 1693 (286).

Condon, J. J. and Harwit, M. 1968, *Physical Review Letters*, 20, 1309; *Physical Review Letters*, 21, 58 (391).

Conklin, E. K. 1966, Glint no. 138 (385).

Conklin, E. K. 1969, PhD Thesis, *Anisotropy and Inhomogeneity in the Cosmic Background Radiation*, Stanford University; *Nature*, 222, 971 (233, 385, 390, 402, 427, 428, 488, 491).

Conklin, E. K. 1972, in IAU Symposium 44, *External Galaxies and Quasi-Stellar Objects*, Dordrecht, Netherlands: Reidel, p. 518 (390, 427, 428, 488).

Conklin, E. K. and Bracewell, R. N. 1967, *Physical Review Letters*, 18, 614; *Nature*, 216, 777 (387, 452, 487).

Corey, B. E. 1978, PhD Thesis, *The Dipole Anisotropy of the Cosmic Microwave Background at a Wavelength of 1.6 cm*, Princeton University (432).

Corey, B. E. and Wilkinson, D. T. 1976, *Bulletin of the American Astronomical Society*, 8, 351 (391, 432, 492).

Cottingham, D. A. 1987, PhD Thesis, *A Sky Temperature Survey at 19.2 GHz Using a Balloon Borne Dicke Radiometer for Anisotropy Tests of the Cosmic Microwave Background*, Princeton University (419).

Covington, A. E. 1950, *Journal of Geophysical Research*, 55, 33 (63).

Cowsik, R. and McClelland, J. 1973, *Astrophysical Journal*, 180, 7 (438).

Crane, P., Hegyi, D. J., Kutner, M. L. and Mandolesi, N. 1989, *Astrophysical Journal*, 346, 136 (416, 479).

Crane, P., Hegyi, D. J., Mandolesi, N. and Danks, A. C. 1986, *Astrophysical Journal*, 309, 822 (479).

Crawford, A. B. and Hogg, D. C. 1956, *Bell System Technical Journal*, 25, 907 (70).

Crawford, A. B., Hogg, D. C. and Hunt, L. E. 1961, *Bell System Technical Journal*, 40, 1005 (71, 160).

Crittenden, R. G. and Turok, N. 1995, *Physical Review Letters*, 75, 2642 (447).

Croft, R. A. C., Weinberg, D. H., Katz, N. and Hernquist, L. 1998, *Astrophysical Journal*, 495, 44 (474).

Crutcher, R. M. 1985, *Astrophysical Journal*, 288, 604 (479).

Dahle, H. 2007, ArXiv Astrophysics e-prints, arXiv:astro-ph/0701598 (474).

Dall'Oglio, G. and de Bernardis, P. 1988, *Astrophysical Journal*, 331, 547 (496).

Dall'oglio, G., Fonti, S., Melchiorri, B., Melchiorri, F., Natale, V., Lombardini, P., Trivero, P. and Sivertsen, S. 1976, *Physical Review D*, 13, 1187 (481).

Danese, L., Burigana, C., Toffolatti, L., de Zotti, G. and Franceschini, A. 1990, in *The Cosmic Microwave Background: 25 Years Later*, ed. N. Mandolesi and N. Vittorio, Dordrecht: Kluwer, 153 (422).

Danese, L. and de Zotti, G. 1978, *Astronomy and Astrophysics*, 68, 157 (480).

Davies, R. D., Lasenby, A. N., Watson, R. A., Daintree, E. J., Hopkins, J., Beckman, J., Sanchez Almeida, J. and Rebolo, R. 1987, *Nature*, 326, 462 (496).

Davis, K. C. 1971, PhD Thesis, *A Measurement of the Isotropy of the Microwave Background Temperature at 6 Centimeter Wavelength*, Princeton University (432).

Davis, M., Efstathiou, G., Frenk, C. S. and White, S. D. M. 1985, *Astrophysical Journal*, 292, 371 (450).

Davis, M. and Peebles, P. J. E. 1983, *Astrophysical Journal*, 267, 465 (457).

de Amici, G., Limon, M., Smoot, G. F., Bersanelli, M., Kogut, A. and Levin, S. 1991, *Astrophysical Journal*, 381, 341 (482).

de Amici, G., Smoot, G. F., Aymon, J., Bersanelli, M., Kogut, A., Levin, S. M. and Witebsky, C. 1988, *Astrophysical Journal*, 329, 556 (482).

de Amici, G., Smoot, G., Friedman, S. D. and Witebsky, C. 1985, *Astrophysical Journal*, 298, 710 (481).

de Amici, G., Witebsky, C., Smoot, G. F. and Friedman, S. D. 1984, *Physical Review D*, 29, 2673 (481).

de Bernardis, P. *et al.* 1990, *Astrophysical Journal Letters*, 360, L31 (498).

de Bernardis, P. *et al.* 1994, *Astrophysical Journal Letters*, 422, L33 (498).

de Bernardis, P. *et al.* 2000, *Nature*, 404, 955 (458, 506).

de Bernardis, P. *et al.* 2002, *Astrophysical Journal*, 564, 559 (506).

De Grasse, R. W., Hogg, D. C., Ohm, E. A. and Scovil, H. E. D. 1959, *Journal of Applied Physics*, 30, 2013; *Proceedings of the National Electronics Conference*, 15, 370, 1959 (49, 65, 71, 145, 167).

De Grasse, R. W., Shulz-Du-Bois, E. D. and Scovil, H. E. D. 1959, *Bell System Technical Journal*, 38, 305 (70, 159).

Delannoy, J., Denisse, J. F., Le Roux, É. and Morlet, B. 1957, *Annales d'Astrophysique*, 20, 222 (64).

Denisse, J.-F., Lequeux, J. and Le Roux, É. 1957, *Comptes Rendus de l'Académie des Sciences, Paris*, 244, 3030 (64).

de Oliveira-Costa, A., Devlin, M. J., Herbig, T., Miller, A. D., Netterfield, C. B., Page, L. A. and Tegmark, M. 1998, *Astrophysical Journal Letters*, 509, L77 (504).

de Oliveira-Costa, A., Tegmark, M., Devlin, M. J., Haffner, L. M., Herbig, T., Miller, A. D., Page, L. A., Reynolds, R. J. and Tufte, S. L. 2000, *Astrophysical Journal Letters*, 542, L5 (504).

de Oliveira-Costa, A., Tegmark, M., Page, L. A. and Boughn, S. P. 1998, *Astrophysical Journal Letters*, 509, L9 (499).

de Sitter, W. 1917, *Monthly Notices of the Royal Astronomical Society*, 78, 3 (53).

de Sitter, W. 1933, *Monthly Notices of the Royal Astronomical Society*, 93, 628 (41).

Devlin, M. J., Clapp, A. C., Gundersen, J. O., Hagmann, C. A., Hristov, V. V., Lange, A. E., Lim, M. A., Lubin, P. M., Mauskopf, P. D., Meinhold, P. R., Richards, P. L., Smoot, G. F., Tanaka, S. T., Timbie, P. T. and Wuensche, C. A. 1994, *Astrophysical Journal Letters*, 430, L1 (500).

Devlin, M. J., de Oliveira-Costa, A., Herbig, T., Miller, A. D., Netterfield, C. B., Page, L. A. and Tegmark, M. 1998, *Astrophysical Journal Letters*, 509, L69 (504).

Dicke, R. H. 1946a, *Astrophysical Journal*, 103, 375 (63).

Dicke, R. H. 1946b, *Review of Scientific Instruments*, 17, 268 (314).

Dicke, R. H. 1964, *The Theoretical Significance of Experimental Relativity*, New York: Gordon and Breach (38).

Dicke, R. H. 1968, *Astrophysical Journal*, 152, 1 (39).

Dicke, R. H., Beringer, R., Kyhl, R. L. and Vane, A. B. 1946, *Physical Review*, 70, 340 (45, 48–49, 63, 168, 177, 188, 307, 324).

Dicke, R. H. and Peebles, P. J. E. 1965, *Space Science Reviews*, 4, 419 (189–192).

Dicke, R. H., Peebles, P. J. E., Roll, P. G. and Wilkinson, D. T. 1965, *Astrophysical Journal*, 142, 414 (155, 172, 192, 205, 215, 287, 293, 298, 306, 325, 342).

Dicker, S. R., Melhuish, S. J., Davies, R. D., Gutiérrez, C. M., Rebolo, R., Harrison, D. L., Davis, R. J., Wilkinson, A., Hoyland, R. J. and Watson, R. A. 1999, *Monthly Notices of the Royal Astronomical Society*, 309, 750 (505).

Dirac, P. A. M. 1938, *Royal Society of London Proceedings, Series A*, 165, 199 (38, 342, 517).

Dodelson, S. 2003, *Modern Cosmology*, Amsterdam: Academic Press (384).

Dodelson, S. and Knox, L. 2000, *Physical Review Letters*, 84, 3523 (455).

Doel, R. E. and McCutcheon, A. 1995, *Journal for the History of Astronomy*, 26, 279 (134).

Doroshkevich, A. G., Khlopov, M. I., Sunyaev, R. A., Szalay, A. S. and Zel'dovich, Ya. B. 1981, *Annals of the New York Academy Sciences*, 375, 32 (439).

Doroshkevich, A. G. and Novikov, I. D. 1964, *Doklady*, 154, 809; *Soviet Physics-Doklady*, 9, 111, 1964 (65, 102–103, 108, 113, 153, 188, 189, 421).

Doroshkevich, A. G. and Sunyaev, R. A. 1969, *Astronomicheskii Zhurnal*, 46, 20; English translation in *Soviet Astronomy*, 13, 15 (110).

Doroshkevich, A. G., Zel'dovich, Ya. B. and Novikov, I. D. 1967, *Astronomicheskii Zhurnal*, 44, 295; English translation in *Soviet Astronomy*, 11, 233 (106).

Doroshkevich, A. G., Zel'dovich, Ya. B. and Sunyaev, R. A. 1978, *Astronomicheskii Zhurnal*, 55, 913; English translation in *Soviet Astronomy*, 22, 523, 1978 (130, 377).

Dragovan, M., Ruhl, J. E., Novak, G., Platt, S. R., Crone, B., Pernic, R. and Peterson, J. B. 1994, *Astrophysical Journal Letters*, 427, L67 (502).

Dunkley, J. *et al.* 2008, ArXiv e-prints, 803, arXiv:0803.0586 (467).

Dyson, F. J. 1992, *From Eros to Gaia*, New York: Pantheon (302).

Eddington, A. S. 1931, *Supplement to Nature*, March 21 (517).

Edge, D. O., Shakeshaft, J. R., McAdam, W. B., Baldwin, J. E. and Archer, S. 1959, *Memoirs of the Royal Astronomical Society*, 68, 37 (158).

Efstathiou, G. and Bond, J. R. 1987, *Monthly Notices of the Royal Astronomical Society*, 227, 33P (445).

Efstathiou, G. and Bond, J. R. 1999, *Monthly Notices of the Royal Astronomical Society*, 304, 75 (378, 446).

Efstathiou, G., Sutherland, W. J. and Maddox, S. J. 1990, *Nature*, 348, 705 (456, 472).

Einstein, A. 1905, *Annalen der Physik*, 17, 891 (390, 528).

Einstein, A. 1915, *Sitzungsberichte der Königlich Preußischen Akademie der Wissenschaften* (Berlin), 844 (519).

Einstein, A. 1917, *Sitzungsberichte der Königlich Preußischen Akademie der Wissenschaften* (Berlin), 142 (9, 13, 15, 39, 515, 515, 517).

Einstein, A. 1922a, *The Meaning of Relativity*, Princeton: Princeton University Press (39).

Einstein, A. 1922b, *Annalen der Physik*, 72, 58 (39).

Einstein, A. 1945, *The Meaning of Relativity*, second edition, Princeton: Princeton University Press, (52).

Einstein, A. and de Sitter, W. 1932, *Proceedings of the National Academy of Science*, 18, 213 (52, 442, 517).

Eisenstein, D. J. *et al.* 2005, *Astrophysical Journal*, 633, 560 (472).

Eke, V. R., Cole, S. and Frenk, C. S. 1996, *Monthly Notices of the Royal Astronomical Society*, 282, 263 (457, 474).

Ellis, G. F. R. 2002, *Nature*, 416, 132 (287).

Ellis, G. F. R. 2006, in *Handbook in Philosophy of Physics*, ed. J. Butterfield and J. Earman, Elsevier, http://arxiv.org/abs/astro-ph/0602280 (383).

Ellis, G. F. R. and Baldwin, J. E. 1984, *Monthly Notices of the Royal Astronomical Society*, 206, 377 (286).

Ellis, G. F. R. and Maartens, R. 2004, *Classical and Quantum Gravity*, 21, 223 (384).

Ellis, G. F. R. and Rothman, T. 1993, *American Journal of Physics*, 61, 883 (381).

Elmegreen, D. M., Elmegreen, B. G., Kaufman, M., Sheth, K., Struck, C., Thomasson, M. and Brinks, E. 2006, *Astrophysical Journal*, 642, 158 (280).

Epstein, E. E. 1967, *Astrophysical Journal Letters*, 148, L157 (233, 488).

Ewing, M. S., Burke, B. F. and Staelin, D. H. 1967, *Physical Review Letters*, 19, 1251 (183, 230, 342, 480).

Fabbri, R., Guidi, I., Melchiorri, F. and Natale, V. 1980b, *Physical Review Letters*, 44, 1563 (438, 493).

Fabbri, R., Guidi, I., Melchiorri, F. and Natale, V. 1982, *Second Marcel Grossman Meeting*, Part B, July 1979 (493).

Fabbri, R., Melchiorri, B., Melchiorri, F., Natale, V., Caderni, N. and Shivanandan, K. 1980a, *Physical Review D*, 21, 2095 (492).

Fall, S. M. 1975, *Monthly Notices of the Royal Astronomical Society*, 172, 23P (457).

Faulkner, J. 1966, *Astrophysical Journal*, 144, 978 (258).

Faulkner, J. 1967, *Astrophysical Journal*, 147, 617 (258).

Faulkner, J. 2003, in *Fred Hoyle's Universe*, ed. C. Wickramasinghe, G. Burbidge and J. Narlikar, Dordrecht: Kluwer Academic Press (246).

Faulkner, J., Hoyle, F. and Narlikar, J. V. 1964, *Astrophysical Journal*, 140, 1100 (253).

Faulkner, J. and Iben, I. J. 1966, *Astrophysical Journal*, 144, 995 (258).

Feldman, H. *et al.* 2003, *Astrophysical Journal Letters*, 596, L131 (473).

Femenia, B., Rebolo, R., Gutierrez, C. M., Limon, M. and Piccirillo, L. 1998, *Astrophysical Journal*, 498, 117 (504).

Field, G. B. and Henry, R. C. 1964, *Astrophysical Journal*, 140, 1002 (279).

Field, G. B., Herbig, G. H. and Hitchcock, J. 1966, *Astronomical Journal*, 71, 161 (77, 173, 196).

Field, G. B. and Hitchcock, J. L. 1966, *Physical Review Letters*, 16, 817; *Astrophysical Journal*, 146, 1 (77, 152, 294, 322, 479).

Fischler, W., Ratra, B. and Susskind, L. 1985, *Nuclear Physics B*, 259, 730 (441).

Fixsen, D. J., Cheng, E. S., Gales, J. M., Mather, J. C., Shafer, R. A. and Wright, E. L. 1996, *Astrophysical Journal*, 473, 576 (421, 483, 512, 528).

Fixsen, D. J., Cheng, E. S. and Wilkinson, D. T. 1983, *Physical Review Letters*, 50, 620 (492, 494).

Fixsen, D. J., Dwek, E., Mather, J. C., Bennett, C. L. and Shafer, R. A. 1998, *Astrophysical Journal*, 508, 123 (420, 421).

Fixsen, D. J., Hinshaw, G., Bennett, C. L. and Mather, J. C. 1997, *Astrophysical Journal*, 486, 623 (420, 499).

Fixsen, D. J., Kogut, A., Levin, S., Limon, M., Lubin, P., Mirel, P., Seiffert, M. and Wollack, E. 2004, *Astrophysical Journal*, 612, 86 (421, 482).

Fok, V. A. 1964, *The Theory of Space, Time and Gravitation*, New York: Macmillan (137).

Fomalont, E. B., Kellermann, K. I. and Wall, J. V. 1984, *Astrophysical Journal Letters*, 277, L23 (286).

Fowler, W. A., Caughlan, G. R. and Zimmerman, B. A. 1967, *Annual Review of Astronomy and Astrophysics*, 5, 525 (261).

Fowler, W. A., Caughlan, G. R. and Zimmerman, B. A. 1975, *Annual Review of Astronomy and Astrophysics*, 13, 69 (261).

Frank-Kamenetskii, D. A. 1959, *Physical Processes in Stars* (in Russian), Moscow: Fizmatgiz (93).

Freedman, W. L., *et al.* 2001, *Astrophysical Journal*, 553, 47 (471).

Freundlich, E. F. 1954, *Philosophical Magazine*, 45, 303; *Proceedings of the Physical Society (London)*, A 67, 192, 1954 (343).

Friedman, S. D., Smoot, G. F., de Amici, G. and Witebsky, C. 1984, *Physical Review D*, 29, 2677 (481).

Friedmann, A. 1922, *Zeitschrift für Physik*, 10, 377 (14, 134, 275).

Friedmann, A. 1924, *Zeitschrift für Physik*, 21, 326 (134).

Friis, H. T. 1971, *Seventy Five Years in an Exciting World*, San Francisco: San Francisco Press (70, 518).

Fu, L. *et al.* 2008, *Astronomy and Astrophysics*, 479, 9 (473).

Fukugita, M. and Peebles, P. J. E. 2004, *Astrophysical Journal*, 616, 643 (18, 470).

Gaier, T., Schuster, J., Gundersen, J., Koch, T., Seiffert, M., Meinhold, P. and Lubin, P. 1992, *Astrophysical Journal Letters*, 398, L1 (455, 499).

Gamow, G. 1942, *Journal of the Washington Academy of Science*, 32, 353 (25, 27).

Gamow, G. 1946, *Physical Review*, 70, 572 (26, 30, 49, 306).

Gamow, G. 1948a, *Physical Review*, 74, 505 (28–31, 35, 49, 60, 66, 193, 306, 436, 438).

Gamow, G. 1948b, *Nature*, 162, 680 (28, 58, 60, 66, 306).

Gamow, G. 1949, *Reviews of Modern Physics*, 21, 367 (33, 34, 59, 66, 190).

Gamow, G. 1952, *The Creation of the Universe*, New York: Viking (52, 62, 63).

Gamow, G. 1953a, in *Symposium on Astrophysics*, University of Michigan, Ann Arbor, June 29 to July 24 (60, 66).

Gamow, G. 1953b, *Danske Matematisk-fysiske Meddelelser*, 27, 10 (61).

Gamow, G. 1954, *Astronomical Journal*, 59, 200 (54).

Gamow, G. 1956, *Vistas in Astronomy*, 2, 1726 (58, 61).

Gamow, G. 1970, *My World Line*, New York: Viking (14, 518).

Gamow, G. and Fleming, J. A. 1942, *Science*, 95, 579 (25).

Gamow, G. and Landau, L. D. 1933, *Nature*, 132, 567 (99).

Gamow, G. and Teller, E. 1939, *Physical Review*, 55, 654 (438).

Ganga, K. 1994, PhD Thesis, *Exploring the Large Scale Anisotropy in the Cosmic Microwave Background Radiation at 170 GHz*, Princeton University (499).

Ganga, K., Cheng, E., Meyer, S. and Page, L. 1993, *Astrophysical Journal Letters*, 410, L57 (453, 497).

Ganga, K., Page, L., Cheng, E. and Meyer, S. 1994, *Astrophysical Journal Letters*, 432, L15 (497).

Ganga, K., Ratra, B., Church, S. E., Sugiyama, N., Ade, P. A. R., Holzapfel, W. L., Mauskopf, P. D. and Lange, A. E. 1997, *Astrophysical Journal*, 484, 517 (503).

Ganga, K., Ratra, B., Lim, M. A., Sugiyama, N. and Tanaka, S. T. 1998, *Astrophysical Journal Supplement*, 114, 165 (500).

Gaustad, J. E., McCullough, P. R. and van Buren, D. 1996, *Publications of the Astronomical Society of the Pacific*, 108, 351 (484).

Gautier, T. N. *et al.* 1984, *Astrophysical Journal Letters*, 278, L57 (338).

Gershtein, S. S. and Zel'Dovich, Ya. B. 1966; English translation in *Soviet Journal of Experimental and Theoretical Physics Letters*, 4, 120 (132).

Giacconi, R., Gorenstein, P., Gursky, H., Usher, P. D., Waters, J. R., Sandage, A., Osmer, P. and Peach, J. V. 1967, *Astrophysical Journal Letters*, 148, L129 (373).

Giacconi, R., Gursky, H., Paolini, F. R. and Rossi, B. B. 1962, *Physical Review Letters*, 9, 439 (341).

Gibson, J., Welch, W. J. and de Pater, I. 2005, *Icarus*, 173, 439 (295).

Ginzburg, V. L. 2001, *The Physics of a Lifetime: Reflections on the Problems and Personalities of 20th Century Physics*, New York: Springer (140).

Ginzburg, V. L. and Ozernoi, L. M. 1966, *Soviet Astronomy*, 9, 726 (132).

Gold, T. and Pacini, F. 1968, *Astrophysical Journal Letters*, 152, L115 (195).

Gorenstein, M. V. and Smoot, G. F. 1981, *Astrophysical Journal*, 244, 361 (490).

Gorski, K. M., Banday, A. J., Bennett, C. L., Hinshaw, G., Kogut, A., Smoot, G. F. and Wright, E. L. 1996, *Astrophysical Journal Letters*, 464, L11 (499).

Gott, J. R., III 1982, *Nature*, 295, 304 (442).

Gough, D. 2005, *The Scientific Legacy of Fred Hoyle*, Cambridge: Cambridge University Press (241, 249, 520).

Grainge, K. *et al.* 2003, *Monthly Notices of the Royal Astronomical Society*, 341, L23 (508).

Grantham, D. D., Rohrbough, S., Salmela, H. A. and Sissenwine, N., 1966, *Air Force Cambridge Research Laboratory Notes on Atmospheric Properties*, #61 (345).

Gratton, R. G., Bragaglia, A., Carretta, E., Clementini, G., Desidera, S., Grundahl, F. and Lucatello, S. 2003, *Astronomy and Astrophysics*, 408, 529 (471).

Gregory, J. 2005, *Fred Hoyle's Universe*, Oxford: Oxford University Press (520).

Gundersen, J. O., Clapp, A. C., Devlin, M., Holmes, W., Fischer, M. L., Meinhold, P. R., Lange, A. E., Lubin, P. M., Richards, P. L. and Smoot, G. F. 1993, *Astrophysical Journal Letters*, 413, L1 (500).

Gundersen, J. O., Lim, M., Staren, J., Wuensche, C. A., Figueiredo, N., Gaier, T. C., Koch, T., Meinhold, P. R., Seiffert, M. D., Cook, G., Segale, A. and Lubin, P. M. 1995, *Astrophysical Journal Letters*, 443, L57 (499).

Gunn, J. E. and Peterson, B. A. 1965, *Astrophysical Journal*, 142, 1633 (122).

Gush, H. P. 1974, *Canadian Journal of Physics*, 52, 554 (416, 481).

Gush, H. P. 1981, *Physical Review Letters*, 47, 745 (417, 481).

Gush, H. P., Halpern, M. and Wishnow, E. H. 1990, *Physical Review Letters*, 65, 537 (17, 125, 198, 351, 359, 413, 418, 421, 482).

Guth, A. H. 1981, *Physical Review D*, 23, 347 (384, 520).

Guth, A. H. 1997, *The Inflationary Universe*, Reading, MA: Addison-Wesley (33, 359, 520).

Guth, A. H. and Pi, S.-Y. 1982, *Physical Review Letters*, 49, 1110 (441).

Halpern, M., Benford, R., Meyer, S., Muehlner, D. and Weiss, R. 1988, *Astrophysical Journal*, 332, 596 (433, 496).

Halverson, N. W., Leitch, E. M., Pryke, C., Kovac, J., Carlstrom, J. E., Holzapfel, W. L., Dragovan, M., Cartwright, J. K., Mason, B. S., Padin, S., Pearson, T. J., Readhead, A. C. S. and Shepherd, M. C. 2002, *Astrophysical Journal*, 568, 38 (508).

Hanany, S. *et al.* 2000, *Astrophysical Journal Letters*, 545, L5 (458, 507).

Hancock, S., Davies, R. D., Lasenby, A. N., Gutierrez de La Cruz, C. M., Watson, R. A., Rebolo, R. and Beckman, J. E. 1994, *Nature*, 367, 333 (496).

Happer, W. and Peebles, P. J. E. 2006, *Proceedings of the American Philosophical Society*, 150, 1 (516).

Harrison, D. L., Rubiño-Martin, J. A., Melhuish, S. J., Watson, R. A., Davies, R. D., Rebolo, R., Davis, R. J., Gutiérrez, C. M. and Macias-Perez, J. F. 2000, *Monthly Notices of the Royal Astronomical Society*, 316, L24 (505).

Harrison, E. R. 1970, *Physical Review D*, 1, 2726 (441, 527).

Harwit, M. 1960, PhD Thesis, *Measurement of Fluctuations in Radiation from a Source in Thermal Equilibrium*, MIT; *Physical Review*, 120, 1551 (332).

Harwit, M. 1961, *Monthly Notices of the Royal Astronomical Society*, 122, 47 (332).

Harwit, M. 1964, in *Les Spectres Infrarouges des Astres*, June 24–26, 1963, Université de Liège, *Mémoires de la Société Royale des Sciences de Liège*, cinquième série, tome IX, 506 (65, 329, 334, 338).

Harwit, M., Houck, J. R. and Fuhrmann, K. 1969, *Applied Optics*, 8, 473 (337).

Haselgrove, C. B. and Hoyle, F. 1959, *Monthly Notices of the Royal Astronomical Society*, 119, 112 (250).

Haslam, C. G. T., Klein, U., Salter, C. J., Stoffel, H., Wilson, W. E., Cleary, M. N., Cooke, D. J. and Thomasson, P. 1981, *Astronomy and Astrophysics*, 100, 209 (429).

Hauser, M. G., Arendt, R. G., Kelsall, T., Dwek, E., Odegard, N., Weiland, J. L., Freudenreich, H. T., Reach, W. T., Silverberg, R. F., Moseley, S. H., Pei, Y. C., Lubin, P., Mather, J. C., Shafer, R. A., Smoot, G. F., Weiss, R., Wilkinson, D. T. and Wright, E. L. 1998, *Astrophysical Journal*, 508, 25 (420).

Hauser, M. G. and Dwek, E. 2001, *Annual Review of Astronomy and Astrophysics*, 39, 249 (65, 188).

Hawking, S. W. 1982, *Physics Letters*, B, 115, 295 (441).

Hawking, S. W. and Ellis, G. F. R. 1965, *Physics Letters*, B17, 246 (380).

Hawking, S. W. and Ellis, G. F. R. 1968, *Astrophysical Journal*, 152, 25 (381, 383).

Hawking, S. W. and Ellis, G. F. R. 1973, *The Large Scale Structure of Space-Time*, London: Cambridge University Press (384).

Hawking, S. W. and Tayler, R. J. 1966, *Nature*, 209, 1278 (261).

Hayashi, C. 1950, *Progress of Theoretical Physics*, 5, 224 (32, 34, 104, 259).

Hayashi, C. and Nishida, M. 1956, *Progress of Theoretical Physics*, 16, 613 (35, 49).

Heeschen, D. S. and Dieter, N. H. 1958, *Proceedings of the Institute of Radio Engineers*, 46, 234 (144).

Hegyi, D. J., Traub, W. A. and Carleton, N. P. 1972, *Physical Review Letters*, 28, 1541 (479).

Hegyi, D. J., Traub, W. A. and Carleton, N. P. 1974, *Astrophysical Journal*, 190, 543 (416, 479).

Henry, P. S. 1971, PhD Thesis, *A Measurement of the Isotropy of the Cosmic Microwave Background at a Wavelength of 3 cm*, Princeton University; *Nature*, 231, 516, 1971 (225, 407, 427, 430, 489, 491).

Herbig, T., de Oliveira-Costa, A., Devlin, M. J., Miller, A. D., Page, L. A. and Tegmark, M. 1998, *Astrophysical Journal Letters*, 509, L73 (504).

Herzberg, G. 1945, *Molecular Spectra and Molecular Structure II. Infrared and Raman Spectra of Polyatomic Molecules*, New York: Van Nostrand (329).

Herzberg, G. 1950, *Molecular Spectra and Molecular Structure I. Spectra of diatomic Molecules*, second edition, New York: Van Nostrand, p. 496 (44, 75, 75, 79, 153, 293, 329, 519).

Hinshaw, G., Banday, A. J., Bennett, C. L., Gorski, K. M., Kogut, A., Smoot, G. F. and Wright, E. L. 1996, *Astrophysical Journal Letters*, 464, L17 (499).

Hinshaw, G. *et al.* 2007, *Astrophysical Journal Supplement*, 170, 288 (530).

Ho, S., Hirata, C. M., Padmanabhan, N., Seljak, U. and Bahcall, N. 2008, *Physical Review D*, 78, 043519 (475).

Hoffmann, W. F., Frederick, C. L. and Emery, R. J. 1971, *Astrophysical Journal Letters*, 164, L23 (74, 338).

Hogg, D. C. 1959, *Journal of Applied Physics*, 30, 1417 (70, 167, 291).

Hogg, D. C. 1968, in *Advances in Microwaves*, ed. L. Young, New York: Academic Press, 3, 1 (73).

Hogg, D. C. and Semplak, R. A. 1961, *Bell System Technical Journal*, 40, 1331 (188).

Hogg, D. C. and Wilson, R. W. 1965, *Bell System Technical Journal*, 44, 1019 (147, 163, 170).

Houck, J. R. and Harwit, M. 1969, *Astrophysical Journal Letters*, 157, L45 (354).

Houck, J. R., Soifer, B. T., Harwit, M. and Pipher, J. L. 1972, *Astrophysical Journal Letters*, 178, L29 (354, 480).

Houck, J. R., Soifer, B. T., Pipher, J. L. and Harwit, M. 1971, *Astrophysical Journal Letters*, 169, L31 (338).

Howell, T. F. and Shakeshaft, J. R. 1966, *Nature*, 210, 1318 (156, 173, 197, 285, 291, 301, 480).

Howell, T. F. and Shakeshaft, J. R. 1967a, *Journal of Atmospheric and Terrestrial Physics*, 29, 1559 (171, 291).

Howell, T. F. and Shakeshaft, J. R. 1967b, *Nature*, 216, 753 (292, 301, 480).

Hoyle, F. 1946, *Monthly Notices of the Royal Astronomical Society*, 106, 255 (249).

Hoyle, F. 1948, *Monthly Notices of the Royal Astronomical Society*, 108, 372 (15, 269).

Hoyle, F. 1953, *Astrophysical Journal*, 118, 513 (277).

Hoyle, F. 1959a, *Monthly Notices of the Royal Astronomical Society*, 119, 124 (250).

Hoyle, F. 1959b, in *Paris Symposium on Radio Astronomy*, ed. R. N. Bracewell, 9, 598 (292).

Hoyle, F. 1965, *Nature*, 208, 111 (57, 272).

Hoyle, F. 1981, *New Scientist*, 92, 521 (62).

Hoyle, F. 1994, *Home is Where the Wind Blows: Chapters from a Cosmologist's Life*, Mill Valley, CA: University Science Books (520).

Hoyle, F. and Burbidge, G. R. 1966, *Nature*, 210, 1346 (57).

Hoyle, F., Burbidge, G. R. and Narlikar, J. V. 1993, *Astrophysical Journal*, 410, 437 (195, 271, 274)

Hoyle, F., Burbidge, G. R. and Narlikar, J. V. 1994, *Monthly Notices of the Royal Astronomical Society*, 267, 1007 (274).

Hoyle, F., Burbidge, G. and Narlikar, J. V. 2000, *A Different Approach to Cosmology: From a Static Universe Through the Big Bang Towards Reality*, New York: Cambridge University Press (274).

Hoyle, F. and Lyttleton, R. A. 1942, *Monthly Notices of the Royal Astronomical Society*, 102, 218 (91).

Hoyle, F. and Narlikar, J. V. 1961, *Monthly Notices of the Royal Astronomical Society*, 123, 133 (269).

Hoyle, F. and Narlikar, J. V. 1962, *The Observatory*, 82, 13 (54).

Hoyle, F. and Narlikar, J. V. 1966, *Royal Society of London Proceedings*, Series A, 290, 162 (41).

Hoyle, F. and Tayler, R. J. 1964, *Nature*, 203, 1108 (37, 60, 187, 190, 192, 192, 243, 257, 259, 262, 272, 522).

Hu, W. and White, M. 1996, *Astrophysical Journal*, 471, 30 (446).

Hubble, E. 1929, *Proceedings of the National Academy of Sciences*, 15, 168 (11, 520).

Hubble, E, 1936, *The Realm of the Nebulae*, New Haven: Yale University Press (31, 54, 55).

Hubble, E, 1937, *The Observational Approach to Cosmology*, Oxford: The Clarendon Press (31).

Hubble, E. and Humason, M. L. 1931, *Astrophysical Journal*, 74, 43 (268).

Huchra, J., Davis, M., Latham, D. and Tonry, J. 1983, *Astrophysical Journal Supplement*, 52, 89 (457).

Hughes, D. J. 1946, *Physical Review*, 70, 106 (28).

Humason, M. L., Mayall, N. U. and Sandage, A. R. 1956, *Astronomical Journal*, 61, 97 (268).

Ichikawa, K., Fukugita, M. and Kawasaki, M. 2005, *Physical Review D*, 71, 043001 (475).

Irvine, W. M. 1961, PhD Thesis, *Local Irregularities in a Universe Satisfying the Cosmological Principle*, Harvard University (522).

Jaffe, A. H. *et al.* 2001, *Physical Review Letters*, 86, 3475 (459).

Jakes, W. C. 1963, *Bell System Technical Journal*, 42, 1421 (49, 72).

Johnson, D. G. and Wilkinson, D. T. 1987, *Astrophysical Journal Letters*, 313, L1 (415, 481).

Jordan, P. 1952, *Schwerkraft und Weltall*, Braunschweig: Vieweg (39).

Jordan, P. 1962, *Reviews of Modern Physics*, 34, 596 (39).

Jungman, G., Kamionkowski, M., Kosowsky, A. and Spergel, D. N. 1996, *Physical Review D*, 54, 1332 (446).

Kaidanovsky, N. L. and Parĭskiĭ, Yu. N. 1987, *Istoriko-Astronomicheskie Issledovaniya "Nauka"*, 59 (106).

Kaiser, M. E. and Wright, E. L. 1990, *Astrophysical Journal Letters*, 356, L1 (479).

Kaiser, N. and Squires, G. 1993, *Astrophysical Journal*, 404, 441 (473).

Kamionkowski, M., Spergel, D. N. and Sugiyama, N. 1994, *Astrophysical Journal Letters*, 426, L57 (445, 446, 455).

Kaufman, M. 1965, *Nature*, 207, 736 (279).

Kaufman, M. 1970, *Astrophysical Journal*, 160, 459 (278).

Kelsall, T. *et al.* 1998, *Astrophysical Journal*, 508, 44 (338).

Kerr, R. P. 1963, *Physical Review Letters*, 11, 237 (266).

Kibble, T. W. B. 1976, *Journal of Physics* A, 9, 1387 (447).

Kipper, A. Ya. 1950, *Astronomicheskii Zhurnal*, 27, 321 (120).

Kislyakov, A. G., Chernyshev, V. I., Lebskii, Y. V., Mal'Tsev, V. A. and Serov, N. V. 1971, *Soviet Astronomy*, 15, 29 (480).

Klein, O. 1958, in Eleventh Solvay Conference, *La Structure et l'Évolution de l'Univers*, Brussels: Stoops, p. 13 (40).

Klypin, A. A., Sazhin, M. V., Strukov, I. A. and Skulachev, D. P. 1987, *Soviet Astronomy Letters*, 13, 104 (132, 495).

Klypin, A. A., Strukov, I. A. and Skulachev, D. P. 1992, *Monthly Notices of the Royal Astronomical Society*, 258, 71 (132).

Knox, L. and Page, L. 2000, *Physical Review Letters*, 85, 1366 (455, 505).

Kogut, A., Banday, A. J., Bennett, C. L., Gorski, K. M., Hinshaw, G., Jackson, P. D., Keegstra, P., Lineweaver, C., Smoot, G. F., Tenorio, L. and Wright, E. L. 1996, *Astrophysical Journal*, 470, 653 (433).

Kogut, A., Bensadoun, M., de Amici, G., Levin, S., Smoot, G. F. and Witebsky, C. 1990, *Astrophysical Journal*, 355, 102 (482).

Kogut, A., Bersanelli, M., de Amici, G., Friedman, S. D., Griffith, M., Grossan, B., Levin, S., Smoot, G. F. and Witebsky, C. 1988, *Astrophysical Journal*, 325, 1 (481).

Kogut, A. *et al.* 2003, *Astrophysical Journal Supplement*, 148, 161 (466).

Kolb, E. W. and Turner, M. S. 1990, *The Early Universe*, Reading, MA: Addison-Wesley (384).

Kompaneets, A. 1956, *Zhurnal Eksperimentalnoi i Teoreticheskoi Fiziki*, 31, 867; English translation in *Soviet Physics JETP*, 4, 730 (111, 122, 136, 422, 528).

Koopmans, L. V. E., Treu, T., Fassnacht, C. D., Blandford, R. D. and Surpi, G. 2003, *Astrophysical Journal*, 599, 70 (471).

Kovac, J. M., Leitch, E. M., Pryke, C., Carlstrom, J. E., Halverson, N. W. and Holzapfel, W. L. 2002, *Nature*, 420, 772 (475).

Kowitt, M. S., Cheng, E. S., Cottingham, D. A., Fixsen, D. J., Inman, C. A., Meyer, S. S., Page, L. A., Puchalla, J. L., Ruhl, J. E. and Silverberg, R. F. 1997, *Astrophysical Journal*, 482, 17 (502).

Kragh, H. 1996, *Cosmology and Controversy*, Princeton, NJ: Princeton University Press (8, 25, 239).

Krauss, L. M. and Chaboyer, B. 2003, *Science*, 299, 65 (471).

Krauss, L. M. and Turner, M. S. 1995, *General Relativity and Gravitation*, 27, 1137 (457).

Kristian, J. and Sachs, R. K. 1966, *Astrophysical Journal*, 143, 379 (369).

Kuhn, J. R., Libbrecht, K. G. and Dicke, R. H. 1988, *Science*, 242, 908 (527).

Kuo, C. L., Ade, P. A. R., Bock, J. J., Cantalupo, C., Daub, M. D., Goldstein, J., Holzapfel, W. L., Lange, A. E., Lueker, M., Newcomb, M., Peterson, J. B., Ruhl, J., Runyan, M. C. and Torbet, E. 2004, *Astrophysical Journal*, 600, 32 (509).

Kurt, V. G. and Sunyaev, R. A. 1970, in IAU Symposium 36, *Ultraviolet Stellar Spectra and Related Ground-Based Observations*, ed. L. Houziaux and H. E. Butler, p. 341 (119).

Lambert, D. 2000, *Un Atome d'Univers: la Vie et l'Oeuvre de Georges Lemaître*, Brussels: Éditions Racines (522).

Landau, L. D. 1932, *Physikalische Zeitschrift Sowjetunion*, 1, 285 (99).

Landau, L. D. 1937, *Akademia Nauk Doklady*, 17, 6; English version in *Nature*, 141, 333, 1938 (99).

Landau, L. D. and Lifshitz, E. M. 1951, *The Classical Theory of Fields*, Reading, MA: Addison-Wesley (185).

Landau, L. D. and Lifshitz, E. M. 1960, *The Theory of Field*, third edition, Moscow (93).

Lange, A. E. *et al.* 2001, *Physical Review D*, 63, 042001 (455, 459).

Lasenby, A. N. and Davies, R. D. 1983, *Monthly Notices of the Royal Astronomical Society*, 203, 1137 (494).

Lastochkin, V. P. and Stankevich, K. S. 1964, *Astronomicheskii Zhurnal*, 41, 769; English translation in *Soviet Astronomy*, 8, 612, 1964 (297).

Lawson, J. L. and Uhlenbeck, G. E. 1950, *Threshold signals*, New York: McGraw-Hill, p. 107 (47).

Layzer, D. 1954, *Astronomical Journal*, 59, 170 (276).

Layzer, D. 1963, *Astrophysical Journal*, 138, 174 (371, 522).

Layzer, D. 1968, *Astrophysics Letters*, 1, 99 (195, 277).

Layzer, D. 1971, in *Astrophysics and General Relativity*, Vol. 2, ed. M. Chrétien, S. Deser and J. Goldstein, New York: Gordon and Breach, 155 (277).

Layzer, D. 1975, in *Galaxies and the Universe*, ed. A. Sandage, M. Sandage and J. Kristian, Chicago: University of Chicago Press, 665 (277).

Layzer, D. 1990, *Cosmogenesis – The Growth of Order in the Universe*, New York: Oxford University Press (277).

Layzer, D. and Hively, R. 1973, *Astrophysical Journal*, 179, 361 (35).

Ledden, J. E., Broderick, J. J., Brown, R. L. and Condon, J. J. 1980, *Astronomical Journal*, 85, 780 (493).

Lee, A. T. *et al.* 2001, *Astrophysical Journal Letters*, 561, L1 (507).

Le Floch, A. and Bretenaker, F. 1991, *Nature*, 352, 198 (65).

Leitch, E. M., Readhead, A. C. S., Pearson, T. J., Myers, S. T., Gulkis, S. and Lawrence, C. R. 2000, *Astrophysical Journal*, 532, 37 (497).

Lemaître, G. 1927, *Annales de la Société Scientifique de Bruxelles*, 47A, 49 (14, 517, 515).

Lemaître, G. 1931, *Nature*, 127, 706 (14, 517, 525).

Lemaître, G. 1933, *Annales de la Société Scientifique de Bruxelles*, 53A, 85 (41).

Levin, S. M., Bensadoun, M., Bersanelli, M., de Amici, G., Kogut, A., Limon, M. and Smoot, G. 1992, *Astrophysical Journal*, 396, 3 (482).

Levin, S. M., Witebsky, C., Bensadoun, M., Bersanelli, M., de Amici, G., Kogut, A. and Smoot, G. F. 1988, *Astrophysical Journal*, 334, 14 (482).

Lewis, A. and Bridle, S. 2002, *Physical Review D*, 66, 103511 (446).

Lifshitz, E. M. 1946, *Zhurnal Eksperimentalnoi i Teoreticheskoi Fiziki*, 16, 587; English translation in *Journal of Physics*, 10, 116, 1946 (130, 136, 276, 277, 367, 370, 375, 435).

Lightman, A. and Brawer, R. 1990, *Origins: The Lives and Worlds of Modern Cosmologists*, Cambridge: Harvard University Press (xi, 4).

Lim, M. A., Clapp, A. C., Devlin, M. J., Figueiredo, N., Gundersen, J. O., Hanany, S., Hristov, V. V., Lange, A. E., Lubin, P. M., Meinhold, P. R., Richards, P. L., Staren, J. W., Smoot, G. F. and Tanaka, S. T. 1996, *Astrophysical Journal Letters*, 469, L69 (500).

Linde, A. D. 1982, *Physics Letters*, B, 108, 389 (520).

Linde, A. D. 1995, *Physics Letters*, B, 351, 99 (443).

Lineweaver, C. H. 1997, in 16th Moriond Astrophysics Meeting, *Microwave Background Anisotropies*, 69 (425).

Lineweaver, C. H., Barbosa, D., Blanchard, A. and Bartlett, J. G. 1997, *Astronomy and Astrophysics*, 322, 365 (457).

Lineweaver, C. H., Hancock, S., Smoot, G. F., Lasenby, A. N., Davies, R. D., Banday, A. J., Gutierrez de La Cruz, C. M., Watson, R. A. and Rebolo, R. 1995, *Astrophysical Journal*, 448, 482 (496).

Longair, M. S. 1966, *Nature*, 211, 949 (57).

Longair, M. S. 2006, *The Cosmic Century: A History of Astrophysics and Cosmology*, Cambridge: Cambridge University Press (56, 239, 422).

Longair, M. S. and Sunyaev, R. A. 1969, *Nature*, 223, 719 (117).

Lorentz, H. A. 1904, *Proceedings of the Academy of Sciences of Amsterdam*, 4, 809; *The Principle of Relativity*, New York: Dover, p. 11, 1952 (390).

Lubimov, V. A., Novikov, E. G., Nozik, V. Z., Tretyakov, E. F. and Kosik, V. S. 1980, *Physics Letters*, B, 94, 266 (439).

Lubin, P. M., Epstein, G. L. and Smoot, G. F. 1983, *Physical Review Letters*, 50, 616 (494).

Lubin, P., Villela, T., Epstein, G. and Smoot, G. 1985, *Astrophysical Journal Letters*, 298, L1 (494).

Lyth, D. H. and Stewart, E. D. 1990, *Physics Letters*, B, 252, 336 (442).

Mach, E. 1883, *Science of Mechanics*, Sixth American Edition 1960, La Salle Illinois: The Open Court Publishing (38) .

Mandolesi, N., Calzolari, P., Cortiglioni, S., Delpino, F. and Sironi, G. 1986a, *Nature*, 319, 751 (495).

Mandolesi, N., Calzolari, P., Cortiglioni, S. and Morigi, G. 1984, *Physical Review D*, 29, 2680 (481).

Mandolesi, N., Calzolari, P., Cortiglioni, S., Morigi, G., Danese, L. and De Zotti, G. 1986b, *Astrophysical Journal*, 310, 561 (481).

Mason, B. S. *et al.* 2003, *Astrophysical Journal*, 591, 540 (507).

Mastenbrook, H. J. 1966, *Naval Research Laboratory Report* 6477 (345).

Mather, J. C. 1974, PhD Thesis, *Far Infrared Spectrometry of the Cosmic Background Radiation*, University of California, Berkeley (360).

Mather, J. C. and Boslough, J. 1996, *The Very First Light: The True Inside Story of the Scientific Journey Back to the Dawn of the Universe*, New York: Basic Books, 1996 (423, 419, 497).

Mather, J. C., Cheng, E. S., Cottingham, D. A., Eplee, R. E., Jr., Fixsen, D. J., Hewagama, T., Isaacman, R. B., Jensen, K. A., Meyer, S. S., Noerdlinger, P. D., Read, S. M., Rosen, L. P., Shafer, R. A., Wright, E. L., Bennett, C. L., Boggess, N. W., Hauser, M. G., Kelsall, T., Moseley, S. H., Jr., Silverberg, R. F., Smoot, G. F., Weiss, R. and Wilkinson, D. T. 1994, *Astrophysical Journal*, (482).

Mather, J. C., Cheng, E. S., Eplee, R. E., Jr., Isaacman, R. B., Meyer, S. S., Shafer, R. A., Weiss, R., Wright, E. L., Bennett, C. L., Boggess, N. W., Dwek, E., Gulkis, S., Hauser, M. G., Janssen, M., Kelsall, T., Lubin, P. M., Moseley, S. H., Jr., Murdock, T. L., Silverberg, R. F., Smoot, G. F. and Wilkinson, D. T. 1990, *Astrophysical Journal Letters*, 354, L37 (17, 125, 198, 295, 359, 413, 418, 421, 482, 514).

Mather, J. C., Fixsen, D. J., Shafer, R. A., Mosier, C. and Wilkinson, D. T. 1999, *Astrophysical Journal*, 512, 511(421, 482).

Mathis, J. S. 1959, *Astrophysical Journal*, 129, 259 (91).

Matsumoto, T., Hayakawa, S., Matsuo, H., Murakami, H., Sato, S., Lange, A. E. and Richards, P. L. 1988, *Astrophysical Journal*, 329, 567 (416, 422, 482).

Mauskopf, P. D. *et al.* 2000, *Astrophysical Journal Letters*, 536, L59 (506).

McKellar, A. 1940, *Publications of the Astronomical Society of the Pacific*, 52, 407 (79, 104).

McKellar, A. 1941, *Publications of the Dominion Astrophysical Observatory*, 7, 251 (44, 74, 104, 270).

McVittie, G. C. 1962, in IAU Symposium 15, *Problems of Extra-Galactic Research*, New York: Macmillan (61).

McVittie, G. C. and Wyatt, S. P. 1959, *Astrophysical Journal*, 130, 1 (65).

Medd, W. J. and Covington, A. E. 1958, *Proceedings of the IRE*, 46, 112 (64).

Meinhold, P., Clapp, A., Devlin, M., Fischer, M., Gundersen, J., Holmes, W., Lange, A., Lubin, P., Richards, P. and Smoot, G. 1993, *Astrophysical Journal Letters*, 409, L1 (500).

Meinhold, P. and Lubin, P. 1991, *Astrophysical Journal Letters*, 370, L11 (499).

Melchiorri, F., Ceccarelli, C., Pietranera, L. and Melchiorri, B. O. 1981, *Astrophysical Journal Letters*, 250, L1 (493).

Meyer, D. M. and Jura, M. 1985, *Astrophysical Journal*, 297, 119 (479).

Meyer, D. M., Roth, K. C. and Hawkins, I. 1989, *Astrophysical Journal Letters*, 343, L1 (416, 479).

Meyer, S. S., Cheng, E. S. and Page, L. A. 1991, *Astrophysical Journal Letters*, 371, L7 (453, 497).

Michie, R. W. 1967, Kitt Peak National Observatory preprint 440 (193).

Millea, M. F., McColl, M., Pedersen, R. J. and Vernon, F. L., Jr. 1971, *Physical Review Letters*, 26, 919 (312, 480).

Miller, A. D., Beach, J., Bradley, S., Caldwell, R., Chapman, H., Devlin, M. J., Dorwart, W. B., Herbig, T., Jones, D., Monnelly, G., Netterfield, C. B., Nolta, M., Page, L. A., Puchalla, J., Robertson, T., Torbet, E., Tran, H. T. and Vinje, W. E. 2002, *Astrophysical Journal Supplement*, 140, 115 (505).

Miller, A. D., Caldwell, R., Devlin, M. J., Dorwart, W. B., Herbig, T., Nolta, M. R., Page, L. A., Puchalla, J., Torbet, E. and Tran, H. T. 1999, *Astrophysical Journal Letters*, 524, L1 (505).

Milne, E. A. 1935, *Relativity, Gravitation and World-Structure*, Oxford: The Clarendon Press (365).

Mitchell, J. L., Keeton, C. R., Frieman, J. A. and Sheth, R. K. 2005, *Astrophysical Journal*, 622, 81 (472).

Mitton, S. 2005, *Fred Hoyle: A Life in Science*, London: Aurum Press (14, 246, 520).

Montgomery, C. G., Dicke, R. H. and Purcell, E. M. 1948, *Principles of Microwave Circuits*, New York: McGraw-Hill (410).

Morgan, W. W., Keenan, P. C. and Kellman, E. 1943, *An Atlas of Stellar Spectra, With an Outline of Spectral Classification*, Chicago: The University of Chicago Press (91).

Morrison, P. 1995, *Masters of Modern Physics* 11, American Institute of Physics, New York: Springer-Verlag. (358).

Mosengeil, K. von, 1907, *Annalen der Physik*, 22, 867 (390, 391).

Muehlner, D. 1977, *Infrared and Submillimeter Astronomy*, 63, 143 (432, 491, 492).

Muehlner, D. and Weiss, R. 1970, *Physical Review Letters*, 24, 742 (354, 413, 491).

Muehlner, D. and Weiss, R. 1973a, *Physical Review D*, 7, 326 (358, 403, 480).

Muehlner, D. and Weiss, R. 1973b, *Physical Review Letters*, 30, 757 (359, 414, 480).

Mukhanov, V. F. and Chibisov, G. V. 1981, *Pis'ma ZhETF*, 33, 549; English translation in *Soviet Journal of Experimental and Theoretical Physics Letters*, 33, 532 (441).

Myers, S. T., Readhead, A. C. S. and Lawrence, C. R. 1993, *Astrophysical Journal*, 405, 8 (497).

Nanos, G. P., Jr. 1974, PhD Thesis, *Polarization of the Blackbody-Radiation at 3.2 cm*, Princeton University (267).

Nanos, G. P., Jr. 1979, *Astrophysical Journal*, 232, 341 (267).

Narlikar, J. V., Edmunds, M. G. and Wickramasinghe, N. C. 1976, in *Far Infrared Astronomy*, ed. M. Rowan-Robinson, New York: Pergamon, 131 (273).

Narlikar, J. V., Vishwakarma, R. G., Hajian, A., Souradeep, T., Burbidge, G. R. and Hoyle, F. 2003, *Astrophysical Journal*, 585, 1 (274).

Netterfield, C. B., Devlin, M. J., Jarolik, N., Page, L. and Wollack, E. J. 1997, *Astrophysical Journal*, 474, 47 (501).

Netterfield, C. B., Jarosik, N., Page, L., Wilkinson, D. and Wollack, E. 1995, *Astrophysical Journal Letters*, 445, L69 (501).

Netterfield, C. B. *et al.* 2002, *Astrophysical Journal*, 571, 604 (506).

Nolta, M. R., Devlin, M. J., Dorwart, W. B., Miller, A. D., Page, L. A., Puchalla, J., Torbet, E. and Tran, H. T. 2003, *Astrophysical Journal*, 598, 97 (505).

Nolta, M. R. *et al.* 2008, ArXiv e-prints, 803, arXiv:0803.0593 (466, 469, 475).

North, J. D. 1965, *The Measure of the Universe: A History of Modern Cosmology*, Oxford: Clarendon Press, 1965 (59).

Novikov. I. D. 1964, *ZhETF* 46, 686; English translation in *Soviet Physics JETP* 19, 467, 1964 (435).

Novikov, I. D. 1990, *Black Holes and the Universe*, Cambridge: Cambridge University Press (100).

Novikov, I. D. 2001, in ASP Conference Series 252, *Historical Development of Modern Cosmology*, ed. V. J. Martínez, V. Trimble and M. J. Pons-Bordería, 43 (100).

O'Dell, C. R. 1963, *Astrophysical Journal*, 138, 1018 (255, 257).

O'Dell, C. R., Peimbert, M. and Kinman, T. D. 1964, *Astrophysical Journal*, 140, 119 (60, 242, 255, 257).

Ohm, E. A. 1961, *Bell System Technical Journal*, 40, 1065 (36, 49–51, 65, 71, 103, 153, 169, 180, 188, 293).

Oort, J. H. 1958, in *La Structure et l'Évolution de l'Univers*, Eleventh Solvay Conference, 1, Brussels: Stoops (40).

Oppenheimer, J. R and Snyder, H. 1939, *Physical Review*, 56, 455 (99).

Oppenheimer, J. R. and Volkoff, G. M. 1939, *Physical Review*, 55, 374 (99).

Osterbrock, D. E. 1989, *Astrophysics of Gaseous Nebulae and Active Galactic Nuclei*, Mill Valley, CA: University Science Books (86).

Osterbrock, D. E. and Ferland, G. J. 2006, *Astrophysics of Gaseous Nebulae and Active Galactic Nuclei*, second edition, Sausalito, CA: University Science Books (86).

Osterbrock, D. E. and Rogerson, J. B. 1961, *Publications of the Astronomical Society of the Pacific*, 73, 129 (59–60, 66, 190, 197, 242, 252).

Ostriker, J. P., Barenbatt, G. I. and Sunyaev, R. A. 1992, *Selected Works of Yakov Borisovich Zel'dovich*, Princeton: Princeton University Press (530).

Ostriker, J. P. and Steinhardt, P. J. 1995, *Nature*, 377, 600 (457).

Ostriker, J. P., Thompson, C. and Witten, E. 1986, *Physics Letters*, B, 180, 231 (447).

O'Sullivan, C., Yassin, G., Woan, G., Scott, P. F., Saunders, R., Robson, M., Pooley, G., Lasenby, A. N., Kenderdine, S., Jones, M., Hobson, M. P. and Duffett-Smith, P. J. 1995, *Monthly Notices of the Royal Astronomical Society*, 274, 861 (503).

Otoshi, T. Y. and Stelzried, C. T. 1975, *IEEE Transactions on Instrumentation and Measurement*, 24, 174 (481).

Overbye, D. 2000, *Einstein in Love*, New York: Viking (517).

Owens, D. K., Muehlner, D. J. and Weiss, R. 1979, *Astrophysical Journal*, 231, 702 (491).

Ozernoi, L. M. and Chernin, A. D. 1968, *Soviet Astronomy*, 11, 907 (125).

Padin, S., Cartwright, J. K., Mason, B. S., Pearson, T. J., Readhead, A. C. S., Shepherd, M. C., Sievers, J., Udomprasert, P. S., Holzapfel, W. L., Myers, S. T., Carlstrom, J. E., Leitch, E. M., Joy, M., Bronfman, L. and May, J. 2001, *Astrophysical Journal Letters*, 549, L1 (507).

Page, L. A., Barnes, C., Hinshaw, G., Spergel, D. N., Weiland, J. L., Wollack, E., Bennett, C. L., Halpern, M., Jarosik, N., Kogut, A., Limon, M., Meyer, S. S., Tucker, G. S. and Wright, E. L. 2003a, *Astrophysical Journal Supplement*, 148, 39 (295).

Page, L. A., Cheng, E. S. and Meyer, S. S. 1990, *Astrophysical Journal Letters*, 355, L1 (497).

Page, L. A., Nolta, M. R., Barnes, C., Bennett, C. L., Halpern, M., Hinshaw, G., Jarosik, N., Kogut, A., Limon, M., Meyer, S. S., Peiris, H. V., Spergel, D. N., Tucker, G. S., Wollack, E. and Wright, E. L. 2003b, *Astrophysical Journal Supplement*, 148, 233 (466, 505).

Page, L. A. *et al.* 2007, *Astrophysical Journal Supplement*, 170, 335 (468).

Pagels, H. R. 1984, *Annals of the New York Academy Sciences*, 422, 15 (439).

Pais, A. 1982, *Subtle is the Lord*, Oxford: Clarendon Press (517).

Palazzi, E., Mandolesi, N., Crane, P., Kutner, M. L., Blades, J. C. and Hegyi, D. J. 1990, *Astrophysical Journal*, 357, 14 (479).

Pariĭskiĭ, Y. N. 1968, *Soviet Astronomy*, 12, 219 (195).

Pariĭskiĭ, Y. N. 1972, *Astronomicheskii Zhurnal*, 49, 1322; English translation in *Soviet Astronomy*, 16, 1048, 1973 (127).

Pariĭskiĭ, Y. N. 1973a, *Soviet Astronomy*, 17, 291 (489).

Pariĭskiĭ, Y. N. 1973b, *Astrophysical Journal Letters*, 180, L47; *Astrophysical Journal Letters*, 188, L113, 1974 (490).

Pariĭskiĭ, Y. N. and Korol'kov, D. 1986, *Soviet Science Review E: Astrophysics and Space Physics*, 5, 39 (486, 489).

Pariiskii, I. N., Petrov, Z. E. and Cherkov, L. N. 1977, *Soviet Astronomy Letters*, 3, 263 (489).

Pariiskii, Y. N. and Pyatunina, T. B. 1971, *Soviet Astronomy*, 14, 1067 (489).

Partridge, R. B. 1969, *American Scientist*, 57, 37 (229, 232, 309)..

Partridge, R. B. 1973, *Nature*, 244, 263 (238).

Partridge, R. B. 1980, *Astrophysical Journal*, 235, 681 (492).

Partridge, R. B. 1995, *3 K: The Cosmic Microwave Background Radiation*, Cambridge: Cambridge University Press (236, 308, 480, 486).

Partridge, R. B. 2004, in *The Cosmological Model*, XXXVII Rencontres de Moriond, ed. Y. Giraud-Héraud, C. Magneville and T. Than Van, Vietnam: The Gioi Publishers (233).

Partridge, R. B. and Wilkinson, D. T. 1967, *Physical Review Letters*, 18, 557 (232, 377, 427, 487, 519).

Patterson, C. C. 1955, *Geochimica et Cosmochimica Acta*, 7, 151 (54).

Pauling, L. 1948, *The Nature of the Chemical Bond*, second edition, Ithaca, NY: Cornell University Press (329).

Pauliny-Toth, I. I. K. and Shakeshaft, J. R. 1962, *Monthly Notices of the Royal Astronomical Society*, 124, 61 (285, 289, 390, 429).

Pearson, T. J. *et al.* 2003, *Astrophysical Journal*, 591, 556 (507).

Peebles, P. J. E. 1962, PhD Thesis, *Observational Tests and Theoretical Problems Relating to the Conjecture that the Strength of the Electromagnetic Interaction may be Variable*, Princeton University (185).

Peebles, P. J. E. 1964, *Astrophysical Journal*, 140, 328 (60, 190).

Peebles, P. J. E. 1965, *Astrophysical Journal*, 142, 1317 (193, 435).

Peebles, P. J. E. 1966, *Physical Review Letters*, 16, 410; *Astrophysical Journal*, 146, 542 (192, 260).

Peebles, P. J. E. 1967, *Astrophysical Journal*, 147, 859 (435).

Peebles, P. J. E. 1968, *Astrophysical Journal*, 153, 1 (121, 194, 436).

Peebles, P. J. E. 1971, *Physical Cosmology*, Princeton, NJ: Princeton University Press (40, 95, 186, 192, 199, 306, 360, 412, 424).

Peebles, P. J. E. 1973, *Astrophysical Journal*, 185, 413 (443).

Peebles, P. J. E. 1981a, *Astrophysical Journal Letters*, 243, L119 (438).

Peebles, P. J. E. 1981b, *Astrophysical Journal*, 248, 885 (472).

Peebles, P. J. E. 1982, *Astrophysical Journal Letters*, 263, L1 (200, 371, 377, 438, 443, 444).

Peebles, P. J. E. 1984, *Astrophysical Journal*, 284, 439 (200, 457).

Peebles, P. J. E. 1986, *Nature*, 321, 27 (200, 457).

Peebles, P. J. E. 1999, *Astrophysical Journal*, 510, 531 (200, 447).

Peebles, P. J. E. and Ratra, B. 1988, *Astrophysical Journal Letters*, 325, L17 (443).

Peebles, P. J. E. and Wilkinson, D. T. 1968, *Physical Review*, 174, 2168 (424).

Peebles, P. J. E. and Yu, J. T. 1970, *Astrophysical Journal*, 162, 815 (67, 194, 377, 441, 445, 527).

Peimbert, M. 1968, *Boletín de los Observatorios Tonantzintla y Tacubaya*, 4, 233 (479).

Pelyushenko, S. A. and Stankevich, K. S. 1969, *Astronomicheskii Zhurnal*, 46, 283; English translation in *Soviet Astronomy*, 13, 223, 1969 (301, 480).

Pen, U.-L., Seljak, U. and Turok, N. 1997, *Physical Review Letters*, 79, 1611 (456).

Penrose, R. 1965, *Physical Review Letters*, 14, 57 (380).

Penzias, A. A. 1964, *Astronomical Journal*, 69, 146 (153).

Penzias, A. A. 1965, *Review of Scientific Instruments*, 36, 68 (148, 164, 293).

Penzias, A. A. 1968, *IEEE Transactions on Microwave Theory Techniques*, 16, 608 (480).

Penzias, A. A. 1979a, *Reviews of Modern Physics*, 51, 425 (103, 155).

Penzias, A. A. 1979b, *Astrophysical Journal*, 228, 430 (157).

Penzias, A. A., Schraml, J. and Wilson, R. W. 1969, *Astrophysical Journal Letters*, 157, L49 (233).

Penzias, A. A. and Wilson, R. W. 1965a, *Astrophysical Journal*, 142, 419 (42, 73, 74, 98, 106, 108, 112, 152, 172, 192, 205, 215, 223, 230, 243, 258, 271, 272, 275, 279, 281, 285, 287, 288, 293, 294, 298, 306, 325, 337, 341, 342, 369, 385, 480, 517).

Penzias, A. A. and Wilson, R. W. 1965b, *Astrophysical Journal*, 142, 1149 (165, 171).

Penzias, A. A. and Wilson, R. W. 1966, *Astrophysical Journal*, 146, 666 (152).

Penzias, A. A. and Wilson, R. W. 1967, *Astronomical Journal*, 72, 315 (156, 173, 197, 301, 480).

Percival, W. J. *et al.* 2007, *Astrophysical Journal*, 657, 51 (473).

Perlmutter, S. *et al.* 1997, *Astrophysical Journal*, 483, 565 (457).

Perlmutter, S. *et al.* 1999, *Astrophysical Journal*, 517, 565 (458, 472).

Peterson, J. B., Griffin, G. S., Newcomb, M. G., Alvarez, D. L., Cantalupo, C. M., Morgan, D., Miller, K. W., Ganga, K., Pernic, D. and Thoma, M. 2000, *Astrophysical Journal Letters*, 532, L83 (506).

Peterson, J. B., Richards, P. L. and Timusk, T. 1985, *Physical Review Letters*, 55, 332 (416, 481).

Piccirillo, L. and Calisse, P. 1993, *Astrophysical Journal*, 411, 529 (500).

Piccirillo, L., Femenia, B., Kachwala, N., Rebolo, R., Limon, M., Gutierrez, C. M., Nicholas, J., Schaefer, R. K. and Watson, R. A. 1997, *Astrophysical Journal Letters*, 475, L77 (504).

Pierce, J. R. 1955, *Jet Propulsion*, 25 (April 1955), 153 (159).

Pierce, J. R. 1979, *Memorial Tributes: National Academy of Engineering*, 1, 159 (521).

Pierce, J. R. and Kompfner, R. 1959, *Proceedings of the Institute of Radio Engineers*, March, 372 (70).

Pipher, J. L. 1971, PhD Thesis, *Rocket Submillimeter Observations of the Galaxy and Background*, Cornell University, pp. 137–143 (338).

Pipher, J. L., Houck, J. R., Jones, B. W. and Harwit, M. 1971, *Nature*, 231, 375 (337, 413).

Platt, S. R., Kovac, J., Dragovan, M., Peterson, J. B. and Ruhl, J. E. 1997, *Astrophysical Journal Letters*, 475, L1 (502).

Podariu, S., Souradeep, T., Gott, J. R., Ratra, B. and Vogeley, M. S. 2001, *Astrophysical Journal*, 559, 9 (486).

Pospieszalski, M. W. 1992, *IEEE MTT-S Digest*, 1369 (455, 519).

Pospieszalski, M. W. *et al.* 1994, *IEEE MTT-S Digest*, 1345 (455, 519).

Pozdnyakov, L. A., Sobol, I. M. and Sunyaev, R. A. 1983, *Soviet Scientific Reviews, E: Astrophysics and Space Physics Reviews* 2, 189 (123).

Puget, J.-L., Abergel, A., Bernard, J.-P., Boulanger, F., Burton, W. B., Desert, F.-X. and Hartmann, D. 1996, *Astronomy and Astrophysics*, 308, L5 (338, 420).

Puzanov, V. I., Salomonovich, A. E. and Stankevich, K. S. 1967, *Astronomicheskii Zhurnal*, 44, 1129; English translation in *Soviet Astronomy*, 11, 905, 1968 (300, 480).

Raghunathan, A. and Subrahmanyan, R. 2000, *Journal of Astrophysics and Astronomy*, 20, 1 (482).

Ratra, B., Ganga, K., Sugiyama, N., Tucker, G. S., Griffin, G. S., Nguyên, H. T. and Peterson, J. B. 1998, *Astrophysical Journal*, 505, 8 (501).

Ratra, B. and Peebles, P. J. E. 1994, *Astrophysical Journal Letters*, 432, L5 (443).

Ratra, B. and Vogeley, M. S. 2008 *Publications of the Astronomical Society of the Pacific*, 120, 235 (466).

Readhead, A. C. S. and Lawrence, C. R. 1992, *Annual Review of Astronomy and Astrophysics*, 30, 653 (496).

Readhead, A. C. S., Lawrence, C. R., Myers, S. T., Sargent, W. L. W., Hardebeck, H. E. and Moffet, A. T. 1989, *Astrophysical Journal*, 346, 566 (497).

Reber, G. 1958, *Proceedings of the Institute of Radio Engineers*, 46, 15 (177).

Rees, M. J. 1968, *Astrophysical Journal Letters*, 153, L1 (267).

Rees, M. J. 2001, *Proceedings of the American Philosophical Society*, 145, 365 (527).

Rees, M. J. and Sciama, D. W. 1968, *Nature*, 217, 511 (267).

Reichardt, C. L. *et al.* 2008, ArXiv e-prints, 801, arXiv:0801.1491 (466).

Reiprich, T. H. and Böhringer, H. 2002, *Astrophysical Journal*, 567, 716 (474).

Riess, A. G., *et al.* 1998, *Astronomical Journal*, 116, 1009 (457, 472).

Riess, A. G. *et al.* 2005, *Astrophysical Journal*, 627, 579 (471).

Roach, F. E. 1964, *Space Science Reviews*, 3, 512 (65).

Robertson, H. P. and Noonan, T. W. 1968, *Relativity and Cosmology*, Philadelphia: Saunders (5).

Robson, E. I. and Clegg, P. E. 1977, *Radio Astronomy and Cosmology*, 74, 319 (481).

Robson, E. I., Vickers, D. G., Huizinga, J. S., Beckman, J. E. and Clegg, P. E. 1974, *Nature*, 251, 591 (414, 481).

Rogerson, J. B. and York, D. G. 1973, *Astrophysical Journal Letters*, 186, L95 (261).

Roll, P. G., Krotkov, R. and Dicke, R. H. 1964, *Annals of Physics*, 26, 442 (213).

Roll, P. G. and Wilkinson, D. T. 1966, *Physical Review Letters*, 16, 405 (154, 173, 196, 223, 285, 294, 299, 311, 325, 342, 480, 519).

Roll, P. G. and Wilkinson, D. T. 1967, *Annals of Physics*, 44, 289 (218, 311, 321, 325).

Roman, N. G. 1950, *Astrophysical Journal*, 112, 554 (91).

Romeo, G., Ali, S., Femenía, B., Limon, M., Piccirillo, L., Rebolo, R. and Schaefer, R. 2001, *Astrophysical Journal Letters*, 548, L1 (460, 504).

Roth, K. C. and Meyer, D. M. 1995, *Astrophysical Journal*, 441, 129 (421, 479).

Roth, K. C., Meyer, D. M. and Hawkins, I. 1993, *Astrophysical Journal Letters*, 413, L67 (479).

Rowan-Robinson, M. 1968, *Monthly Notices of the Royal Astronomical Society*, 138, 445 (264).

Rudnick, L. 1978, *Astrophysical Journal*, 223, 37 (491).

Ruhl, J. E., Dragovan, M., Platt, S. R., Kovac, J. and Novak, G. 1995, *Astrophysical Journal Letters*, 453, L1 (502).

Ruhl, J. E. *et al.* 2003, *Astrophysical Journal*, 599, 786 (506).

Ryle, M. 1955, *The Observatory*, 75, 137 (55).

Sachs, R. K. and Wolfe, A. M. 1967, *Astrophysical Journal*, 147, 73 (67, 194, 364, 370, 373, 375, 396, 443, 444, 475, 527).

Sandage, A. 1958, *Astrophysical Journal*, 127, 513 (54).

Sandage, A. 1961, *Astrophysical Journal*, 133, 355 (54, 381, 471).

Sandage, A. and Tammann, G. A. 1964–65, *Annual Report of the Director, Mount Wilson and Palomar Observatories*, Pasadena, CA: The Observatories, p. 35 (65).

Sandage, A., Tammann, G. A., Saha, A., Reindl, B., Macchetto, F. D. and Panagia, N. 2006, *Astrophysical Journal*, 653, 843 (471).

Sakharov, A. D. 1965, *Zhurnal Eksperimental'noi i Teoreticheskoi Fiziki*, 49, 345; English translation in *Soviet Physics JETP*, 22, 241, 1966 (131).

Sakharov, A. D. 1967, *Soviet Journal of Experimental and Theoretical Physics Letters*, 5, 24 (132).

Sakharov, A. 1968, *Symmetry of the Universe* (in Russian), in *The Future of Science*, Second Annual Issue, Moscow: Znanie Press (95).

Sakharov, A. 1990, *Memoirs*, New York: Chekhov Press; English translation by R. Lourie, London: Hutchinson (96).

Sazonov, S. Y. and Sunyaev, R. A. 2000, *Astrophysical Journal*, 543, 28 (128).

Scheuer, P. A. G. 1957, *Proceedings of the Cambridge Philosophical Society*, 53, 764 (290).

Scheuer, P. A. G. 1975, in *Galaxies and the Universe*, ed. A. Sandage, M. Sandage and J. Kristian, Chicago: University of Chicago Press, p. 725 (291).

Schmidt, M. 1959, *Astrophysical Journal*, 129, 243 (91).

Schmidt, M. 1968, *Astrophysical Journal*, 151, 393 (264).

Schuster, J., Gaier, T., Gundersen, J., Meinhold, P., Koch, T., Seiffert, M., Wuensche, C. A. and Lubin, P. 1993, *Astrophysical Journal Letters*, 412, L47 (499).

Schwarzschild, M. 1946, *Astrophysical Journal*, 104, 203 (249).

Sciama, D. W. 1959, *The Unity of the Universe*, Garden City NY: Doubleday (38, 258, 262).

Sciama, D. W. 1966, *Nature*, 211, 277 (195).

Sciama, D. W. 1971, *Modern Cosmology*, Cambridge: Cambridge University Press (95).

Sciama, D. W. 2001, *Astrophysics and Space Science*, 276, 151 (197).

Sciama, D. W. and Rees, M. J. 1966, *Nature*, 211, 1283 (57, 264).

Scott, P. F., Saunders, R., Pooley, G., O'Sullivan, C., Lasenby, A. N., Jones, M., Hobson, M. P., Duffett-Smith, P. J. and Baker, J. 1996, *Astrophysical Journal Letters*, 461, L1 (292, 503).

Scott, P. F. *et al.* 2003, *Monthly Notices of the Royal Astronomical Society*, 341, 1076 (508).

Seielstad, G. A., Masson, C. R. and Berge, G. L. 1981, *Astrophysical Journal*, 244, 717 (493).

Seljak, U., Slosar, A. and McDonald, P. 2006, *Journal of Cosmology and Astro-Particle Physics*, 10, 14 (474).

Seljak, U. and Zaldarriaga, M. 1996, *Astrophysical Journal*, 469, 437 (446).

Shakeshaft, J. R. 1954, *Philosophical Magazine*, 45, 1136 (65).

Sheldon, E. S. *et al.* 2007, ArXiv e-prints, 709, arXiv:0709.1162 (473).

Shepley, L. C. 1965, *Proceedings of the National Academy of Sciences*, 52, 1403 (380).

Shivanandan, K., Houck, J. R. and Harwit, M. O. 1968, *Physical Review Letters*, 21, 1460 (84, 232, 329, 337, 338, 341, 354).

Shklovsky, I. S. 1960, *Astronomicheskii Zhurnal*, 37, 256; English translation in *Soviet Astronomy*, 4, 243, 1960 (297).

Shklovsky, I. S., 1966, *Astronomical Circular*, 364, Soviet Academy of Science (105, 152, 173).

Shmaonov, T., 1957, *Pribori i Tekhnika Experimenta* (in Russia), 1, 83 (64, 105).

Shulman, S., Bortolot, V. J. and Thaddeus, P. 1974, *Astrophysical Journal*, 193, 97 (85).

Shvartsman, V. F. 1969, *Soviet Journal of Experimental and Theoretical Physics Letters*, 9, 184 (132).

Silk, J. 1967, *Nature*, 215, 1155 (67, 194, 375, 437).

Silk, J. 1968a, PhD Thesis, *The Formation of Galaxies*, Harvard University (374).

Silk, J. 1968b, *Astrophysical Journal*, 151, 459 (67, 128, 194, 376).

Silk, J. 1968c, *Nature*, 218, 453 (128).

Silk, J. 2006, *Infinite Cosmos*, Oxford: Oxford University Press (371).

Silk, J. and Wilson, M. L. 1981, *Astrophysical Journal Letters*, 244, L37 (377).

Simonetti, J. H., Dennison, B. and Topasna, G. A. 1996, *Astrophysical Journal Letters*, 458, L1 (484).

Singal, J., Fixsen, D. J., Kogut, A., Levin, S., Limon, M., Lubin, P., Mirel, P., Seiffert, M. and Wollack, E. J. 2006, *Astrophysical Journal*, 653, 835 (482).

Sironi, G. and Bonelli, G. 1986, *Astrophysical Journal*, 311, 418 (481).

Sironi, G., Bonelli, G. and Limon, M. 1991, *Astrophysical Journal*, 378, 550 (482).

Sironi, G., Inzani, P. and Ferrari, A. 1984, *Physical Review D*, 29, 2686 (481).

Sironi, G., Limon, M., Marcellino, G., Bonelli, G., Bersanelli, M., Conti, G. and Reif, K. 1990, *Astrophysical Journal*, 357, 301 (482).

Smirnov, Yu. N. 1964, *Astronomicheskii Zhurnal*, 41, 1084; English translation in *Soviet Astronomy AJ*, 8, 864, 1965 (36, 60, 94).

Smith, M. G. and Partridge, R. B. 1970, *Astrophysical Journal*, 159, 737 (232).

Smith, S. 1936, *Astrophysical Journal*, 83, 23 (31).

Smoot, G. F., Bensadoun, M., Bersanelli, M., de Amici, G., Kogut, A., Levin, S. and Witebsky, C. 1987, *Astrophysical Journal Letters*, 317, L45 (415, 481).

Smoot, G. F., de Amici, G., Friedman, S. D., Witebsky, C., Mandolesi, N., Partridge, R. B., Sironi, G., Danese, L. and de Zotti, G. 1983, *Physical Review Letters*, 51, 1099 (481).

Smoot, G. F., de Amici, G., Friedman, S. D., Witebsky, C., Sironi, G., Bonelli, G., Mandolesi, N., Cortiglioni, S., Morigi, G., Partridge, R. B., Danese, L. and De Zotti, G. 1985, *Astrophysical Journal Letters*, 291, L23 (230, 481).

Smoot, G. F., Gorenstein, M. V. and Muller, R. A. 1977, *Physical Review Letters*, 39, 898 (225, 391, 433, 452, 490).

Smoot, G. F. and Lubin, P. M. 1979, *Astrophysical Journal Letters*, 234, L83 (433, 490).

Smoot, G. F. and Scott, D. 1998, *The European Physical Journal* C3, 1 (486).

Smoot, G. F. *et al.* 1991, *Astrophysical Journal Letters*, 371, L1 (499).

Smoot, G. F. *et al.* 1992, *Astrophysical Journal Letters*, 396, L1 (132, 371, 376, 452, 499, 527).

Soifer, B. T., Houck, J. R. and Harwit, M. 1971, *Astrophysical Journal Letters*, 168, L73 (337).

Soifer, B. T., Pipher, J. L. and Houck, J. R. 1972, *Astrophysical Journal*, 177, 315 (338).

Spergel, D. N. and Zaldarriaga, M. 1997, *Physical Review Letters*, 79, 2180 (466).

Spergel, D. N. *et al.* 2003, *Astrophysical Journal Supplement*, 148, 175 (466).

Spergel, D. N. *et al.* 2007, *Astrophysical Journal Supplement*, 170, 377 (260, 378, 446, 466).

Staggs, S. T., Jarosik, N. C., Meyer, S. S. and Wilkinson, D. T. 1996b, *Astrophysical Journal Letters*, 473, L1 (421, 482).

Staggs, S. T., Jarosik, N. C., Wilkinson, D. T. and Wollack, E. J. 1996a, *Astrophysical Journal*, 458, 407 (482).

Stankevich, K. S. 1974, *Soviet Astronomy*, 18, 126 (490).

Stankevich, K. S., Lastochkin, V. P. and Torkhov, V. A. 1967, *Radiofizika*, 10, 1758; English translation in *Radiophysics and Quantum Electronics*, 10, 984, 1967 (299, 480).

Stankevich, K. S., Wielebinski, R. and Wilson, W. E. 1970, *Australian Journal of Physics*, 23, 529 (302).

Staren, J., Meinhold, P., Childers, J., Lim, M., Levy, A., Lubin, P., Seiffert, M., Gaier, T., Figueiredo, N., Villela, T., Wuensche, C. A., Tegmark, M. and de Oliveira-Costa, A. 2000, *Astrophysical Journal*, 539, 52 (505).

Starobinsky, A. A. 1982, *Physics Letters*, B, 117, 175 (441).

Steigman, G. 2007, *Annual Review of Nuclear and Particle Science*, 57, 463 (37, 470).

Steinhardt, P. J. and Turner, M. S. 1984, *Physical Review D*, 29, 2162 (468).

Steinhardt, P. J. and Turok, N. 2007, *Endless Universe*, New York: Doubleday (42).

Stoicheff, B. 2002, *Gerhard Herzberg, an Illustrious Life in Science*, Ottawa: NRC Press (519).

Stokes, R. A., Partridge, R. B. and Wilkinson, D. T. 1967, *Physical Review Letters*, 19, 1199, 1360 (156, 183, 228–231, 295, 309, 312, 326, 480).

Strukov, I. A., Brukhanov, A. A., Skulachev, D. P. and Sazhin, M. V. 1992, *Monthly Notices of the Royal Astronomical Society*, 258, 37P (495).

Strukov, I. A. and Skulachev, D. P. 1984, *Soviet Astronomy Letters*, 10, 1 (495).

Strukov, I. A. and Skulachev, D. P. 1988, *Astrophysics and Space Physics Reviews*, 6, 147 (132).

Strukov, I. A. and Skulachev, D. P. 1991, *Advances in Space Research*, 11, 255 (131).

Strukov, I. A., Skulachev, D. P., Boyarskii, M. N. and Tkachev, A. N. 1987, *Soviet Astronomy Letters*, 13, 65 (132).

Struve, O. 1950, *Stellar Evolution, an Exploration from the Observatory*, Princeton NJ: Princeton University Press (276).

Sugiyama, N. and Gouda, N. 1992, *Progress of Theoretical Physics*, 88, 803 (446).

Sugiyama, N. and Silk, J. 1994, *Physical Review Letters*, 73, 509 (378).

Sullivan, W. 1965, *New York Times*, May 21, section 1, p. 1 (77, 152, 172).

Sullivan, W. 1969, *New York Times*, June 18, section 2, p. 1 (390).

Sunyaev, R. A. 1968, *Akademiia Nauk SSSR Doklady*, 179, 45; English translation in *Soviet Physics Doklady*, 13, 183 (121).

Sunyaev, R. A. 1969, *Astronomicheskii Zhurnal*, 46, 929; English translation in *Soviet Astronomy*, 13, 729, 1970 (120).

Sunyaev, R. A. 1980, *Soviet Astronomy Letters*, 6, 213 (128).

Sunyaev, R. A. 2004, in *Zel'dovich reminiscences*, ed. R. A. Sunyaev, Boca Raton, FL: Chapman and Hall (97, 98, 133, 530).

Sunyaev, R. A. and Zel'dovich, Ya. B. 1970a, *Astrophysics and Space Science*, 7, 20 (124).

Sunyaev, R. A. and Zel'dovich, Ya. B. 1970b, *Comments on Astrophysics and Space Physics*, 2, 66 (125).

Sunyaev, R. A. and Zel'dovich, Ya. B. 1970c, *Astrophysics and Space Science*, 7, 3 (67, 130, 194, 377).

Sunyaev, R. A. and Zel'dovich, Ya. B. 1972, *Comments on Astrophysics and Space Physics*, 4, 173 (127).

Sunyaev, R. A. and Zel'dovich, Ya. B. 1980, *Monthly Notices of the Royal Astronomical Society*, 190, 413 (127).

Szalay, A. S. and Marx, G. 1974, *Acta Physica Hungarica*, 35, 113 (438).

Szalay, A. S. and Marx, G. 1976, *Astronomy and Astrophysics*, 49, 437 (438).

Tabor, W. J. and Sibilia, J. T. 1963, *Bell System Technical Journal*, 42, 1963 (163).

Tanaka, H. 1979, *Shizen*, January, p 100 (63).

Tanaka, H., Kakinuma, T., Shindo, H. and Takayanagi, T. 1951, *Kuhden Kenkyu-jo Houkoku*, Nagoya University, 2, 121; English translation in Haruo Tanaka and Takakiyo Kakinuma, *Proceedings of the Research Institute of Atmospherics*, Nagoya University, 1, 85, 1953 (63).

Tanaka, S. T., Clapp, A. C., Devlin, M. J., Figueiredo, N., Gundersen, J. O., Hanany, S., Hristov, V. V., Lange, A. E., Lim, M. A., Lubin, P. M., Meinhold, P. R., Richards, P. L., Smoot, G. F. and Staren, J. 1996, *Astrophysical Journal Letters*, 468, L81 (500).

Tayler, R. J. 1990, *Quarterly Journal of the Royal Astronomical Society*, 31, 371 (61).

Tegmark, M., de Oliveira-Costa, A., Staren, J. W., Meinhold, P. R., Lubin, P. M., Childers, J. D., Figueiredo, N., Gaier, T., Lim, M. A., Seiffert, M. D., Villela, T. and Wuensche, C. A. 2000, *Astrophysical Journal*, 541, 535 (505).

Tegmark, M. *et al.* 2004, *Astrophysical Journal*, 606, 702 (472).

Teller, E. 1948, *Physical Review*, 73, 801 (38).

ter Haar, D. 1950, *Reviews of Modern Physics*, 22, 119 (33, 67).

Terrell, J. 1964, *Science*, 145, 918 (57).

Thaddeus, P. 1972, *Annual Review of Astronomy and Astrophysics*, 10, 305 (84, 103, 322, 416, 479).

Thaddeus, P. and Clauser, J. F. 1966, *Physical Review Letters*, 16, 819 (84, 152, 173, 294, 322).

Thorne, K. S. 1967, *Astrophysical Journal*, 148, 51 (261).

Timbie, P. T. and Wilkinson, D. T. 1990, *Astrophysical Journal*, 353, 140 (497).

Tolman, R. C. 1931, *Physical Review*, 37, 1639 (17, 42).

Tolman, R. C. 1934, *Relativity, Thermodynamics and Cosmology*, Oxford: The Clarendon Press (41, 185, 410, 472).

Torbet, E., Devlin, M. J., Dorwart, W. B., Herbig, T., Miller, A. D., Nolta, M. R., Page, L., Puchalla, J. and Tran, H. T. 1999, *Astrophysical Journal Letters*, 521, L79 (505).

Trimble, V. 2006, *New Astronomy Review*, 50, 844 (62).

Tsukerman, V. A. and Azarkh, Z. M. 1994, *People and Explosions* (in Russian), p. 118, published in the September, October, and November issues of the journal *Zvedza*; English translation by T. Sergay: *Arzamas-16: Soviet Scientists in the Nuclear Age: A Memoir*, 1999, p. 143, Nottingham: Bamcote Press (97).

Tucker, G. S., Griffin, G. S., Nguyen, H. T. and Peterson, J. B. 1993, *Astrophysical Journal Letters*, 419, L45 (501).

Tucker, G. S., Gush, H. P., Halpern, M., Shinkoda, I. and Towlson, W. 1997, *Astrophysical Journal Letters*, 475, L73 (503).

Tukey, J. W. 1966, in *Spectral Analysis of Time Series*, ed. B. Harris, New York: Wiley (452).

Turok, N. 1989, *Physical Review Letters*, 63, 2625 (447).

Turok, N. 1996, *Physical Review Letters*, 77, 4138 (466).

Uson, J. M. and Wilkinson, D. T. 1982, *Physical Review Letters*, 49, 1463 (377, 493).

Uson, J. M. and Wilkinson, D. T. 1984a, *Astrophysical Journal*, 283, 471(493).

Uson, J. M. and Wilkinson, D. T. 1984b, *Nature*, 312, 427 (493).

Varshalovich, D. A. and Sunyaev, R. A. 1968, *Astrophysics*, 4, 140 (120).

Vilenkin, A. 1985, *Physics Reports*, 121, 263 (447).

Vittorio, N. and Silk, J. 1984, *Astrophysical Journal Letters*, 285, L39 (378, 450).

Vittorio, N. and Silk, J. 1985, *Astrophysical Journal Letters*, 297, L1 (378).

von Weizäcker, C. F. 1938, *Physikalische Zeitschrift*, 39, 633 (25, 28, 40).

Wagoner, R. V. 1967, *Science*, 155, 1369 (243, 261).

Wagoner, R. V. 1973, *Astrophysical Journal*, 179, 343 (261).

Wagoner, R. V. 1990, in *Modern Cosmology in Retrospect*, ed. R. Bertotti, R. Balbinot, S. Bergia and A. Messina, Cambridge: Cambridge University Press, 159 (259).

Wagoner, R. V., Fowler, W. A. and Hoyle, F. 1966, *Science*, 152, 677 (260).

Wagoner, R. V., Fowler, W. A. and Hoyle, F. 1967, *Astrophysical Journal*, 148, 3 (192, 243, 260, 272).

Wall, J. V., Chu, T. Y. and Yen, J. L. 1970, *Australian Journal of Physics*, 23, 45 (285).

Wall, J. V., Perley, R., Liang, R., Silk, J. and Taylor, A. 2006, private communication (287).

Watson, R. A., Gutierrez de La Cruz, C. M., Davies, R. D., Lasenby, A. N., Rebolo, R., Beckman, J. E. and Hancock, S. 1992, *Nature*, 357, 660 (496).

Weber, J. 1951, PhD Thesis, *Microwave Technique in Chemical Kinetics*, The Catholic University of America (62).

Webster, A. S. 1974, *Monthly Notices of the Royal Astronomical Society*, 166, 355 (488).

Weinberg, S. 1962, *Physical Review*, 128, 1457 (41).

Weinberg, S. 1977, *The First Three Minutes*, New York: Bantam Books (8, 105).

Weiss, R. 1980, *Annual Review of Astronomy and Astrophysics*, 18, 489 (480, 486).

Weiss, R. and Grodzins, L. 1962, *Physics Letters*, 1, 342 (342).

Welch, W. J., Keachie, S., Thornton, D. D. and Wrixon, G. 1967, *Physical Review Letters*, 18, 1068 (230, 294, 300, 480).

Westerhout, G. and Oort, J. H. 1951, *Bulletin of the Astronomical Institutes of the Netherlands*, 11, 323 (288).

Wetterich, C. 1988, *Nuclear Physics* B, 302, 668 (443).

Weymann, R. 1965, *Physics of Fluids*, 8, 2112 (123, 195).

Weymann, R. 1966, *Astrophysical Journal*, 145, 560 (195, 291, 301).

Wheeler, J. A. 1958 (with his students) in Eleventh Solvay Conference, *La Structure et l'Évolution de l'Univers*, Brussels: Stoops (41).

Wheeler J. A. 1964, *Geometrodynamics and the Issue of the Final State*, in *Relativity, Groups, and Topology*, ed. C. DeWitt and B. S. DeWitt, New York: Gordon and Breach (380).

Wheeler, J. A. and Feynman, R. P. 1945, *Reviews of Modern Physics*, 17, 157 (238).

White, S. D. M. and Frenk, C. S. 1991, *Astrophysical Journal*, 379, 52 (456, 474).

White, S. D. M., Navarro, J. F., Evrard, A. E. and Frenk, C. S. 1993, *Nature*, 366, 429 (456, 474).

Wilkinson, D. T. 1962, PhD Thesis, *A Precision Measurement of the G-Factor of the Free Electron*, University of Michigan (326).

Wilkinson, D. T. 1967, *Physical Review Letters*, 19, 1195 (156, 229, 231, 322, 326, 480).

Wilkinson, D. T. 1980, *Physica Scripta*, 21, 606 (312).

Wilkinson, D. T. and Partridge, R. B. 1967, *Nature*, 215, 719 (225, 487).

Wilkinson, D. T. and Peebles, P. J. E. 1983, in *Serendipitous Discoveries in Radio Astronomy*, ed. K. I. Kellermann and B. Sheets, Greenbank, WV: National Radio Astronomical Observatory (23).

Williamson, K. D., Blair, A. G., Catlin, L. L., Hiebert, R. D., Loyd, E. G. and Romero, H. V. 1973, *Nature*, 241, 79 (480).

Wilson, G. W., Knox, L., Dodelson, S., Coble, K., Cheng, E. S., Cottingham, D. A., Fixsen, D. J., Goldin, A. B., Inman, C. A., Kowitt, M. S., Meyer, S. S., Page, L. A., Puchalla, J. L., Ruhl, J. E. and Silverberg, R. F. 2000, *Astrophysical Journal*, 532, 57 (502).

Wilson, M. L. 1983, *Astrophysical Journal*, 273, 2 (377).

Wilson, M. L. and Silk, J. 1981, *Astrophysical Journal*, 243, 14 (377).

Wilson, R. W. 1963, *Astrophysical Journal*, 137, 1038 (158).

Wilson, R. W. 1979, *Reviews of Modern Physics*, 51, 433 (148).

Wilson, R. W. and Bolton, J. G. 1960, *Publications of the Astronomical Society of the Pacific*, 72, 331 (158).

Wilson, R. W., Jefferts, K. B. and Penzias, A. A. 1970, *Astrophysical Journal Letters*, 161, L43 (175).

Wilson, R. W. and Penzias, A. A. 1967, *Science*, 156, 1100 (174, 426, 488).

Wilson, R. W., Penzias, A. A., Jefferts, K. B. and Solomon, P. M. 1973, *Astrophysical Journal Letters*, 179, L107 (157).

Witebsky, C., Smoot, G., de Amici, G. and Friedman, S. D. 1986, *Astrophysical Journal*, 310, 145 (481).

Wolfe, A. M. and Burbidge, G. R. 1969, *Astrophysical Journal*, 156, 345 (195, 232, 273).

Wollack, E. J., Jarosik, N. C., Netterfield, C. B., Page, L. A. and Wilkinson, D. T. 1993, *Astrophysical Journal Letters*, 419, L49 (501).

Woody, D. P., Mather, J. C., Nishioka, N. S. and Richards, P. L. 1975, *Physical Review Letters*, 34, 1036 (481).

Woody, D. P. and Richards, P. L. 1979, *Physical Review Letters*, 42, 925 (481).

Wood-Vasey, W. M. *et al.* 2007, *Astrophysical Journal*, 666, 694 (472).

Wright, E. L., Bennett, C. L., Gorski, K., Hinshaw, G. and Smoot, G. F. 1996, *Astrophysical Journal Letters*, 464, L21 (499).

Wright, E. L., Smoot, G. F., Bennett, C. L. and Lubin, P. M. 1994a, *Astrophysical Journal*, 436, 443 (443).

Wright, E. L., Smoot, G. F., Kogut, A., Hinshaw, G., Tenorio, L., Lineweaver, C., Bennett, C. L. and Lubin, P. M. 1994b, *Astrophysical Journal*, 420, 1 (499).

Wright, E. L. *et al.* 1992, *Astrophysical Journal Letters*, 396, L13 (452, 527).

Xu, Y., Tegmark, M., de Oliveira-Costa, A., Devlin, M. J., Herbig, T., Miller, A. D., Netterfield, C. B. and Page, L. 2001, *Physical Review D*, 63, 103002 (504).

Yakubov, V. B. 1964, *Astronomicheskii Zhurnal*, 41, 884; English translation in *Soviet Astronomy*, 8, 708, 1965 (42).

Yates, K. W. and Wielebinski, R. 1967, *Astrophysical Journal*, 149, 439 (284).

Yodh, G. B. and Wallis, R. F. 2001, *Physics Today*, 54, 070000 (529).

Yu, J. T. 1968, PhD Thesis, *Clusters of Galaxies – Their Statistics and Formation*, Princeton University (198, 363).

Yu, J. T. and Peebles, P. J. E. 1969, *Astrophysical Journal*, 158, 103 (198, 363, 443).

Zaldarriaga, M., Spergel, D. N. and Seljak, U. 1997, *Astrophysical Journal*, 488, 1 (446).

Zatsepin, G. T. and Kuz'min, V. A. 1966, *Soviet Journal of Experiment and Theoretical Physics Letters*, 4, 78 (132).

Zel'dovich, Ya. B. 1962, *Zhurnal Eksperimentalnoi i Teoreticheskoi Fiziki*, 43, 1561; English translation in *Soviet Physics JETP*, 16, 1102, 1963 (35, 100, 189, 276, 277).

Zel'dovich, Ya. B. 1963a, *Atomnaya Energiya*, 14, 92; English translation in *Soviet Atomic Energy*, 14, 83, 1964 (35, 66, 94, 259).

Zel'dovich, Ya. B. 1963b, *Uspekhi Fizicheskikh Nauk*, 80, 357; English translation in *Soviet Physics-Uspekhi*, 6, 475, 1964 (35, 95).

Zel'dovich, Ya. B. 1965, *Advances in Astronomy and Astrophysics*, 3, 241 (35, 36, 104, 111, 189).

Zel'dovich, Ya. B. 1966, *Uspekhi Fizicheskikh Nauk*, 89, 647; English translation in *Soviet Physics-Uspekhi*, 9, 602, 1966 (95, 97, 111, 436).

Zel'dovich, Ya. B., 1968, *Uspekhi Fizicheskikh Nauk*, 95, 209; English translation in *Soviet Physics-Uspekhi*, 11, 381 (111).

Zel'dovich, Ya. B. 1972, *Monthly Notices of the Royal Astronomical Society*, 160, 1P (441, 527).

Zel'dovich, Ya. B. 1980, *Monthly Notices of the Royal Astronomical Society*, 192, 663 (447).

Zel'dovich Ya. B. 1985, *Selected Proceedings: Particles, Nuclei and the Universe*, Moscow: Nauka Press (97).

Zel'dovich, Ya. B. 1993, *Selected works of Yakov Borisovich Zel'dovich*, ed. J. P. Ostriker, G. I. Barenblatt and R. A. Sunyaev, Princeton NJ: Princeton University Press (111).

Zel'dovich, Ya. B., Einasto, J. and Shandarin, S. F. 1982, *Nature*, 300, 407 (439).

Zel'dovich, Ya. B., Kurt, V. G. and Sunyaev, R. A. 1968, *Zhurnal Eksperimental noi i Teoreticheskoi Fiziki*, 55, 278; English translation in *Soviet Physics JETP*, 28, 146, 1969 (121, 436).

Zel'dovich, Ya. B. and Novikov, I. D. 1967, *Relativistic Astrophysics*, Moscow: Izdatel'stvo Nauka; enlarged English translation in *Relativistic Astrophysics 1. Stars and Relativity*, Chicago: University of Chicago Press (109, 111).

Zel'dovich, Ya. B. and Novikov, I. D. 1983, *Relativistic Astrophysics 2. The Structure and Evolution of the Universe*, Chicago: University of Chicago Press, translated and enlarged from Zel'dovich and Novikov (1975) (100).

Zel'dovich, Ya. B. and Raizer, Yu. P. 1966, *Physics of Shock Waves and High Temperature Hydrodynamic Phenomena*, Moscow: Nauka Press (115).

Zel'dovich, Ya. B. and Raizer, Yu. P. 1968, *Elements of Gas Dynamics and the Classical Theory of Shock Waves*, New York: Academic Press (143).

Zel'dovich, Ya. B., Rakhmatulina, A. K. and Sunyaev, R. A. 1972, *Radiophysics and Quantum Electronics*, 15, 121 (130, 443).

Zel'dovich, Ya. B. and Shakura, N. I. 1969, *Astronomicheskii Zhurnal*, 46, 225; English translation in *Soviet Astronomy*, 13, 175 (124).

Zel'dovich, Ya. B. and Sunyaev, R. A. 1969, *Astrophysics and Space Science*, 4, 301 (123, 195, 422, 528).

Zwicky, F. 1929, *Proceedings of the National Academy of Sciences*, 15, 773 (52, 529).

Zwicky, F. 1933, *Helvetica Physica Acta*, 6, 110 (20, 31).

Zwicky, F., Herzog, E. and Wild, P. 1961–68, *Catalogue of Galaxies and Clusters of Galaxies*, Pasadena: California Institute of Technology (362).

Index